Meshfree and Particle Methods

Meshfree and Particle Methods

Fundamentals and Applications

Ted Belytschko
Northwestern University
USA

J. S. Chen
University of California
USA

Michael Hillman
The Pennsylvania State University
USA

This edition first published 2024

Registered Office
John Wiley & Sons, Inc., 111 River Street, Hoboken, NJ 07030, USA
John Wiley & Sons Ltd, The Atrium, Southern Gate, Chichester, West Sussex, PO19 8SQ, UK

For details of our global editorial offices, customer services, and more information about Wiley products visit us at www.wiley.com.

Wiley also publishes its books in a variety of electronic formats and by print-on-demand. Some content that appears in standard print versions of this book may not be available in other formats.

Library of Congress Cataloging-in-Publication Data
Names: Belytschko, Ted, 1943–2014, author. | Chen, J. S., 1958– author. | Hillman, Michael, author.
Title: Meshfree and particle methods : fundamentals and applications / Ted Belytschko, J. S Chen, Michael Hillman.
Description: Hoboken, NJ : Wiley, 2024. | Includes bibliographical references and index.
Identifiers: LCCN 2022041519 (print) | LCCN 2022041520 (ebook) | ISBN 9780470848005 (hardback) | ISBN 9781119811152 (adobe pdf) | ISBN 9781119811138 (epub)
Subjects: LCSH: Meshfree methods (Numerical analysis) | Particle methods (Numerical analysis)
Classification: LCC QA297 .B379 2023 (print) | LCC QA297 (ebook) | DDC 518/.2–dc23/eng20221219
LC record available at https://lccn.loc.gov/2022041519
LC ebook record available at https://lccn.loc.gov/2022041520

Cover Design: Wiley
Cover Image: Courtesy of Jonghyuk Baek, J. S. Chen, and Michael Hillman

Set in 9.5/12.5pt STIXTwoText by Straive, Pondicherry, India

Contents

Preface

We dedicate this book "Meshfree and Particle Methods: Fundamentals and Applications" to the late Professor Ted Belytschko for his vision, leadership, and remarkable contributions in this field. The book project was initiated quite a few years before Ted's illness. In the beginning and before Ted's passing in 2014, the efforts by the first two authors were on the fundamental formulation of meshfree methods, and the progress was initially slow, while the topics in the book simultaneously evolved due to the active research in meshfree methods and other related fields. Through the addition of the third author, this book project was finally brought to completion. A two-dimensional MATLAB implementation of the reproducing kernel particle method for solving linear elasticity (RKPM2D) was also attached to this book to illustrate the programming of meshfree methods.

History seems to repeat itself indeed. Like finite difference and finite element methods, meshfree and particle methods originated as fundamental research topics in academia, and eventually found their way into industrial applications. The first workshops, titled "Workshop on Meshfree Methods," sponsored by the National Science Foundation and organized by the University of Iowa, were held in 2000 and 2001. Afterward, workshops called "Workshop on Meshless Methods, Generalized Finite Element Methods, and Related Approaches" were held at the University of Maryland from 2005 to 2009. Around the same time, the biennial "International Workshop on Meshfree Methods for Partial Differential Equations" was initiated by the University of Bonn, Germany, in 2001 and then held every odd year. The biennial US version, "USACM Thematic Conference on Meshfree and Particle Methods" (slightly different names were given to each), started in 2014 and has been held every even year since.

This book is an attempt to present both the fundamentals of meshfree and particle methods, as well as the state-of-the-art of several topics that we feel are very practical for engineering applications. It does not reflect the breadth and amount of research activities across the field but instead focuses on subjects where these methods are most advantageous. It would, of course, be impossible to cover the entirety of the state of research in one book.

Our intent then is to provide a comprehensive discussion on the fundamental concepts, basic formulations, numerical algorithms, computer implementation, and the application of meshfree and particle methods to challenging engineering and scientific problems. We expect it to be a useful introduction and reference for engineers and scientists across academia, industry, and government, and suitable for the instruction of graduate courses at universities. We present the basic formulation of meshfree methods through the moving least squares and reproducing kernel approximations, to demonstrate the unique approximation and discretization properties for solving both diffusion and linear and nonlinear mechanics problems.

The Utility of Meshfree Methods

For the reader who is not intimately familiar with these approaches, we first describe why they are useful. We begin by defining meshfree methods: a class of numerical techniques that do not rely on any mesh, grid, or structured discretization, aside from a set of points. That is, the connectivity between the points (called nodes) does not have to be dictated *a priori* in an adjoining fashion, and only needs to satisfy some minimum requirements. The Galerkin class of meshfree methods was designed to inherit the main advantages of the finite element method, such as the compact support of shape functions, good approximation properties, and mathematical foundations in variational and related principles. At the same time, they overcome the main disadvantages of the finite element method, such as the strong tie between mesh quality and approximation quality, difficulties in constructing discontinuous or highly continuous approximations, tedious adaptive refinement, solution sensitivity to mesh distortion, and solution divergence due to mesh entanglement in large deformation problems.

The nodes, which form patches of supports, only need to cover the domain of interest. This feature obviates the conforming requirement in the finite element method. Therefore, constructing a model for engineering analysis is much less burdensome than the traditional mesh-based approach: one does not need to be concerned with "high quality" elements. When the Galerkin class of these meshfree methods spawned an explosion of research in the 1990s, this feature was highly celebrated. The tedious and time-consuming task of generating a mesh suitable for analysis can be circumvented entirely.

Several other features of these methods are quite remarkable and perhaps even more appealing. First, the order of completeness in the approximation is not only arbitrary but uncoupled from the order of continuity. This is in contrast to most formulations based on conforming polynomials, where to increase continuity, one must also increase the order. Thus low-order methods with high-order smoothness are possible, and vice versa. This feature is very practical for solving problems in mechanics, where the governing equations can involve high-order derivatives such as thin shells. It is also no longer necessary to employ the weak formulation to reduce the order of differentiation to accommodate the low order of global continuity of traditional finite elements. The most significant advantage of meshfree methods is this flexibility in customizing the approximation functions for desired regularity and ability to capture essential physics and features of particular problems of interest by embedding special functions. Adaptivity and multiple-scale solution strategies also can be implemented with relative ease.

The last unique aspect that we will highlight here is that during the simulation, significant distortions (even fluid-like material flow), fracture, and surface closure, are easily accommodated since no mesh is employed. Historically, much of the method development has been driven by large-deformation plasticity problems (metal forming and earthmoving were among the first industrial applications of the Galerkin version), high-rate defense simulations, and elastomeric devices. To this day, these remain the primary domains where these methods are applied.

Over the years, it has become clear that meshfree methods provide considerable advantages over the conventional finite element methods in solving problems involving moving discontinuities, evolving interfaces, multiple-scale phenomena, large material distortion and structural deformation, and fracture and damage processes. The overall extreme versatility has opened up seemingly limitless possibilities in method development, and there appears to be an ever-present interest in these methods despite nearly three decades of development.

What is Unique About this Book

A handful of books have been published on meshfree and particle methods, so we would like to highlight some unique aspects of this book:

- Detailed descriptions of essential issues and how to address them, not covered in detail in other books, organized over several dedicated chapters: essential boundary condition enforcement, numerical integration, and nonlinear meshfree methods.
- Up-to-date and complete information about the state-of-the-art in Galerkin and collocation meshfree and particle methods, covering the fundamental theories and applications.
- The inclusion of many meshfree methods, such as the Galerkin type, collocation type, partition of unity methods, and kernel estimate of conservation equations (smoothed particle hydrodynamics).
- The topics are integrated with an open-source code, with a chapter describing the code in detail that cross-references the methods described in the book.

Another key feature is that it can serve as both an introduction and a valuable reference to students, engineers, and scientists who either want to learn about meshfree methods or are working in this area already.

Level and Background

This book is designed for readers without prior experience with meshfree and particle methods, but it still requires some basic knowledge of numerical analysis and mechanics. In particular, readers will greatly benefit from an understanding of the linear and nonlinear finite element method, and have a deeper understanding and appreciation for the materials presented. An introductory course in mechanics or elasticity covering indicial and tensor notation is a prerequisite.

The primary audience includes practitioners and researchers in the mechanical, aerospace, civil, and structural engineering industries. A secondary audience is graduate students in these fields, and students of applied mathematics. This book provides fundamental theories, mathematical formulations, numerical algorithms, and code implementation steps to learn the fundamentals and help develop meshfree codes for performing research and analysis.

Content and Structure of this Book

The first six chapters of this book have been compiled with the help of lecture notes (in particular the example problems) from SE 279 "Meshfree Methods for Linear and Nonlinear Mechanics" at The University of California, San Diego, and CE 597 "Meshfree Methods and Advanced Computational Solid Mechanics" at The Pennsylvania State University, as well as short courses offered by the second and third authors. They serve as the main introduction to these methods, focusing on linear problems. The remaining chapters do not necessarily contain more advanced materials but detail the application of these methods to unique problem domains (Chapter 7), the variety of meshfree methods formulated over the years (Chapter 8), leveraging smoothness to solve the strong form directly (Chapter 9), and the computer implementation of these methods (Chapter 10).

This book is organized as follows. We first present the history of method development, the definition of a meshfree method, the key approximation properties, a demonstration of meshfree modeling, and classes of meshfree methods in Chapter 1, "Introduction to Meshfree and Particle Methods." Next, the strong forms and weak forms (variational equations) of diffusion and elasticity problems are given in Chapter 2, "Preliminaries: Strong and Weak Forms of Diffusion, Elasticity, and Solid Continua," with particular emphasis on the imposition of Dirichlet boundary conditions since meshfree approximations are generally kinematically inadmissible. Here, we note that the readers with intimate knowledge of strong and weak forms, constrained variational principles like the Lagrange multiplier method, and the governing equations of continua may prefer to skip this chapter, but they may still find it useful for reference later in the book. In Chapter 3, "Meshfree Approximations," complete derivations of the popularly used moving-least squares (MLS) approximation employed in the element-free Galerkin (EFG) method, and the reproducing kernel (RK) approximation used in the reproducing kernel particle method (RKPM) are provided. Their associated properties and various methods to compute derivatives needed for solving PDEs are also presented. Although convergence and stability of various meshfree formulations are discussed with numerical demonstrations provided, the proofs of the mathematical properties are not described, but appropriate references are given for further information and details. Complete procedures for the discretization of PDEs using meshfree approximations with the associated enforcement of the Dirichlet boundary conditions are discussed in Chapter 4, "Solving PDEs with the Galerkin Meshfree Methods." Various methods to construct kinematically admissible meshfree approximations are introduced in Chapter 5, "Construction of Kinematically Admissible Shape Functions," as well as their incorporation with consistent weak formulations to achieve optimal convergence when higher-order bases are used. In Chapter 6, "Quadrature in Meshfree Methods," the concept of integration constraints (ICs) is first introduced as the sufficient conditions to achieve optimal convergence, and various stabilized nodal integration methods with variationally consistent corrections to meet the ICs are presented. Lagrangian and semi-Lagrangian meshfree approximations and discretizations for solving large (finite) deformation solid mechanics problems are discussed in Chapter 7, "Nonlinear Meshfree Methods," along with their stability conditions in transient problems and their extension to contact problems. Chapter 8, "Other Galerkin Meshfree Methods," presents other approximations commonly used in the weak formulation, including smoothed particle hydrodynamics (SPH), the partition of unity method, *h-p* clouds, and Sibson and non-Sibson interpolation used in the natural element method. Utilizing the smoothness of the meshfree approximations to discretize PDEs directly, Chapter 9, "Strong Form Collocation Meshfree Methods," introduces the alternative approach of using collocation rather than weak forms. The radial basis collocation method (RBCM), the reproducing kernel collocation method (RKCM), the gradient reproducing kernel collocation method (GRKCM), and the application of these approaches to solving problems with heterogeneity and discontinuities are discussed. Chapter 9 includes content republished by permission from Springer Nature: *Advances in Computational Plasticity: A Book in Honour of D. Roger J. Owen*, Performance Comparison of Nodally Integrated Galerkin Meshfree Methods and Nodally Collocated Strong Form Meshfree Methods, Hillman, M.; Chen, J.S., 2017. Finally, Chapter 10, "RKPM2D: A Two-dimensional Implementation of RKPM," details the implementation aspects of meshfree methods, with a description of an open-source software RKPM2D with preprocessing, solver, and postprocessing integrated under the MATLAB environment. The computer code should help readers understand the programming of meshfree and particle methods and allow the implementation of advanced meshfree algorithms and extensions to other related applications. Chapter 10 includes content republished by permission from Springer Nature: *Computational Particle Mechanics*, An Open-Source Implementation of Nodally Integrated Reproducing Kernel Particle Method for Solving Partial Differential Equations, Tsung-Hui Huang, et al., 2019.

Suggested Use in Instruction

A one-semester graduate course on linear meshfree methods can cover Chapters 2–6, while a one-quarter course can cover most topics in these chapters. An instructor may also choose to cover selected topics from Chapters 7–9, such as the strong-form collocation method, or a brief introduction to other meshfree methods, as we have done in the past. A more advanced course could also be compiled from select topics from the book. A few examples are given below:

One-semester course: "Meshfree and Particle Methods for Linear Mechanics."

Topics: Linear Galerkin meshfree methods (Chapters 2–6) and strong form collocation meshfree methods (Chapter 9).

One-quarter course: "Meshfree and Particle Methods for Linear Mechanics."

Topics: Strong and weak formulations for linear problems (Sections 2.1–2.2), meshfree approximations (Sections 3.1–3.4), solving diffusion and elasticity (Sections 4.1–4.2), kinematically admissible meshfree methods (Sections 5.1–5.5), quadrature (Sections 6.1–6.6), and introduction to strong form collocation methods (Sections 9.1–9.3).

One-quarter or one-semester course: "Meshfree and Particle Methods for Linear and Nonlinear Mechanics."

Topics: Select materials from Chapters 2–6, Chapter 7 (Sections 7.1–7.5), and select materials from Chapters 8–9.

A short course would undoubtedly need to be much more selective in the material but could refer to the book for more details.

Acknowledgements

We would like to sincerely thank the following former and current students and postdocs of the second and third authors who have contributed to proofreading chapters and generating figures for this book; their efforts are greatly appreciated:

- Jonghyuk Baek, The University of California, San Diego
- Sheng-Wei Chi, The University of Illinois, Chicago
- Andy Groeneveld, The Pennsylvania State University, University Park
- Xiaolong He, The University of California, San Diego
- Tsung-Hui Huang, National Tsing Hua University, Taiwan
- Siavash Jafarzadeh, The Pennsylvania State University, University Park
- Feihong Liu, The Pennsylvania State University, University Park
- Ryan Schlinkman, The University of California, San Diego
- Kristen Susuki, The University of California, San Diego
- Karan Taneja, The University of California, San Diego
- Hui-Ping Wang, General Motors Global Research and Development
- Jiarui Wang, The Pennsylvania State University, University Park
- Yanran Wang, The Pennsylvania State University, University Park

The English proofing by Jennifer Dougal is also appreciated.

J. S. Chen
University of California, San Diego
Michael Hillman
The Pennsylvania State University, University Park

Glossary of Notation

Diffusion Equation

Π_{D}: Functional for the standard variational principle
$\Pi_{\mathrm{D}}^{\mathrm{P}}$: Functional for the penalty approach
$\Pi_{\mathrm{D}}^{\mathrm{L}}$: Functional for the Lagrange multiplier approach
$\Pi_{\mathrm{D}}^{\mathrm{N}}$: Functional for the Nitsche approach
$\Pi_{\mathrm{D}}^{\mathrm{Nit}}$: Nitsche's contributions to the functional
Ω: Domain
Γ: Total boundary
Γ_u: Essential boundary
Γ_q: Natural boundary
$\overline{u}$: Prescribed scalar value of u on the essential boundary
$\overline{q}$: Prescribed scalar value of normal flux on the natural boundary
$\mathbf{q}$ (q_i): Flux vector
$\mathbf{k}$ (k_{ij}): Diffusivity, can also be a scalar k
s: Source term
u: Unknown field or trial functions
δu: Variation on the unknown field or test functions
v: Test functions (usually equivalent to δu)
u^h: Galerkin solution
δu^h: Discrete test function for the Galerkin equation
u_I: Unknown coefficients in the Galerkin equation
δu_I: Arbitrary coefficients in the Galerkin equation

Elasticity

Π_{E}: Functional for the standard variational principle
$\Pi_{\mathrm{E}}^{\mathrm{P}}$: Functional for the penalty approach
$\Pi_{\mathrm{E}}^{\mathrm{L}}$: Functional for the Lagrange multiplier approach
$\Pi_{\mathrm{E}}^{\mathrm{N}}$: Functional for the Nitsche approach
$\Pi_{\mathrm{E}}^{\mathrm{Nit}}$: Nitsche's contributions to the functional
Ω: Domain

Γ: Total boundary
Γ_u: Essential boundary
Γ_t: Natural boundary
$\bar{\mathbf{u}}$ $(\bar{u}_i)$: Prescribed vector value of $\mathbf{u}$ on the essential boundary
$\bar{\mathbf{t}}$ $(\bar{t}_i)$: Prescribed vector traction $\mathbf{t}$ on the natural boundary
$\boldsymbol{\sigma}$ (σ_{ij}): Stress tensor
$\boldsymbol{\varepsilon}$ (ε_{ij}): Strain tensor
$\mathbf{C}$ (C_{ijkl}): Fourth-order elasticity tensor
$\mathbf{b}$ (b_i): Body force
$\mathbf{u}$ (u_i): Displacement, or trial functions
$\delta\mathbf{u}$ (δu_i): Variation on displacement or test functions
$\mathbf{v}$ (v_i): Test functions (usually equivalent to $\delta\mathbf{u}$)
$\mathbf{u}^h$ (u_i^h): Galerkin solution
$\delta\mathbf{u}^h$ (δu_i^h): Discrete test function for the Galerkin equation
$\mathbf{u}_I$ (u_{iI}): Unknown coefficients in the Galerkin equation
$\delta\mathbf{u}_I$ (δu_{iI}): Arbitrary coefficients in the Galerkin equation

Boundary Value Problems and Variational Methods

L_2: Sobolov space of degree zero, $\mathrm{L}_2 = \mathrm{L}_2(\Omega) = \{f | \int_\Omega f^2 d\Omega < \infty\}$

H^s: Sobolov space of degree s, $\mathrm{H}^s = \mathrm{H}^s(\Omega) = \left\{ f \mid f \in \mathrm{L}_2,\ f_{,i} \in \mathrm{L}_2, \ldots, \underbrace{f_{,ij\ldots k}}_{s\ \text{indices}} \in \mathrm{L}_2 \right\}$

$\|f\|_{\mathrm{H}^s}$: Sobolev norm of a function f of degree s,

$$\|f\|_{\mathrm{H}^s} = \left(\int_\Omega f^2 + f_{,i} f_{,i} + \ldots + \underbrace{f_{,ij\ldots k}}_{s\ \text{indices}} \underbrace{f_{,ij\ldots k}}_{s\ \text{indices}} d\Omega \right)^{1/2}$$

$|f|_{\mathrm{H}^s}$: Sobolev semi-norm of a function f of degree s, $|f|_{\mathrm{H}^s} = \left(\int_\Omega \underbrace{f_{,ij\ldots k}}_{s\ \text{indices}} \underbrace{f_{,ij\ldots k}}_{s\ \text{indices}} d\Omega \right)^{1/2}$

U: Set of functions in H^1 with essential boundary conditions satisfied (admissible trial function space)
U_0: Set of functions in H^1 with homogenous boundary conditions satisfied (admissible test function space)
C^k: Functions with kth order continuity; the function exists and is continuous up to its kth order derivatives

Meshfree Approximations

NP: Number of points
$w_I(\mathbf{x}) \equiv w_a(\mathbf{x} - \mathbf{x}_I)$: Weight function at node I
$\phi_I(\mathbf{x}) \equiv \phi_a(\mathbf{x} - \mathbf{x}_I)$: Kernel function at node I

$C(\mathbf{x}; \mathbf{x}-\mathbf{x}_I)$:	Correction function at node I
a_I:	Support at node I
a:	Constant support for the entire domain
N_I:	Shape functions
Ψ_I:	MLS and RK shape functions; meshfree shape functions in general
$\hat{\Psi}_I$:	Kinematically admissible MLS and RK shape functions
$\Psi_I^{(\boldsymbol{\alpha})}$:	Diffuse or implicit derivative of MLS and RK shape functions
$\mathbf{u}_x^h$ or $\mathbf{u}_y^h$:	Implicit gradient approximation of the derivative in the x- and y-directions
S:	Set of all meshfree nodes that discretize the domain and boundaries
S_x or $S_{\mathbf{x}}$:	Set of nodes with non-zero kernel or weight over a point x or $\mathbf{x}$
S_u:	Set of Lagrange multiplier nodes associated with the essential boundary
S_{F}:	Set of free nodes
S_{C}:	Set of constrained nodes
$\mathbf{p}^{\mathrm{T}} = [1 \;\; x \;\; \dots \;\; x^n]$:	Vector of basis functions in the moving least squares approximation
$\mathbf{H}^{\mathrm{T}}(x-s) = \left[1 \;\; x-s \;\; (x-s)^2 \;\; \dots \;\; (x-s)^n\right]$:	Vector of basis functions in the reproducing kernel approximation
δ_{IJ}:	Kronecker delta function
Ns:	Number of source points (discrete points or nodes) in the strong form collocation method
Nc:	Number of collocation points in the strong form collocation method
$\mathbf{x}_{\hat{J}}$:	Collocation points in the strong form collocation method

Quadrature and Patch Tests

W_L:	Lth weight in a domain quadrature rule
S_L:	Lth weight in a boundary quadrature rule
$\tilde{\mathbf{B}}_I$:	Smoothed gradient associated with node I
$a(\cdot,\cdot)$:	Bilinear form
$a\langle\cdot,\cdot\rangle$:	Quadrature version of the bilinear form $a(\cdot,\cdot)$
$a_{\mathrm{D}}\langle\cdot,\cdot\rangle$:	Direct nodal integration of the bilinear form $a(\cdot,\cdot)$
$a_{\mathrm{S}}\langle\cdot,\cdot\rangle$:	Stabilized conforming nodal integration (SCNI) for the bilinear form $a(\cdot,\cdot)$
$a_{\mathrm{M}}\langle\cdot,\cdot\rangle$:	"Modified-type" stabilization
$a_{\mathrm{N}}\langle\cdot,\cdot\rangle$:	"Natural-type" stabilization

$a_{\text{SM}}\langle\cdot,\cdot\rangle$: $a_{\text{SM}}\langle\cdot,\cdot\rangle = a_{\text{S}}\langle\cdot,\cdot\rangle + a_{\text{M}}\langle\cdot,\cdot\rangle$ for modified SCNI (MSCNI)
$a_{\text{DN}}\langle\cdot,\cdot\rangle$: $a_{\text{DN}}\langle\cdot,\cdot\rangle = a_{\text{D}}\langle\cdot,\cdot\rangle + a_{\text{N}}\langle\cdot,\cdot\rangle$ for naturally stabilized nodal integration (NSNI)
$a_{\text{SN}}\langle\cdot,\cdot\rangle$: $a_{\text{SN}}\langle\cdot,\cdot\rangle = a_{\text{S}}\langle\cdot,\cdot\rangle + a_{\text{N}}\langle\cdot,\cdot\rangle$ for naturally stabilized conforming nodal integration (NSCNI)
$\boldsymbol{\varepsilon}_L$: Nodal strain computed with direct differentiation
$\tilde{\boldsymbol{\varepsilon}}_L$: Nodal strain computed with smoothing
$\boldsymbol{\varepsilon}_L^K$: Nodal sub-cell strain
$\mathbf{u}^{\text{P}}$ (u_i^{P}): Arbitrary vector polynomial solution in the elasticity problem patch test
$\boldsymbol{\sigma}^{\text{P}}$ (σ_{ij}^{P}): Stress tensor associated with arbitrary polynomial solution $\mathbf{u}^{\text{P}}$ in the elasticity problem patch test
$\bar{\mathbf{t}}^{\text{P}}$ ($\bar{t}_i^{\text{P}}$): Prescribed vector traction $\bar{\mathbf{t}}$ in the arbitrary polynomial patch test
$\bar{\mathbf{u}}^{\text{P}}$ ($\bar{u}_i^{\text{P}}$): Prescribed vector value of displacement boundary condition $\bar{\mathbf{u}}$ in the arbitrary polynomial patch test
$\bar{\mathbf{b}}^{\text{P}}$ ($\bar{b}_i^{\text{P}}$): Prescribed vector value of body force $\mathbf{b}$ in the arbitrary polynomial patch test

Matrix Forms

$\mathbf{B}_I$: Strain-displacement (gradient) matrix for node I in elasticity
$\mathbf{D}$: Matrix form of the fourth-order elasticity tensor $\mathbf{C}$ in elasticity
$\boldsymbol{\eta}$: Matrix form of the unit normals on the boundary in elasticity
$\mathbf{f}$: Right-hand side force in the Galerkin equation for diffusion (f_I) and elasticity ($\mathbf{f}_I$)
$\mathbf{K}$: Matrix (stiffness) in the Galerkin equation for diffusion (K_{IJ}) and elasticity ($\mathbf{K}_{IJ}$)
$\mathbf{d}$: Displacement (generalized) vector in the Galerkin equation for diffusion (u_I) and elasticity ($\mathbf{u}_I$)
$\boldsymbol{\Sigma} = \left[\sigma_{11}\ \sigma_{22}\ \sigma_{12}\right]^{\text{T}}$: Stress vector (2D) for the matrix form of elasticity (Voigt notation)
$\mathcal{E} = \left[\varepsilon_{11}\ \varepsilon_{22}\ 2\varepsilon_{12}\right]^{\text{T}}$: Strain vector (2D) for the matrix form of elasticity (Voigt notation)
$\mathbf{G}$: Lagrange multiplier contribution to system matrix for diffusion ($G_{I\alpha}$) and elasticity ($\mathbf{G}_{I\alpha}$)
$\mathbf{A}$: Penalty contribution to stiffness matrix for diffusion (A_{IJ}) and elasticity ($\mathbf{A}_{IJ}$)
$\boldsymbol{\Lambda}$: Transformation matrix

Nonlinear Continua

Π_{C}: Functional for nonlinear continua
$\Pi_{\text{C}}^{\text{P}}$: Functional for the penalty approach in nonlinear continua
$\Pi_{\text{C}}^{\text{L}}$: Functional for the Lagrange multiplier approach in nonlinear continua

Π_C^N: Functional for the Nitsche approach in nonlinear continua
$\boldsymbol{\tau}$ (τ_{ij}): Cauchy stress
$\mathbf{P}$ (P_{ij}): Nominal stress (transpose of the first Piola-Kirchhoff stress)
$\mathbf{S}$ (S_{ij}): Second Piola-Kirchhoff stress
Ω: Domain of the deformed configuration Ω
$\mathbf{x} = \boldsymbol{\varphi}(\mathbf{X}, t)$: Coordinate vector for a material point in the deformed configuration Ω
Γ: Total boundary of the deformed configuration Ω
Γ_u: Essential boundary of the deformed configuration Ω
Γ_t: Natural boundary of the deformed configuration Ω
ρ: Density at the current state
b_i: Body force defined in the deformed domain Ω
h_i: Surface traction defined on the deformed traction boundary Γ_t
Ω^0: Domain of the initial configuration
$\mathbf{X}$: Coordinate vector for a material point in the initial configuration Ω^0
Γ^0: Total boundary of the initial configuration
Γ_u^0: Essential boundary of the initial configuration
Γ_t^0: Natural boundary of the initial configuration
ρ^0: Initial density
b_i^0: Body force defined in the initial domain Ω^0
h_i^0: Surface traction mapped onto the initial traction boundary Γ_t^0
$\mathbf{T} = [\tau_{11}\ \tau_{22}\ \tau_{12}]^T$: Cauchy stress vector (2D) for the matrix form (Voigt notation)
$\mathbf{P} = [P_{11}\ P_{22}\ P_{12}\ P_{11}]^T$: Nominal stress vector (2D) for the matrix form
$\Psi_I^0(\mathbf{X})$: Lagrangian RK shape functions
$\Psi_I(\mathbf{x})$: Semi-Lagrangian RK shape functions
Ψ_I^C: RK shape functions defined on the contact surface
$\mathbf{B}_I^0$: Strain-displacement (gradient) matrix for node I for the deformation gradient
$\mathbf{B}_I$: Strain-displacement (gradient) matrix for node I for the symmetric displacement gradient with respect to the current coordinate
$\mathcal{M}$ ($\mathcal{M}_{IJ}$): Consistent mass matrix
$\mathcal{M}^L$ ($\mathcal{M}_{IJ}^L$): Lumped mass matrix
$\mathbf{N}$ ($\mathbf{N}_{IJ}$): Convective transport matrix
$\mathbf{f}^{ext}$ ($\mathbf{f}_I^{ext}$): External force vector
$\mathbf{f}^{int}$ ($\mathbf{f}_I^{int}$): Internal force vector

1

Introduction to Meshfree and Particle Methods

Meshfree methods have several origins: on one genealogical tree are the particle methods originated by Lucy [1] and Gingold and Monaghan [2], then refined and extended by Monaghan and his coworkers [3, 4]. On the other are the generalized finite difference methods, originated by Jensen [5] and refined by Perrone, Kao, Liszka, and Orkisz [6, 7]. One of the most important papers in the emergence of these methods was the work of Nayroles, Touzot, and Villon [8]. They called the method the diffuse element method. As we shall see, there are many similarities between these methods, though the initial viewpoints appear to be decidedly different.

Remarkably, a watershed in the development of these methods occurred when they were named "meshfree" (or meshless) methods, which shows that an attractive name goes a long way. The advantage of the name meshfree is that it highlights the most compelling attribute of these methods: the absence of a mesh of elements interconnected by nodes. The generation of finite element meshes for three-dimensional problems of bodies with a variety of features is still very challenging, especially when the model must be remeshed as the solution evolves, as in solidification and dynamic fracture problems. Adaptive refinement using finite elements is also cumbersome due to compatibility requirements along element boundaries.

Both names have persisted: Larry Libersky pointed out at an early specialty meeting on these methods that the name meshfree is more marketable than meshless: we don't call foods "fatless" or soft drinks "sugarless"; they are called "fat-free" and "sugar-free"! Thus, meshfree is a more attractive name for these methods.

1.1 Definition of Meshfree Method

For the purposes of this book, we define a meshfree method as any method that constructs the approximation in terms of nodal values, where the connectivity of the nodes need not be specified explicitly, and the arrangement of the nodes is arbitrary. In fact, information like connectivity is usually extracted in meshfree methods on the fly. Thus, the major distinction from finite element methods (FEMs) is that elements are not used to construct the approximation functions. For traditional finite difference methods, the arrangement of nodes is also not arbitrary. Nodes are arranged in a highly structured manner so that their indices can identify adjacent nodes required in constructing the equations. The difference between these two approaches is crystallized in Figure 1.1 by observing how a finite element discretization contrasts with a meshfree discretization: the approximation is generally associated only with nodes in the latter. The meshfree approximation function associated with node I is denoted by Ψ_I, and the subdomain over which it is nonzero, the

Meshfree and Particle Methods: Fundamentals and Applications, First Edition.
Ted Belytschko, J. S. Chen, and Michael Hillman.

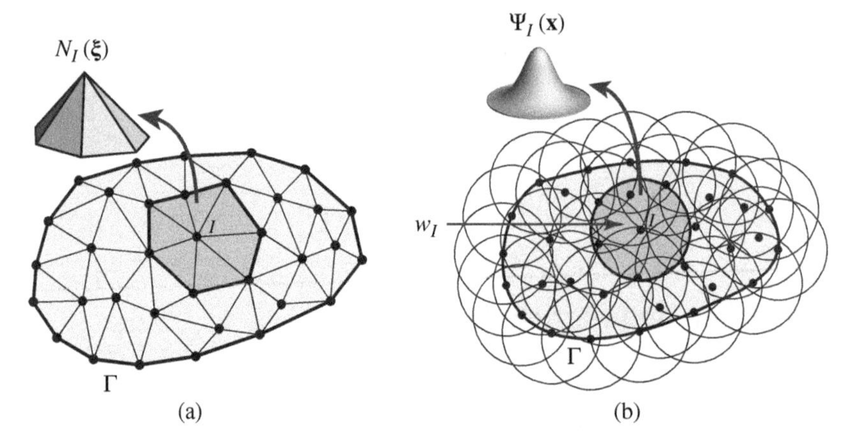

Figure 1.1 (a) Patching of finite element shape functions from local element domains and (b) meshfree approximation functions constructed directly at the nodes in the global coordinate, employing circular supports. The boundary of the domain is denoted as Γ. *Source:* Adapted from Chen and Belytschko [9], figure 1, p. 887. Reproduced with permission of Springer Nature.

support of Ψ_I, is denoted by w_I. It can be seen that the support naturally defines connectivity rather than fixed connectivity dictated by a grid or a mesh of elements. Thus, any field variable in differential equations can be approximated using a point cloud without any particular structure.

1.2 Key Approximation Characteristics

Meshfree methods possess several key characteristics that make them a highly unique class of numerical methods for solving differential equations. First, by the definition we have laid out, the approximation is formed without the need for a mesh or grid, and the connectivity is instead defined naturally. Meshfree approximations can also be constructed with arbitrary smoothness, and they can be much smoother or rougher than the finite element shape functions. In fact, the order of continuity and completeness can be made independent from one another in most meshfree methods, as opposed to finite element or isogeometric analysis, where increasing continuity requires increasing the polynomial order (completeness).

With smoother approximations, quantities involving derivatives such as strains in continuum mechanics and fluxes in heat transfer are also much smoother. This attribute increases the accuracy of the solution when the approximated function is smooth. Since solutions of diffusion equations and elasticity are usually very smooth when the material coefficients are continuous, this advantage applies to many useful classes of problems. Several of these properties, and other special properties are detailed in Chapter 3 "Meshfree Approximations."

Utilizing the smoothness of the approximation, it is also possible to solve the strong form of a problem directly without resorting to the weak formulation. Second-order derivatives are often required; these are simple to compute based on the smoothness of the meshfree shape functions and can also be approximated very efficiently. Chapter 9 "Strong Form Collocation Meshfree Methods" describes these approaches. This advantage of smoothness also applies to the weak formulation of thin beams, plates, and shells, where the global continuity required is easily achieved without the substantial effort required in FEM.

The ease of adaptive refinement is another attractive feature of meshfree methods. In finite elements, approximation functions are constructed in a local parent domain, and thus, compatibility is required along the element boundaries in local adaptive refinement. On the other hand, meshfree methods only rely on nodal locations in the global coordinates to construct the approximation, and enforcing compatibility is avoided in adaptive refinement.

A major distinction between meshfree and finite element approximations is that meshfree shape functions are usually not interpolants. Finite elements are usually interpolatory, and in fact, they are Lagrange interpolants for most of the commonly used elements. On the other hand, meshfree approximation functions are usually not, and we will instead call them simply approximation functions or shape functions. They are sometimes called interpolation functions, but this terminology ignores the fundamental definition of interpolants. This property of meshfree approximations makes the direct application of boundary conditions on the primary variable, often called essential boundary conditions or Dirichlet boundary conditions, more difficult. Chapter 2 "Preliminaries: Strong and Weak Forms of Diffusion, Elasticity and Solid Continua" lays out constrained variational principles to weakly enforce Dirichlet conditions such as the Lagrange multiplier method, the penalty method, and Nitsche's method. Special techniques can also be introduced to construct meshfree shape functions so that the traditional strong approach can be employed, which are described in Chapter 5 "Construction of Kinematically Admissible Shape Functions."

Another fundamental difference between meshfree and finite element methods is the relationship between the support of the shape functions and the subdomains over which quadrature is carried out. The element and quadrature domains are the same in finite elements, which unifies the approximation of field variables and numerical integration. This is clearly not the case for meshfree methods as the supports are defined simply by the nodes themselves, as seen in Figure 1.1b. At the same time, this offers some unique possibilities for meshfree quadrature, yet implementing a practical approach is nontrivial. Numerical integration for meshfree methods is discussed in Chapter 6, "Quadrature in Meshfree Methods."

The absence of a mesh circumvents element distortion and entanglement issues, making meshfree methods quite suitable for arbitrarily large deformations and distortions in Lagrangian continuum mechanics. Chapter 7 "Nonlinear Meshfree Methods" summarizes their implementation in this class of problems, leveraging several of the special properties of meshfree approximations.

1.3 Meshfree Computational Model

A meshfree computational model consists of a set of nodes and a description of the model's surfaces. We will call the set of nodes and the boundary description the *grid*, so as to avoid the word "mesh" in a "meshfree method." It can be seen in Figure 1.1b that the arrangement of the nodes can be quite arbitrary, although uniform nodal distributions are often used when domain geometry is regular. The boundaries can be described by a variety of methods, for example, level set functions and CAD methods. It is noteworthy that in contrast to FEM, the geometric description of the body must usually be given, rather than the geometry defined directly by the mesh topology.

The generation of a meshfree grid for computational analysis is shown in Figure 1.2. Starting with a geometric definition of the model, points are generated inside the volumes while the surface definitions are retained. Nodes can be generated to fill the volume in a variety of ways. A simple way is to simply use a triangulation of the associated domain, retaining only the vertices. In fact, any finite element meshing technology can be employed, with only the nodes retained from the mesh

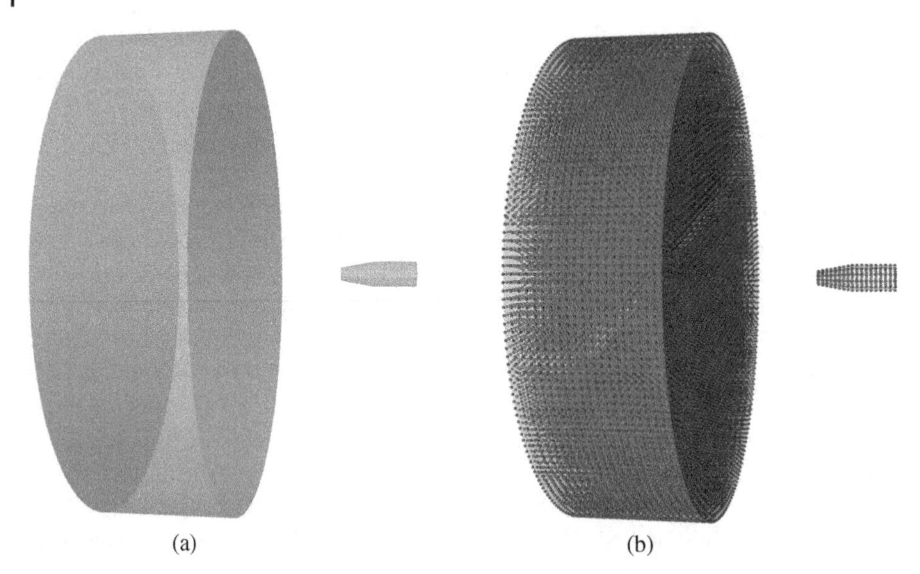

Figure 1.2 Constructing a meshfree computational grid for a bullet penetration analysis: (a) CAD geometry and (b) the meshfree grid, which fills the CAD volumes with nodes and retains the surfaces for applying boundary conditions.

generation. On the other hand, relying on meshing technology is not necessary or even desirable, as meshfree methods obviate meshing which is a key advantage. A wide variety of sphere and ellipsoidal packing algorithms exist [10], which can be used to avoid meshing altogether. In addition, triangulation algorithms or Voronoi diagrams can be used to obtain nodal volumes, and thus the only requirement for discretization is the generation of nodes.

1.4 A Demonstration of Meshfree Analysis

Figure 1.3 demonstrates several key characteristics of a meshfree solution. A penetration analysis associated with a meshfree grid (see Figure 1.2) is performed for a concrete target. The computation proceeds without the difficulties of element distortion and entanglement. The material break-up and ensuing multi-body contact with free surface formation can be handled by node-based algorithms (see Chapter 7). As the analysis proceeds, a meshfree error detector [11] embedded in the meshfree approximation indicates where additional accuracy is needed, and points are added in these areas on-the-fly with multiple levels of adaptive refinement. The refinement is performed simply by inserting additional points without issues such as "hanging nodes" as in FEM. Thus, the framework of meshfree analysis provides a flexible tool for challenging scientific and engineering problems.

1.5 Classes of Meshfree Methods

Meshfree methods have been developed under two general branches of formulations which are covered in this book:

1) **The Galerkin meshfree methods based on the weak form of partial differential equations (PDEs).** While no mesh is needed in the construction of the approximation, domain integration

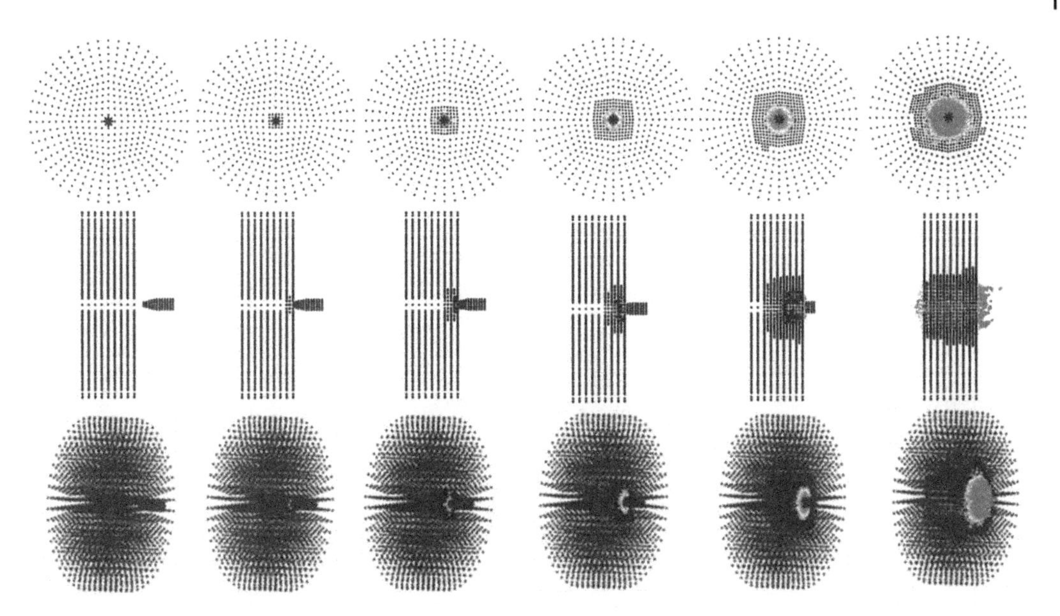

Figure 1.3 Meshfree simulation of a bullet penetration event with adaptive model refinement. Top to bottom: back view, side view, and a front perspective; left to right: as the simulation proceeds.

is required, and special techniques to enforce essential boundary conditions are needed. Domain integration and enforcement of boundary conditions are discussed in Chapters 4–6; and

2) **The collocation meshfree methods based on the strong form of PDEs.** Because of the ease in constructing smooth meshfree approximations, PDEs can be solved using the strong form directly at collocation points without special domain integration and essential boundary condition procedures, as will be presented in Chapter 9.

Table 1.1 shows the wide variety of the meshfree methods that have been proposed; Table 1.2 lists the common acronyms for these methods. Essentially, one may classify these methods based on the approximation and how it is employed to solve the governing equations at hand. As can be seen, meshfree methods have generally been developed using the classical weak formulation.

In this book, we do not go into too much detail on the meshfree implementation of Peridynamics [27] or optimal transportation meshfree methods [48], which are based on governing equations other than the classical differential ones. We also do not include the Lagrangian–Eulerian type methods of particle in cell [56, 57] and material point methods [12, 13], which employ Lagrangian points with an Eulerian computational mesh. Instead, we refer the interested reader to the review by Li and Liu [58] for a broad overview that includes meshfree methods that do not strictly fall under the categories of Galerkin or collocation.

We will focus primarily on meshfree methods of the Galerkin type that employ shape functions with compact support, which are covered in Chapters 3–8. Some collocation methods and approximation functions with both global and compact supports will be discussed in Chapter 9. A compact support is essential if the governing equations are to be sparse; large systems of equations that are not sparse are extremely expensive. This observation also applies to solutions by explicit time integration. When the supports are not compact, the acceleration at any node depends on the nodal values of all nodes of the model. This makes the computation of the acceleration very expensive.

Table 1.1 Various Galerkin and collocation-based meshfree methods.

			Solution scheme (Discretization)	
			Weak form	**Strong form**
Approximation	Local Polynomial	Eulerian mesh with Lagrangian points	MPM [12, 13], PFEM-2 [14, 15]	
		Reconstructed	PFEM [16, 17]	
		Discontinuous/meshless	FPM [18]	
	Moving least squares and reproducing kernel	Direct derivatives	EFG [19], RKPM [20, 21]	FP [22], RKCM [23, 24]
		Diffuse/implicit derivatives	DEM [8]	GRKCM [25, 26], GFD [5, 7]
		Non-local derivatives		PD [27], ULPH [28], RKPD [29]
		Smoothed derivatives	SCNI [30], RKGS [31]	GSCM [32]
		Enriched	XEFG [33, 34]	
		Petrov–Galerkin	MLPG [35], VCI [36]	
		Eulerian mesh with Lagrangian points	Improved MPM [37]	
		Reconstructed meshfree	SLRKPM [38, 39]	
	Partition of unity	Polynomial enrichment	*hp*C [40, 41], MFS [42]	
		General enrichment	PUM [43, 44], PPU [45]	
	MaxEnt	Lagrangian	MaxEnt [46, 47]	
		Reconstructed	OTM [48]	
	Natural neighbor	Lagrangian	NEM [49, 50]	
		Reconstructed	MFEM [51]	
	Radial basis functions		RPIM [52]	RBCM [53, 54]
	Kernel approximation		SPH [1, 2]	MPS [55]

Table 1.2 Common acronyms of meshfree methods.

DEM	Diffuse element method
EFG	Element-free Galerkin
FPM	Fragile points method
FP	Finite point
GFD	Generalized finite difference
GRKCM	Gradient reproducing kernel collocation method
GSCM	Gradient smoothing collocation method
*hp*C	*h-p* clouds
MaxEnt	Maximum entropy
MFEM	Meshless finite element method
MFS	Method of finite spheres
MLPG	Meshless local Petrov–Galerkin
MPM	Material point method
MPS	Moving particle semi-implicit
NEM	Natural element method
OTM	Optimal transportation meshfree
PD	Peridynamics
PFEM	Particle finite element method
PFEM-2	Particle finite element method, second generation
PPU	Particle partition of unity
PU	Partition of unity
RBCM	Radial basis collocation method
RKCM	Reproducing kernel collocation method
RKGS	Reproducing kernel gradient smoothing
RKPD	Reproducing kernel peridynamics
RKPM	Reproducing kernel particle method
RPIM	Radial point interpolation method
SCNI	Stabilized conforming nodal integration
SLRKPM	Semi-Lagrangian reproducing kernel particle method
ULPH	Updated Lagrangian particle hydrodynamics
VCI	Variationally consistent integration
XEFG	Extended element-free Galerkin

The situation is akin to molecular dynamics. Potentials that depend on all atoms in a system are very slow, whereas potentials that only involve the nearest neighbors are very fast.

1.6 Applications of Meshfree Methods

Meshfree methods are particularly well-suited for large deformation applications where FEM fails due to mesh entanglement and other mesh-related issues. They have been applied to many different solid mechanics problems such as large deformation of hyperelastic materials [59, 60], metal forming [61, 62], geotechnical analysis [63, 64], earthmoving [38], explosive and magnetic welding [65], and additive manufacturing [66]. Figure 1.4 demonstrates a meshfree simulation of a landslide due to the 1989 Loma Prieta earthquake with substantial fluid-like motion of material undergoing plastic flow. Figure 1.5 shows a simulation of the explosive welding process (which is featured on the cover of this book), with a comparison of the experimentally observed steady-state interfacial wave.

Fracture mechanics is another area where meshfree methods offer a unique strength. Early on, it was recognized that meshfree formulations such as the element-free Galerkin method could offer an effective alternative to finite elements in modeling fracture by cutting the particle influence across a crack and further provide easy adaptive refinement to attain accuracy near crack tips [68, 69]. The cracking particle method [70] offers one such unique implementation. Alternatively, enrichment of the approximation functions for crack tip singularities can be embedded in the approximation [71, 72], or added extrinsically [71, 73] as in the extended finite element method. Figure 1.6 shows different fracture mechanisms developed in the meshfree simulation of a concrete unit-cell specimen subjected to tension and shear.

The naturally conforming properties of meshfree approximations allow adaptivity to be performed in a much more effective manner than conventional FEM. Nodes can be inserted or removed with ease, and error indicators have been formulated to guide adaptive refinement. For example, the multiresolution reproducing kernel particle method [75, 76] enables the scale decomposition of the numerical solution by using the meshfree approximation as a filter. The high-scale solution has also been used as the error indicator for adaptive h-refinement [11].

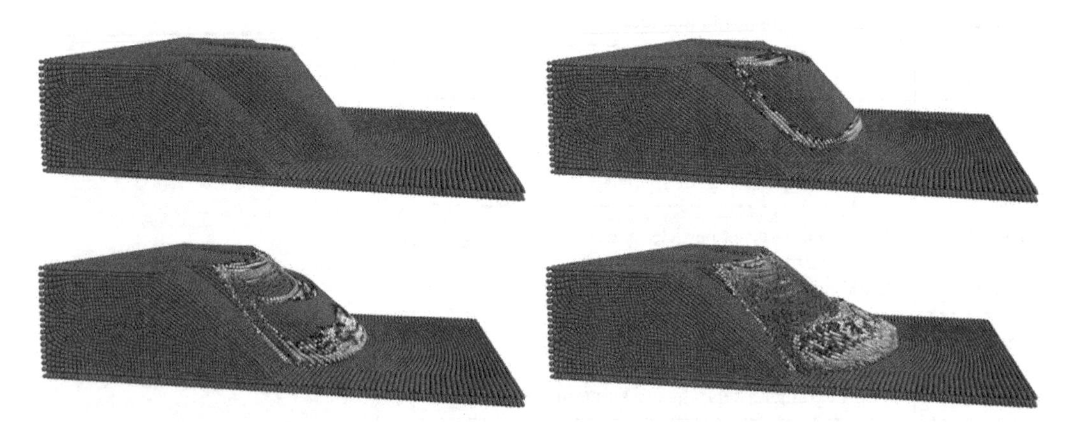

Figure 1.4 Simulation of a landslide triggered by the 1989 Loma Prieta earthquake using meshfree methods. *Source:* Chen et al. [67], figure 26, p. 29 / With permission of ASCE.

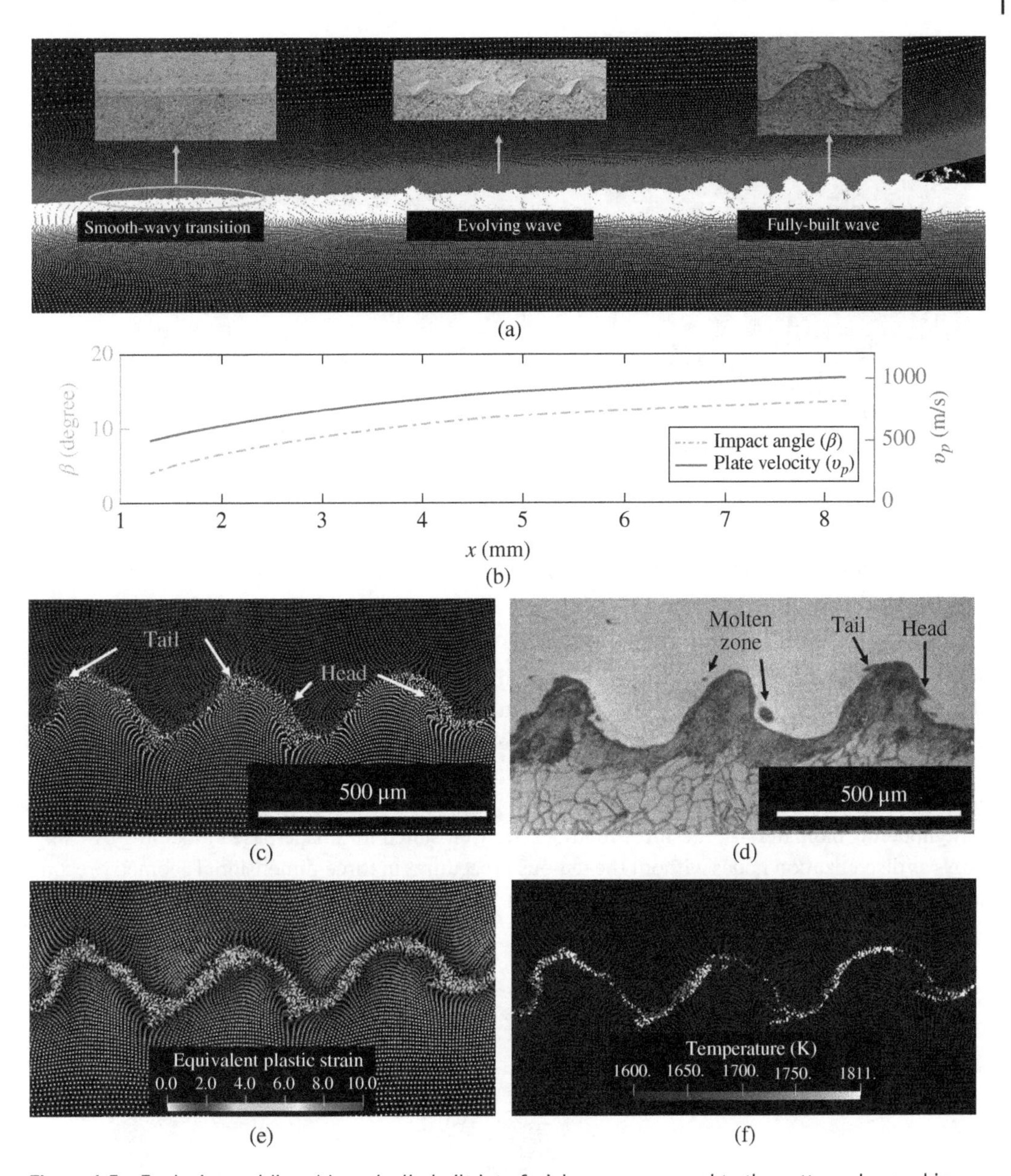

Figure 1.5 Explosive welding: (a) gradually-built interfacial wave compared to the pattern observed in experimental work, (b) the gradually increasing impact angle and vertical plate velocity, (c) numerical wave at steady state, (d) experimental wave at steady state, (e) equivalent plastic strain distribution, and (f) temperature distribution. *Source:* Baek et al. [65], Springer Nature, CC BY 4.0.

While p-adaptivity is not so straightforward, h-p clouds allow the bases to vary throughout the domain such that higher order accuracy can be obtained where needed [40].

Researchers have employed the flexibility of meshfree methods for regularization in localization problems [77–79] to circumvent ambiguous boundary conditions in gradient methods. Methods have also been developed and applied successfully to localization problems difficult for FEM [80, 81].

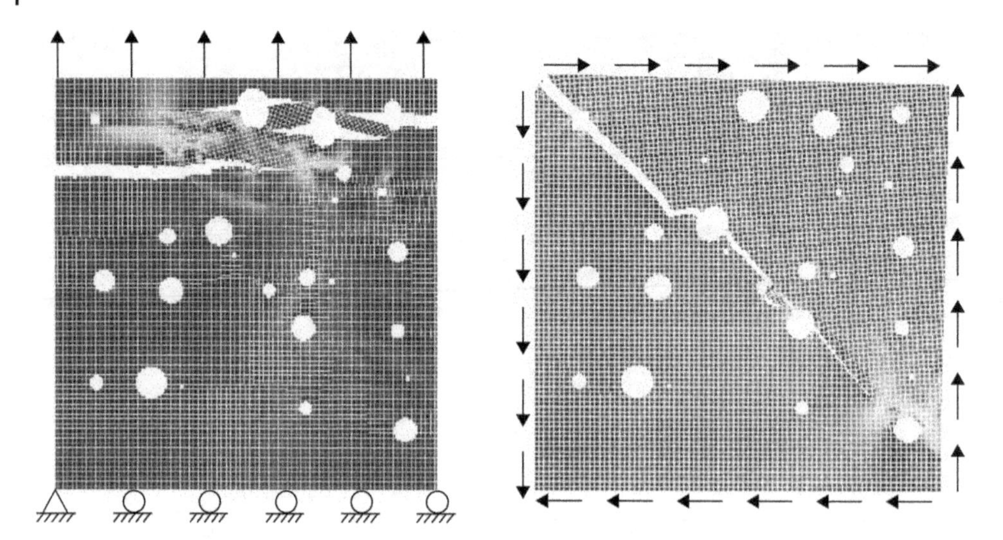

Figure 1.6 Fractures developed in a concrete unit-cell specimen subjected to tension and shear. *Source:* Adapted from Liang et al. [74], figure 11, p 10, and figure 12, p. 10 / With permission of Springer Nature.

By employing the arbitrary smoothness in meshfree approximation functions, a smooth contact algorithm has been proposed that allows continuum-based contact formulations with the full tangent [62]. This, in turn, allows optimal convergence in contact iterations in contrast to finite element-based contact with low continuity and enables robust analysis in applications such as deep drawing of sheet metal with large sliding contact, as shown in Figure 1.7, where C^0 contact fails to converge.

In modeling biomaterials, meshfree methods are well suited for image-based modeling by using pixels as discretization nodes without the tedious procedures in three-dimensional geometry reconstruction from the images with mesh generation [83]. Meshfree methods can also represent the

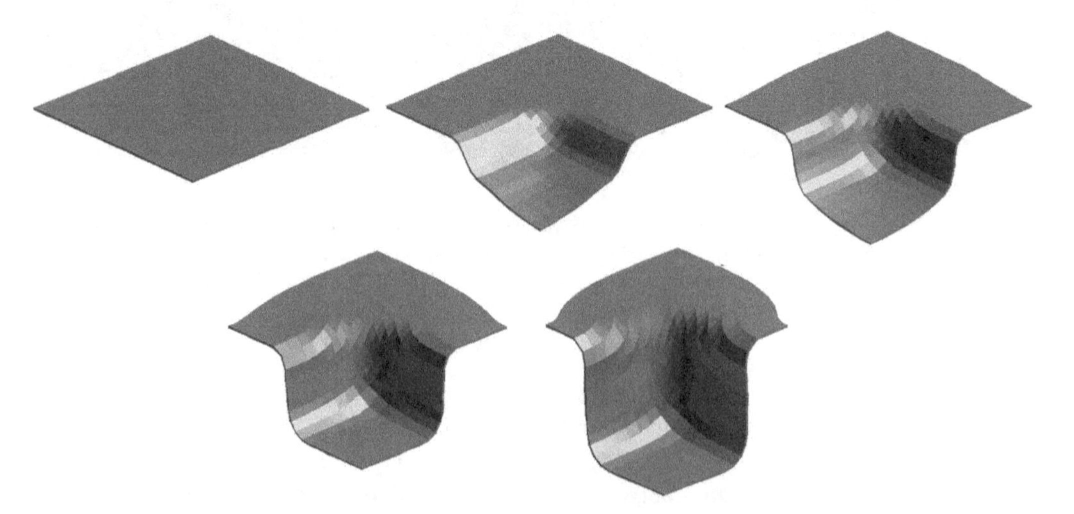

Figure 1.7 Progressive deformation in deep drawing of metal. *Source:* Adapted from Chen et al. [82], Figure 23, p. 38.

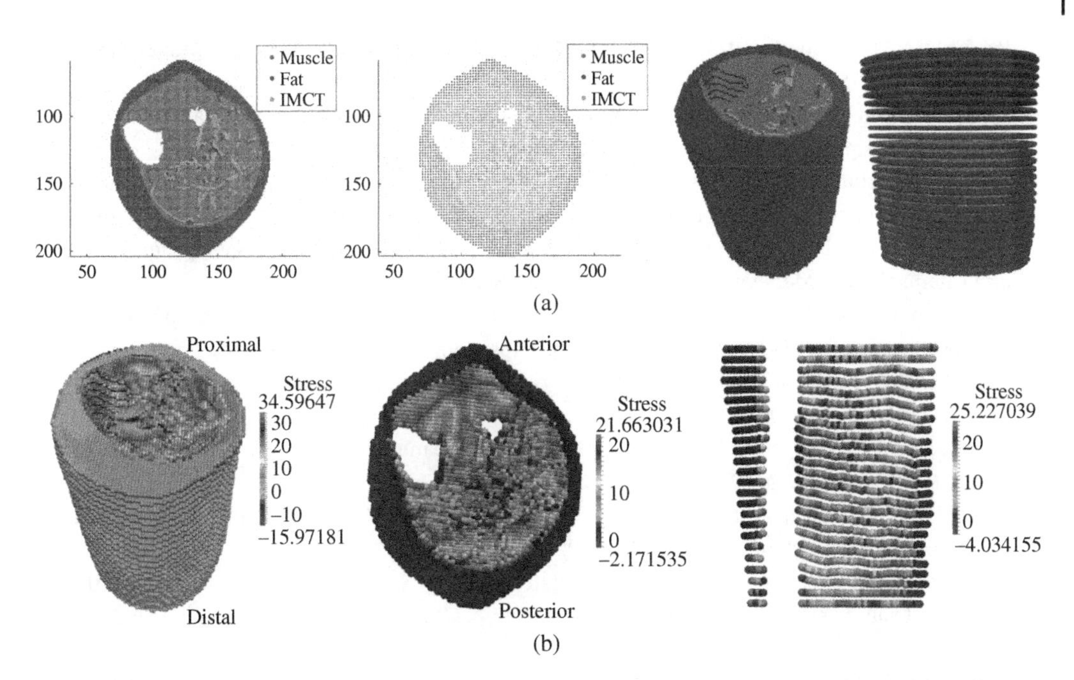

Figure 1.8 Image-based meshfree modeling of a skeletal muscle (a) pixel points as the meshfree discrete points (with muscle, fat, and intramuscular connective tissue (IMCT); fine and course models), perspective views, and (b) maximum principal Cauchy stress (N/cm^2) distribution at pixel points. *Source:* Chen et al. [83], figures 9 and 13, p. 9 and 12 / With permission of Taylor & Francis (https://www.tandfonline.com/).

smooth transition of material properties across material interfaces in biomaterials. Figure 1.8 shows how a skeletal muscle is modeled by meshfree methods. Image pixel points can be used directly as the discretization. The associated stress distribution is computed using material properties and fiber orientations defined at the pixel points.

Another good application of meshfree methods is for problems that involve higher-order differentiation in the PDE, such as thin plate and shell problems [84–86], where meshfree approximation functions with higher-order continuity can be employed with virtually no additional effort. For shape optimization, meshfree methods can avoid mesh distortion in the iterative process [87, 88]. Finally, the smooth meshfree approximation functions are well-suited for optimal employment of Petrov–Galerkin stabilization techniques in convection-dominated problems [89].

References

1 Lucy, L.B. (1977). A numerical approach to the testing of the fission hypothesis. *Astron. J.* 82: 1013–1024.

2 Gingold, R.A. and Monaghan, J.J. (1977). Smoothed particle hydrodynamics: theory and application to non-spherical stars. *Mon. Not. R. Astron. Soc.* 181 (3): 375–389.

3 Monaghan, J.J. (1982). Why particle methods work. *SIAM J. Sci. Stat. Comput.* 3 (4): 422–433.

4 Monaghan, J.J. (1988). An introduction to SPH. *Comput. Phys. Commun.* 48 (1): 89–96.

5 Jensen, P.S. (1972). Finite difference techniques for variable grids. *Comput. Struct.* 2 (1–2): 17–29.

6 Perrones, N. and Kao, R. (1974). A general-finite difference method for arbitrary meshes. *Comput. Struct.* 5 (1): 45–57.

7 Liszka, T.J. and Orkisz, J. (1980). The finite difference method at arbitrary irregular grids and its application in applied mechanics. *Comput. Struct.* 11 (1–2): 83–95.

8 Nayroles, B., Touzot, G., and Villon, P. (1992). Generalizing the finite element method: diffuse approximation and diffuse elements. *Comput. Mech.* 10 (5): 307–318.

9 Chen, J.-S. and Belytschko, T. (2011). Meshless and meshfree methods. In: *Encyclopedia of Applied and Computational Mathematics* (ed. B. Engquist), 886–894. Springer.

10 Hifi, M. and M'hallah, R. (2009). A literature review on circle and sphere packing problems: models and methodologies. *Adv. Oper. Res.* 2009: 22pp. https://doi.org/10.1155/2009/150624.

11 You, Y., Chen, J.-S., and Lu, H. (2003). Filters, reproducing kernel, and adaptive meshfree method. *Comput. Mech.* 31 (3): 316–326.

12 Sulsky, D.L., Chen, Z., and Schreyer, H.L. (1994). A particle method for history-dependent materials. *Comput. Methods Appl. Mech. Eng.* 118 (1–2): 179–196.

13 Sulsky, D.L., Zhou, S.J., and Schreyer, H.L. (1995). Application of a particle-in-cell method to solid mechanics. *Comput. Phys. Commun.* 87 (1–2): 236–252.

14 Idelsohn, S.R., Nigro, N., Limache, A., and Oñate, E. (2012). Large time-step explicit integration method for solving problems with dominant convection. *Comput. Methods Appl. Mech. Eng.* 217–220: 168–185.

15 Idelsohn, S.R., Marti, J., Becker, P., and Oñate, E. (2014). Analysis of multifluid flows with large time steps using the particle finite element method. *Int. J. Numer. Methods Fluids* 75 (9): 621–644.

16 Idelsohn, S.R., Oñate, E., and Del Pin, F. (2003). A Lagrangian meshless finite element method applied to fluid–structure interaction problems. *Comput. Struct.* 81 (8–11): 655–671.

17 Idelsohn, S.R., Oñate, E., and Del Pin, F. (2004). The particle finite element method: a powerful tool to solve incompressible flows with free-surfaces and breaking waves. *Int. J. Numer. Methods Eng.* 61 (7): 964–989.

18 Dong, L., Yang, T., Wang, K., and Atluri, S.N. (2019). A new Fragile Points Method (FPM) in computational mechanics, based on the concepts of Point Stiffnesses and Numerical Flux Corrections. *Eng. Anal. Bound. Elem.* 107: 124–133.

19 Belytschko, T., Lu, Y.Y., and Gu, L. (1993). Element-free Galerkin methods. *Int. J. Numer. Methods Eng.* 37: 229–256.

20 Liu, W.K., Jun, S., and Zhang, Y.F. (1995). Reproducing kernel particle methods. *Int. J. Numer. Methods Fluids* 20 (8–9): 1081–1106.

21 Chen, J.-S., Pan, C., Wu, C.-T., and Liu, W.K. (1996). Reproducing Kernel Particle Methods for large deformation analysis of non-linear structures. *Comput. Methods Appl. Mech. Eng.* 139 (1–4): 195–227.

22 Oñate, E., Idelsohn, S.R., Zienkiewicz, O.C., and Taylor, R.L. (1996). A finite point method in computational mechanics. Applications to convective transport and fluid flow. *Int. J. Numer. Methods Eng.* 39 (22): 3839–3866.

23 Aluru, N.R. (2000). A point collocation method based on reproducing kernel approximations. *Int. J. Numer. Methods Eng.* 47 (6): 1083–1121.

24 Hu, H.-Y., Chen, J.-S., and Hu, W. (2011). Error analysis of collocation method based on reproducing kernel approximation. *Numer. Methods Partial Differ. Equ.* 27 (3): 554–580.

25 Chi, S.-W., Chen, J.-S., Hu, H.-Y., and Yang, J.P. (2013). A gradient reproducing kernel collocation method for boundary value problems. *Int. J. Numer. Methods Eng.* 93 (13): 1381–1402.

26 Mahdavi, A., Chi, S.-W., and Zhu, H. (2019). A gradient reproducing kernel collocation method for high order differential equations. *Comput. Mech.* 64 (5): 1421–1454.

27 Silling, S.A. and Askari, E. (2005). A meshfree method based on the peridynamic model of solid mechanics. *Comput. Struct.* 83 (17–18): 1526–1535.

28 Yan, J., Li, S., Kan, X. et al. (2020). Higher-order nonlocal theory of Updated Lagrangian Particle Hydrodynamics (ULPH) and simulations of multiphase flows. *Comput. Methods Appl. Mech. Eng.* 368: 113176.

29 Hillman, M., Pasetto, M., and Zhou, G. (2020). Generalized reproducing kernel peridynamics: unification of local and non-local meshfree methods, non-local derivative operations, and an arbitrary-order state-based peridynamic formulation. *Comput. Part. Mech.* 7: 435–469.

30 Chen, J.-S., Wu, C.-T., and Yoon, S. (2001). A stabilized conforming nodal integration for Galerkin mesh-free methods. *Int. J. Numer. Methods Eng.* 50 (2): 435–466.

31 Wang, D. and Wu, J. (2019). An inherently consistent reproducing kernel gradient smoothing framework toward efficient Galerkin meshfree formulation with explicit quadrature. *Comput. Methods Appl. Mech. Eng.* 349: 628–672.

32 Qian, Z., Wang, L., Gu, Y., and Zhang, C. (2021). An efficient meshfree gradient smoothing collocation method (GSCM) using reproducing kernel approximation. *Comput. Methods Appl. Mech. Eng.* 374: 113573.

33 Rabczuk, T. and Areias, P.M.A. (2006). A meshfree thin shell for arbitrary evolving cracks based on an extrinsic basis. *CMES - Comput. Model. Eng. Sci.* 16 (2): 115–130.

34 Rabczuk, T., Bordas, S.P.A., and Zi, G. (2007). A three-dimensional meshfree method for continuous multiple-crack initiation, propagation and junction in statics and dynamics. *Comput. Mech.* 40 (3): 473–495.

35 Atluri, S.N. and Zhu, T.L. (1998). A new Meshless Local Petrov-Galerkin (MLPG) approach in computational mechanics. *Comput. Mech.* 22 (2): 117–127.

36 Chen, J.-S., Hillman, M., and Rüter, M. (2013). An arbitrary order variationally consistent integration for Galerkin meshfree methods. *Int. J. Numer. Methods Eng.* 95 (5): 387–418.

37 Sulsky, D. and Gong, M. (2016). Improving the material-point method. In: *Innovative Numerical Approaches for Multi-Field and Multi-Scale Problem* (ed. K. Weinberg and A. Pandolfi), 217–240. Springer.

38 Guan, P.C., Chen, J.-S., Wu, Y. et al. (2009). Semi-Lagrangian reproducing kernel formulation and application to modeling earth moving operations. *Mech. Mater.* 41 (6): 670–683.

39 Guan, P.C., Chi, S.W., Chen, J.-S. et al. (2011). Semi-Lagrangian reproducing kernel particle method for fragment-impact problems. *Int. J. Impact Eng.* 38 (12): 1033–1047.

40 Duarte, C.A.M. and Oden, J.T. (1996). An *h-p* adaptive method using clouds. *Comput. Methods Appl. Mech. Eng.* 139 (1–4): 237–262.

41 Duarte, C.A.M. and Oden, J.T. (1996). *H-p* clouds—an *h-p* meshless method. *Numer. Methods Partial Differ. Equ.* 12 (6): 673–705.

42 De, S. and Bathe, K.J. (2000). The method of finite spheres. *Comput. Mech.* 25 (4): 329–345.

43 Melenk, J.M. and Babuška, I. (1996). The partition of unity finite element method: basic theory and applications. *Comput. Methods Appl. Mech. Eng.* 139 (1–4): 289–314.

44 Babuška, I. and Melenk, J.M. (1997). The partition of unity method. *Int. J. Numer. Methods Eng.* 40 (4): 727–758.

45 Griebel, M. and Schweitzer, M.A. (2000). A particle-partition of unity method for the solution of elliptic, parabolic, and hyperbolic PDEs. *SIAM J. Sci. Comput.* 22 (3): 853–890.

46 Arroyo, M. and Ortiz, M. (2006). Local maximum-entropy approximation schemes: a seamless bridge between finite elements and meshfree methods. *Int. J. Numer. Methods Eng.* 65 (13): 2167–2202.

47 Sukumar, N. (2004). Construction of polygonal interpolants: a maximum entropy approach. *Int. J. Numer. Methods Eng.* 61 (12): 2159–2181.

48 Li, B., Habbal, F., and Ortiz, M. (2010). Optimal transportation meshfree approximation schemes for fluid and plastic flows. *Int. J. Numer. Methods Eng.* 83 (12): 1541–1579.

49 Sukumar, N. and Belytschko, T. (1998). The natural element method in solid mechanics. *Int. J. Numer. Methods Eng.* 43 (5): 839–887.

50 Braun, J. and Sambridge, M. (1995). A numerical method for solving partial differential equations on highly irregular evolving grids. *Nature* 376 (6542): 655–660.

51 Idelsohn, S.R., Oñate, E., Calvo, N., and Del Pin, F. (2003). The meshless finite element method. *Int. J. Numer. Methods Eng.* 58 (6): 893–912.

52 Wang, J.G. and Liu, G.-R. (2002). A point interpolation meshless method based on radial basis functions. *Int. J. Numer. Methods Eng.* 54 (11): 1623–1648.

53 Kansa, E.J. (1990). Multiquadrics – a scattered data approximation scheme with applications to computational fluid-dynamics – I surface approximations and partial derivative estimates. *Comput. Math. Appl.* 19 (8): 127–145.

54 Kansa, E.J. (1990). Multiquadrics – a scattered data approximation scheme with applications to computational fluid-dynamics – II solutions to parabolic, hyperbolic and elliptic partial differential equations. *Comput. Math. Appl.* 19 (8): 147–161.

55 Koshizuka, S. and Oka, Y. (1996). Moving-particle semi-implicit method for fragmentation of incompressible fluid. *Nucl. Sci. Eng.* 123 (3): 421–434.

56 Brackbill, J.U. and Ruppel, H.M. (1986). FLIP: a method for adaptively zoned, particle-in-cell calculations of fluid flows in two dimensions. *J. Comput. Phys.* 65 (2): 314–343.

57 Brackbill, J.U., Kothe, D.B., and Ruppel, H.M. (1988). Flip: a low-dissipation, particle-in-cell method for fluid flow. *Comput. Phys. Commun.* 48 (1): 25–38.

58 Li, S. and Liu, W.K. (2002). Meshfree and particle methods and their applications. *Appl. Mech. Rev.* 1 (55): 1–34.

59 Chen, J.-S., Pan, C., and Wu, C.-T. (1997). Large deformation analysis of rubber based on a reproducing kernel particle method. *Comput. Mech.* 19 (3): 211–227.

60 Chen, J.-S., Yoon, S., Wang, H.-P., and Liu, W.K. (2000). An improved reproducing kernel particle method for nearly incompressible finite elasticity. *Comput. Methods Appl. Mech. Eng.* 181: 117–145.

61 Chen, J.-S., Pan, C., Roque, C.M.O.L., and Wang, H.-P. (1998). A Lagrangian reproducing kernel particle method for metal forming analysis. *Comput. Mech.* 22 (3): 289–307.

62 Wang, H.-P., Wu, C.-T., and Chen, J.-S. (2014). A reproducing kernel smooth contact formulation for metal forming simulations. *Comput. Mech.* 54 (1): 151–169.

63 Bui, H.H., Fukagawa, R., Sako, K., and Ohno, S. (2008). Lagrangian meshfree particles method (SPH) for large deformation and failure flows of geomaterial using elastic-plastic soil constitutive model. *Int. J. Numer. Anal. Methods Geomech.* 32 (12): 1537–1570.

64 Wei, H., Chen, J.-S., Beckwith, F., and Baek, J. (2020). A naturally stabilized semi-Lagrangian meshfree formulation for multiphase porous media with application to landslide modeling. *J. Eng. Mech.* 146 (4): 4020012.

65 Baek, J., Chen, J.-S., Zhou, G. et al. (2021). A semi-Lagrangian reproducing kernel particle method with particle-based shock algorithm for explosive welding simulation. *Comput. Mech.* 67 (6): 1601–1627.

66 Fan, Z. and Li, B. (2019). Meshfree simulations for additive manufacturing process of metals. *Integr. Mater. Manuf. Innov.* 8 (2): 144–153.

67 Chen, J.-S., Hillman, M., and Chi, S.W. (2017). Meshfree methods: progress made after 20 years. *J. Eng. Mech.* 143 (4): 04017001. 38pp.

68 Belytschko, T., Gu, L., and Lu, Y.Y. (1994). Fracture and crack growth by element free Galerkin methods. *Model. Simul. Mater. Sci. Eng.* 2 (3A): 519–534.

69 Belytschko, T., Lu, Y.Y., and Gu, L. (1995). Crack propagation by element-free Galerkin methods. *Eng. Fract. Mech.* 51 (2): 295–315.

70 Rabczuk, T. and Belytschko, T. (2004). Cracking particles: a simplified meshfree method for arbitrary evolving cracks. *Int. J. Numer. Methods Eng.* 61 (13): 2316–2343.

71 Fleming, M., Chu, Y.A., Moran, B., and Belytschko, T. (1997). Enriched element-free Galerkin methods for crack tip fields. *Int. J. Numer. Methods Eng.* 40 (8): 1483–1504.

72 Belytschko, T., Krongauz, Y., Fleming, M. et al. (1996). Smoothing and accelerated computations in the element free Galerkin method. *J. Comput. Appl. Math.* 74 (1–2): 111–126.

73 Ventura, G., Xu, J.X., and Belytschko, T. (2002). A vector level set method and new discontinuity approximations for crack growth by EFG. *Int. J. Numer. Methods Eng.* 54 (6): 923–944.

74 Liang, S., Chen, J.-S., Li, J. et al. (2017). Numerical investigation of statistical variation of concrete material properties between scales. *Int. J. Fract.* 208 (1): 97–113.

75 Liu, W.K., Chen, Y., Chang, C.T., and Belytschko, T. (1996). Advances in multiple scale kernel particle methods. *Comput. Mech.* 18 (2): 73–111.

76 Liu, W.K., Hao, W., Chen, Y. et al. (1997). Multiresolution reproducing kernel particle methods. *Comput. Mech.* 20 (4): 295–309.

77 Chen, J.-S., Wu, C.-T., and Belytschko, T. (2000). Regularization of material instabilities by meshfree approximations with intrinsic length scales. *Int. J. Numer. Methods Eng.* 47 (7): 1303–1322.

78 Chen, J.-S., Zhang, X., and Belytschko, T. (2004). An implicit gradient model by a reproducing kernel strain regularization in strain localization problems. *Comput. Methods Appl. Mech. Eng.* 193 (27–29): 2827–2844.

79 Wei, H. and Chen, J.-S. (2018). A damage particle method for smeared modeling of brittle fracture. *Int. J. Multiscale Comput. Eng.* 16 (4): 303–324.

80 Liu, W.K. and Jun, S. (1998). Multiple-scale reproducing kernel particle methods for large deformation problems. *Int. J. Numer. Methods Eng.* 41 (7): 1339–1362.

81 Li, S., Liu, W.K., Qian, D., and Guduru, P.R. (2001). Dynamic shear band propagation and micro-structure of adiabatic shear band. *Comput. Methods Appl. Mech. Eng.* 191: 73–92.

82 Chen, J.-S., Liu, W.K., Hillman, M. et al. (2018). Reproducing kernel particle method for solving partial differential equations. In: *Encyclopedia of Computational Mechanics, Volume 2*, 2e (eds. E. Stein, R. de Borst and T. Hughes), London: Wiley.

83 Chen, J.-S., Basava, R.R., Zhang, Y. et al. (2015). Pixel-based meshfree modelling of skeletal muscles. *Comput. Methods Biomech. Biomed. Eng. Imaging Vis.* 4 (2): 73–85.

84 Chen, J.-S. and Wang, D. (2006). A constrained reproducing kernel particle formulation for shear deformable shell in Cartesian coordinates. *Int. J. Numer. Methods Eng.* 68 (2): 151–172.

85 Wang, D. and Chen, J.-S. (2008). A Hermite reproducing kernel approximation for thin-plate analysis with sub-domain stabilized conforming integration. *Int. J. Numer. Methods Eng.* 74 (3): 368–390.

86 Behzadinasab, M., Alaydin, M., Trask, N., and Bazilevs, Y. (2022). A general-purpose, inelastic, rotation-free Kirchhoff–Love shell formulation for peridynamics. *Comput. Methods Appl. Mech. Eng.* 389: 114422.

87 Kim, N.H., Choi, K.K., and Chen, J.-S. (2001). Die shape design optimization of sheet metal stamping process using meshfree method. *Int. J. Numer. Methods Eng.* 51 (12): 1385–1406.

88 Kim, N.H., Choi, K.K., Chen, J.-S., and Park, Y.H. (2000). Meshless shape design sensitivity analysis and optimization for contact problem with friction. *Comput. Mech.* 25 (2–3): 157–168.

89 Huerta, A. and Fernández-Méndez, S. (2003). Time accurate consistently stabilized mesh-free methods for convection dominated problems. *Int. J. Numer. Methods Eng.* 56 (9): 1225–1242.

2

Preliminaries: Strong and Weak Forms of Diffusion, Elasticity, and Solid Continua

In this chapter, the strong forms of the equations of diffusion, elasticity, and nonlinear continuum mechanics will be given. Variational principles and weak forms will be presented for each of these, emphasizing various ways of treating the essential boundary conditions. The latter is important in meshfree methods because it is difficult to construct trial and test functions that satisfy the standard conditions on essential boundaries, unlike finite element methods. The term *strong form* refers to the actual partial differential equations and their boundary conditions. The *weak form* is an equivalent integral equation that usually has weaker requirements for the continuity of the solutions.

We will use both tensor and indicial notation. Boldface symbols are tensors or vectors; when indicial notation is used, the subscript refers to the scalar component, and repeated indices are summed over their range (Einstein notation).

2.1 Diffusion Equation

2.1.1 Strong Form of the Diffusion Equation

We will first give the strong and weak forms of the linear diffusion equation. Consider a domain Ω with boundary Γ as shown in Figure 2.1. The boundary Γ is subdivided into two parts Γ_u and Γ_q, where the primary variable u and its flux q are prescribed, respectively. The two parts of the boundary are complementary, i.e., $\Gamma_u \cap \Gamma_q = \emptyset$ and $\Gamma_u \cup \Gamma_q = \Gamma$. The boundary is a curve in two dimensions and a surface in three dimensions, but we will often refer to it simply as a surface.

The partial differential equation and boundary conditions for the steady-state isotropic linear diffusion equation are given by the problem of finding the scalar field $u(\mathbf{x})$ that satisfies

$$\nabla \cdot (k\nabla u) + s = 0 \quad \text{in } \Omega, \quad \text{or} \quad (ku_{,i})_{,i} + s = 0 \quad \text{in } \Omega, \tag{2.1}$$

$$u = \overline{u} \text{ on } \Gamma_u, \qquad \text{(essential, Dirichlet)} \tag{2.2}$$

$$k\nabla u \cdot \mathbf{n} = -\mathbf{q} \cdot \mathbf{n} = \overline{q} \quad \text{on } \Gamma_q, \quad \text{or} \quad ku_{,i}n_i = -q_i n_i = \overline{q} \quad \text{on } \Gamma_q, \qquad \text{(natural, Neumann)} \tag{2.3}$$

where $(\cdot)_{,i} \equiv \partial(\cdot)/\partial x_i$ denotes differentiation with respect to a coordinate, and the flux is $q_i = -ku_{,i}$.

We have given the equations in both vector and indicial notation form. Any steady diffusion problem with isotropic diffusivity can be put in this form. For heat conduction, the variable u is the

Meshfree and Particle Methods: Fundamentals and Applications, First Edition.
Ted Belytschko, J. S. Chen, and Michael Hillman.

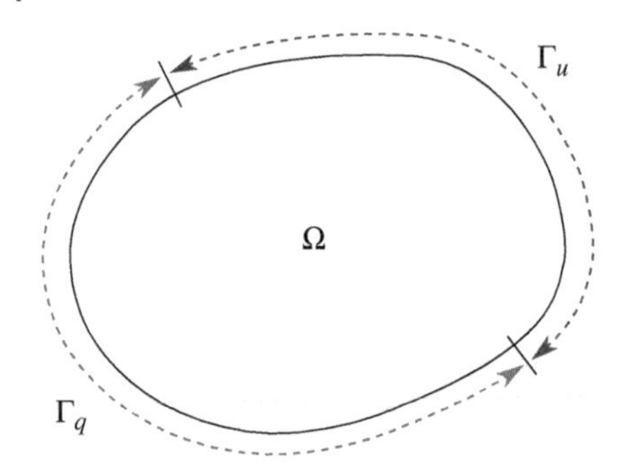

Figure 2.1 Problem domain Ω, essential boundary Γ_u, and natural boundary Γ_q for the diffusion equation.

temperature, s is the heat source, and k is the conductivity. For other diffusion problems, such as the diffusion of interstitials and vacancies in metals, u is the concentration, and k is the diffusion coefficient (or diffusivity). The fact that k is a scalar is a result of the assumption that the medium is isotropic (thus u diffuses the same in all directions).

Equations (2.2) and (2.3) are the boundary conditions, which can be applied to either the function or its gradient. We have given two alternate names for the boundary conditions on the function and its normal gradient. The conditions on the function are usually called *essential boundary conditions* in variational and finite element methods; this name originates from the fact that these conditions must be met by the approximate trial solutions in Galerkin and Rayleigh–Ritz methods. They are called *Dirichlet boundary conditions* in the finite difference literature.

The conditions on the normal gradient of the function are called *natural boundary conditions* in the literature on variational and finite element methods and *Neumann boundary conditions* in the finite difference literature. The name natural boundary condition emanates from the fact that they are enforced naturally in Galerkin methods.

Recall that these two parts of the boundary are complementary:

$$\Gamma_u \cup \Gamma_q = \Gamma, \ \Gamma_u \cap \Gamma_q = \emptyset. \tag{2.4}$$

This means any point on the boundary must either be a natural boundary or an essential boundary, but it cannot be both. There are good underlying physical arguments for this. For example, one cannot prescribe both the temperature and the heat flux at a point since the temperature drives the flux. The combination of the governing equation and the boundary conditions will be called the strong form.

In some cases, such as in acoustic problems, boundary conditions involve both the function and its derivatives. These are called Robin boundary conditions and are written in the form

$$cu + k\nabla u \cdot \mathbf{n} = \bar{q}, \tag{2.5}$$

where c is a constant. For nonlinear, possibly anisotropic time-dependent diffusion problems, where the solution is now a function of both position and time, denoted as $u(\mathbf{x}, t)$, the governing equation can be expressed in terms of the flux $\mathbf{q}$ by

$$-\nabla \cdot \mathbf{q} + s = \alpha \frac{\partial u}{\partial t}. \tag{2.6}$$

For heat conduction, α is the product of the material's density and specific heat capacity. For conciseness, we will combine the time derivative with the source term in many of the developments. In that case, the governing equation (2.6) becomes

$$-\nabla \cdot \mathbf{q} + \tilde{s} = 0 \quad \text{where} \quad \tilde{s} = s - \alpha \frac{\partial u}{\partial t}. \tag{2.7}$$

The flux $\mathbf{q}$ in the above is generally a function of the gradient of the primary variable $u(\mathbf{x}, t)$. This relation takes on various names for various physical problems: in heat conduction, it is called

Fourier's law for the linear case; in other diffusion equations, it is called Fick's law. Sometimes the relation between the flux and the gradient is called a constitutive equation, although this name is primarily used for relations between stresses and strains for solids and fluids.

We can write the general nonlinear relation between flux and the gradient of the field as

$$\mathbf{q} = \mathbf{q}(\nabla u). \tag{2.8}$$

For a linear relationship between the flux and the gradient, which is known as Fourier's law in heat conduction, the equation corresponding to (2.8) is

$$\mathbf{q} = -\mathbf{k} \cdot \nabla u, \quad \text{or} \quad q_i = -k_{ij} u_{,j}, \tag{2.9}$$

where in heat conduction, k_{ij} is symmetric and positive definite due to physical considerations; heat must flow from hot to cold areas. For isotropic diffusion, $k_{ij} = k\delta_{ij}$. Substituting $k_{ij} = k\delta_{ij}$ into (2.9), and substituting the result into (2.6), and setting $\alpha = 0$ yields (2.1). Therefore, (2.1) can be viewed as a special case of (2.6) and (2.9) for linear, isotropic diffusion at steady-state conditions (the solution does not depend on time, as the time-derivative term drops).

For anisotropic diffusion under steady-state conditions (setting $\alpha = 0$), one obtains

$$\nabla \cdot (\mathbf{k} \cdot \nabla u) + s = 0, \quad \text{or} \quad \left(k_{ij} u_{,j}\right)_{,i} + s = 0, \tag{2.10}$$

and for constant, isotropic $\mathbf{k}$ ($k_{ij} = k\delta_{ij}$) we obtain the well-known Poisson equation

$$k\nabla^2 u + s = 0, \quad \text{or} \quad k u_{,ii} + s = 0, \tag{2.11}$$

where $\nabla^2 \equiv \nabla \cdot \nabla$ is the Laplace operator.

A diffusion problem is said to have a potential W when

$$\mathbf{q} = -\frac{\partial W}{\partial(\nabla u)}, \quad \text{or} \quad q_i = -\frac{\partial W}{\partial u_{,i}}. \tag{2.12}$$

In the following, variational principles and weak forms are given for the diffusion equation. For steady-state diffusion problems with a potential, either a variational approach or a weak form can be used, depending on which is more convenient. But the variational form only applies to steady-state problems in which $u(\mathbf{x})$ can be obtained by a potential. Weak forms, on the contrary, can easily be developed for more general flux-gradient relationships and time-dependent problems as in (2.6).

2.1.2 The Variational Principle for the Diffusion Equation

Here, we will give the standard variational principle that corresponds to (2.1)–(2.3). Variational principles and variational equations are used to select the best solution (usually minimizing the error in the derivatives) from a space of *trial solutions,* also called *candidate solutions*. We first consider the variational principle for linear, isotropic diffusion that applies to trial solutions that satisfy the essential boundary conditions. We will then give the variational equation and show that it implies the strong form.

It is, however, non-trivial to construct such approximations for meshfree methods, so this variational principle is often not applicable. Therefore, we will also give principles that are applicable when the essential boundary conditions are not satisfied by the trial solutions.

We will first state the variational principle for the diffusion equation and then give the variational equation that emanates from it; it will be seen later that this variational equation is identical to the weak form that is obtained by the method of weighted residuals. The method of weighted residuals is a more general technique that can be applied to partial differential equations for which variational principles are not available. Discretizations of partial differential equations can be obtained either directly from the variational equation or from the corresponding weak form.

2.1.2.1 The Standard Variational Principle

The standard variational principle for the isotropic diffusion equation (2.1) considers that the trial solutions satisfy the essential boundary conditions. This principle for the diffusion equation states the following: the solution $u(\mathbf{x})$ of (2.1) with boundary conditions (2.2) and (2.3) is the minimizer of the functional $\Pi_D(u(\mathbf{x}))$ for $u(\mathbf{x}) \in \mathrm{U}$ where

$$\Pi_D(u(\mathbf{x})) = \int_\Omega \left(\frac{1}{2} k u_{,i} u_{,i} - su\right) d\Omega - \int_{\Gamma_q} u\bar{q} d\Gamma, \tag{2.13}$$

where the subscript "D" indicates that it pertains to the diffusion equation and

$$\mathrm{U} = \{u(\mathbf{x}) | u(\mathbf{x}) \in \mathrm{H}^1 \text{ and } u(\mathbf{x}) = \bar{u} \text{ on } \Gamma_u\}. \tag{2.14}$$

The space of functions U, as indicated by (2.14), requires functions to be smooth enough (with C^0 continuity) with square-integrable derivatives, and satisfy the essential boundary condition. Functions that satisfy these conditions, i.e., the ones that are in the space U, are called *admissible*. In more mathematical terms, it is said that the trial functions are in the Sobolev space of degree one (H^1), which is the space of functions where the derivative and the function itself are square integrable (see the Glossary of Notation for a full definition).

A partial differential equation can be solved via a variational principle either by expressing the functional (2.13) in terms of the parameters of the trial solutions and then minimizing the function that corresponds to the functional $\Pi_D(u(\mathbf{x}))$ or by developing a variational equation from (2.13) directly and obtaining the solution from that equation. The latter is employed in Chapter 4 for meshfree approximations.

2.1.2.2 The Variational Equation

To derive the equation that corresponds to the variational principle, it is important to bear in mind that $\Pi_D(u(\mathbf{x}))$ is a functional, meaning that it is a function of functions. Its variation, i.e., its change, is denoted by $\delta\Pi_D$ and given by

$$\delta\Pi_D(u(\mathbf{x})) = \Pi_D(u(\mathbf{x}) + \delta u(\mathbf{x})) - \Pi_D(u(\mathbf{x})), \tag{2.15}$$

where $\delta u(\mathbf{x})$ are variations of the trial solution; we will also call them test functions in the context of developing variational equations, for they correspond to test functions in the method of weighted residuals. The variations $\delta u(\mathbf{x})$ of the trial solutions must be constructed such that the variations vanish on the essential boundaries and also be in H^1. This space of variations of the trial solutions is denoted by U_0 and given as

$$\mathrm{U}_0 = \left\{\delta u(\mathbf{x}) \middle| \delta u(\mathbf{x}) \in \mathrm{H}^1 \text{ and } \delta u(\mathbf{x}) = 0 \text{ on } \Gamma_u\right\}. \tag{2.16}$$

The operation of taking a variation of a functional is identical in form to taking the derivative of a function. Thus, standard rules for derivatives can be used to evaluate variations.

The *variational equations*, i.e., the equations for the minimizer of $\Pi_D(u(\mathbf{x}))$, are found by taking the variation of the functional and, since the minimizer is a stationary point, setting the variation to zero. This gives

$$\delta\Pi_D = \int_\Omega (ku_{,i}\,\delta u_{,i} - s\delta u)d\Omega - \int_{\Gamma_q} \delta u\bar{q}d\Gamma = 0. \tag{2.17}$$

The variational equation corresponding to the strong form is then:

$$\text{Find } u \in \mathrm{U} \text{ such that } \delta\Pi_D = 0 \ \forall \ \delta u \in \mathrm{U}_0. \tag{2.18}$$

When a variational equation such as (2.17) is used to obtain an approximate solution by constructing discrete counterparts of $u(\mathbf{x})$ and $\delta u(\mathbf{x})$, the functions are often called *trial functions* and *test functions*, respectively.

2.1.2.3 Equivalence of the Variational Equation and the Strong Form

We next show that (2.18) is equivalent to the governing equation (2.1) including the boundary conditions (2.2) and (2.3). The main step in this demonstration is an integration by parts of the first term in Eq. (2.17). The integration by parts formula is an important tool for relating weak (or variational) forms to strong forms, so we briefly develop it. To start, recall that the derivative using the product rule gives

$$(\delta u(ku_{,i}))_{,i} = \delta u_{,i}ku_{,i} + \delta u(ku_{,i})_{,i}. \tag{2.19}$$

If we rearrange the order of the above terms and integrate them over the domain, we obtain

$$\int_\Omega \delta u(ku_{,i})_{,i}d\Omega = \int_\Omega (\delta u(ku_{,i}))_{,i}d\Omega - \int_\Omega \delta u_{,i}ku_{,i}d\Omega. \tag{2.20}$$

The *divergence theorem* (also known as Gauss's theorem) states that for any C^1 vector field $\mathbf{g}(\mathbf{x})$

$$\begin{gathered}\int_\Omega \nabla\cdot\mathbf{g}d\Omega = \int_\Gamma \mathbf{g}\cdot\mathbf{n}\,d\Gamma \ \text{ if } \mathbf{g} \text{ is } C^1,\\ \text{or}\\ \int_\Omega g_{i,i}d\Omega = \int_\Gamma g_i n_i\,d\Gamma \ \text{ if } g_i \text{ is } C^1,\end{gathered} \tag{2.21}$$

where n_i is the unit normal to the boundary. For a scalar field f, (2.21) reduces to

$$\begin{gathered}\int_\Omega \nabla f d\Omega = \int_\Gamma f\mathbf{n}\,d\Gamma \ \text{ if } f \text{ is } C^1,\\ \text{or}\\ \int_\Omega f_{,i}\,d\Omega = \int_\Gamma fn_i\,d\Gamma \ \text{ if } f \text{ is } C^1.\end{gathered} \tag{2.22}$$

The divergence theorem and product rule can be used to develop *integration by parts* in multiple dimensions, which states

$$\begin{aligned}&\int_{\Omega} v\nabla\cdot\mathbf{g}d\Omega = \int_{\Gamma} v\mathbf{g}\cdot\mathbf{n}d\Gamma - \int_{\Omega}\nabla v\cdot\mathbf{g}d\Omega \text{ if } v \text{ and } \mathbf{g} \text{ are in } C^1,\\ &\text{or}\\ &\int_{\Omega} vg_{i,i}d\Omega = \int_{\Gamma} vg_i n_i d\Gamma - \int_{\Omega} v_{,i}g_i d\Omega \text{ if } v \text{ and } g_i \text{ are in } C^1.\end{aligned} \tag{2.23}$$

Using (2.21) on the first term on the right-hand side of (2.20), with $g_i = \delta u(ku_{,i})$, gives the following integration by parts formula after rearranging

$$\int_{\Omega} ku_{,i}\delta u_{,i}d\Omega = \int_{\Gamma}\delta uku_{,n}d\Gamma - \int_{\Omega}\delta u(ku_{,i})_{,i}d\Omega, \tag{2.24}$$

where $u_{,n} \equiv u_{,i}n_i$ is the gradient normal to the boundary. The above could have also been obtained directly by the divergence theorem. Since δu vanishes on Γ_u, it follows from the above that

$$\int_{\Omega} ku_{,i}\delta u_{,i}d\Omega = \int_{\Gamma_q}\delta uku_{,n}d\Gamma - \int_{\Omega}\delta u(ku_{,i})_{,i}d\Omega. \tag{2.25}$$

Now, substituting (2.25) into the first term of (2.17) and collecting terms gives (after a change of sign)

$$\int_{\Omega}\delta u\left((ku_{,i})_{,i} + s\right)d\Omega - \int_{\Gamma_q}\delta u(ku_{,n} - \overline{q})d\Gamma = 0. \tag{2.26}$$

Since the above must hold for arbitrary $\delta u(\mathbf{x})$, it follows that the coefficients of $\delta u(\mathbf{x})$ in the above vanish, i.e., $(ku_{,i})_{,i} + s = 0$ in Ω and $ku_{,n} - \overline{q} = 0$ on Γ_q (this is called the fundamental lemma of the calculus of variations, termed *the principle of variations* here for simplicity). These are, respectively, the governing equation (2.1) and the natural boundary condition (2.3). Any trial solutions must meet the essential boundary conditions (2.2) because they must be in the space U, as seen in (2.14). Thus, the solution to the variational equation satisfies the governing equation and all boundary conditions.

Furthermore, because the variational equation is equivalent to the strong form, we can conclude that the minimization of the functional also corresponds to the strong form.

One crucial proviso to the above development is that we have taken the space of trial functions to be in H^1, i.e., they are C^0 functions. However, the divergence theorem requires more smoothness; as stated in (2.21), the integrand must be C^1. Therefore, the above development of the weak form does not account for weak or strong discontinuities, where *weak discontinuities* are discontinuities in the derivatives and *strong discontinuities* are discontinuities in the functions themselves. For consideration of arbitrary weak and strong discontinuities, we will need to generalize the formulas in (2.20)–(2.25).

To extend the divergence theorem and integration by parts to less smooth functions, we first need to develop an integration by parts formula for functions that are C^0, with weak discontinuities on a set of surfaces, or even C^{-1}, with discontinuities in the functions themselves. Nevertheless, we must still insist that the functions are C^1 everywhere else.

A general scenario is shown in Figure 2.2: let the interior boundaries of n_s subdomains Ω_i with strong and/or weak discontinuities be denoted by Γ_i, $i = 1$ to n_s, with the union of all these interior interfaces denoted by Γ_{int}. Later, we will distinguish between the interfaces with weak and strong discontinuities as $\Gamma_{\text{int}}^{\text{W}}$ and $\Gamma_{\text{int}}^{\text{S}}$, respectively. Note that each boundary Γ_i will be shared by two domains, which defines an interface (see Figure 2.2).

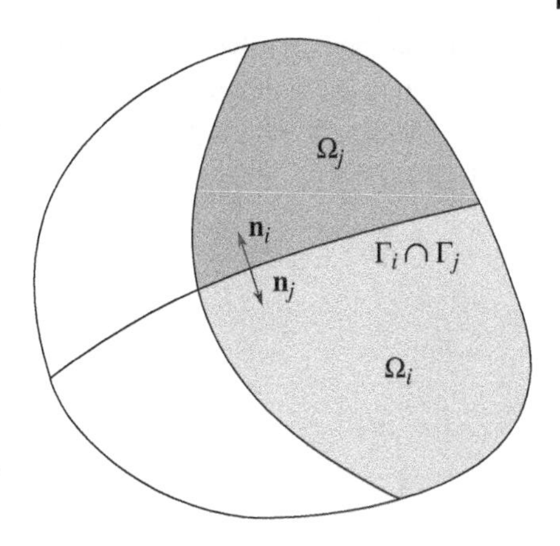

Figure 2.2 Strong and weak discontinuities on interfaces between subdomains.

First, we decompose the integral (2.22) on the left-hand side into the contributions from each of the interior domains Ω_i, where the functions and its derivatives are continuous (where $f \in C^1$):

$$\int_{\Omega} \nabla f d\Omega = \sum_{i=1}^{n_s} \int_{\Omega_i} \nabla f d\Omega. \tag{2.27}$$

Next, applying (2.22) to each of the subdomains in (2.27), we obtain the following:

$$\int_{\Omega} \nabla f d\Omega = \int_{\Gamma} \mathbf{n} f d\Omega + \sum_{i=1}^{n_s} \int_{\Gamma_i} \mathbf{n}_i f_i \, d\Gamma, \tag{2.28}$$

where $\mathbf{n}_i$ is the outward unit normal to the interior boundary Γ_i and $f_i(\mathbf{x}) = \lim_{\varepsilon \to 0} f(\mathbf{x} - \varepsilon \mathbf{n}_i)$, i.e., the value of f on the side of Ω_i.

Now, consider two neighboring subdomains Ω_i and Ω_j, with a shared boundary $\Gamma_i \cap \Gamma_j$ as shown in Figure 2.2: there are two contributions with equal and opposite normals $\mathbf{n}_i$ and $\mathbf{n}_j$, respectively. Using this fact, the last expression in (2.28) can be reduced to integration along the union of all interior interfaces Γ_{int}:

$$\int_{\Omega} \nabla f d\Omega = \int_{\Gamma} \mathbf{n} f d\Omega + \int_{\Gamma_{\text{int}}} [f] \mathbf{n} d\Gamma \text{ when } f \in C^{-1}, \tag{2.29}$$

where for two given subdomains Ω_i and Ω_j that share an interior surface, $[f] = f_i - f_j$ if $\mathbf{n}$ is chosen to be $\mathbf{n}_i$ on Γ_{int}, and $[f] = f_j - f_i$ if $\mathbf{n}$ is chosen to be $\mathbf{n}_j$ on Γ_{int}. In either case, it can be seen that $[f]$ is the jump in the function $f(\mathbf{x})$ across the discontinuity. For simplicity, we can denote a "positive side" and a "negative side" to avoid notational complexity: $[f] = f^+ - f^-$ if $\mathbf{n}$ is $\mathbf{n}^+$ and $[f] = f^- - f^+$ if $\mathbf{n}$ is $\mathbf{n}^-$. Incidentally, it is also often less confusing to simply use a jump of the form $[f]\mathbf{n} = f^+\mathbf{n}^+ + f^-\mathbf{n}^-$ to evaluate the product of the jump and the normal.

The above can be applied similarly to a vector function. The result is called the *generalized divergence theorem* and is given by

$$\int_{\Omega} \nabla \cdot \mathbf{g} d\Omega = \int_{\Gamma} \mathbf{g} \cdot \mathbf{n} d\Gamma + \int_{\Gamma_{\text{int}}} [\mathbf{g}] \cdot \mathbf{n} d\Gamma \quad \text{if } \mathbf{g} \text{ is } C^{-1},$$

or

$$\int_{\Omega} g_{i,i} d\Omega = \int_{\Gamma} g_i n_i d\Gamma + \int_{\Gamma_{\text{int}}} [g_i] n_i d\Gamma \quad \text{if } g_i \text{ is } C^{-1}. \tag{2.30}$$

As with the standard divergence theorem, Eq. (2.30) can also be extended to higher-order tensors.

Now, when the standard integration by parts formula (2.23) is applied to the product of a derivative involving discontinuities, as in the derivation of (2.30), we obtain a *generalized integration by parts*

$$\int_\Omega v\nabla\cdot\mathbf{g}d\Omega = \int_\Gamma v\mathbf{g}\cdot\mathbf{n}d\Gamma + \int_{\Gamma_{\text{int}}}[v\mathbf{g}]\cdot\mathbf{n}d\Gamma - \int_\Omega \nabla v\cdot\mathbf{g}d\Omega \text{ if } v\mathbf{g} \text{ is in } C^{-1},$$

or

$$\int_\Omega vg_{i,i}d\Omega = \int_\Gamma vg_in_i d\Gamma + \int_{\Gamma_{\text{int}}}[vg_i]n_i d\Gamma - \int_\Omega v_{,i}g_i d\Omega \text{ if } vg_i \text{ is in } C^{-1}. \tag{2.31}$$

An immediate result that applies to the diffusion problem is found by setting $v = \delta u$ and $\mathbf{g} = k\nabla u$:

$$\int_\Omega \delta u\nabla\cdot k\nabla u d\Omega = \int_\Gamma \delta uk\nabla u\cdot\mathbf{n}d\Gamma + \int_{\Gamma_{\text{int}}}[\delta uk\nabla u]\cdot\mathbf{n}d\Gamma - \int_\Omega \nabla\delta u\cdot k\nabla u d\Omega,$$

or

$$\int_\Omega \delta u(ku_{,i})_{,i}d\Omega = \int_\Gamma \delta uku_{,i}n_i d\Gamma + \int_{\Gamma_{\text{int}}}[\delta uku_{,i}]n_i d\Gamma - \int_\Omega k\delta u_{,i}u_{,i}d\Omega. \tag{2.32}$$

When the solution or approximation contains weak or strong discontinuities, the term $\delta\Pi_{\text{D}}$ for the variational equation for diffusion in (2.18) becomes

$$\delta\Pi_{\text{D}}(u(\mathbf{x})) = \int_\Omega\left(\frac{1}{2}k\delta u_{,i}u_{,i} - su\right)d\Omega - \int_{\Gamma_q}\delta u\bar{q}d\Gamma - \int_{\Gamma^{\text{S}}_{\text{int}}}[\delta u]\bar{q}_{\text{int}}d\Gamma, \tag{2.33}$$

where $\bar{q}_{\text{int}}$ is the cohesive flux term on $\Gamma^{\text{S}}_{\text{int}}$ for solutions with strong discontinuities, with

$$\begin{aligned}\bar{q}_{\text{int}} &= \bar{q}^+_{\text{int}} \text{ if } [u] = u^+ - u^-,\\ \bar{q}_{\text{int}} &= \bar{q}^-_{\text{int}} \text{ if } [u] = u^- - u^+,\end{aligned} \tag{2.34}$$

where $\bar{q}^+_{\text{int}} = (k\nabla u)^+\cdot\mathbf{n}^+$ and $\bar{q}^-_{\text{int}} = (k\nabla u)^-\cdot\mathbf{n}^-$ are prescribed fluxes on the "+" and "−" sides, respectively, with $\bar{q}^+_{\text{int}} + \bar{q}^-_{\text{int}} = 0$.

Using (2.32), it can be shown that (2.33) attests to the strong form of the steady-state isotropic diffusion problem with essential and natural boundary conditions (2.1)–(2.3), as well as the following additional conditions on the interfaces:

$$[k\nabla u]\cdot\mathbf{n} = 0 \text{ on } \Gamma^{\text{W}}_{\text{int}}, \tag{2.35}$$

$$\begin{aligned}(k\nabla u)^+\cdot\mathbf{n}^+ &= -(k\nabla u)^-\cdot\mathbf{n}^- = \bar{q}^+_{\text{int}} = \bar{q}_{\text{int}} \quad \text{on } \Gamma^{\text{S}}_{\text{int}} \text{ if } [u] = u^+ - u^-,\\ (k\nabla u)^-\cdot\mathbf{n}^- &= -(k\nabla u)^+\cdot\mathbf{n}^+ = \bar{q}^-_{\text{int}} = \bar{q}_{\text{int}} \quad \text{on } \Gamma^{\text{S}}_{\text{int}} \text{ if } [u] = u^- - u^+.\end{aligned} \tag{2.36}$$

In arriving at the above, we have used the fact that δu is continuous along $\Gamma^{\text{W}}_{\text{int}}$, but discontinuous on $\Gamma^{\text{S}}_{\text{int}}$. The first condition (2.35) states the fluxes are equal and opposite along $\Gamma^{\text{W}}_{\text{int}}$, and the second condition (2.36) gives the prescribed fluxes on each side of $\Gamma^{\text{S}}_{\text{int}}$.

The last term in (2.33) could drop out for a solution that contains only weak discontinuities. However, care should be taken to respect the condition (2.35) on weak interfaces if smooth approximations are used, such as those generally constructed with meshfree shape functions.

2.1.3 Constrained Variational Principles for the Diffusion Equation

For most meshfree approximations, it is non-trivial to construct trial functions that satisfy the essential boundary conditions in the required set U in (2.14). To develop variational principles for these situations, it is necessary to impose the essential boundary conditions by constraints. We will consider three methods for imposing essential boundary conditions:

1) The penalty method
2) The Lagrange multiplier method
3) Nitsche's method

2.1.3.1 The Penalty Method

In the penalty method, the square of the constraint is added to the functional, so that when the constraint is satisfied, the penalty vanishes. In this case, the constraint is the essential boundary condition. The trial solutions then no longer need to satisfy the essential boundary conditions *a priori*.

For the diffusion equation, the functional with the essential boundary conditions imposed by a penalty is given by

$$\Pi_{\mathrm{D}}^{\mathrm{P}} = \Pi_{\mathrm{D}} + \frac{1}{2}\int_{\Gamma_u} \beta(u(\mathbf{x}) - \overline{u}(\mathbf{x}))^2 d\Gamma, \tag{2.37}$$

where β is the penalty parameter and Π_{D} is given by (2.13). The penalty parameter is usually chosen to be a large positive constant; it is discussed further in Section 4.1.1. Note that in (2.37) the penalty term is the square of the difference between the boundary value of $u(\mathbf{x})$ and what it is prescribed to be, $\overline{u}(\mathbf{x})$, integrated over the essential boundary. This penalty integral vanishes only when the essential boundary condition is satisfied on the entire essential boundary. Therefore, if the penalty parameter is sufficiently large with respect to the constant(s) in the standard variational principle, which is the diffusivity k in the diffusion problem for Π_{D}, the minimizer of $\Pi_{\mathrm{D}}^{\mathrm{P}}$ will almost satisfy the boundary condition along the entire boundary. Still, it will not exactly satisfy the essential boundary condition, as shown later.

The variational principle for the penalty method is

$$\text{Find } u \in \mathrm{H}^1 \text{ such that } \delta\Pi_{\mathrm{D}}^{\mathrm{P}} = 0 \;\; \forall \;\; \delta u \in \mathrm{H}^1. \tag{2.38}$$

The variational solution is thus the minimizer of (2.37) without any conditions on the test or trial solutions for the essential boundaries. We will show that this functional does not exactly satisfy the essential boundary conditions regardless of the magnitude of the penalty parameter β, so this variational principle is often modified. Another disadvantage of the penalty method is that it requires the selection of the penalty parameter, which is often difficult and may result in ill conditioning of the discrete equations. Further guidance on the selection of this parameter is summarized in Section 4.1.1.

Next, we demonstrate that the minimizer of (2.37) satisfies the governing equation and the natural boundary conditions but does not satisfy the essential boundary conditions exactly, although the accuracy in the essential boundary condition often suffices. For conciseness, we only consider trial functions in H^1. Taking the variation of Π_D^P gives

$$\delta\Pi_D^P = \int_{\Omega} (ku_{,i}\delta u_{,i} - s\delta u)d\Omega - \int_{\Gamma_q} \delta u\overline{q}d\Gamma + \int_{\Gamma_u} \beta\delta u(u - \overline{u})d\Gamma = 0. \tag{2.39}$$

Note that in contrast to the standard variational principle, the third term appears because $\delta u(\mathbf{x})$ does not vanish on the essential boundaries. Using integration by parts (2.24) and changing the sign of the equation then gives

$$\delta\Pi_D^P = \int_{\Omega} \delta u((ku_{,i})_{,i} + s)d\Omega - \int_{\Gamma_q} \delta u(ku_{,n} - \overline{q})d\Gamma - \int_{\Gamma_u} \delta uku_{,n}d\Gamma - \int_{\Gamma_u} \beta\delta u(u - \overline{u})d\Gamma = 0. \tag{2.40}$$

Applying the principle of variations, the diffusion equation and the natural boundary condition can be obtained from the first two terms of Eq. (2.40). The third and fourth terms need to be combined to extract the condition on the essential boundary. This gives

$$\int_{\Gamma_u} \delta u(ku_{,n} + \beta(u - \overline{u}))d\Gamma = 0. \tag{2.41}$$

Therefore

$$u - \overline{u} = -\frac{1}{\beta}ku_{,n} \text{ on } \Gamma_u. \tag{2.42}$$

It can be seen that the larger the penalty parameter β with respect to k, the more exactly the essential boundary condition $u = \overline{u}$ is enforced. However, there will always be a slight discrepancy due to the term $\frac{1}{\beta}ku_{,n}$ in (2.42).

Nitsche's method, which is described later, adds another term to the functional (2.37) so that the integral of $ku_{,n}$ over the essential boundary is eliminated in Eq. (2.40), which also eliminates this term in (2.42). Consequently, the essential boundary conditions can be satisfied exactly in the variational equation.

2.1.3.2 The Lagrange Multiplier Method

In the Lagrange multiplier method, the essential boundary condition is enforced by adding the product of the essential boundary condition and a Lagrange multiplier to the original functional (2.13). This gives a modified equation

$$\Pi_D^L = \Pi_D + \int_{\Gamma_u} \lambda(\mathbf{x})(u(\mathbf{x}) - \overline{u}(\mathbf{x}))\,d\Gamma, \tag{2.43}$$

where $\lambda(\mathbf{x})$ is the Lagrange multiplier field.

The variational principle is

$$\text{Find } u \in \mathrm{H}^1,\ \lambda \in \mathrm{L}_2 \text{ such that } \delta\Pi_\mathrm{D}^\mathrm{L} = 0\ \forall\ \delta u \in \mathrm{H}^1,\ \delta\lambda \in \mathrm{L}_2, \tag{2.44}$$

where $\mathrm{L}_2 \equiv \mathrm{H}^0$ is the Sobolov space of degree zero (see the Glossary of Notation). Note that when the essential conditions are enforced by a Lagrange multiplier, the solution is no longer a minimizer, but is instead a saddle point of the functional, but the variational equation is obtained by the same procedure, with $\delta\Pi_\mathrm{D}^\mathrm{L} = 0$. As in the penalty method, the trial functions do not need to satisfy the essential boundary conditions, i.e., the trial functions need not be in U. Instead, it is only necessary to be sufficiently integrable, so that they are in H^1. The Lagrange multipliers must be square-integrable, i.e., be in L_2, so C^{-1} functions are suitable. They also need to satisfy the Babuška–Brezzi condition [1, 2] for the stability of the solution; in the discrete case, the choice of approximations of the two fields is not entirely trivial. More discussion on this aspect of this approach is given in Section 4.1.2.

The variational equations are obtained by taking a variation of $\Pi_\mathrm{D}^\mathrm{L}$, which by referring to (2.13) and (2.43) can be seen to be

$$\delta\Pi_\mathrm{D}^\mathrm{L} = \delta\Pi_\mathrm{D} + \int_{\Gamma_u} \delta\lambda(u-\bar{u})\,d\Gamma + \int_{\Gamma_u} \delta u\,\lambda\,d\Gamma = 0, \tag{2.45}$$

or

$$\delta\Pi_\mathrm{D}^\mathrm{L} = \int_\Omega (ku_{,i}\delta u_{,i} - s\delta u)\,d\Omega - \int_{\Gamma_q} \delta u\bar{q}\,d\Gamma + \int_{\Gamma_u} \delta\lambda(u-\bar{u})\,d\Gamma + \int_{\Gamma_u} \delta u\,\lambda\,d\Gamma = 0. \tag{2.46}$$

We next show that the variational equation for the Lagrange multiplier method, Eq. (2.46), implies the strong form. For this purpose, we use integration by parts applied to the diffusion equation (2.24), to rewrite the first term in (2.46) as

$$\int_\Omega \delta u_{,i} k u_{,i}\,d\Omega = \int_{\Gamma_u} (\delta u(ku_{,i}))n_i\,d\Gamma + \int_{\Gamma_q} (\delta u(ku_{,i}))n_i\,d\Gamma - \int_\Omega \delta u(ku_{,i})_{,i}\,d\Omega. \tag{2.47}$$

Substituting (2.47) into (2.46), we obtain

$$\begin{aligned}
\delta\Pi_\mathrm{D}^\mathrm{L} = &- \int_\Omega \delta u\left((ku_{,i})_{,i} + s\right)d\Omega - \int_{\Gamma_q} \delta u\bar{q}\,d\Gamma + \int_{\Gamma_u} (\delta u(ku_{,n}))\,d\Gamma \\
&+ \int_{\Gamma_q} (\delta u(ku_{,n}))\,d\Gamma + \int_{\Gamma_u} \delta\lambda(u-\bar{u})\,d\Gamma + \int_{\Gamma_u} \delta u\,\lambda\,d\Gamma \\
= &- \int_\Omega \delta u\left((ku_{,i})_{,i} + s\right)d\Omega + \int_{\Gamma_q} \delta u(ku_{,n} - \bar{q})\,d\Gamma + \int_{\Gamma_u} (\delta u(ku_{,n} + \lambda))\,d\Gamma \\
&+ \int_{\Gamma_u} \delta\lambda(u-\bar{u})\,d\Gamma = 0.
\end{aligned} \tag{2.48}$$

From the arbitrariness of the variations δu and $\delta\lambda$, it follows that

$$(ku_{,i})_{,i} + s = 0 \text{ in } \Omega,\ \ ku_{,n} = \bar{q} \text{ on } \Gamma_q, u = \bar{u} \text{ on } \Gamma_u,\ ku_{,n} + \lambda = 0 \text{ on } \Gamma_u. \tag{2.49}$$

The first three equations in (2.49) are the governing equation, the natural boundary condition, and the essential boundary condition corresponding to (2.1) through (2.3). The last equation is an extraneous condition that arises from the Lagrange multiplier formulation and expresses the Lagrange multiplier in terms of the flux, or the normal gradient of u.

2.1.3.3 Nitsche's Method

To obtain Nitsche's method, we use the last equation in (2.49) to express λ in terms of $ku_{,n}$, $\lambda = -ku_{,n}$ on Γ_u, insert this relationship into (2.43) and add a penalty identical to that in (2.37). This yields the following functional

$$\Pi_{\mathrm{D}}^{\mathrm{N}} = \Pi_{\mathrm{D}} - \int_{\Gamma_u} ku_{,n}(u-\bar{u})\,d\Gamma + \frac{1}{2}\int_{\Gamma_u} \beta(u-\bar{u})^2 d\Gamma. \tag{2.50}$$

Nitsche's variational principle states that [3]:

$$\text{Find } u \in \mathrm{H}^1 \text{ such that } \delta\Pi_{\mathrm{D}}^{\mathrm{N}} = 0\ \forall\ \delta u \in \mathrm{H}^1. \tag{2.51}$$

The above variational form does not require the test or trial functions to satisfy the essential boundary conditions, and there is no need for a separate Lagrange multiplier field. We will show that in contrast to the penalty method, the essential boundary conditions are satisfied exactly even when the trial functions do not. This is not true, however, for finite-dimensional solution spaces.

The variational equation, or weak form, can be obtained by taking the variation of (2.50) (also using (2.17)), which yields

$$\begin{aligned}\delta\Pi_{\mathrm{D}}^{\mathrm{N}} = &\int_{\Omega}(ku_{,i}\delta u_{,i} - s\delta u)d\Omega - \int_{\Gamma_q}\delta u\bar{q}d\Gamma \\ &- \int_{\Gamma_u}k\delta u_{,n}(u-\bar{u})\,d\Gamma - \int_{\Gamma_u}\delta uku_{,n}\,d\Gamma + \int_{\Gamma_u}\beta\delta u(u-\bar{u})\,d\Gamma = 0.\end{aligned} \tag{2.52}$$

We stress that Nitsche's variational principle is easily derived by simply solving for the Lagrange multiplier, as was done above, inserting that into the Lagrange multiplier for the variational principle and adding a penalty. This is how Lu et al. [4] rediscovered the method, although no penalty was added in that paper.

We next show that Nitsche's variational principle is equivalent to the strong form. Using integration by parts, we obtain

$$\begin{aligned}\delta\Pi_{\mathrm{D}}^{\mathrm{N}} = &-\int_{\Omega}\delta u\left((ku_{,i})_{,i} + s\right)d\Omega - \int_{\Gamma_q}\delta u\bar{q}d\Gamma + \int_{\Gamma_q}\delta uku_{,n}d\Gamma + \int_{\Gamma_u}\delta uku_{,n}d\Gamma \\ &- \int_{\Gamma_u}\delta uku_{,n}d\Gamma - \int_{\Gamma_u}\delta u_{,n}k(u-\bar{u})d\Gamma + \beta\int_{\Gamma_u}\delta u(u-\bar{u})d\Gamma = 0.\end{aligned} \tag{2.53}$$

The fourth and fifth terms cancel, and if we combine the rest, we have

$$\begin{aligned}\delta\Pi_{\mathrm{D}}^{\mathrm{N}} = &- \int_{\Omega} \delta u\left(\left(ku_{,i}\right)_{,i} + s\right) d\Omega + \int_{\Gamma_q} \delta u(ku_{,n} - \overline{q}) d\Gamma \\ &- \int_{\Gamma_u} \delta u_{,n} k(u - \overline{u}) d\Gamma + \beta \int_{\Gamma_u} \delta u(u - \overline{u}) d\Gamma = 0.\end{aligned} \tag{2.54}$$

The first integral gives the diffusion equation and the second yields the natural boundary condition. The last two give the essential boundary conditions twice, once with the normal gradient as a test function, the second time with the function itself. As can be surmised, the penalty term is unnecessary from a strong consistency standpoint, yet it now plays the role of ensuring the stability of the resulting discrete equations.

It is interesting to observe that Nitsche's variational principle introduces a term, the second term on the right-hand side of (2.50), that cancels with the fourth integral in (2.53) that causes the discrepancy in the essential boundary condition in the penalty method. Therefore, Nitsche's method compensates for this and makes the variational principle consistent with the strong form.

2.1.4 Weak Form of the Diffusion Equation by the Method of Weighted Residuals

The method of weighted residuals can be applied to a far wider class of problems than variational principles. For example, it can be applied to nonlinear and time-dependent problems. Nonlinear problems only possess variational forms if the flux is derived from a potential. We will show that the weak forms are identical to the variational equations for the class of problems governed by variational principles. Furthermore, as we will see, variational methods give valuable guidance on how to construct constrained weak forms.

The standard way to develop a weak form is to multiply the governing equation by an arbitrary test function $v(\mathbf{x})$ and integrate over the domain of the problem. In this case, we consider the diffusion equation in terms of the flux

$$\int_{\Omega} v(\mathbf{x})\left(-q_{i,i}(u(\mathbf{x},t)) + \tilde{s}(u(\mathbf{x},t))\right) d\Omega = 0. \tag{2.55}$$

Note that the source $\tilde{s}$ subsumes any time-dependent terms, so the above will apply to non-steady-state, i.e., time-dependent diffusion problems. The flux can be an arbitrary nonlinear function of the gradient of the field $u(\mathbf{x})$ and does not need to come from a potential, as was the case in (2.12).

Since we would like this weak form to satisfy the natural boundary condition as well, we also multiply the natural boundary condition on Γ_q by the test functions and integrate over that boundary, so that

$$\begin{aligned}&\int_{\Gamma_q} v(\mathbf{x})(\overline{q}(\mathbf{x},t) + q_n(u(\mathbf{x},t))) d\Gamma = 0, \\ &q_n = q_i n_i = -ku_{,i} n_i = -ku_{,n}.\end{aligned} \tag{2.56}$$

The above equation can be added directly to (2.55) for a weak formulation consistent with the governing equation and natural boundary condition:

$$\int_{\Omega} v(\mathbf{x})\left(-q_{i,i}(u(\mathbf{x},t)) + \tilde{s}(u(\mathbf{x},t))\right) d\Omega + \int_{\Gamma_q} v(\mathbf{x})(\overline{q}(\mathbf{x},t) + q_n(u(\mathbf{x},t))) d\Gamma = 0. \tag{2.57}$$

However, as will be seen, it is more convenient to invoke the natural boundary conditions during the construction of the weak form if integration by parts is introduced for (2.55). In the cases where integration by parts is not introduced in constructing the weak form, such as the strong form collocation method to be discussed in Chapter 9, (2.57) can be used to impose equilibrium and the natural boundary condition, together with the weighted essential boundary condition.

Equation (2.57) is equivalent to the governing equation, i.e., implies (2.1)–(2.3) and vice versa, if certain integrability conditions are imposed on the test and trial functions along with essential boundary conditions. However, the solution by (2.57) requires that the trial functions be C^1 functions, which is often awkward for most approximations (but notably, not meshfree). Furthermore, implementing the weak form (2.57) generally leads to nonsymmetric discrete equations; this can be foreseen by noting that the weak form is not symmetric in the test and trial functions. The test function appears directly while the trial function eventually appears as a second derivative. As it turns out, this is much more problematic than once thought [5].

Using integration by parts applied to the diffusion equation in (2.24), we can transform the first integral on the left-hand side of (2.55) to obtain

$$\begin{aligned} \int_{\Omega} v q_{i,i} d\Omega &= \int_{\Gamma} v q_n d\Gamma - \int_{\Omega} v_{,i} q_i d\Omega \\ &= \int_{\Gamma_u} v q_n d\Gamma + \int_{\Gamma_q} v q_n d\Gamma - \int_{\Omega} v_{,i} q_i d\Omega, \end{aligned} \tag{2.58}$$

where in the second line of the above we have used the complementarity of the boundaries to express the boundary integral in the first line as the sum of integrals over the essential and natural boundaries.

Substituting the resulting (2.58) into (2.55) then gives

$$\int_{\Omega} (v_{,i} q_i + v\tilde{s})\, d\Omega - \int_{\Gamma_u} v q_n d\Gamma - \int_{\Gamma_q} v q_n d\Gamma = 0. \tag{2.59}$$

In standard constructions of the weak form, the boundary integral over Γ_u is dropped by constructing the test functions $v(\mathbf{x})$ so that they vanish on Γ_u, so we will do this in this part and set the last term to zero:

$$\int_{\Omega} (v_{,i} q_i + v\tilde{s})\, d\Omega - \int_{\Gamma_q} v q_n d\Gamma = 0. \tag{2.60}$$

Later, in the development of weak forms consistent with meshfree methods in Chapter 5, this cannot be done.

If we now invoke the natural boundary condition (2.3) directly, or equivalently add the weak form of the natural boundary condition (2.56) to (2.60), we obtain the weak form of the diffusion equation with natural boundary conditions

$$\int_{\Omega} (v_{,i} q_i + v\tilde{s})\, d\Omega + \int_{\Gamma_q} v\bar{q} d\Gamma = 0, \tag{2.61}$$

i.e., the above is the weak form of (2.1) and (2.3). We have called the expression on the left-hand side $\delta\Pi_{\mathrm{D}}$ in the variational equations.

Note that if we restrict ourselves to the steady-state case ($\tilde{s} = s$), with the flux-gradient relation as the linear law (2.9) with isotropic diffusion, then the weak form is

$$\int_{\Omega} (-v_{,i} k u_{,i} + vs)\, d\Omega + \int_{\Gamma_q} v\bar{q}\, d\Gamma = 0. \tag{2.62}$$

If we let $v = \delta u$, then the above is identical (except for a change of sign) to the variational equation (2.17). So, the method of weighted residuals gives the same weak form as the variational method. This is true for any self-adjoint system of partial differential equations, or ones that possess a potential.

To obtain a weak form which is equivalent to the strong form, it is necessary to construct trial functions that satisfy the essential boundary conditions and test functions that vanish on the essential boundaries; the latter was used in going from (2.59) to (2.61). Furthermore, the test and trial functions must be such that the integrals in (2.61) can be evaluated. The critical term is $v_{,i}q_i$, which becomes the product of derivatives $v_{,i}ku_{,i}$ for a linear system as in (2.62). Generally, the same class of functions is employed for the test and trial functions, so this requires that the derivatives of the functions be square-integrable, i.e., the functions must be in H^1, exactly the same as in the variational approach. If the functions are not singular, well-behaved functions such as polynomials that are C^0, are usually in H^1.

So, to obtain an equivalent weak form, we construct two function spaces (i.e., sets of functions):

$$\mathrm{U} = \{u(\mathbf{x}) | u(\mathbf{x}) \in \mathrm{H}^1 \text{ and } u(\mathbf{x}) = \bar{u} \text{ on } \Gamma_u\}, \tag{2.63}$$

$$\mathrm{U}_0 = \{v(\mathbf{x}) | v(\mathbf{x}) \in \mathrm{H}^1 \text{ and } v(\mathbf{x}) = 0 \text{ on } \Gamma_u\}. \tag{2.64}$$

Note that the function space U consists of functions that satisfy the essential boundary conditions and are sufficiently smooth so that the weak form is integrable. This would entail that the functions be piecewise and continuously differentiable, which are often designated as C^0 functions. The function space U_0 consists of functions with the same smoothness that vanish on the essential boundaries. Note that the two spaces are *exactly the same* as those required for the variational approach.

The weak form can then be stated as

$$\text{Find } u \in \mathrm{U} \text{ such that (2.61) holds } \forall\, v \in \mathrm{U}_0. \tag{2.65}$$

Since the weak form is identical to the variational equation (2.17), identical steps can be used to show that it implies the strong form, and we will not repeat that.

As can be seen from the fact that the weak form is directly developed in terms of the flux, it holds for any relation between the flux and the gradient of the field: these relations may be nonlinear and quite discontinuous. Furthermore, the weak form applies to an arbitrary source term that can be a function of $u(\mathbf{x}, t)$ and its time derivatives.

A desirable feature of the weak form (2.62) is that the first term is symmetric in the test function and trial function. It will be seen later that, as a consequence, this weak form yields symmetric discrete equations.

We next develop the weak form for the diffusion equation by the weighted residual approach for trial solutions that do not satisfy the essential boundary conditions. In other words, the test and trial functions are in H^1, but the trial function $u(\mathbf{x})$ does not satisfy the essential boundary conditions and the test function $v(\mathbf{x})$ does not vanish on the essential boundary. Here, we consider the Lagrange multiplier approach; other methods can be developed similarly.

The derivation of a Lagrange multiplier weak form under the weighted residual approach is difficult without referring to the variational methods. A naive approach would be to add the essential boundary condition constraint

$$\int_{\Gamma_u} \lambda(u-\bar{u})d\Gamma = 0 \quad \forall \lambda(\mathbf{x}), \tag{2.66}$$

where $\lambda(\mathbf{x})$ is a Lagrange multiplier field. However, the correct strong form cannot be recovered.

The easiest way to obtain the correct weak form is to use the variational principle to provide guidance as to its form. If we denote the variations in the Lagrange multiplier field by $\gamma(\mathbf{x})$ and the variation $\delta u(\mathbf{x})$ by $v(\mathbf{x})$, then the term corresponding to the variation of the Lagrange multiplier constraint in (2.45) is

$$\int_{\Gamma_u} \gamma(u-\bar{u})d\Gamma + \int_{\Gamma_u} \lambda v d\Gamma. \tag{2.67}$$

Subtracting the above term from (2.61), and repeating the same steps that were used to obtain the weak form gives

$$\text{Find } u(\mathbf{x}) \in \mathrm{H}^1 \text{ and } \lambda \in \mathrm{L}_2 \text{ such that}$$
$$\int_{\Omega} (v_{,i}q_i + v\tilde{s})d\Omega + \int_{\Gamma_q} v\bar{q}d\Gamma - \int_{\Gamma_u} \gamma(u-\bar{u})d\Gamma - \int_{\Gamma_u} v\lambda d\Gamma = 0 \quad \forall v \in \mathrm{H}^1 \text{ and } \forall \gamma \in \mathrm{L}_2. \tag{2.68}$$

If we let $v = \delta u$, $\gamma = \delta\lambda$, and $\tilde{s} = s$, and substitute the Fourier law for the flux, this gives the same variational equation as (2.46) with a change of sign.

2.2 Elasticity

2.2.1 Strong Form of Elasticity

We will now give the strong form for linear elasticity. The variational principles will then be given, and the variational equations derived. Since much attention has been given to the diffusion equation, the presentation will be more condensed and some derivations will be left to the reader.

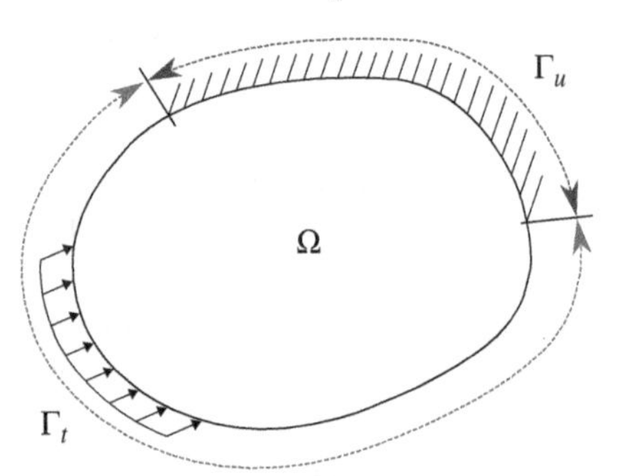

Figure 2.3 Problem domain Ω, essential boundary Γ_u, and natural boundary Γ_t for the elasticity equation.

We consider a body Ω with boundary Γ as shown in Figure 2.3. The boundary Γ consists of a prescribed displacement boundary Γ_u and a prescribed traction boundary Γ_t. The two boundaries are complementary in that $\Gamma_u \cup \Gamma_t = \Gamma$ and $\Gamma_u \cap \Gamma_t = \emptyset$. One can also subdivide the boundary according to components, where the ith displacement component is prescribed on Γ_u^i and the ith traction component is prescribed on Γ_t^i. The boundaries associated with each component are then complementary: $\Gamma_u^i \cup \Gamma_t^i = \Gamma^i$ and $\Gamma_u^i \cap \Gamma_t^i = \emptyset$. However, this form introduces substantial notational complexity, so we limit these derivations to the previous case in this book. The more general case can be found in [6].

The strong form for linear elasticity consists of the equilibrium equations and the essential and natural boundary conditions, which are given, respectively, by

$$\begin{aligned} &\nabla \cdot \boldsymbol{\sigma} + \mathbf{b} = \mathbf{0} \quad \text{in } \Omega, \\ &\text{or} \\ &\sigma_{ij,j} + b_i = 0 \quad \text{in } \Omega, \end{aligned} \tag{2.69}$$

$$\begin{aligned} &\mathbf{u} = \bar{\mathbf{u}} \quad \text{on } \Gamma_u, \\ &\text{or} \\ &u_i = \bar{u}_i \quad \text{on } \Gamma_u, \end{aligned} \tag{2.70}$$

$$\begin{aligned} &\boldsymbol{\sigma} \cdot \mathbf{n} = \bar{\mathbf{t}} \quad \text{on } \Gamma_t, \\ &\text{or} \\ &\sigma_{ij} n_j = \bar{t}_i \quad \text{on } \Gamma_t, \end{aligned} \tag{2.71}$$

where $\mathbf{b}$ is the body force and $\bar{\mathbf{u}}$ and $\bar{\mathbf{t}}$ are the prescribed boundary displacement and traction.

Equation (2.69) is the equilibrium equation. The following two equations are the boundary conditions. Equation (2.70) is the prescribed displacement boundary condition, often called the essential boundary condition or Dirichlet boundary condition. Equation (2.71) is the traction boundary condition, often called the natural or Neumann boundary condition.

For elasticity with strong or weak discontinuities, there are additional conditions in the strong and corresponding weak form that need to be considered. The variational principles in this case are given in the Appendix.

In addition to the above, a constitutive equation that relates stress to strain and ultimately to the displacement gradient is needed. In linear elasticity, the stress is related to the strain by the Hookean tensor by

$$\begin{aligned} &\boldsymbol{\sigma} = \mathbf{C} : \boldsymbol{\varepsilon}, \\ &\text{or} \\ &\sigma_{ij} = C_{ijkl} \varepsilon_{kl}, \end{aligned} \tag{2.72}$$

where the strain is given in terms of the displacements by

$$\begin{aligned} &\boldsymbol{\varepsilon} = \nabla_s \mathbf{u}, \\ &\text{or} \\ &\varepsilon_{ij} = \frac{1}{2}\left(u_{i,j} + u_{j,i}\right) \equiv u_{(i,j)}, \end{aligned} \tag{2.73}$$

where $\nabla_s(\cdot) = 1/2(\nabla \otimes (\cdot) + (\cdot) \otimes \nabla)$, and its index notation is shown in (2.73), where it is called the symmetric part of the displacement gradient. The elastic constants have the minor symmetries $C_{ijkl} = C_{jikl} = C_{ijlk}$ and the major symmetries $C_{ijkl} = C_{klij}$.

Substituting (2.72) and (2.73) into (2.69) gives

$$\begin{aligned} &\nabla \cdot (\mathbf{C} : \nabla_s \mathbf{u}) + \mathbf{b} = \mathbf{0} \quad \text{in } \Omega, \\ &\text{or} \\ &\left(C_{ijkl} u_{(k,l)}\right)_{,j} + b_i = 0 \quad \text{in } \Omega. \end{aligned} \tag{2.74}$$

In the indicial form of the above, we have taken advantage of the minor symmetry of $\mathbf{C}$ to write the expression simply in terms of the gradient of the displacement rather than the symmetric part of the gradient.

2.2.2 The Variational Principle for Elasticity

The variational principle for elasticity is called the theorem of minimum potential energy. It states that the solution $u_i(\mathbf{x}) \in \mathrm{U}$ is the minimizer of

$$\Pi_{\mathrm{E}} = W^{\mathrm{int}} - W^{\mathrm{ext}}, \tag{2.75}$$

where

$$W^{\mathrm{int}} = \frac{1}{2}\int_{\Omega} \boldsymbol{\varepsilon} : \mathbf{C} : \boldsymbol{\varepsilon}\, d\Omega, \quad \text{or} \quad W^{\mathrm{int}} = \frac{1}{2}\int_{\Omega} \varepsilon_{ij} C_{ijkl}\, \varepsilon_{kl}\, d\Omega, \tag{2.76}$$

$$W^{\mathrm{ext}} = \int_{\Omega} \mathbf{u}\cdot\mathbf{b}\, d\Omega + \int_{\Gamma_t} \mathbf{u}\cdot\bar{\mathbf{t}}\, d\Gamma, \quad \text{or} \quad W^{\mathrm{ext}} = \int_{\Omega} u_i b_i\, d\Omega + \int_{\Gamma_t} u_i \bar{t}_i\, d\Gamma, \tag{2.77}$$

$$\mathrm{U} = \left\{ u_i(\mathbf{x}) \middle| u_i(\mathbf{x}) \in \mathrm{H}^1 \text{ and } u_i(\mathbf{x}) = \bar{u}_i \text{ on } \Gamma_u \right\}, \tag{2.78}$$

where the subscript "E" refers to elasticity.

The term W^{int} is the internal energy, also called the strain energy or stored energy; W^{ext} is the external work. The two terms in (2.77) are the work done by the body forces and the applied tractions.

Since the solution is the minimizer of Π_{E}, as in the diffusion equation, we obtain the variational equation by taking the variation of Π_{E} and setting it to zero. That gives

$$\delta\Pi_{\mathrm{E}} = \int_{\Omega} \delta u_{(i,j)} C_{ijkl} u_{(k,l)}\, d\Omega - \int_{\Omega} \delta u_i b_i\, d\Omega - \int_{\Gamma_t} \delta u_i \bar{t}_i\, d\Gamma = 0, \tag{2.79}$$

for all $\delta u_i \in \mathrm{U}_0$ with

$$\mathrm{U}_0 = \left\{ \delta u_i(\mathbf{x}) \middle| \delta u_i(\mathbf{x}) \in \mathrm{H}^1 \text{ and } \delta u_i(\mathbf{x}) = 0 \text{ on } \Gamma_u \right\}. \tag{2.80}$$

We next show that the variational equation (2.79) gives the strong form of the elasticity problem. Applying integration by parts to the first integral in (2.79) yields

$$\int_{\Omega} \delta u_{(i,j)} C_{ijkl} u_{(k,l)}\, d\Omega = \int_{\Gamma} \delta u_i \sigma_{ij} n_j\, d\Gamma - \int_{\Omega} \delta u_i \left(C_{ijkl} u_{(k,l)} \right)_{,j} d\Omega. \tag{2.81}$$

Combining (2.81) and (2.79), and using $\delta\mathbf{u}(\mathbf{x}) = \mathbf{0}$ on Γ_u (from the definition of U_0), we then have

$$\delta\Pi_{\mathrm{E}} = -\int_{\Omega} \delta u_i \left(\left(C_{ijkl} u_{(k,l)} \right)_{,j} + b_i \right) d\Omega + \int_{\Gamma_t} \delta u_i \left(\sigma_{ij} n_j - \bar{t}_i \right) d\Gamma = 0. \tag{2.82}$$

By the principle of variations, each of the parenthesized terms in (2.82) must vanish. This gives, from left to right, the equilibrium equation and the natural boundary condition. The essential boundary conditions are satisfied intrinsically by the definition of U in (2.78). Note that in order to arrive at this equivalence, the definition of U_0 in (2.80) must be invoked. That is, the homogenous condition on the essential boundary for the test function in the standard variational principle is necessary to attest to the strong form.

2.2.3 Constrained Variational Principles for Elasticity

For trial solutions that do not satisfy the essential boundary conditions, as with the diffusion equation, variational principles can be constructed by adding the essential boundary conditions as constraints. The following methods will be described as before:

1) The penalty method
2) The Lagrange multiplier method
3) Nitsche's method

Again, the spaces are not required to be admissible; the test and trial functions only need to be in H^1.

2.2.3.1 The Penalty Method

In the penalty method, the essential boundary condition is added as a penalty to the standard functional, so the functional is

$$\Pi_{\mathrm{E}}^{\mathrm{P}} = \Pi_{\mathrm{E}} + \frac{1}{2}\beta \int_{\Gamma_u} (\mathbf{u} - \bar{\mathbf{u}}) \cdot (\mathbf{u} - \bar{\mathbf{u}}) d\Gamma. \tag{2.83}$$

The variational equation is obtained by taking the variation of (2.83) and setting it to zero. The variational principle corresponding to the strong form (2.74) is to find $u_i \in \mathrm{H}^1$ such that

$$\delta\Pi_{\mathrm{E}}^{\mathrm{P}} = \delta\Pi_{\mathrm{E}} + \beta \int_{\Gamma_u} \delta\mathbf{u} \cdot (\mathbf{u} - \bar{\mathbf{u}}) d\Gamma = 0, \tag{2.84}$$

for all $\delta u_i \in \mathrm{H}^1$. This variational principle again does not attest exactly to the strong form; the essential boundary conditions are not satisfied as in the diffusion equation with the penalty method. The conditions here are analogous to (2.42).

2.2.3.2 The Lagrange Multiplier Method

In the Lagrange multiplier method, the variational principle is constructed by appending the product of Lagrange multipliers and the essential boundary condition and integrating over the essential boundary, so that the functional is

$$\Pi_{\mathrm{E}}^{\mathrm{L}} = \Pi_{\mathrm{E}} + \int_{\Gamma_u} \boldsymbol{\lambda} \cdot (\mathbf{u} - \bar{\mathbf{u}}) d\Gamma, \tag{2.85}$$

where the Lagrange multiplier $\boldsymbol{\lambda}$ is now a vector to enforce the constraints in each component of $\mathbf{u}$.

The variational principle corresponding to the strong form (2.74) is: find $u_i \in \mathrm{H}^1$ and $\lambda_i \in \mathrm{L}_2$ such that

$$\delta\Pi_{\mathrm{E}}^{\mathrm{L}} = \int_{\Omega} \delta u_{(i,j)} C_{ijkl} u_{(k,l)} d\Omega - \int_{\Omega} \delta u_i b_i d\Omega - \int_{\Gamma_t} \delta u_i \bar{t}_i d\Gamma + \int_{\Gamma_u} (\delta\lambda_i (u_i - \bar{u}_i) + \lambda_i \delta u_i) d\Gamma = 0, \tag{2.86}$$

for all $\delta u_i \in \mathrm{H}^1$, $\delta\lambda_i \in \mathrm{L}_2$.

The strong form is obtained by integration by parts of the first term in (2.86) and invoking the arbitrariness of δu_i and $\delta\lambda_i$ (this is left as an exercise), which gives

$$\left(C_{ijkl} u_{(k,l)}\right)_{,j} + b_i = 0 \quad \text{in } \Omega, \quad \text{or} \quad \sigma_{ij,j} + b_i = 0 \quad \text{in } \Omega, \qquad \text{(equilibrium equation)} \tag{2.87}$$

$$C_{ijkl} u_{(k,l)} n_j = \bar{t}_i \quad \text{on } \Gamma_t, \quad \text{or} \quad \sigma_{ij} n_j = \bar{t}_i \quad \text{on } \Gamma_t, \qquad \text{(traction boundary condition)} \tag{2.88}$$

$$u_i = \bar{u}_i \quad \text{on } \Gamma_u, \qquad \text{(displacement boundary condition)} \tag{2.89}$$

$$\lambda_i = -t_i = -\sigma_{ij} n_j \quad \text{on } \Gamma_u, \tag{2.90}$$

where we have introduced the traction t_i computed from the stress $t_i = \sigma_{ij} n_j$ on Γ_u. As can be seen from the above, the Lagrange multiplier variational equation implies the equilibrium equation, and the traction and displacement boundary conditions. As in the case of the diffusion equation, an additional equation (2.90) is implied. This equation identifies the Lagrange multiplier to be the negative traction on the displacement boundary.

2.2.3.3 Nitsche's Method

Nitsche's method can be constructed directly from the Lagrange multiplier form (2.85) by replacing the Lagrange multiplier with the relation given in (2.90) and adding a penalty. This gives

$$\Pi_{\mathrm{E}}^{\mathrm{N}} = \Pi_{\mathrm{E}} - \int_{\Gamma_u} \mathbf{t} \cdot (\mathbf{u} - \bar{\mathbf{u}}) d\Gamma + \frac{1}{2} \int_{\Gamma_u} \beta (\mathbf{u} - \bar{\mathbf{u}}) \cdot (\mathbf{u} - \bar{\mathbf{u}}) d\Gamma, \tag{2.91}$$

where β is a penalty parameter. The penalty parameter in Nitsche's method can be significantly smaller than in a standard penalty method and serves as a stabilization parameter.

The variational equation for Nitsche's method is: find $u_i \in \mathrm{H}^1$

$$\delta\Pi_{\mathrm{E}}^{\mathrm{N}} = \delta\Pi_{\mathrm{E}} - \int_{\Gamma_u} \delta\mathbf{t} \cdot (\mathbf{u} - \bar{\mathbf{u}}) d\Gamma - \int_{\Gamma_u} \delta\mathbf{u} \cdot \mathbf{t} d\Gamma + \beta \int_{\Gamma_u} \delta\mathbf{u} \cdot (\mathbf{u} - \bar{\mathbf{u}}) d\Gamma = 0, \tag{2.92}$$

for all $\delta u_i \in \mathrm{H}^1$. It is left as an exercise to show that the Lagrange multiplier and Nitsche methods imply the strong form of the problem (2.69)–(2.71).

Exercise 2.1 Derive the relationship between the prescribed value of displacement ($\bar{\mathbf{u}}$) and the true displacement ($\mathbf{u}$) for the penalty method for elasticity in (2.84), i.e. the analogy to the conditions in the diffusion equation (2.42).

Exercise 2.2 Using integration by parts, derive the conditions (2.87)–(2.90) from (2.86), i.e., show that the Lagrange multiplier method attests to the strong form for elasticity, plus the additional condition.

Exercise 2.3 Using integration by parts show that Nitsche's method in (2.92) attests to the strong form for elasticity (2.69)–(2.71).

Exercise 2.4 Derive the weak form of the standard variational principle (2.79) by using the method of weighted residuals. Start with (2.74) and integrate the equation over the domain with an arbitrary test function $\mathbf{v}$; note that since (2.74) represents d unknowns, the dimension of the arbitrary test function must also have d components.

2.3 Nonlinear Continuum Mechanics

2.3.1 Strong Form for General Continua

The governing equations in a Lagrangian description of a general continuum employ the material (Lagrangian) coordinates $\mathbf{X} \in \Omega^0$ as independent variables in the undeformed configuration with domain Ω^0 and boundary Γ^0. The motion is described by a mapping

$$\mathbf{x} = \boldsymbol{\varphi}(\mathbf{X}, t), \tag{2.93}$$

where $\mathbf{x} \in \Omega$ are the spatial (Eulerian) coordinates in the deformed configuration with domain Ω and boundary Γ, and t is the time, see Figure 2.4.

Two important variables in describing the deformation resulting from the motion are the deformation gradient $\mathbf{F}$ given by

$$\mathbf{F} = \nabla_0 \phi, \quad \text{or} \quad F_{ij} = \frac{\partial \phi_i}{\partial X_j}, \tag{2.94}$$

where ∇_0 is the gradient taken with respect to the undeformed configuration, and the Jacobian determinant J, given by

$$J = \det(\mathbf{F}), \tag{2.95}$$

where $\det(\cdot)$ denotes the determinant.

The governing equations are listed in Box 2.1. In the top of the Box, the equations are given for a Lagrangian type of stress, the nominal stress $\mathbf{P}$, which is the transpose of the first Piola–Kirchhoff

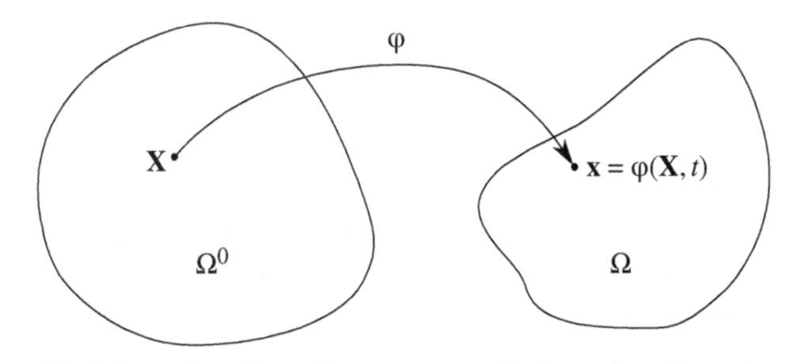

Figure 2.4 Mapping of a material point $\mathbf{X}$ in the undeformed configuration to point $\mathbf{x} = \phi(\mathbf{X}, t)$ in the deformed configuration.

(PK) stress. In Box 2.2, the equations are given in terms of an Eulerian type of stress, the Cauchy stress $\boldsymbol{\tau}$, or true stress, along with physical quantities in the deformed configuration. The physical basis of these two sets of equations is identical, see [6].

In addition, a constitutive equation is required. The constitutive equation should initially be developed in terms of frame invariant rates. The rate of the nominal stress $\dot{\mathbf{P}}$ is not frame invariant, nor is the rate of the deformation gradient $\dot{\mathbf{F}}$. However, frame invariant constitutive equations can always be transformed into equations relating $\dot{\mathbf{P}}$ to $\dot{\mathbf{F}}$, so this form will be assumed where convenient. For reversible materials, the nominal stress can be expressed in terms of a potential by

$$\mathbf{P} = \frac{\partial W^{\text{int}}}{\partial \mathbf{F}^{\text{T}}} \tag{2.96}$$

where W^{int} is the strain energy density function. Materials that can undergo large reversible deformations are called hyperelastic. A well-known example is rubber.

Box 2.1 Conservation Equations in Terms of Lagrangian Variables

Mass conservation

$$\rho(\mathbf{X}, t)J(\mathbf{X}, t) = \rho^0,$$

or

$$\rho J = \rho^0.$$

Linear momentum conservation

$$\rho_0 \frac{\partial^2 \mathbf{u}(\mathbf{X}, t)}{\partial t^2} = \nabla_0 \cdot \mathbf{P}^{\text{T}} + \rho^0 \mathbf{b}^0,$$

or

$$\rho^0 \frac{\partial^2 u_i}{\partial t^2} = \frac{\partial P_{ji}}{\partial X_j} + \rho^0 b_i^0.$$

Angular momentum conservation

$$\mathbf{F} \cdot \mathbf{P} = \mathbf{P}^{\text{T}} \cdot \mathbf{F}^{\text{T}},$$

or

$$F_{ik} P_{kj} = P_{ik}^{\text{T}} F_{kj}^{\text{T}} = F_{jk} P_{ki}.$$

Conservation of internal energy w^{int}

$$\rho^0 \frac{\partial w^{\text{int}}(\mathbf{X}, t)}{\partial t} = \dot{\mathbf{F}}^{\text{T}} : \mathbf{P} - \nabla_0 \cdot \mathbf{q}^0 + \rho^0 s^0,$$

or

$$\rho^0 \frac{\partial w^{\text{int}}}{\partial t} = \dot{F}_{ji} P_{ij} - \frac{\partial q_i^0}{\partial X_i} + \rho^0 s^0.$$

where $(\dot{\ })$ denotes differentiation with respect to time, s^0 is the source term in heat conduction in the undeformed configuration, and $\mathbf{q}^0 = -\mathbf{K} \cdot \nabla_0 \theta$ is the heat flux in the undeformed configuration with Fourier's law; θ is the temperature change.

Boundary conditions

$$u_i = \overline{u}_i \quad \text{on} \quad \Gamma_u^0, \qquad P_{ji} n_i^0 = \overline{t}_j^0 \quad \text{on} \quad \Gamma_t^0.$$

Additional boundary conditions if the energy equation is active

$$\theta = \overline{\theta} \quad \text{on} \quad \Gamma_\theta, \qquad q_i^0 = \overline{q}_i^0 \quad \text{on} \quad \Gamma_q.$$

Box 2.2 Conservation Equations in terms of Eulerian Variables

Mass conservation

$$\frac{\partial \rho(\mathbf{X},t)}{\partial t} + \rho(\mathbf{X},t)\left\{\nabla\cdot\frac{\partial \mathbf{u}(\mathbf{X},\ t)}{\partial t}\right\} = 0,$$

or

$$\frac{\partial \rho}{\partial t} + \rho \dot{u}_{i,i} = 0.$$

Linear momentum conservation

$$\rho\frac{\partial^2 \mathbf{u}(\mathbf{X},t)}{\partial t^2} = \nabla\cdot\boldsymbol{\tau} + \rho\mathbf{b},$$

or

$$\rho\frac{\partial^2 u_i}{\partial t^2} = \tau_{ij,j} + \rho b_i,$$

where ρ and $\mathbf{b}$ are the density and body force per unit volume in the current (deformed) state, and $\nabla \equiv \partial/\partial\mathbf{x}$.

Angular momentum conservation

$$\boldsymbol{\tau} = \boldsymbol{\tau}^{\mathrm{T}},$$

or

$$\tau_{ij} = \tau_{ji}.$$

Energy equation

$$\rho\frac{\partial w^{\mathrm{int}}(\mathbf{X},t)}{\partial t} = (\nabla_s\dot{\mathbf{u}}) : \boldsymbol{\tau} - \nabla\cdot\mathbf{q} + \rho s,$$

or

$$\rho\frac{\partial w^{\mathrm{int}}}{\partial t} = \dot{u}_{(i,j)}\tau_{ij} - \frac{\partial q_i}{\partial x_i} + \rho s,$$

where $\dot{u}_{(i,j)}$ denotes the symmetric part of $\dot{u}_{i,j}$, s is the current source term in heat conduction, and $\mathbf{q}$ is the current heat flux.

2.3.2 Principle of Stationary Potential Energy

Variational principles can only be developed for reversible materials, i.e., those with a potential. The theorem is called the theorem of stationary potential energy, because in contrast with linear elasticity, where a minimum potential energy exists for a well-posed problem, in nonlinear elasticity, stationary points such as saddle points often occur. The potential energy then takes the form:

$$\Pi_{\mathrm{C}} = \int_{\Omega^0} W^{\mathrm{int}} d\Omega - \int_{\Omega^0} \mathbf{u}\cdot\rho^0\mathbf{b}^0 d\Omega - \int_{\Gamma_t^0} \mathbf{u}\cdot\bar{\mathbf{t}}^0 d\Gamma, \tag{2.97}$$

where the subscript "C" stands for the potential for a general solid continuum, ρ^0 is the initial density, $\mathbf{b}^0$ is the body force per unit volume defined in the undeformed domain, and Γ_t^0 is the portion of the boundary of the undeformed domain where the natural boundary conditions are prescribed; here we also introduce Γ_u^0, where the essential boundary conditions are prescribed.

It is assumed that the body force $\mathbf{b}^0$ and the applied traction $\bar{\mathbf{t}}^0(\mathbf{X})$, $\mathbf{X} \in \Gamma_t^0$ are fixed, i.e., independent of the displacements. Any cohesive work at discontinuities has also been omitted here (refer to the Appendix for the treatment of discontinuities). The principle of stationary potential energy then states

$$\text{For } u_i(\mathbf{x}) \in \mathrm{U},\ \delta\Pi_\mathrm{C} = 0\ \forall\, u_i(\mathbf{x}) \in \mathrm{U}_0 \text{ is equivalent to the strong form.} \tag{2.98}$$

The spaces U and U_0 here consider essential and homogenous conditions, respectively, on the essential boundary of the undeformed body Γ_u^0:

$$\begin{aligned} \mathrm{U} &= \left\{u_i(\mathbf{x}) \middle| u_i(\mathbf{x}) \in \mathrm{H}^1 \text{ and } u_i(\mathbf{x}) = \bar{u}_i \text{ on } \Gamma_u^0\right\}, \\ \mathrm{U}_0 &= \left\{\delta u_i(\mathbf{x}) \middle| \delta u_i(\mathbf{x}) \in \mathrm{H}^1 \text{ and } \delta u_i(\mathbf{x}) = 0 \text{ on } \Gamma_u^0\right\}. \end{aligned} \tag{2.99}$$

The variational equation is obtained by taking the variation of (2.97) and setting it to zero, which is the so-called total Lagrangian formulation [7]:

$$\delta\Pi_\mathrm{C} = \int_{\Omega^0} \delta W^{\mathrm{int}} d\Omega - \int_{\Omega^0} \delta\mathbf{u} \cdot \rho^0 \mathbf{b}^0 d\Omega - \int_{\Gamma_t^0} \delta\mathbf{u} \cdot \bar{\mathbf{t}}^0 d\Gamma = 0. \tag{2.100}$$

We now use the chain rule on the first integrand, which gives

$$\delta W^{\mathrm{int}} = \frac{\partial W^{\mathrm{int}}}{\partial F_{ij}} \delta F_{ij} = P_{ji} \delta F_{ij} = P_{ji} \frac{\partial(\delta u_i)}{\partial X_j}. \tag{2.101}$$

Substituting (2.101) into (2.100) and using integration by parts on the first integral gives

$$\delta\Pi_\mathrm{C} = \int_{\Gamma_t^0} \delta u_i P_{ji} n_j^0 d\Gamma - \int_{\Omega^0} \delta u_i \frac{\partial P_{ji}}{\partial X_j} d\Omega - \int_{\Omega^0} \delta u_i \rho^0 b_i^0 d\Omega - \int_{\Gamma_t^0} \delta u_i \bar{t}_i^0 d\Gamma = 0. \tag{2.102}$$

where n_j^0 is the outward unit normal to Γ_t^0. In the above, we have also used the fact that $\delta\mathbf{u}$ vanishes on Γ_u^0 (as in elasticity), so the integral is restricted to the complementary domain Γ_t^0 in the first term. Furthermore, we have used our assumption that $\mathbf{b}^0$ and $\bar{\mathbf{t}}^0$ are independent of the displacement; otherwise, additional terms would emanate from the last two integrals and the strong form could not be derived. Collecting terms gives

$$\delta\Pi_\mathrm{C} = \int_{\Gamma_t^0} \delta u_i \left(P_{ji} n_j^0 - \bar{t}_i^0\right) d\Gamma - \int_{\Omega^0} \delta u_i \left(\frac{\partial P_{ji}}{\partial X_j} + \rho^0 b_i^0\right) d\Omega = 0. \tag{2.103}$$

Using the principle of variations, the first integral gives the traction boundary condition on the natural boundary, while the second integral gives the equilibrium equation.

2.3.3 Standard Weak Form for Nonlinear Continua

In the absence of heat conduction, only the linear momentum conservation equation remains to be solved [6], so we will first give the weak forms of that equation.

The derivation of the weak form for trial functions that satisfy the essential boundary conditions is similar to elasticity. We first multiply the governing equation and the natural boundary condition by the test function $\mathbf{v}(\mathbf{X})$ and integrate over the domains where they hold. This gives

$$\int_{\Omega^0} \mathbf{v} \cdot \left(\nabla_0 \cdot \mathbf{P} + \rho^0 \tilde{\mathbf{b}}^0\right) d\Omega = 0, \tag{2.104}$$

$$\int_{\Gamma_t^0} \mathbf{v} \cdot \left(\mathbf{t} - \bar{\mathbf{t}}^0\right) d\Gamma = 0, \tag{2.105}$$

where $\nabla_0 \equiv \partial/\partial\mathbf{X}$ denotes the gradient with respect to the undeformed coordinates, and $\tilde{\mathbf{b}}^0 = \mathbf{b}^0 - \dfrac{\partial^2 \mathbf{u}(\mathbf{X}, t)}{\partial t^2}$.

We now apply integration by parts to the first term in (2.104) which gives

$$\begin{aligned}\int_{\Omega^0} v_i \frac{\partial P_{ji}}{\partial X_j} d\Omega &= \int_{\Omega^0} \frac{\partial}{\partial X_j}(v_i P_{ji}) d\Omega - \int_{\Omega^0} P_{ji} \frac{\partial v_i}{\partial X_j} d\Omega \\ &= \int_{\Gamma_u^0 \cup \Gamma_t^0} v_i P_{ji} n_j^0 d\Gamma - \int_{\Omega^0} P_{ji} \frac{\partial v_i}{\partial X_j} d\Omega.\end{aligned} \tag{2.106}$$

Combining (2.106) with (2.104)–(2.105) and letting $u_i(\mathbf{x}) \in \mathrm{U}$ and $v_i(\mathbf{x}) \in \mathrm{U}_0$ (the same sets as the stationary principle in (2.99)) gives

$$\int_{\Omega^0} (\nabla_0 \mathbf{v})^{\mathrm{T}} : \mathbf{P} d\Omega - \int_{\Omega^0} \rho^0 \mathbf{v} \cdot \tilde{\mathbf{b}}^0 d\Omega - \int_{\Gamma_t^0} \mathbf{v} \cdot \bar{\mathbf{t}}^0 d\Gamma = 0. \tag{2.107}$$

Note that in the development of this weak form, we have not restricted the body force and applied tractions to be independent of the displacement $\mathbf{u}$, so this weak form holds for more general cases than the principle of stationary potential energy. Furthermore, it includes the inertial term, so it applies to dynamic problems. For a more general case with discontinuities, see the Appendix.

Guided by the above, we can obtain the weak form for the case when the essential boundary conditions are imposed by Lagrange multipliers. We add the displacement boundary condition similar to (2.85) in elasticity and take the variation to yield

$$\delta\Pi_{\mathrm{C}}^{\mathrm{L}} = \delta\Pi_{\mathrm{C}} + \int_{\Gamma_u^0} \delta\boldsymbol{\lambda} \cdot (\mathbf{u} - \bar{\mathbf{u}}) d\Gamma + \int_{\Gamma_u^0} \delta\mathbf{u} \cdot \boldsymbol{\lambda} d\Gamma = 0. \tag{2.108}$$

As for the penalty method, following (2.84), we have

$$\delta\Pi_{\mathrm{C}}^{\mathrm{P}} = \delta\Pi_{\mathrm{C}} + \beta \int_{\Gamma_u^0} \delta\mathbf{u} \cdot (\mathbf{u} - \bar{\mathbf{u}}) d\Gamma = 0. \tag{2.109}$$

Similarly, for Nitsche's method, it is

$$\delta\Pi_{\mathrm{C}}^{\mathrm{N}} = \delta\Pi_{\mathrm{C}} - \int_{\Gamma_u^0} \delta\mathbf{t} \cdot (\mathbf{u} - \bar{\mathbf{u}}) d\Gamma - \int_{\Gamma_u^0} \mathbf{t} \cdot \delta\mathbf{u} d\Gamma + \int_{\Gamma_u^0} \beta\delta\mathbf{u} \cdot (\mathbf{u} - \bar{\mathbf{u}}) d\Gamma. \tag{2.110}$$

2.A Appendix

2.A.1 Elasticity with Discontinuities

Wherever the material constants are discontinuous, such as at interfaces between materials, the displacement field will have weak discontinuities and equilibrium requires the tractions to be in equilibrium on these surfaces $\Gamma_{\text{int}}^{\text{W}}$

$$\mathbf{n}^+ \cdot \boldsymbol{\sigma}^+ + \mathbf{n}^- \cdot \boldsymbol{\sigma}^- = \mathbf{n}^+ \cdot \mathbf{C}^+ : \boldsymbol{\varepsilon}^+ + \mathbf{n}^- \cdot \mathbf{C}^- : \boldsymbol{\varepsilon}^- = 0 \text{ on } \Gamma_{\text{int}}^{\text{W}}. \tag{2.111}$$

Note that by using $\mathbf{n}^+ = -\mathbf{n}^-$ the traction equilibrium yields zero stress jump in the form of $[\boldsymbol{\sigma}] \cdot \mathbf{n} = \mathbf{0}$ on the weak discontinuity interface $\Gamma_{\text{int}}^{\text{W}}$ (with $\mathbf{n} = \mathbf{n}^+$ or $\mathbf{n} = \mathbf{n}^-$). In entities such as cracks and dislocations in the body, it is necessary to also add strong discontinuities in the displacement field. Recall we have

$$\delta\Pi_{\text{E}} = \delta W^{\text{int}} - \delta W^{\text{ext}}, \tag{2.112}$$

where δW^{ext} now must include a contour integral on the surfaces with strong discontinuities $\Gamma_{\text{int}}^{\text{S}}$

$$\delta W^{\text{ext}} = \int_{\Omega} \delta\mathbf{u} \cdot \mathbf{b} d\Omega + \int_{\Gamma_t} \delta\mathbf{u} \cdot \bar{\mathbf{t}} d\Gamma + \int_{\Gamma_{\text{int}}^{\text{S}}} [\delta\mathbf{u}] \cdot \mathbf{t}^{\text{coh}} d\Gamma,$$

or

$$\delta W^{\text{ext}} = \int_{\Omega} \delta u_i b_i^0 d\Omega + \int_{\Gamma_t} \delta u_i \bar{t}_i d\Gamma + \int_{\Gamma_{\text{int}}^{\text{S}}} [\delta u_i] t_i^{\text{coh}} d\Gamma, \tag{2.113}$$

and U is now defined as

$$\mathrm{U} = \{u_i(\mathbf{x}) | u_i(\mathbf{x}) \in \mathrm{H}^1 \text{ except when } u_i(\mathbf{x}) \text{ is discontinuous on } \Gamma_{\text{int}}^{\text{S}}, \text{ and } u_i(\mathbf{x}) = \bar{u}_i \text{ on } \Gamma_u \}. \tag{2.114}$$

The last term in (2.113) is the work done across the strong discontinuity on $\Gamma_{\text{int}}^{\text{S}}$, which depends on the cohesive traction across the discontinuity $\mathbf{t}^{\text{coh}}$. Note that when $\mathbf{t}^{\text{coh}}$ vanishes, the last term in (2.113) also vanishes. The conditions on the strong discontinuity can be thought of as natural boundary conditions.

Like the diffusion equation, here t_i^{coh} is specified with respect to the jump operator as

$$t_i^{\text{coh}} = \left(t_i^{\text{coh}}\right)^+ \text{ if } [u_i] = u_i^+ - u_i^-,$$
$$t_i^{\text{coh}} = \left(t_i^{\text{coh}}\right)^- \text{ if } [u_i] = u_i^- - u_i^+, \tag{2.115}$$

where $\left(t_i^{\text{coh}}\right)^+ = \boldsymbol{\sigma}^+ \cdot \mathbf{n}^+$ and $\left(t_i^{\text{coh}}\right)^- = \boldsymbol{\sigma}^- \cdot \mathbf{n}^-$ are prescribed fluxes on the "+" and "−" sides of the interface, with $\left(t_i^{\text{coh}}\right)^+ + \left(t_i^{\text{coh}}\right)^- = 0$.

We can now obtain the variational equation by taking the variation of Π_{E} in (2.112)–(2.114) and setting it equal to zero:

$$\delta\Pi_{\text{E}} = \int_{\Omega} \delta u_{(i,j)} C_{ijkl} u_{(k,l)} d\Omega - \int_{\Omega} \delta u_i b_i d\Omega - \int_{\Gamma_t} \delta u_i \bar{t}_i d\Gamma - \int_{\Gamma_{\text{int}}^{\text{S}}} [\delta u_i] t_i^{\text{coh}} d\Gamma = 0. \tag{2.116}$$

We next show that the variational equation (2.116) gives the strong form of the elasticity problem plus discontinuities. For this purpose, we use integration by parts on the first term of the right-hand side and break the strong discontinuity interface into two surfaces, with strong discontinuities on Γ^{S}_{int} and weak discontinuities on Γ^{W}_{int}. We intend this derivation to hold for functions in H^1, so we must also consider the surfaces of weak discontinuities. Applying the generalized integration by parts formula then gives

$$\begin{aligned}\int_{\Omega} \delta u_{(i,j)} C_{ijkl} u_{(k,l)} d\Omega = &\int_{\Gamma} \delta u_i \sigma_{ij} n_j d\Gamma + \int_{\Gamma^{W}_{int}} [\delta u_i \sigma_{ij}] n_j d\Gamma \\ &+ \int_{\Gamma^{S}_{int}} [\delta u_i \sigma_{ij}] n_j d\Gamma - \int_{\Omega} \delta u_i \left(C_{ijkl} u_{(k,l)}\right)_{,j} d\Omega.\end{aligned} \tag{2.117}$$

Since $\delta\mathbf{u}$ is continuous on Γ^{W}_{int}, we can combine the jump terms on the two surfaces. Then using the definition of the jump, we have

$$\begin{aligned}\int_{\Omega} \delta u_{(i,j)} \left(C_{ijkl} u_{(k,l)}\right) d\Omega = &\int_{\Gamma_t} \delta u_i \sigma_{ij} n_j d\Gamma + \int_{\Gamma^{W}_{int}} \delta u_i \left(\sigma^{+}_{ij} n^{+}_j + \sigma^{-}_{ij} n^{-}_j\right) d\Gamma + \int_{\Gamma^{S}_{int}} \left(\delta u^{+}_i \sigma^{+}_{ij} n^{+}_j + \delta u^{-}_i \sigma^{-}_{ij} n^{-}_j\right) d\Gamma \\ &- \int_{\Omega} \delta u_i \left(C_{ijkl} u_{(k,l)}\right)_{,j} d\Omega.\end{aligned} \tag{2.118}$$

Note that $\delta\mathbf{u}$ is continuous on Γ^{W}_{int}, whereas it is not continuous on Γ^{S}_{int}. Also, the domain of the first integral has been changed to Γ_t because $\delta\mathbf{u}$ vanishes on Γ_u.

Substituting the resulting form of (2.118) into (2.116) then gives

$$\begin{aligned}\delta\Pi_E = &\int_{\Gamma_t} \delta u_i \sigma_{ij} n_j d\Gamma + \int_{\Gamma^{W}_{int}} \delta u_i \left(\sigma^{+}_{ij} n^{+}_j + \sigma^{-}_{ij} n^{-}_j\right) d\Gamma + \int_{\Gamma^{S}_{int}} \left(\delta u^{+}_i \sigma^{+}_{ij} n^{+}_j + \delta u^{-}_i \sigma^{-}_{ij} n^{-}_j\right) d\Gamma \\ &- \int_{\Omega} \delta u_i \left(C_{ijkl} u_{(k,l)}\right)_{,j} d\Omega - \int_{\Omega} \delta u_i b_i d\Omega - \int_{\Gamma_t} \delta u_i \bar{t}_i d\Gamma - \int_{\Gamma^{S}_{int}} [\delta u_i] t^{coh}_i d\Gamma.\end{aligned} \tag{2.119}$$

Rearranging the terms then gives

$$\begin{aligned}\delta\Pi_E = &-\int_{\Omega} \delta u_i \left(\left(C_{ijkl} u_{(k,l)}\right)_{,j} + b_i\right) d\Omega + \int_{\Gamma_t} \delta u_i \left(\sigma_{ij} n_j - \bar{t}_i\right) d\Gamma + \int_{\Gamma^{W}_{int}} \delta u_i \left(\sigma^{+}_{ij} n^{+}_j + \sigma^{-}_{ij} n^{-}_j\right) d\Gamma \\ &+ \int_{\Gamma^{S}_{int}} \left(\delta u^{+}_i \left(\sigma^{+}_{ij} n^{+}_j - t^{coh}_i\right) + \delta u^{-}_i \left(\sigma^{-}_{ij} n^{-}_j + t^{coh}_i\right)\right) d\Gamma.\end{aligned} \tag{2.120}$$

In addition to the equations obtained from the standard variational principle without discontinuities, the third term on the right-hand side gives the jump conditions in (2.111), while the last integral shows that on surfaces with strong discontinuities, the tractions on both sides are equal to t^{coh}_i with different signs:

$$\begin{aligned}\sigma^{+}_{ij} n^{+}_j &= -\sigma^{-}_{ij} n^{-}_j = t^{coh}_i, \text{ on } \Gamma^{S}_{int} \text{ if } [u_i] = u^{+}_i - u^{-}_i, \\ \sigma^{-}_{ij} n^{-}_j &= -\sigma^{+}_{ij} n^{+}_j = t^{coh}_i, \text{ on } \Gamma^{S}_{int} \text{ if } [u_i] = u^{-}_i - u^{+}_i.\end{aligned} \tag{2.121}$$

2.A.2 Continuum Mechanics with Discontinuities

For continuum mechanics, consideration of discontinuities on contours Γ^0_{int} follows elasticity. We first supplement the conditions (2.104) and (2.105) with

$$\int_{\Gamma^0_{\text{int}}} [\![\mathbf{v}]\!] \cdot \left(\mathbf{t} - \mathbf{t}^{\text{coh}}\right) d\Gamma_0 = 0. \tag{2.122}$$

We now apply integration by parts to the first term in (2.104), which gives

$$\begin{aligned}\int_{\Omega^0} v_j \frac{\partial P_{ij}}{\partial X_i} d\Omega &= \int_{\Omega^0} \frac{\partial}{\partial X_i}\left(v_j P_{ij}\right) d\Omega - \int_{\Omega^0} P_{ij} \frac{\partial v_j}{\partial X_i} d\Omega \\ &= \int_{\Gamma^0_u \cup \Gamma^0_t} v_i P_{ji} n^0_j d\Gamma + \int_{\Gamma^0_{\text{int}}} [\![v_j P_{ij} n^0_i]\!] d\Gamma - \int_{\Omega^0} P_{ij} \frac{\partial v_j}{\partial X_i} d\Omega.\end{aligned} \tag{2.123}$$

The contour integral must include all interior surfaces over which either the test function $\mathbf{v}(\mathbf{X})$ or the normal component $\mathbf{n}^0 \cdot \mathbf{P}$ has discontinuities.

Combining (2.123) with (2.104)–(2.105) and (2.122) gives

$$\int_{\Omega^0} (\nabla_0 \mathbf{v})^{\text{T}} : \mathbf{P} d\Omega_0 - \int_{\Omega^0} \rho^0 \mathbf{v} \cdot \tilde{\mathbf{b}}^0 d\Omega - \int_{\Gamma^0_t} \mathbf{v} \cdot \bar{\mathbf{t}}^0 d\Gamma - \int_{\Gamma^0_{\text{int}}} [\![\mathbf{v}]\!] \cdot \mathbf{t}^{\text{coh}} d\Gamma = 0. \tag{2.124}$$

Note that using the weighted residual procedure here, $\mathbf{t}^{\text{coh}}$ can be assumed to be a function of the displacement which is generally the case.

References

1 Babuška, I. (1973). The finite element method with Lagrangian multipliers. *Numer. Math.* 20 (3): 179–192.
2 Brezzi, F. (1974). On the existence, uniqueness and approximation of saddle-point problems arising from Lagrangian multipliers. *Publ. Mathématiques Inform. Rennes.* S4: 1–26.
3 Nitsche, J. (1971). Über ein Variationsprinzip zur Lösung von Dirichlet-Problemen bei Verwendung von Teilräumen, die keinen Randbedingungen unterworfen sind. *Abhandlungen Aus Dem Math. Semin. Der Univ. Hambg.* 36: 9–15.
4 Lu, Y.Y., Belytschko, T., and Gu, L. (1994). A new implementation of the element free Galerkin method. *Comput. Methods Appl. Mech. Eng.* 113 (3–4): 397–414.
5 Wang, J. and Hillman, M. (2022). Temporal stability of collocation, Petrov–Galerkin, and other non-symmetric methods in elastodynamics and an energy conserving time integration. *Comput. Methods Appl. Mech. Eng.* 393: 114738.
6 Belytschko, T., Liu, W.K., Moran, B., and Elkhodary, K. (2013). *Nonlinear Finite Elements for Continua and Structures.* John Wiley & Sons.
7 Bathe, K. (2006). *Finite Element Procedures.* Prentice Hall.

3

Meshfree Approximations

A key task in the development of a meshfree method is the construction of the approximation functions, i.e., the trial and test functions in a Galerkin method. Meshfree approximations should preferably have compact supports. Furthermore, it is desirable that they be quite smooth in solving problems with smooth solutions. For example, solving elasticity using approximations with at least C^1 continuity yields continuous stresses and strains. This arbitrary order of smoothness (continuity) is easy to achieve in the approximations to be discussed in this chapter. The standard C^0 finite element for elasticity, on the other hand, generates discontinuous stresses and strains across element boundaries.

The moving least squares (MLS) and the reproducing kernel (RK) meshfree approximations are the most widely used. Many others are similar, so we will first describe MLS in Section 3.1. We will then introduce the RK approximation in Section 3.2 from a different lineage in its construction, based on the kernel estimate (KE) that originated from smoothed particle hydrodynamics (SPH) in 1977 by Lucy [1] and Gingold and Monaghan [2]. For the application of meshfree approximations to the solution of partial differential equations, derivatives are to be computed; Section 3.3 gives an overview of differentiation of the associated shape functions to be employed. The properties common to both the MLS and RK approximations are then discussed in Section 3.4, and various other methods to compute meshfree derivatives of functions are discussed in Section 3.5.

3.1 MLS Approximation

MLS was originally introduced in the surface fitting of a scattered set of points in a three-dimensional space by Lancaster and Salkauskas in 1981 [3]; Liszka and Orkisz also derived a similar method independently around the same time [4]. MLS was first discovered by the computational mechanics community by Nayroles, Touzot, and Villon in 1992 [5] in the diffuse element method (DEM), which employs MLS in conjunction with the Galerkin method to formulate a meshfree computational approach.

To begin, the MLS approximation will be described in one dimension for a scalar variable. The construction of vector and scalar approximations in multiple dimensions will then be discussed.

Suppose that on an interval $c \leq x \leq d$, we have a set of nodes $\{x_I\}_{I=1}^{NP}$. The basic idea in the MLS approximation is to approximate a function $u(x)$ in terms of *variable coefficients* $b_i(x)$ by

$$u^h(x) = b_0(x) + b_1(x)x + b_2(x)x^2 + \dots \tag{3.1}$$

Meshfree and Particle Methods: Fundamentals and Applications, First Edition.
Ted Belytschko, J. S. Chen, and Michael Hillman.

where the superscript h is used to indicate an approximation. Equation (3.1) is very similar to the functions used in the Rayleigh–Ritz method with $u^h = b_0x + b_1x^2 + \ldots$. The key difference is that the coefficients b_i are assumed to be variable, i.e., functions of the spatial position x.

There are several ways to explain the construction of these types of approximations. We will first describe the approach based on a "weighted" fit to nodal values by means of a weight function $w_I(x)$. Each weight function is associated with a node I; we will sometimes use the notation $w_I(x) \equiv w_a(x - x_I)$ to indicate that they are often translates of each other. They are also chosen to be functions with compact support denoted by "a"; i.e., they are only nonzero over a small subdomain $(x - a,\ x + a)$. An example of a weight function with a compact support is shown in Figure 3.1. Usually, the weight functions for all the nodes are identical except for the sizes of the supports, which are denoted by a_I; in most applications, a_I will depend on the local spacing of the nodes which we will generally denote h. The variations in the support sizes can be induced by scaling the arguments of the weight function by a_I, which can be written as $w_I(x) = w_a((x - x_I)/a_I)$. But carrying around all of this notation is quite awkward, so we will use either $w_I(x)$ or $w_a(x - x_I)$.

To generalize the presentation somewhat, we write the MLS approximation in the form

$$u^h(x) = \sum_{i=0}^{n} b_i(x)p_i(x) = \mathbf{p}^{\mathrm{T}}(x)\,\mathbf{b}(x). \tag{3.2}$$

In the above, $b_i(x)$ and $p_i(x)$ are the functions used in constructing the approximation. They are often called the basis functions or simply the basis. For the one dimensional approximation in (3.2), if only the constant and linear terms are retained, the basis is called a linear basis and the vectors in (3.2) are given by

$$\mathbf{p}^{\mathrm{T}} = [1\ \ x], \quad \mathbf{b}^{\mathrm{T}} = [\,b_0(x) \quad b_1(x)\,] \text{ (linear basis)}, \tag{3.3}$$

while, if we retain up to quadratic terms, the basis is

$$\mathbf{p}^{\mathrm{T}} = \left[1 \quad x \quad x^2\right], \quad \mathbf{b}^{\mathrm{T}} = [\,b_0(x) \quad b_1(x) \quad b_2(x)\,] \text{ (quadratic basis)}. \tag{3.4}$$

In the MLS approximation, the coefficients at each point x are found by minimizing a weighted norm J_w given by

$$J_w(x) = \sum_{I=1}^{NP} w_I(x)\left(\mathbf{p}^{\mathrm{T}}(x_I)\mathbf{b}(x) - u_I\right)^2, \tag{3.5}$$

where $u_I \equiv u(x_I)$ is nodal value, and $w_I(x)$ is a weight function with compact support (see Figure 3.1). For clarity, and this is usually also done in practice, we let the weight functions be a function of the

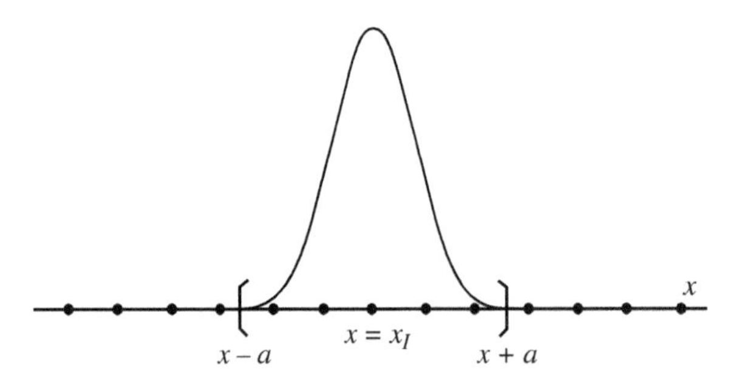

Figure 3.1 Weight function $w_I(x) \equiv w_a(x - x_I)$.

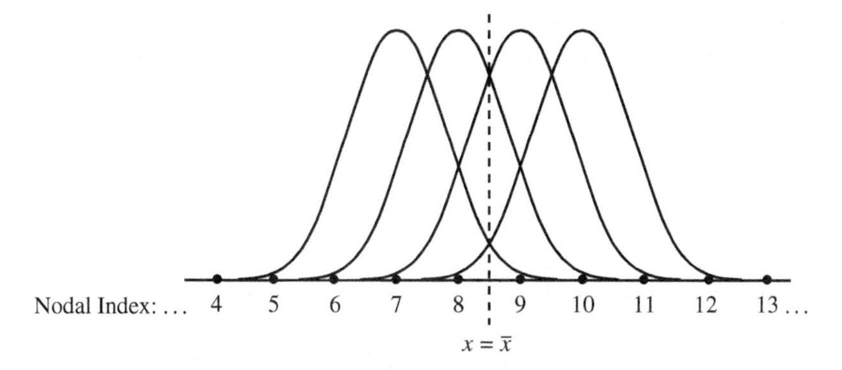

Figure 3.2 Illustration of nonzero contributions to the sums in constructing a meshfree approximation.

distance between x and x_I, i.e., $w_I(x) = w_a(|x - x_I|)$. Let the support of $w_I(x)$ be denoted by a_I, so $w_a(|x - x_I|) = 0$ for $|x - x_I| > a_I$. It follows that only the nodes I with $|x - x_I| \le a_I$ make a nonzero contribution to the sum (3.5). We denote this set of points by $S_x = \{I \mid x \in \text{supp}\, w_I(x)\}$, where "supp" denotes the support of a function. Note that the set S_x changes with x and consists of a set of nodes that are in the vicinity of point x; the members of the set depend on the size of the supports and the spacing of the neighboring nodes. For example, for the point $x = \bar{x}$ in Figure 3.2, the set of nodes in the domain of influence of x is given by $S_{\bar{x}} = \{7, 8, 9, 10\}$; in this one-dimensional grid with the nodes numbered sequentially, the nodes in the set are also sequential, but this is usually not the case in multiple dimensions.

The summation in (3.5) can then be replaced by

$$J_w(x) = \sum_{I \in S_x} w_I(x)\left(\mathbf{p}^{\mathrm{T}}(x_I)\mathbf{b}(x) - u_I\right)^2. \tag{3.6}$$

For convenience in the subsequent algebraic manipulations, we let $w_I \equiv w_I(x)$, $\mathbf{p}_I \equiv \mathbf{p}(x_I)$, $\mathbf{b} \equiv \mathbf{b}(x)$. Writing (3.6) in this notation we have

$$J_w = \sum_{I \in S_x} w_I\left(\mathbf{p}_I^{\mathrm{T}}\mathbf{b} - u_I\right)^2. \tag{3.7}$$

The minimum value of J_w is now found by taking its derivative with respect to $\mathbf{b}$ and setting the result to zero:

$$\frac{\partial J_w}{\partial \mathbf{b}} = 2\sum_{I \in S_x} w_I \mathbf{p}_I\left(\mathbf{p}_I^{\mathrm{T}}\mathbf{b} - u_I\right) = \mathbf{0}. \tag{3.8}$$

Here, we have used the identity $\partial\left(\mathbf{p}_I^{\mathrm{T}}\mathbf{b}\right)/\partial\mathbf{b} = \partial\left(\mathbf{b}^{\mathrm{T}}\mathbf{p}_I\right)/\partial\mathbf{b} = \mathbf{p}_I$. Dropping the common factor of two, the above can be written as

$$\left(\sum_{I \in S_x} w_I \mathbf{p}_I \mathbf{p}_I^{\mathrm{T}}\right)\mathbf{b} = \sum_{I \in S_x} w_I \mathbf{p}_I u_I. \tag{3.9}$$

This can be seen to be a set of linear algebraic equations in the unknowns $\mathbf{b}$. The number of equations depends on the number of basis functions.

The form in (3.9) can be used directly when the MLS approximants are used for fitting data. However, for applications to partial differential equations, it is useful to obtain the MLS approximation in the form

$$u^h(x) = \sum_{I=1}^{NP} \Psi_I(x) u_I, \tag{3.10}$$

where u_I are nodal parameters and $\Psi_I(x)$ are called the shape functions; this is the same terminology as in finite element methods, and we will usually use this term to avoid misusing the term interpolants. They are also called the approximants.

To obtain the shape function from (3.10), (3.2), and (3.9), we will first define the following matrix

$$\mathbf{A}(x) = \sum_{I \in S_x} w_I(x) \mathbf{p}(x_I) \mathbf{p}^{\mathrm{T}}(x_I). \tag{3.11}$$

The matrix **A** is often called the moment matrix; the origin of this name for a polynomial basis will be clarified shortly. It is also called a Gram matrix of the basis functions $\mathbf{p}(x)$ with respect to $w_I(x)$.

Equation (3.9) can then be written as

$$\mathbf{A}(x)\mathbf{b}(x) = \sum_{I \in S_x} w_I(x) \mathbf{p}(x_I) u_I \quad \Rightarrow \quad \mathbf{b}(x) = \mathbf{A}^{-1}(x) \sum_{I \in S_x} w_I(x) \mathbf{p}(x_I) u_I, \tag{3.12}$$

where the functional dependence of the terms has been reintroduced. As noted, **A** is a function of x, so the equations for **b** must be solved at every location where the approximation $u^h(x)$ is computed.

Substituting (3.12) into (3.2), we obtain

$$u^h(x) = \mathbf{p}^{\mathrm{T}}(x)\mathbf{b}(x) = \mathbf{p}^{\mathrm{T}}(x)\mathbf{A}^{-1}(x) \sum_{I \in S_x} w_I(x) \mathbf{p}(x_I) u_I. \tag{3.13}$$

Comparing equations (3.10) and (3.13), it can be seen that the shape functions $\Psi_I(x)$ are given by

$$\Psi_I(x) = \mathbf{p}^{\mathrm{T}}(x)\mathbf{A}^{-1}(x)\mathbf{p}(x_I) w_I(x). \tag{3.14}$$

Note that the sum in (3.10) is over all nodes, whereas the sum in (3.13) is only over the nodes in the set S_x. This means that only those nodes with weight functions $w_I(x)$ that are nonzero at x make contributions to the sum in (3.10); the remaining nodes have no effect on the value of the approximation $u^h(x)$. In other words, only those nodes whose supports include the location x affect the approximation at x, $u^h(x)$, that is

$$u^h(x) = \sum_{I=1}^{NP} \Psi_I(x) u_I = \sum_{I \in S_x} \Psi_I(x) u_I. \tag{3.15}$$

This will be seen to be important in ensuring the sparsity of the discretizations for partial differential equations.

The above approximation is called a *moving* least square approximation because the norm J_w depends on the position at which the coefficients **b** are to be evaluated. By contrast, a classical, non-MLS method uses a weight function that is constant in space, so only a single set of coefficients needs to be evaluated. Such a non-MLS fit is called a regression function in statistics, see Lancaster and Salkauskas [3]. The notion of moving approximations is often confused with local and global approximations, but is quite unrelated.

Exercise 3.1 Consider a domain Ω discretized by a set of NP points $\{x_I\}_{I=1}^{NP}$, with the nodal value of a function u at each point x_I given as $u(x_I) = u_I$, $I = 1,2,\ldots, NP$. The approximation of the function u, denoted as u^h, using a set of basis functions $\mathbf{p}^{\mathrm{T}}(x) = [p_0(x) \quad p_1(x) \quad \cdots \quad p_n(x)]$, can be obtained using the minimization of the following residuals:

1) Least squares (LS):
   The LS approximation of u is given as

$$u^h(x) = \mathbf{p}(x)^{\mathrm{T}}\mathbf{b},$$

   with the residual defined as

$$J = \left(u - u^h, u - u^h\right),$$

   where

$$(u, v) = \sum_{I=1}^{NP} u(x_I)v(x_I).$$

   The LS approximation of u is obtained by minimizing J. Note that the coefficients in the LS approximation are *constant.*
2) Weighted least squares (WLS):
   Consider a local approximation of u near $x = \overline{x}$ as

$$u_{\overline{x}}^h(x) = \mathbf{P}^{\mathrm{T}}(x)\mathbf{b}(\overline{x}),$$

   with a weighted residual defined as

$$J_{w_{\overline{x}}} = \left(u - u^h, u - u^h\right)_{\overline{x}},$$

   where

$$(u, v)_{\overline{x}} = \sum_{I \in S_{\overline{x}}} u(x_I)v(x_I)w_a(\overline{x} - x_I).$$

   The WLS approximation of u is obtained by minimizing $J_{w_{\overline{x}}}$. Note that the coefficients now are functions of the location $\overline{x}$.
3) Moving least squares (MLS)
   i) Consider a WLS approximation of u near $x = \overline{x}$: $u_{\overline{x}}^h(x)$.
   ii) The MLS approximation of u is then obtained by setting $\overline{x} \to x$ in $u_{\overline{x}}^h(x)$,

$$u^h(x) = u_{\overline{x}\to x}^h(x).$$

Show that the LS, WLS, and MLS approximations of u using the above procedures yield the results shown in Table 3.1. This exercise demonstrates an alternative way of deriving the MLS approximation.

Table 3.1 Definition of least squares (LS), weighted least squares (WLS), and moving least squares (MLS) approximations.

	Approximation	Least squares residual	Minimization of residual
LS	$u^h(x) = \mathbf{p}^T(x)\mathbf{b}$	$J = (u - u^h, u - u^h)$ $(u, v) = \sum_{I=1}^{NP} u(x_I)v(x_I)$	$u^h(x) = \mathbf{p}^T(x)\mathbf{A}^{-1}(x)\sum_{I=1}^{NP}\mathbf{p}(x_I)u_I$ $\mathbf{A}(x) = \sum_{I=1}^{NP}\mathbf{p}(x_I)\mathbf{p}^T(x_I)$
WLS	$u^h_{\bar{x}}(x) = \mathbf{p}^T(x)\mathbf{b}(\bar{x})$	$J_{w_{\bar{x}}} = (u - u^h, u - u^h)_{\bar{x}}$ $(u, v)_{\bar{x}} = \sum_{I\in S_{\bar{x}}} u(x_I)v(x_I)w_a(\bar{x} - x_I)$	$u^h_{\bar{x}} = \mathbf{p}^T(x)\mathbf{A}^{-1}(\bar{x})\sum_{I\in S_{\bar{x}}}\mathbf{p}(x_I)w_a(\bar{x} - x_I)u_I$ $\mathbf{A}(\bar{x}) = \sum_{I\in S_{\bar{x}}}\mathbf{p}(x_I)\mathbf{p}^T(x_I)w_a(\bar{x} - x_I)$
MLS	$u^h(x) = u^h_{\bar{x}\to x}(x)$	$J_{w_x} = (u - u^h, u - u^h)_x$ $(u, v)_x = \sum_{I\in S_x} u(x_I)v(x_I)w_a(x - x_I)$	$u^h(x) = u^h_{\bar{x}\to x}(x)$ $= \mathbf{p}^T(x)\mathbf{A}(x)^{-1}\sum_{I\in S_x}\mathbf{p}(x_I)w_a(x - x_I)u_I$ $\mathbf{A}(x) = \sum_{I\in S_x}\mathbf{p}(x_I)\mathbf{p}^T(x_I)w_a(x - x_I)$

3.1.1 Weight Functions

The most widely used weight functions are given in Table 3.2. In one dimension, we have given only symmetric weight functions, i.e., weight functions for which

$$w_a(x - x_I) = w_a(x_I - x). \tag{3.16}$$

Table 3.2 Most widely used weight functions.

Weight function	Expression	Order of continuity
Box function	$w(z) = \begin{cases} 1, & 0 \le z \le 1 \\ 0, & 1 < z \end{cases}, \quad z = \frac{\lvert x - x_I \rvert}{a}$	C^{-1}
Tent (or hat) function	$w(z) = \begin{cases} 1 - z, & 0 \le z \le 1 \\ 0, & 1 < z \end{cases}, \quad z = \frac{\lvert x - x_I \rvert}{a}$	C^0
Quadratic B-spline function	$w(z) = \begin{cases} 1 - 2z^2, & 0 \le z \le \frac{1}{2} \\ 2 - 4z + 2z^2, & \frac{1}{2} < z \le 1, \\ 0, & 1 < z \end{cases} \quad z = \frac{\lvert x - x_I \rvert}{a}$	C^1
Cubic B-spline function	$w(z) = \begin{cases} \frac{2}{3} - 4z^2 + 4z^3, & 0 \le z \le \frac{1}{2} \\ \frac{4}{3} - 4z + 4z^2 - \frac{4}{3}z^3, & \frac{1}{2} < z \le 1, \\ 0, & 1 < z \end{cases} \quad z = \frac{\lvert x - x_I \rvert}{a}$	C^2
Truncated exponential function	$w(z) = \begin{cases} \frac{\exp(-(z/c)^2) - \exp(-(1/c)^2)}{(1 - \exp(-(1/c)^2))}, & 0 \le z \le 1 \\ 0, & 1 < z \end{cases}, \quad z = \frac{\lvert x - x_I \rvert}{a}$ c : scaling factor	C^∞ *in* $0 \le z < 1$, C^0 *at* $z = 1$

The following conditions are usually imposed in constructing a weight function

1) It should have compact support, i.e.,

$$w_a(x - x_I) \neq 0 \text{ only for } |x - x_I| \leq a, \quad a \sim O(h). \tag{3.17}$$

By $O(h)$, we mean from approximately h to at most about $5h$. The compact support keeps the stiffness matrix sparse; the dependence of the support on the nodal spacing h yields a bandwidth that will not increase with the refinement of h, and this definition is also necessary for convergence (see Section 3.4).

2) It should be positive in its support

$$w_a(x - x_I) > 0 \text{ for } |x - x_I| < a. \tag{3.18}$$

This property helps ensure the invertibility of the matrix involved in constructing the meshfree approximation, see Section 3.1.4.

3) It should be a monotonically decreasing function of $|x - x_I|$; this ensures an evaluation point x has a greater influence on the approximation when it is closer to the sampling point x_I.
4) It must have the same continuity as desired in the approximation.

Various weight functions with different orders of continuities are shown in Figure 3.3. The cubic B-spline, as indicated in Table 3.2, is C^2. Its advantage is that its computation is fast. The continuity

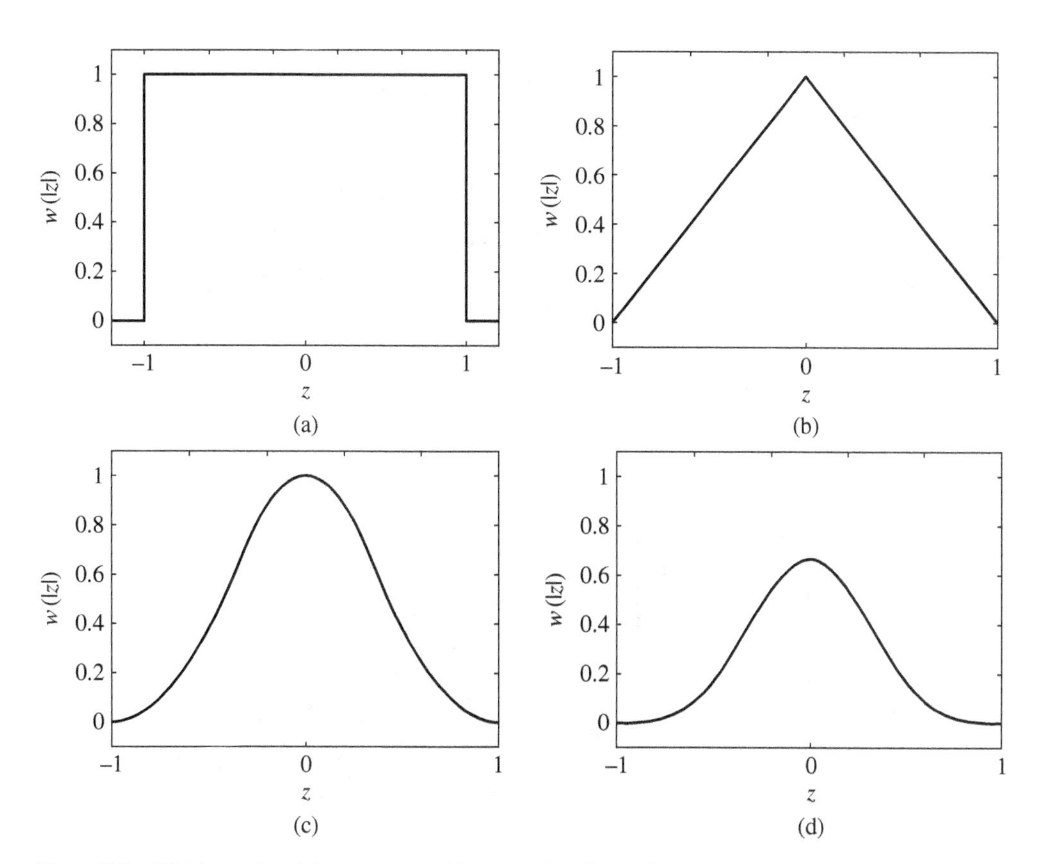

Figure 3.3 Widely used weight functions: (a) box function, (b) hat function, (c) quadratic B-spline, and (d) cubic B-spline.

of the exponential weight function is a little puzzling. Strictly speaking, its support is infinite since the function never vanishes, and in that case, it is infinitely continuously differentiable. However, truncating the function renders the support to be finite and seems to have little effect on its accuracy. The resulting MLS approximation function also behaves more smoothly than a C^1 weight despite the truncation.

The weight functions in multiple dimensions are usually generated from the one-dimensional weight functions by replacing $|x - x_I|$ with the distance r to the node I, which is given by

$$r = \begin{cases} ((x-x_I)^2 + (y-y_I)^2)^{\frac{1}{2}}, & \text{two dimensions} \\ ((x-x_I)^2 + (y-y_I)^2 + (z-z_I)^2)^{\frac{1}{2}}. & \text{three dimensions} \end{cases} \tag{3.19}$$

We denote the weight function of node I by $w_I(r)$. The weight functions $w_I(r)$ are radially symmetric, and they can be considered isotropic, although anisotropic ellipsoidal kernels can be constructed and are sometimes desirable. In any case, such functions are easily constructed with high orders of continuity from one-dimensional weight functions. The support of a radial weight function $w_I(r)$ in two dimensions is a circle centered at node I, as shown in Figure 3.4a.

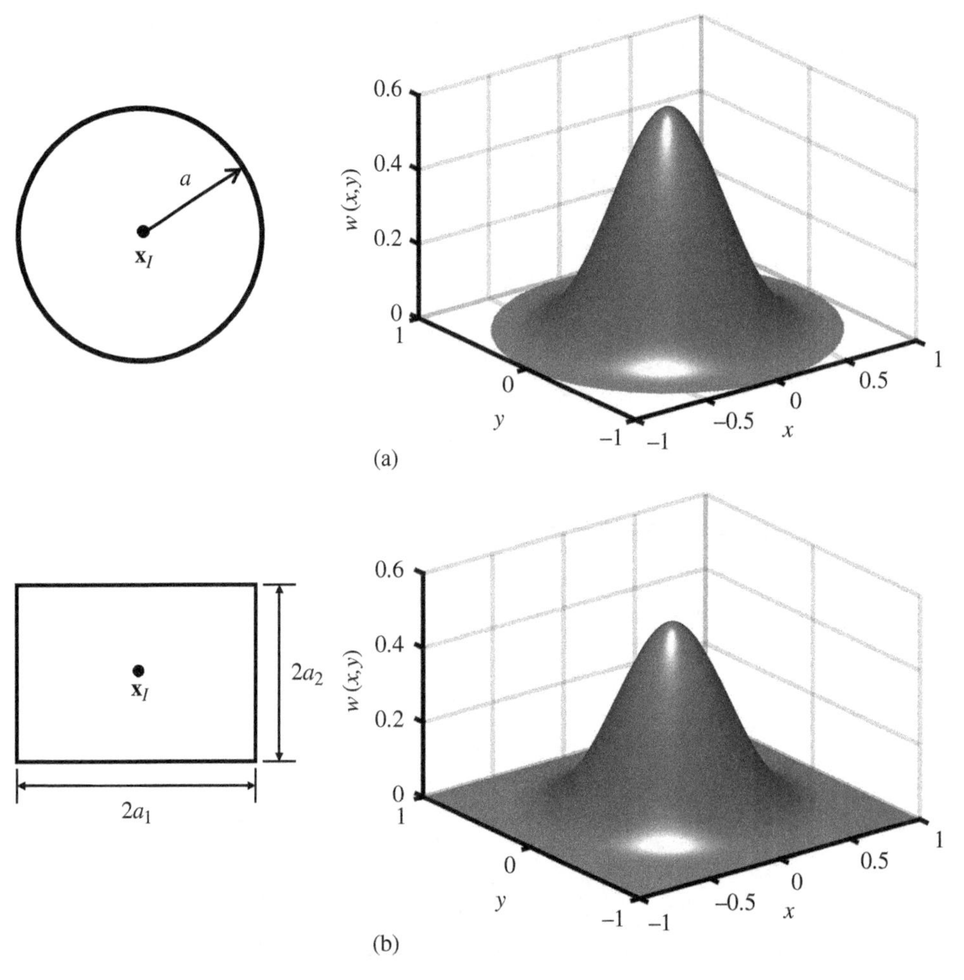

Figure 3.4 Two-dimensional weight functions and their supports generated by (a) the distance function (3.19), and (b) the product rule (3.20).

Weight functions in multiple dimensions can also be constructed by tensor product procedures, in which the weight function is taken to be a product of one-dimensional weight functions in the base spatial coordinates. For example, a tensor product function based on the one-dimensional weight can be obtained as

$$\begin{aligned} w_a(x-x_I, y-y_I) &= w_{a_1}(x-x_I)w_{a_2}(y-y_I), && \text{two dimensions} \\ w_a(x-x_I, y-y_I, z-z_I) &= w_{a_1}(x-x_I)w_{a_2}(y-y_I)w_{a_3}(z-z_I). && \text{three dimensions} \end{aligned} \tag{3.20}$$

The support of a tensor product weight function in two dimensions is a rectangle with dimensions $2a_1 \times 2a_2$ as shown in Figure 3.4b, and in three dimensions, it is a brick with dimensions $2a_1 \times 2a_2 \times 2a_3$. Note that different support sizes for $w_{a_1}(x-x_I)$, $w_{a_2}(y-y_I)$, and $w_{a_3}(z-z_I)$ can be selected to allow the use of different nodal densities in different directions (anisotropic node distributions) in certain applications.

The one-dimensional shape functions corresponding to a cubic B-spline weight are shown in Figure 3.5 for constant, linear, and quadratic bases. The two-dimensional shape functions corresponding to a tensor product cubic B-spline weight for these bases are shown in Figure 3.6. The MLS functions with a constant basis are called Shepard functions.

Exercise 3.2 Following Exercise 3.1, use the least squares, weighted least squares, and moving least squares methods for fitting the data given in Table 3.3.

1) Use quadratic bases $\mathbf{p}^{\mathrm{T}}(\mathbf{x}) = \begin{bmatrix} 1 & x & x^2 \end{bmatrix}$ for LS, WLS, and MLS methods, and employ the cubic B-spline weight function shown in Table 3.2 for WLS ($\bar{x} = 0.5$, $a = 0.25$) and MLS ($a = 0.25$) methods. Compare the fitting of the three methods with the original data in a figure.
2) Vary the support "a" in WLS and MLS and study the effect of a in the data fitting.

3.1.2 MLS Approximation of Vectors in Multiple Dimensions

The approximation of vector and tensor functions in multiple dimensions is almost identical to the one-dimensional approximation. Consider a domain Ω; we denote the spatial coordinates by the vector $\mathbf{x}$, where $\mathbf{x} = [x \quad y]^{\mathrm{T}}$ in two dimensions and $\mathbf{x} = [x \quad y \quad z]^{\mathrm{T}}$ in three dimensions. Scattered within the domain Ω is a set of nodes $\{\mathbf{x}_I\}_{I=1}^{NP}$. The conditions on the placement of these nodes are given later.

We wish to approximate a vector function $u_i(\mathbf{x})$, where the subscripts on u denote the components of the vector. We denote the approximation by $u_i^h(\mathbf{x})$; the procedure for constructing a scalar approximation in multiple dimensions is identical to the following procedure and can be extracted from it by just dropping the index on u.

We start with a monomial expansion in terms of the basis functions as in (3.2)

$$u_i^h(\mathbf{x}) = \sum_{\alpha=0}^{n_i} b_{i\alpha}(\mathbf{x})p_{i\alpha}(\mathbf{x}) = \mathbf{p}_i^{\mathrm{T}}(\mathbf{x})\mathbf{b}_i(\mathbf{x}). \tag{3.21}$$

Note that for each vector component u_i, we have a different set of bases $\mathbf{p}_i(\mathbf{x})$ and coefficients $\mathbf{b}_i(\mathbf{x})$. It is possible to employ different basis functions for each component. If one chooses to use the same basis for each component of $u_i(\mathbf{x})$, the subscript i is dropped in $\mathbf{p}_i(\mathbf{x})$ and the coefficients $\mathbf{b}_i(\mathbf{x})$. Taking the same or different basis function in a different direction i, would not complicate the development much.

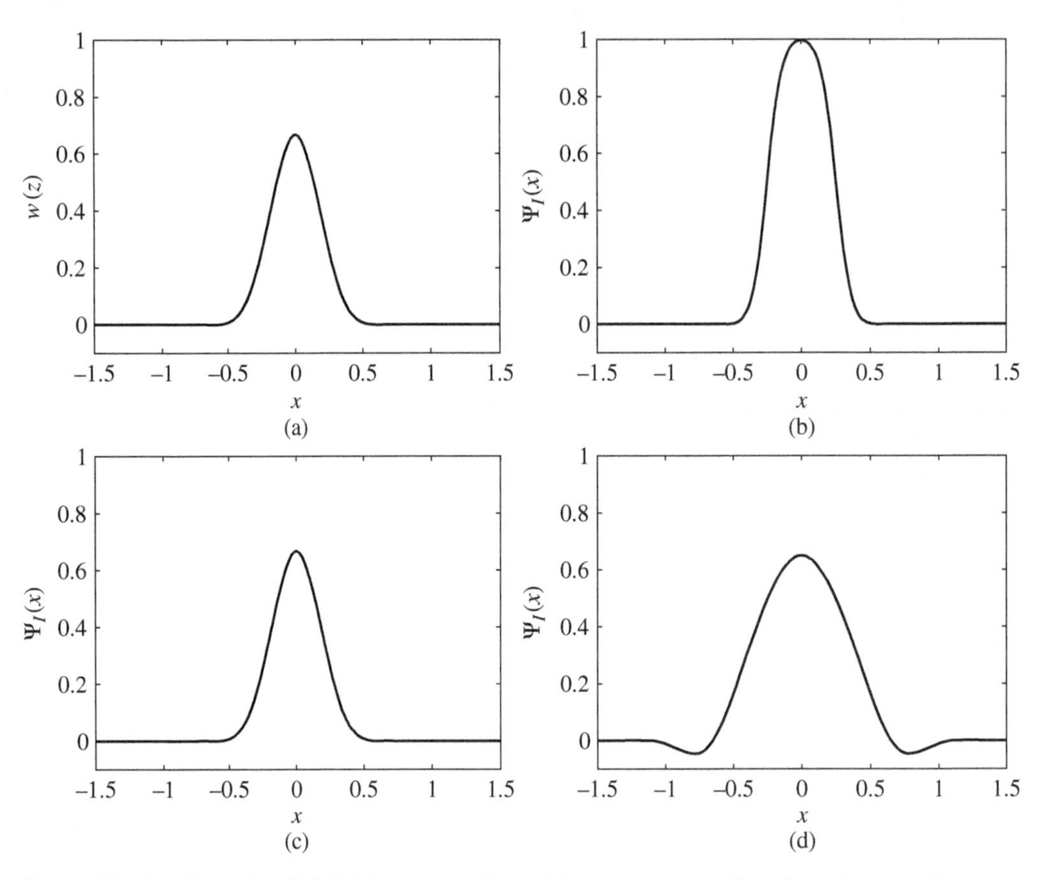

Figure 3.5 One-dimensional MLS: (a) cubic B-spline weight, and corresponding shape functions for (b) constant basis, (c) linear basis, and (d) quadratic basis.

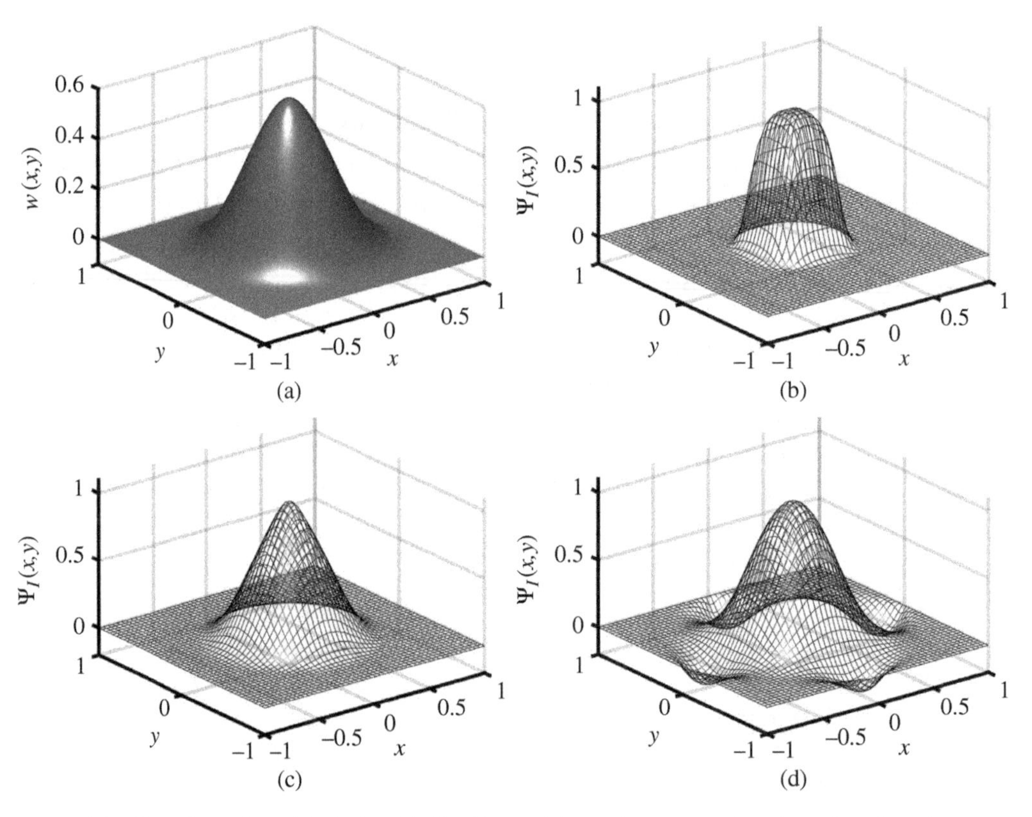

Figure 3.6 Two-dimensional MLS: (a) cubic B-spline weight, and corresponding shape functions for (b) constant basis, (c) linear basis, and (d) quadratic basis.

Table 3.3 Definition of LS, WLS, and MLS approximation data.

x_I	0.000	0.120	0.190	0.310	0.430	0.480	0.610	0.720	0.790	0.910	1.000
u_I	1.000	1.496	1.771	2.191	2.513	2.614	2.781	2.825	2.816	2.763	2.718

The coefficients $\mathbf{b}_i(x)$ are found at any point $\mathbf{x}$ by minimizing a weighted norm

$$J_w^i(\mathbf{x}) = \sum_{I \in S_\mathbf{x}} w_I(\mathbf{x}) \left(\mathbf{b}_i^\mathrm{T}(\mathbf{x})\mathbf{p}_i(\mathbf{x}_I) - u_{iI}\right) \left(\mathbf{b}_i^\mathrm{T}(\mathbf{x})\mathbf{p}_i(\mathbf{x}_I) - u_{iI}\right), \tag{3.22}$$

where $S_\mathbf{x} = \{I \mid \mathbf{x} \in \text{supp}\, w_I(\mathbf{x})\}$ is the set of nodes with the associated weight functions that cover $\mathbf{x}$, and u_{iI} is the value of the ith component at node I. The minimum of each J_w^i is found by setting its derivatives with respect to $\mathbf{b}_i$ to zero, which gives

$$\frac{\partial J_w^i}{\partial \mathbf{b}_i} = 2 \sum_{I \in S_\mathbf{x}} w_I \mathbf{p}_{iI} \left(\mathbf{p}_{iI}^\mathrm{T} \mathbf{b}_i - u_{iI}\right) = 0, \tag{3.23}$$

where $\mathbf{p}_{iI} \equiv \mathbf{p}_i(\mathbf{x}_I)$, $\mathbf{b}_i \equiv \mathbf{b}_i(\mathbf{x})$, and $w_I \equiv w_I(\mathbf{x})$. The algebra involved in developing these equations is identical to that between (3.5) and (3.8), except that subscripts i must now be carried on $\mathbf{p}_i$, $\mathbf{b}_i$, and u_i. We define similar matrices $\mathbf{A}_i$ as in the one-dimensional development

$$\mathbf{A}_i(\mathbf{x}) = \sum_{I \in S_\mathbf{x}} w_I \mathbf{p}_{iI} \mathbf{p}_{iI}^\mathrm{T} = \mathbf{P}_i^\mathrm{T} \mathbf{W} \mathbf{P}_i, \tag{3.24}$$

$$\mathbf{P}_i = \begin{bmatrix} \mathbf{p}_i^\mathrm{T}\left(\mathbf{x}_{S_\mathbf{x}(1)}\right) \\ \vdots \\ \mathbf{p}_i^\mathrm{T}\left(\mathbf{x}_{S_\mathbf{x}(K)}\right) \end{bmatrix}, \tag{3.25}$$

$$\mathbf{W} = \begin{bmatrix} w_{S_\mathbf{x}(1)}(\mathbf{x}) & & 0 \\ & \ddots & \\ 0 & & w_{S_\mathbf{x}(K)}(\mathbf{x}) \end{bmatrix}, \tag{3.26}$$

where $K = \text{card}(S_\mathbf{x})$ is the number of members in $S_\mathbf{x}$ and $S_\mathbf{x}(i)$ is the node number of ith member in $S_\mathbf{x}$. The remaining steps follow exactly the same as the one-dimensional case in (3.9)–(3.14). The final MLS multi-dimensional shape function reads

$$u_i^h(\mathbf{x}) = \mathbf{p}_i^\mathrm{T}(\mathbf{x})\mathbf{b}_i(\mathbf{x}) = \mathbf{p}_i^\mathrm{T}(\mathbf{x})\mathbf{A}_i^{-1}(\mathbf{x}) \sum_{I \in S_\mathbf{x}} w_I(\mathbf{x})\mathbf{p}_i(\mathbf{x}_I) u_{iI} = \sum_{I \in S_\mathbf{x}} \Psi_{iI}(\mathbf{x}) u_{iI}, \tag{3.27}$$

where

$$\Psi_{iI}(\mathbf{x}) = \mathbf{p}_i^\mathrm{T}(\mathbf{x})\mathbf{A}_i^{-1}(\mathbf{x})\mathbf{p}_i(\mathbf{x}_I) w_I(\mathbf{x}). \tag{3.28}$$

In most applications, the approximations of u_i are selected to be independent of the directional index i. That is, basis functions are independent of i:

$$u_i^h(\mathbf{x}) = \sum_{\alpha=0}^{n_l} b_{i\alpha}(\mathbf{x}) p_\alpha(\mathbf{x}) = \mathbf{p}^\mathrm{T}(\mathbf{x})\mathbf{b}_i(\mathbf{x}). \tag{3.29}$$

The coefficients $\mathbf{b}_i(\mathbf{x})$ are found at any location $\mathbf{x}$ by minimizing a weighted norm

$$J_w(\mathbf{x}) = \sum_{I \in S_x} w_I(\mathbf{x}) \left(\mathbf{b}_i^{\mathrm{T}}(\mathbf{x})\mathbf{p}(\mathbf{x}_I) - u_{iI}\right) \left(\mathbf{b}_i^{\mathrm{T}}(\mathbf{x})\mathbf{p}(\mathbf{x}_I) - u_{iI}\right). \tag{3.30}$$

The minimum of each J_w is found by setting its derivatives with respect to $\mathbf{b}_i$ to zero, which gives

$$\frac{\partial J_w}{\partial \mathbf{b}_i} = 2\sum_{I \in S_x} w_I \mathbf{p}_I \left(\mathbf{p}_I^{\mathrm{T}} \mathbf{b}_i - u_{iI}\right) = 0. \tag{3.31}$$

Solving $\mathbf{b}_i$ from (3.31) and substituting it into (3.29) yields

$$u_i^h(\mathbf{x}) = \mathbf{p}^{\mathrm{T}}(\mathbf{x})\mathbf{A}^{-1}(\mathbf{x})\sum_{I \in S_x} w_I(\mathbf{x})\mathbf{p}(\mathbf{x}_I)u_{iI} = \sum_{I \in S_x} \Psi_I(\mathbf{x})u_{iI}, \tag{3.32}$$

where $\Psi_I(\mathbf{x})$ is the MLS shape function of u_i, and is found to be independent of the component i,

$$\Psi_I(\mathbf{x}) = \mathbf{p}^{\mathrm{T}}(\mathbf{x})\mathbf{A}^{-1}(\mathbf{x})\mathbf{p}(\mathbf{x}_I)w_I(\mathbf{x}), \tag{3.33}$$

$$\mathbf{A} = \sum_{I \in S_x} w_I(\mathbf{x})\mathbf{p}_I\mathbf{p}_I^{\mathrm{T}} = \mathbf{P}^{\mathrm{T}}\mathbf{W}\mathbf{P}, \tag{3.34}$$

$$\mathbf{P} = \begin{bmatrix} \mathbf{p}^{\mathrm{T}}\left(\mathbf{x}_{S_x(1)}\right) \\ \vdots \\ \mathbf{p}^{\mathrm{T}}\left(\mathbf{x}_{S_x(K)}\right) \end{bmatrix}. \tag{3.35}$$

The properties of the MLS shape functions $\Psi_I(\mathbf{x})$ are the same as those of $\Psi_{iI}(\mathbf{x})$, with only one difference: the basis functions $\mathbf{p}(\mathbf{x})$ in $\Psi_I(\mathbf{x})$ are the same for all components of u_i.

3.1.3 Reproducing Properties

The reproducing properties of an approximation refer to its ability to exactly reproduce a given function. These are also referred to as an approximation's *completeness* when referring to polynomial approximations and *consistency* when solving partial differential equations (PDEs). For example, if a set of approximation functions $N_I(\mathbf{x})$ are able to reproduce $p(\mathbf{x})$ using nodal information about $p(\mathbf{x})$, this requires

$$\sum_I p(\mathbf{x}_I)N_I(\mathbf{x}) = p(\mathbf{x}). \tag{3.36}$$

Also, if the approximants $N_I(\mathbf{x})$ are able to reproduce the constant function exactly, then setting $p(\mathbf{x}) = 1$ we have

$$\sum_I N_I(\mathbf{x}) = 1. \tag{3.37}$$

This is called *zeroth-order completeness*. Functions that satisfy the above property are *partitions of unity*, which is important in developing certain meshfree approximations that converge, as discussed in Chapter 8.

Linear completeness requires the approximation to reproduce a linear function exactly. This condition can be expressed as

$$\sum_I \mathbf{x}_I N_I(\mathbf{x}) = \mathbf{x} \quad \text{or} \quad \sum_I x_{iI} N_I(\mathbf{x}) = x_i. \tag{3.38}$$

It can be shown that the MLS shape functions satisfy the reproducing properties in (3.36) when the given function $p(\mathbf{x})$ is contained in the basis vector $\mathbf{p}$. This can be verified by examining how the MLS approximation is constructed by minimizing the least squares measures in (3.30) via (3.31). In (3.31), one finds that when setting $u_{iI} = p_i(\mathbf{x}_I)$, $\mathbf{b}_i$ with its αth component $b_{\alpha i} = \delta_{\alpha i}$ yields $\partial J_w / \partial \mathbf{b}_i = \mathbf{0}$ and thus minimizes the least-square measure J_w. In fact, it leads to $J_w = 0$. That is, MLS shape functions exactly reproduce all the basis functions in the bases vector $\mathbf{p}$:

$$\sum_I \mathbf{p}(\mathbf{x}_I)\Psi_I(\mathbf{x}) = \mathbf{p}(\mathbf{x}). \tag{3.39}$$

For example, if $\Psi_I(x)$ is constructed with quadratic basis functions $\mathbf{p}^{\mathrm{T}} = \begin{bmatrix} 1 & x & x^2 \end{bmatrix}$ in one dimension, then the MLS shape functions have the following reproducing properties:

$$\sum_{I \in S_x} \Psi_I(x) = 1, \; \sum_{I \in S_x} x_I \Psi_I(x) = x, \; \sum_{I \in S_x} x_I^2 \Psi_I(x) = x^2. \tag{3.40}$$

3.1.4 Continuity of Shape Functions

For a polynomial basis, the shape functions have the same continuity as the weight or kernel functions, so if $w_I(\mathbf{x})$ is C^k, then $\Psi_I(\mathbf{x})$ is C^k. For example, two-dimensional shape functions for linear and quadratic basis are shown in Figure 3.6: in both cases, a C^2 cubic B-spline weight function was used; the shape functions also have C^2 continuity.

This can be seen as follows. The polynomial basis $\mathbf{p}(\mathbf{x})$ is infinitely continuously differentiable, so the product of the terms is C^k; but to complete the argument, it must be shown that $\mathbf{A}^{-1}$ is at least C^k. This follows from the fact that $\mathbf{A}$ is C^k, and that the inverse of a matrix (if it exists) has the same continuity as the matrix, as long as it is invertible, which will be discussed next.

In the following, we will show that the moment matrix is positive definite under certain conditions. Positive definiteness guarantees that the moment matrix $\mathbf{A}$ can be inverted. To be more specific, we show that the matrix $\mathbf{A}$ is positive definite if and only if the vectors $\mathbf{p}_I$ are linearly independent. The proof is as follows. Recall that a matrix $\mathbf{A}$ is positive definite if

$$\mathbf{z}^{\mathrm{T}}\mathbf{A}\mathbf{z} > 0 \;\; \forall \mathbf{z} \neq 0. \tag{3.41}$$

Substituting the definition of $\mathbf{A}$ given in (3.24) into the above yields

$$\mathbf{z}^{\mathrm{T}}\mathbf{A}\mathbf{z} = \mathbf{z}^{\mathrm{T}}\mathbf{P}^{\mathrm{T}}\mathbf{W}\mathbf{P}\mathbf{z} = \mathbf{y}^{\mathrm{T}}\mathbf{W}\mathbf{y} \text{ where } \mathbf{y} = \mathbf{P}\mathbf{z}. \tag{3.42}$$

Since the weight function is positive within its support, it can be seen from the definition of $\mathbf{W}$ in (3.26) that $\mathbf{y}^{\mathrm{T}}\mathbf{W}\mathbf{y}$ is positive for any nonzero $\mathbf{y}$ provided that the weight's support is sufficiently large. Note for $\mathbf{y}$ to be nonzero for any nonzero vector $\mathbf{z}$, the basis vectors $\mathbf{p}_I$ are required to be linearly independent. Further, since $\mathbf{A}(\mathbf{x})$ is a summation of rank-one matrices, for $\mathbf{A}(\mathbf{x})$ to be nonsingular, a necessary condition is for any $\mathbf{x}$, there are at least $m = \binom{n+d}{d} = (n+d)!/(n!d!)$ nonzero terms in the summation in (3.34) [6], with n being the order of the complete polynomial in the basis vector $\mathbf{p}$, and d the space dimension. This requires that any $\mathbf{x}$ be covered by at least m weight functions. An additional necessary condition is that the nodal points $\mathbf{x}_I$ are not located on a surface that can be represented by the basis functions. In two dimensions with linear basis, this entails that the points are not collinear. In three dimensions with linear basis, this requires that the points not be coplanar.

3.2 Reproducing Kernel Approximation

3.2.1 Continuous Reproducing Kernel Approximation

The earliest and simplest approximation method introduced in a meshfree framework is the KE employed in 1977 by Lucy [1] and Gingold and Monaghan [2] in SPH. The RK approximation can be viewed as an enhancement of the kernel estimate in SPH. In one dimension, the KE of a function $u(x)$, denoted as $u^{\mathrm{k}}(x)$, is expressed as the convolution between the function $u(x)$ and a kernel function $\phi_a(x)$ in a domain Ω as [1, 2, 7, 8]:

$$u^{\mathrm{k}}(x) = \int_{\Omega} \phi_a(x-s)u(s)ds, \tag{3.43}$$

where $\phi_a(x)$ is called the kernel function with a compact support size a, which plays a similar role as the weight function in the moving least squares approximation. For example, a cubic B-spline kernel function in Table 3.2, shown in Figure 3.3(d) is

$$\phi_a(x-s) \equiv \phi_a(z) = \begin{cases} \frac{2}{3} - 4z^2 + 4z^3, & 0 \leq z \leq \frac{1}{2} \\ \frac{4}{3} - 4z + 4z^2 - \frac{4}{3}z^3, & \frac{1}{2} \leq z \leq 1, \\ 0, & z > 1 \end{cases} \qquad z = \frac{|x-s|}{a}. \tag{3.44}$$

The kernel functions generally have the same properties discussed in Section 3.1.1 imposed, for the same reasons.

Note that $u^{\mathrm{k}}(x)$ approaches $u(x)$ if $\phi_a(x-s)$ approaches $\delta(x-s)$, and the domain is infinite, that is

$$u^{\mathrm{k}}(x) = \int_{-\infty}^{\infty} \delta(x-s)u(s)ds = u(x). \tag{3.45}$$

The error in (3.43) is due to the finite domain Ω and the kernel where $\phi_a(x-s) \neq \delta(x-s)$.

Now consider a Taylor expansion of $u(s)$ in (3.43) as

$$u(s) = \sum_{i=0}^{\infty} \frac{(s-x)^i}{i!} u^{(i)}(x), \tag{3.46}$$

where $u^{(i)}(x) \equiv d^i u(x)/dx^i$. Here, we have assumed that u is infinitely differentiable. Substituting (3.46) into (3.43), we have

$$u^{\mathrm{k}}(x) = m_0(x)u(x) + \sum_{i=1}^{\infty} \frac{(-1)^i}{i!} m_i(x)u^{(i)}(x), \tag{3.47}$$

where $m_i(x)$ is the ith moment of ϕ_a defined by

$$m_i(x) = \int_{\Omega} (x-s)^i \phi_a(x-s)ds. \tag{3.48}$$

For $u^{\mathrm{k}}(x) = u(x)$, according to (3.47), the required conditions are

$$\begin{aligned} m_0(x) &= 1, \\ m_i(x) &= 0, \ \text{for} \ 0 < i < \infty. \end{aligned} \tag{3.49}$$

The only kernel function that meets the requirements in (3.49) is $\phi_a(x-s) = \delta(x-s), x \in \Omega$, which is not very useful for numerical purposes.

Instead of requiring the conditions in (3.49), a weakened consideration is to require $u^{\mathrm{k}}(x) = u(x)$ *if $u(x)$ is a polynomial of degree n*. In this case, according to (3.47), the kernel function has to satisfy the following conditions:

$$\begin{aligned} m_0(x) &= 1, \\ m_i(x) &= 0, \ \text{for} \ 1 \leq i \leq n. \end{aligned} \tag{3.50}$$

The conditions in (3.50) are called the reproducing conditions. Not all kernel functions satisfy the above reproducing conditions, but a kernel function can be modified to meet them. The RK approximation, denoted as $u^{\mathrm{r}}(x)$, introduces a correction to the KE as follows [9]:

$$u^{\mathrm{r}}(x) = \int_{\Omega} \overline{\phi}_a(x; x-s)u(s)ds, \tag{3.51}$$

where $\overline{\phi}_a(x; x-s)$ is a modified kernel function, also called the *reproducing kernel* function, given by

$$\overline{\phi}_a(x; x-s) = C(x; x-s)\phi_a(x-s), \tag{3.52}$$

where $C(x; x-s)$ is called the correction function, expressed as

$$C(x; x-s) = \sum_{i=0}^{n} (x-s)^i b_i(x) \equiv \mathbf{H}^{\mathrm{T}}(x-s)\mathbf{b}(x), \tag{3.53}$$

and $\mathbf{H}(x-s)$ is the vector of basis functions:

$$\mathbf{H}^{\mathrm{T}}(x-s) = \begin{bmatrix} 1 & x-s & (x-s)^2 & \dots & (x-s)^n \end{bmatrix}, \tag{3.54}$$

and $\mathbf{b}^{\mathrm{T}}(x) = \begin{bmatrix} b_0(x) & b_1(x) & b_2(x) & \dots & b_n(x) \end{bmatrix}$ is the coefficient vector. By introducing a Taylor expansion on $u(x)$ in (3.51), we have

$$u^{\mathrm{r}}(x) = \overline{m}_0 u(x) + \sum_{i=1}^{\infty} \frac{(-1)^i}{i!} \overline{m}_i u^{(i)}(x), \tag{3.55}$$

where $\overline{m}_i(x)$ is the ith moment of $\overline{\phi}_a$ and is defined as

$$\begin{aligned} \overline{m}_i(x) &= \int_{\Omega} (x-s)^i \overline{\phi}_a(x; x-s)ds = \int_{\Omega} (x-s)^i \left(\sum_{k=0}^{n} b_k(x)(x-s)^k \right) \phi_a(x-s)ds \\ &= \sum_{k=0}^{n} b_k(x) m_{i+k}(x), \end{aligned} \tag{3.56}$$

where $m_i(x)$ is defined in (3.48). The nth-order reproducing conditions in the RK approximation in (3.55) are now:

$$\overline{m}_0(x) = 1, \\ \overline{m}_i(x) = 0, \text{ for } 1 \le i \le n. \tag{3.57}$$

Using (3.56), the equations in (3.57) can be expressed as the following system of equations

$$\mathbf{M}(x)\mathbf{b}(x) = \mathbf{H}(0), \tag{3.58}$$

where $\mathbf{M}(x)$ is called the moment matrix with respect to the kernel $\phi_a(x)$ and is expressed as

$$\mathbf{M}(x) = \begin{bmatrix} m_0(x) & m_1(x) & \cdots & m_n(x) \\ m_1(x) & m_2(x) & \cdots & m_{n+1}(x) \\ \vdots & \vdots & \ddots & \vdots \\ m_n(x) & m_{n+1}(x) & \cdots & m_{2n}(x) \end{bmatrix}. \tag{3.59}$$

It can be easily shown that $\mathbf{M}(x)$ can be expressed by the vector of basis functions $\mathbf{H}(x)$ as

$$\mathbf{M}(x) = \int_\Omega \mathbf{H}(x-s)\mathbf{H}^{\mathrm{T}}(x-s)\phi_a(x-s)ds. \tag{3.60}$$

By obtaining the coefficient vector $\mathbf{b}(x)$ from (3.58), and substituting it back into (3.51), the RK approximation in one dimension can now be expressed as

$$u^{\mathrm{r}}(x) = \mathbf{H}^{\mathrm{T}}(0)\mathbf{M}^{-1}(x)\int_\Omega \mathbf{H}(x-s)\phi_a(x-s)u(s)ds. \tag{3.61}$$

The derivation above demonstrates the basic construction of the RK approximation. This construction, however, becomes somewhat tedious in multiple dimensions. An alternative approach can be used to obtain the RK approximation following [10]. Consider the imposition of the nth-order polynomial reproducing conditions on the RK approximation in (3.51):

$$\int_\Omega \overline{\phi}_a(x; x-s)s^i ds = x^i, \;\; i = 0,\ldots,n, \tag{3.62}$$

or,

$$\int_\Omega C(x; x-s)\phi_a(x-s)s^i ds = x^i, \; i = 0,\ldots,n. \tag{3.63}$$

Equivalently, we have

$$\int_\Omega C(x; x-s)\phi_a(x-s)(x-s)^i ds = \delta_{i0}, \; i = 0,\ldots,n. \tag{3.64}$$

The above equation can be expressed further as

$$\int_\Omega C(x; x-s)\phi_a(x-s)\mathbf{H}(x-s)ds = \mathbf{H}(0). \tag{3.65}$$

Substituting correction function $C(x; x-s)$ in (3.53) into (3.65) yields the same linear system as that in (3.58) in solving for the coefficient vector $\mathbf{b}(x)$, and consequently the same final form of the RK approximation in (3.61).

With this approach, the multi-dimensional (in d dimensions) RK approximation can be written as

$$u^{\mathrm{r}}(\mathbf{x}) = \int_{\Omega} C(\mathbf{x}; \mathbf{x}-\mathbf{s})\phi_a(\mathbf{x}-\mathbf{s})u(\mathbf{s})d\mathbf{s}, \tag{3.66}$$

where $\mathbf{x} \equiv [x_1 \ \ldots \ x_d]$, and $\mathbf{s} \equiv [s_1 \ \ldots \ s_d]$. The correction function $C(\mathbf{x}; \mathbf{x}-\mathbf{s})$ in multiple dimensions is expressed as the linear combination of monomial bases:

$$\begin{aligned} C(\mathbf{x}; \mathbf{x}-\mathbf{s}) &= \sum_{i+j+k=0}^{n} b_{ijk}(\mathbf{x})(x_1-s_1)^i(x_2-s_2)^j(x_3-s_3)^k, \quad n \geq 0 \\ &\equiv \mathbf{H}^{\mathrm{T}}(\mathbf{x}-\mathbf{s})\mathbf{b}(\mathbf{x}). \end{aligned} \tag{3.67}$$

Here $\mathbf{H}(\mathbf{x}-\mathbf{s})$ is the vector of basis functions:

$$\mathbf{H}^{\mathrm{T}}(\mathbf{x}-\mathbf{s}) = \begin{bmatrix} 1 & x_1-s_1 & x_2-s_2 & \ldots & (x_3-s_3)^n \end{bmatrix}. \tag{3.68}$$

The coefficient vector $\mathbf{b}^{\mathrm{T}}(\mathbf{x}) = \begin{bmatrix} b_{000}(\mathbf{x}) & b_{100}(\mathbf{x}) & b_{010}(\mathbf{x}) & \cdots & b_{003}(\mathbf{x}) \end{bmatrix}$ is then easily solved for by enforcing nth-order monomial reproducibility as in (3.36):

$$\int_{\Omega} C(\mathbf{x}; \mathbf{x}-\mathbf{s})\phi_a(\mathbf{x}-\mathbf{s})s_1^i s_2^j s_3^k d\mathbf{s} = x_1^i x_2^j x_3^k, \quad 0 \leq i+j+k \leq n. \tag{3.69}$$

Equation (3.69) can be rewritten as

$$\int_{\Omega} C(\mathbf{x}; \mathbf{x}-\mathbf{s})\phi_a(\mathbf{x}-\mathbf{s})(x_1-s_1)^i(x_2-s_2)^j(x_3-s_3)^k d\mathbf{s} = \delta_{i0}\delta_{j0}\delta_{k0}, \quad 0 \leq i+j+k \leq n, \tag{3.70}$$

or,

$$\int_{\Omega} C(\mathbf{x}; \mathbf{x}-\mathbf{s})\phi_a(\mathbf{x}-\mathbf{s})\mathbf{H}(\mathbf{x}-\mathbf{s})d\mathbf{s} = \mathbf{H}(\mathbf{0}). \tag{3.71}$$

Introducing the correction function in (3.67) into (3.71) yields the following equation:

$$\left(\int_{\Omega} \mathbf{H}(\mathbf{x}-\mathbf{s})\mathbf{H}^{\mathrm{T}}(\mathbf{x}-\mathbf{s})\phi_a(\mathbf{x}-\mathbf{s})d\mathbf{s} \right) \mathbf{b}(\mathbf{x}) = \mathbf{H}(\mathbf{0}). \tag{3.72}$$

By solving for $\mathbf{b}(\mathbf{x})$ from (3.72), the multi-dimensional RK approximation is obtained

$$u^{\mathrm{r}}(\mathbf{x}) = \mathbf{H}^{\mathrm{T}}(\mathbf{0})\mathbf{M}^{-1}(\mathbf{x}) \int_{\Omega} \mathbf{H}(\mathbf{x}-\mathbf{s})\phi_a(\mathbf{x}-\mathbf{s})u(\mathbf{s})d\mathbf{s}, \tag{3.73}$$

where

$$\mathbf{M}(\mathbf{x}) = \int_{\Omega} \mathbf{H}(\mathbf{x}-\mathbf{s})\mathbf{H}^{\mathrm{T}}(\mathbf{x}-\mathbf{s})\phi_a(\mathbf{x}-\mathbf{s})d\mathbf{s}. \tag{3.74}$$

Here $\mathbf{M}(\mathbf{x})$ is the multi-dimensional moment matrix, and $\phi_a(\mathbf{x}-\mathbf{s})$ is the multi-dimensional kernel function that can be constructed by using the kernel function in one dimension, by either of the two following ways. Similar to the construction of the weight function in the MLS approximation in (3.19) and (3.20), the multi-dimensional kernel can be obtained as

$$(1) \quad \phi_a(\mathbf{x}-\mathbf{s}) \equiv \phi_a(z), \quad z = \frac{\|\mathbf{x}-\mathbf{s}\|}{a}, \quad \text{(Spherical support)} \tag{3.75}$$

or

$$(2) \quad \phi_a(\mathbf{x}-\mathbf{s}) = \prod_{i=1}^{3} \phi_{a_i}(x_i - s_i) = \phi_{a_1}(x_1 - s_1)\phi_{a_2}(x_2 - s_2)\phi_{a_3}(x_3 - s_3). \quad \text{(Brick support)} \tag{3.76}$$

The kernel function defined in (3.75) has a spherical support, while the one defined in (3.76) has a cuboid (brick-shaped) support. These two types of kernels are depicted in Figure 3.4a and b for the two-dimensional case, where they are circles and rectangles, respectively.

Note that $\mathbf{M}(\mathbf{x})$ is the Gram matrix of the basis functions $\mathbf{H}(\mathbf{x}-\mathbf{s})$ with respect to $\phi_a(\mathbf{x}-\mathbf{s})$. $\mathbf{M}(\mathbf{x})$ is positive definite if the basis functions in $\mathbf{H}(\mathbf{x}-\mathbf{s})$ are linearly independent and $\phi_a(\mathbf{x}-\mathbf{s}) > 0$ for $\|\mathbf{x}-\mathbf{s}\| < a$. On the other hand, for the discrete RK approximation introduced in the next section, a sufficient number of nodal supports have to cover the point $\mathbf{x}$ for $\mathbf{M}(\mathbf{x})$ to be positive definite.

To construct the approximation functions for a finite-dimensional solution of partial differential equations based on the RK approximation, discretization of the continuous RK function in (3.73) is needed. However, this can lead to the violation of the reproducing conditions, and consequently monomials are not exactly reproduced [11]. To address this issue, here a discrete RK approximation is formulated directly following [12].

3.2.2 Discrete RK Approximation

Consider a domain of interest $\overline{\Omega} = \Omega \cup \Gamma$ discretized by a set of points $S = \{\mathbf{x}_1, \mathbf{x}_2, \cdots, \mathbf{x}_{NP} \mid \mathbf{x}_I \in \overline{\Omega}\}$, and let the approximation of a function u, denoted by u^h, be constructed using the information at discrete points in the set S as follows:

$$u^h(\mathbf{x}) = \sum_{I=1}^{NP} \Psi_I(\mathbf{x}) u_I, \tag{3.77}$$

where $\Psi_I(\mathbf{x})$ is the shape function of node I associated with $\mathbf{x}_I$, and u_I is the coefficient sought. In the discrete approximation, the shape functions are based entirely on point data. The RK shape functions are constructed as follows [11, 13] with consideration of discrete reproducing conditions [12]:

$$\Psi_I(\mathbf{x}) = C(\mathbf{x}; \mathbf{x}-\mathbf{x}_I)\phi_a(\mathbf{x}-\mathbf{x}_I), \tag{3.78}$$

where $\phi_a(\mathbf{x}-\mathbf{x}_I)$ is the kernel function with compact support measured by the support dimension a, and $C(\mathbf{x}; \mathbf{x}-\mathbf{x}_I)$ is the correction function composed of monomial bases:

$$C(\mathbf{x}; \mathbf{x}-\mathbf{x}_I) = \sum_{i+j+k=0}^{n} b_{ijk}(\mathbf{x})(x_1 - x_{1I})^i (x_2 - x_{2I})^j (x_3 - x_{3I})^k, \quad n \geq 0, \tag{3.79}$$

where n represents the degree of monomial bases, and $b_{ijk}(\mathbf{x})$ are coefficients obtained by the following discrete reproducing conditions [12]:

$$\sum_{I=1}^{NP} \Psi_I(\mathbf{x}) x_{1I}^i x_{2I}^j x_{3I}^k = x_1^i x_2^j x_3^k, \quad i+j+k = 0, 1, \ldots, n. \tag{3.80}$$

Here it can be seen that the completeness conditions (3.39) are explicit in the construction.

Obtaining $b_{ijk}(\mathbf{x})$ from (3.80) yields the following RK shape function:

$$\Psi_I(\mathbf{x}) = \mathbf{H}^{\mathrm{T}}(\mathbf{0})\mathbf{M}^{-1}(\mathbf{x})\mathbf{H}(\mathbf{x}-\mathbf{x}_I)\phi_a(\mathbf{x}-\mathbf{x}_I), \tag{3.81}$$

$$\mathbf{M}(\mathbf{x}) = \sum_{I=1}^{NP} \mathbf{H}(\mathbf{x}-\mathbf{x}_I)\mathbf{H}^{\mathrm{T}}(\mathbf{x}-\mathbf{x}_I)\phi_a(\mathbf{x}-\mathbf{x}_I), \tag{3.82}$$

$$\mathbf{H}^{\mathrm{T}}(\mathbf{x}-\mathbf{x}_I) = \begin{bmatrix} 1 & x_1 - x_{1I} & x_2 - x_{2I} & \ldots & (x_3 - x_{3I})^n \end{bmatrix}. \tag{3.83}$$

Note that in practice, sums in (3.77) and (3.82) should be carried out over the set $S_{\mathbf{x}} = \{I \mid \mathbf{x} \in \text{supp } \phi_a(\mathbf{x} - \mathbf{x}_I)\}$, and the supports are usually defined node-wise:

$$u^h(\mathbf{x}) = \sum_{I \in S_{\mathbf{x}}} \Psi_I(\mathbf{x}) u_I, \tag{3.84}$$

$$\mathbf{M}(\mathbf{x}) = \sum_{I \in S_{\mathbf{x}}} \mathbf{H}(\mathbf{x}-\mathbf{x}_I)\mathbf{H}^{\mathrm{T}}(\mathbf{x}-\mathbf{x}_I)\phi_{a_I}(\mathbf{x}-\mathbf{x}_I). \tag{3.85}$$

Finally, the approximation of a vector function follows the MLS derivation in Section 3.1.2. When the basis functions are the same in all directions we have

$$u_i^h(\mathbf{x}) = \sum_{I \in S_{\mathbf{x}}} \Psi_I(\mathbf{x}) u_{iI}, \quad \text{or} \quad \mathbf{u}^h(\mathbf{x}) = \sum_{I \in S_{\mathbf{x}}} \Psi_I(\mathbf{x}) \mathbf{u}_I, \tag{3.86}$$

where $\Psi_I(\mathbf{x})$ is constructed the same as in the scalar case (3.81)–(3.83).

In the RK approximation, the kernel function ϕ_a has a compact support a, and determines the continuity of the approximation; for example, the cubic B-spline function is C^2 continuous, and yields a C^2 approximation. The degree of monomial basis functions n in the correction function $C(\mathbf{x}; \mathbf{x} - \mathbf{x}_I)$ controls the order of completeness in the approximation, which is related to the order of consistency when introducing the RK approximation for solving partial differential equations. For the moment matrix $\mathbf{M}(\mathbf{x})$ to be nonsingular, the compact support of the kernel function has to be sufficiently large to keep the reproducing conditions in (3.80) linearly independent [11], which follows the requirements for the matrix $\mathbf{A}(\mathbf{x})$ in MLS in Section 3.1.4. A typical RK discretization using circular supports is shown in Figure 3.7a where the support of node I is shaded in grey. The corresponding shape function over the node's compact support is shown in Figure 3.7b. The algorithm for constructing the RK approximation is given in **Box 3.1**. Further discussion is provided in Section 10.3.3. Scripts for obtaining and plotting one and two-dimensional shape functions are also provided with the RKPM2D code described in Chapter 10.

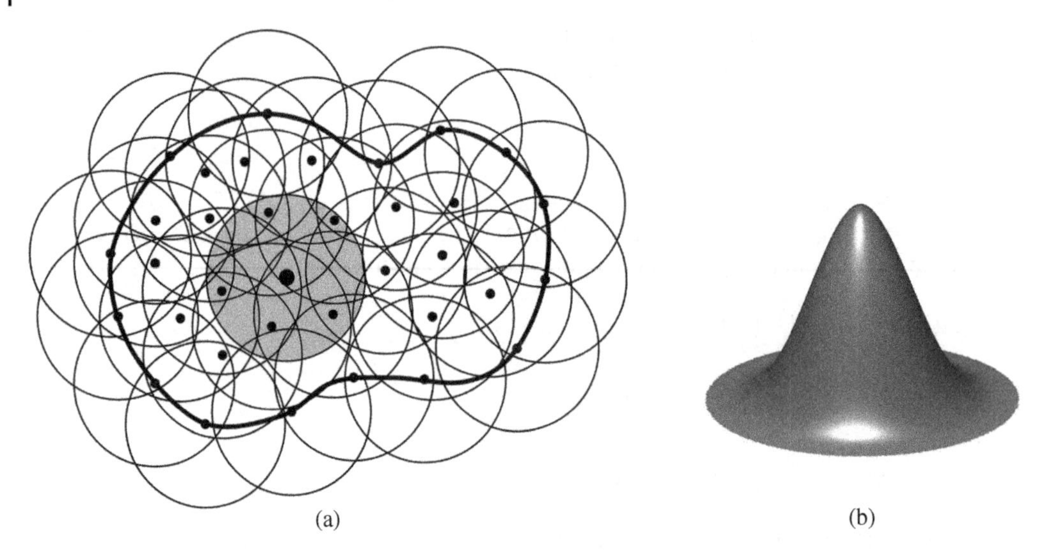

Figure 3.7 (a) RK discretization using circular supports with the support of a node I shaded in grey, and (b) RK shape function at node I.

Box 3.1 Construction of RK Shape Functions

Given: the set of indices $S_{\mathbf{x}}$, kernel function, set of nodal supports $\{a_I\}_{I=1}^{NP}$, and order of basis n.

Step 1. Moment matrix. Loop over the indices of the nodes in $S_{\mathbf{x}}$ with index I:

- Compute $\mathbf{H}(\mathbf{x}-\mathbf{x}_I)$ and $\phi_{a_I}(\mathbf{x}-\mathbf{x}_I)$ and store in memory.
- Add the contributions to the moment matrix:

$$\mathbf{M}(\mathbf{x}) = \sum_{I \in S_{\mathbf{x}}} \mathbf{H}(\mathbf{x}-\mathbf{x}_I)\mathbf{H}^{\mathrm{T}}(\mathbf{x}-\mathbf{x}_I)\phi_{a_I}(\mathbf{x}-\mathbf{x}_I).$$

Step 2. Get correction function. Solve for $\mathbf{b}(\mathbf{x})$ from $\mathbf{M}(\mathbf{x})\mathbf{b}(\mathbf{x}) = \mathbf{H}(\mathbf{0})$.

Step 3. Shape function construction. Loop over the indices of the nodes in $S_{\mathbf{x}}$ with index I to obtain the shape functions with non-zero contribution to the approximation at $\mathbf{x}$:

$$\Psi_I(\mathbf{x}) = \mathbf{b}^{\mathrm{T}}(\mathbf{x})\mathbf{H}(\mathbf{x}-\mathbf{x}_I)\phi_{a_I}(\mathbf{x}-\mathbf{x}_I).$$

It should be noted that the RK and MLS approximations discussed generally do not possess the Kronecker delta property, and hence are not kinematically admissible unless special techniques are introduced, such as coupling with finite elements near the essential boundary [14, 15], or modifications to the approximation [16–19]. Without such methods, the essential boundary conditions have to be imposed by techniques such as the penalty method [20, 21], Nitsche's method [22–24], the modified variational principle [25], or the Lagrange multiplier method [26]. These approaches will be discussed in Chapters 4 and 5, respectively; the associated variational principles were given in Chapter 2.

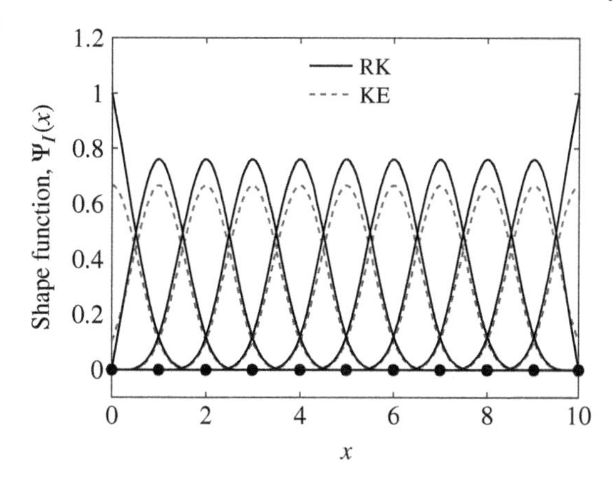

Figure 3.8 KE and RK shape functions for a set of equally spaced nodes.

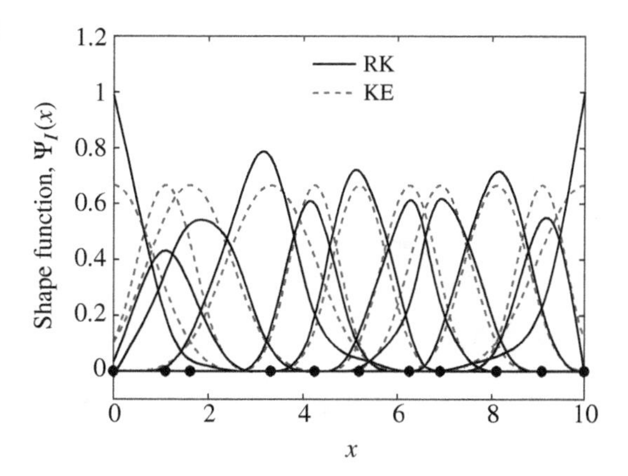

Figure 3.9 KE and RK shape functions for a set of unevenly spaced nodes.

Figure 3.8 shows a comparison of the one-dimensional KE shape function $\Psi_I(x) = \phi_a(x - x_I)$ and RK shape function $\Psi_I(x) = C(x; x - x_I)\phi_a(x - x_I)$ constructed based on a cubic spline kernel function and a set of equally spaced nodes. In this example, the linear basis functions are introduced to construct the RK shape functions, i.e., $\mathbf{H}^{\mathrm{T}}(x - x_I) = [1 \quad x - x_I]$. A similar comparison is shown in Figure 3.9 where the shape functions are constructed based on a set of unevenly spaced nodes. As can be seen in these two figures, the correction function in the RK approximation plays an important role to "correct" the KE shape functions according to the the nodal spacing and proximity to the boundary, so that the zeroth and first-order completeness conditions, $\sum_{I=1}^{NP} \Psi_I(x) = 1$ and $\sum_{I=1}^{NP} \Psi_I(x) x_I = x$, are satisfied, as shown in Figure 3.10. The KE shape functions, on the other hand, do not satisfy the completeness conditions. The two-dimensional reproduction of zeroth and first-order monomials in the RK approximation and KE are compared in Figure 3.11, with derivative completeness conditions shown in Figure 3.12 (satisfaction of $\sum_{I=1}^{NP} \Psi_{I,x}(x) = 1_{,x} = 0$ and $\sum_{I=1}^{NP} \Psi_{I,x}(x) x_I = x_{,x} = 1$). The latter property will be expanded upon further in the next section. Detailed discussions can be found in Belytschko et al. [27] and Chen et al. [12].

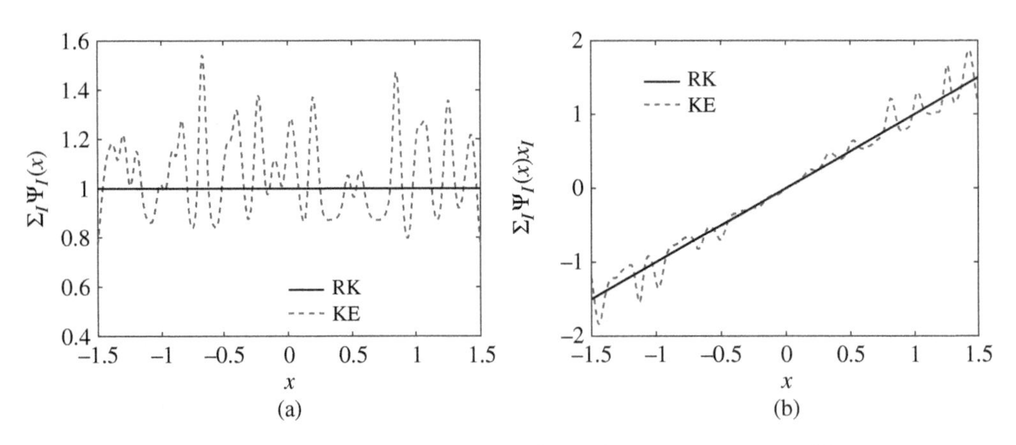

Figure 3.10 Comparison of (a) the zeroth and (b) the first order completeness conditions.

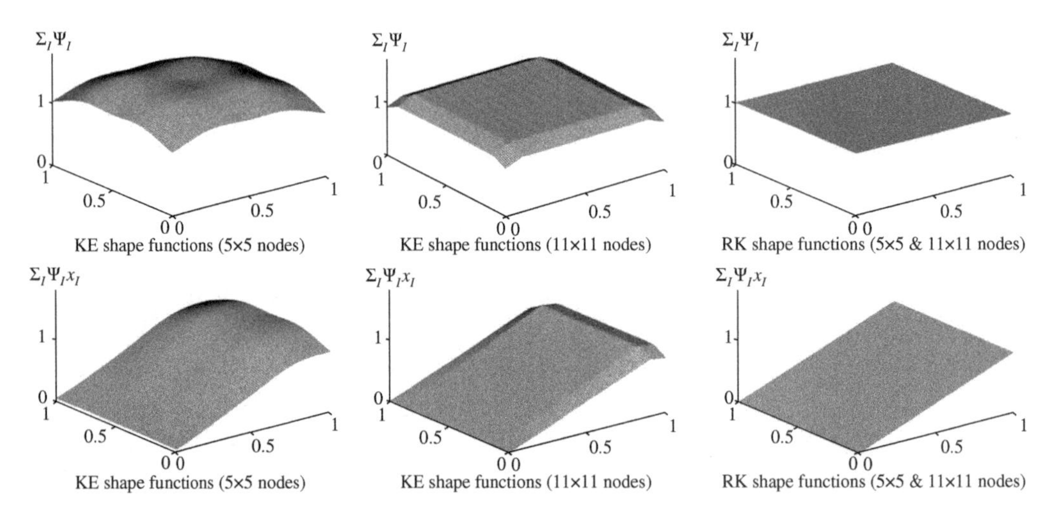

Figure 3.11 Constant and linear completeness conditions of KE and RK shape functions in 2D.

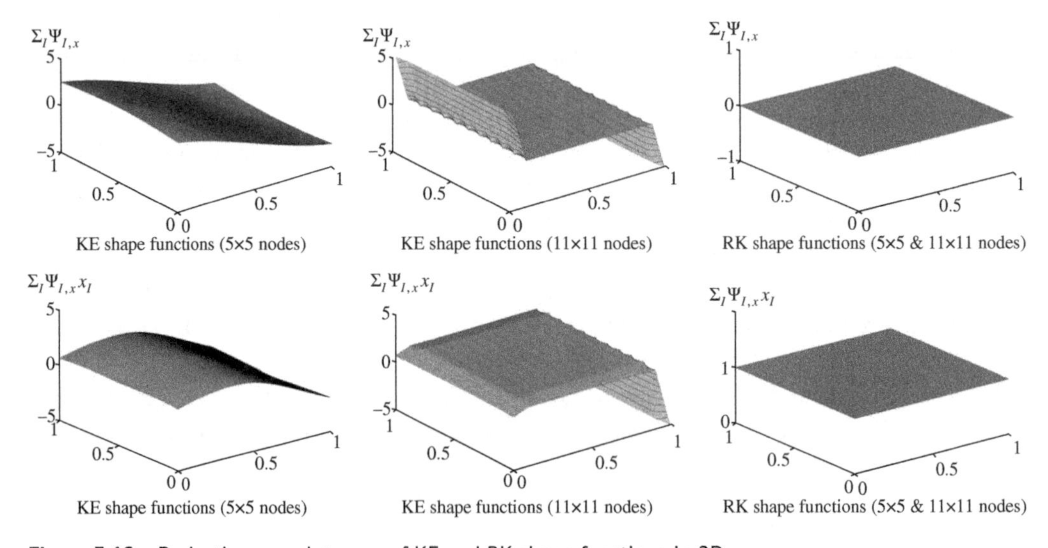

Figure 3.12 Derivative completeness of KE and RK shape functions in 2D.

3.3 Differentiation of Meshfree Shape Functions and Derivative Completeness Conditions

Solving a partial differential equation involves differentiation of the approximation function at hand either in the weak form or strong-form collocation. For finite element shape functions, this involves the chain rule to obtain derivatives with respect to the Cartesian coordinate, since they are a function of the element parent coordinates. In contrast, meshfree shape functions are constructed directly in the Cartesian coordinate, which actually simplifies matters somewhat. Nevertheless there are several terms in the MLS approximation (3.33) and RK approximation (3.81), and care must be taken when constructing direct derivatives.

First, using the product rule, the first-order derivative of the MLS approximation (3.33) with respect to the coordinate x_i is

$$\Psi_{I,i}(\mathbf{x}) = \mathbf{p}_{,i}^{\mathrm{T}}(\mathbf{x})\mathbf{A}^{-1}(\mathbf{x})\mathbf{p}(\mathbf{x}_I)w_I(\mathbf{x}) + \mathbf{p}^{\mathrm{T}}(\mathbf{x})\{\mathbf{A}^{-1}(\mathbf{x})\}_{,i}\mathbf{p}(\mathbf{x}_I)w_I(\mathbf{x}) + \mathbf{p}^{\mathrm{T}}(\mathbf{x})\mathbf{A}^{-1}(\mathbf{x})\mathbf{p}(\mathbf{x}_I)w_{I,i}(\mathbf{x}). \tag{3.87}$$

For the RK shape function (3.81), we have

$$\begin{aligned}\Psi_{I,i}(\mathbf{x}) = \mathbf{H}^{\mathrm{T}}(\mathbf{0})\Big\{&\{\mathbf{M}^{-1}(\mathbf{x})\}_{,i}\mathbf{H}(\mathbf{x}-\mathbf{x}_I)\phi_a(\mathbf{x}-\mathbf{x}_I) + \mathbf{M}^{-1}(\mathbf{x})\mathbf{H}_{,i}(\mathbf{x}-\mathbf{x}_I)\phi_a(\mathbf{x}) \\ &+ \mathbf{M}^{-1}(\mathbf{x})\mathbf{H}_{,i}(\mathbf{x}-\mathbf{x}_I)\phi_{a,i}(\mathbf{x})\}.\end{aligned} \tag{3.88}$$

Here, we place special emphasis that the terms $\{\mathbf{A}^{-1}(\mathbf{x})\}_{,i}$ and $\{\mathbf{M}^{-1}(\mathbf{x})\}_{,i}$, which are the derivatives of a matrix inverse, are not the same as the inverse of a matrix derivative, i.e., $\{\mathbf{B}^{-1}(\mathbf{x})\}_{,i} \neq \{\mathbf{B}_{,i}(\mathbf{x})\}^{-1}$.

To obtain this term, we start with the identity $\mathbf{B}^{-1}(\mathbf{x})\mathbf{B}(\mathbf{x}) = \mathbf{I}$, where $\mathbf{I}$ is the identity matrix. Differentiation of both sides of the equation yields

$$\{\mathbf{B}^{-1}(\mathbf{x})\}_{,i}\mathbf{B}(\mathbf{x}) + \mathbf{B}^{-1}(\mathbf{x})\mathbf{B}_{,i}(\mathbf{x}) = \mathbf{0} \rightarrow \{\mathbf{B}^{-1}(\mathbf{x})\}_{,i} = -\mathbf{B}^{-1}(\mathbf{x})\mathbf{B}_{,i}(\mathbf{x})\mathbf{B}^{-1}(\mathbf{x}). \tag{3.89}$$

The terms $\mathbf{p}_{,i}^{\mathrm{T}}(\mathbf{x})$ and $\mathbf{H}_{,i}(\mathbf{x}-\mathbf{x}_I)$ are simply the direct differentiation of the basis functions. For example, for second-order bases in one dimension, we have

$$\begin{aligned}\mathbf{p}_{,x}^{\mathrm{T}}(\mathbf{x}) &= \begin{bmatrix}1 & x & x^2\end{bmatrix}_{,x} = [0 \quad 1 \quad 2x], \\ \mathbf{H}_{,x}^{\mathrm{T}}(\mathbf{x}-\mathbf{x}_I) &= \begin{bmatrix}1 & x-x_I & (x-x_I)^2\end{bmatrix}_{,x} = [0 \quad 1 \quad 2(x-x_I)].\end{aligned} \tag{3.90}$$

Finally, when weight and kernel functions are given in terms of $z = \|\mathbf{x}-\mathbf{x}_I\|/a$, the chain rule can be used to obtain

$$w_{I,i} = \frac{\partial w_I}{\partial z}\frac{\partial z}{\partial x_i}, \quad \varphi_{a,i} = \frac{\partial \varphi_a}{\partial z}\frac{\partial z}{\partial x_i}. \tag{3.91}$$

The term $\partial z/\partial x_i = (x_i - x_{iI})/za^2$ is singular since as $x_i \rightarrow x_{iI}$, $z \rightarrow 0$. In order to avoid this issue, it is possible to use $\partial z/\partial x_i \approx (x_i - x_{iI})/(za^2 + \mathrm{eps})$, or $\partial z/\partial x_i = 0$ when $|x_i - x_{iI}| < \mathrm{eps}$, where eps is on the order of machine precision, with negligible error. Algorithms to construct direct derivatives are described in Section 10.3.3.

The derivatives of the shape functions (3.87) and (3.88) can be easily shown to satisfy the following *derivative reproducing conditions* (or *derivative completeness conditions* when the functions are monomials) by differentiating both sides of (3.39):

$$\sum_I \mathbf{p}(\mathbf{x}_I)\Psi_{I,i}(\mathbf{x}) = \mathbf{p}_{,i}(\mathbf{x}). \tag{3.92}$$

For example, in one dimension with quadratic basis functions $\mathbf{p}^T = \begin{bmatrix} 1 & x & x^2 \end{bmatrix}$, the MLS and RK shape functions satisfy

$$\sum_{I \in S_x} \Psi_{I,x}(x) = 0, \sum_{I \in S_x} x_I \Psi_{I,x}(x) = 1, \sum_{I \in S_x} x_I^2 \Psi_{I,x}(x) = 2x. \tag{3.93}$$

The property $\sum_{I \in S_x} \Psi_{I,i}(x) = 0$ for each $i = 1, \ldots, d$ is called the *partition of nullity*.

3.4 Properties of the MLS and Reproducing Kernel Approximations

The shape funtions of MLS in equation (3.33) and RK in equation (3.81), possess the following properties with consideration of nth-order complete monomials in the basis functions:

1) The smoothness of the shape functions $\Psi_I(\mathbf{x})$ is determined by the smoothness of the kernel functions $\phi_I(\mathbf{x})$ (simplified notation for the kernel function associated with node I) in the RK approximation or the weight functions $w_I(\mathbf{x})$ in the MLS approximation. For instance, if $\phi_I(\mathbf{x}) \in C^k$, then $\Psi_I(\mathbf{x}) \in C^k$. The smoothness of the MLS functions was discussed in Section 3.1.4, and the proof for the RK shape functions follows directly.

   Two very noteworthy consequences of this property are

   a) The continuity of MLS and RK can be increased without increasing the order of approximation, and vice versa, in contrast to finite elements and isogeometric approximations where they are coupled. A good example is one-dimensional finite elements. For C^0 continuity needed in elasticity (a rod/truss), shape functions can be linear polynomials. Yet to attain the continuity needed for an Euler–Bernoulli beam, the shape functions must be cubic polynomials. In general, to attain kth-order continuity for the finite element method, the polynomial order n required is $n \geq 2k + 1$, and for isogeometric analysis, it is $n \geq k + 1$. For MLS and RK, increasing the order is not necessary.
   b) Approximations must be at least C^{m-1} *globally* for weak forms containing derivatives order m. While finite elements can achieve this level of smoothness in one dimension by increasing the polynomial order to $n = 2m - 1$, this is very difficult to achieve in multiple dimensions for $m > 1$, e.g. for thin plates and shells, where global C^1 continuity is required. In contrast, the MLS and RK approximations can effortlessly solve these types of problems by selecting the proper kernel.

2) For $\mathbf{M}(\mathbf{x})$ and $\mathbf{A}(\mathbf{x})$ to be nonsingular, any position $\mathbf{x} \in \Omega$ needs to be covered by at least $m = (n + d)!/(n!d!)$ nodes, with the restriction on placement discussed in Section 3.1.4. Generally, choosing $a = h \times (n + 1)$ is sufficient to satisfy these two requirements.
3) The shape functions are not guaranteed to satisfy the Kronecker delta property, i.e., $\Psi_I(\mathbf{x}_J) \neq \delta_{IJ}$, and thus $u^h(\mathbf{x}_I) \neq u_I$. The approximation does possess this property however when K (the cardinality of $S_\mathbf{x}$) is equal to the number of basis functions m [5], but this is difficult to achieve in arbitrary geometries so it is not very practical for things like essential boundary condition enforcement. In general, $K > m$ and thus $\Psi_I(\mathbf{x}_J) \neq \delta_{IJ}$.

4) The construction of the shape functions $\Psi_I(\mathbf{x})$ does not rely on a mesh.
5) There are no conforming issues in adaptive refinement.
6) Both of the shape functions satisfy the *reproducing conditions* at nodal points, $\Sigma_I \mathbf{p}(\mathbf{x}_I)\Psi_I(\mathbf{x}) = \mathbf{p}(\mathbf{x})$, as discussed for MLS in Section 3.1.3, and in the RK approximation, this is explicit in the construction by (3.80). One consequence is that the partition of unity is inherently satisfied, and the shape functions can be used to patch together special enrichment functions which can greatly enhance the solution accuracy in PDEs. The special functions can also be included in the basis using the reproducing conditions. More details are provided in Chapter 8.
7) Derivatives of the shape functions satisfy *derivative reproducing conditions* as discussed in Section 3.3, which is an important property in solving PDEs. For further discussion of this, and other derivative approximations that also satisfy these conditions, see Section 3.5.
8) The shape functions with nth-order completeness satisfy the following interpolation estimate for the approximation of a function u when it is sufficiently smooth, and the particle distribution is subordinate to open cover of the domain with sufficient neighbors for each point in the domain [6, 28]:

$$\left\|u - u^h\right\|_{\mathrm{H}^s} \leq c_s a^{n+1-s} |u|_{\mathrm{H}^{n+1}}, \quad 0 \leq s \leq n, \tag{3.94}$$

where $\|\cdot\|_{\mathrm{H}^s}$ and $|\cdot|_{\mathrm{H}^s}$ denote Sobolov norms and semi-norms of degree s, respectively (see the Glossary of Notation), a is the maximum support dimension of the approximation functions and c_s is independent of a. For example, for the L_2 norm $s = 0$:

$$\left\|u - u^h\right\|_{\mathrm{L}_2} \leq c_0 a^{n+1} |u|_{\mathrm{H}^{n+1}}. \tag{3.95}$$

For the H^1 norm, we have

$$\left\|u - u^h\right\|_{\mathrm{H}^1} \leq c_1 a^{n} |u|_{\mathrm{H}^{n+1}}. \tag{3.96}$$

Here, we note that the rates in the H^s *semi*-norms are the same as those in the full H^s norms. While the convergence rate in meshfree approximations with nth-order completeness yields the same rate of convergence compared to that of nth-order finite elements, the constant c_s can be made smaller in meshfree methods with proper selection of smoothness in the meshfree approximation. Another feature of (3.94) is that the approximation only converges with the refinement of the nodal spacing h when the measure a decreases as well, but typically a is always $O(h)$.

As a demonstration of the last property, convergence of a linear ($n = 1$) RK approximation with uniform nodes is shown in Figure 3.13a for various kernels and $a = 2h$. Here, an L_2 least-squares fit of data generated by $u_I = \sin(x_I)$ over the one-dimensional domain [0, 10] is employed. It can be seen that the optimal rate of two ($n + 1 = 2$) is obtained, and smoothness can help decrease the constant c_s in (3.94). Figure 3.13b shows the same solution with cubic B-spline kernels and various choices of dilation (normalized support size); all yield the optimal rate, but some larger dilations can increase the accuracy of the approximation and lower the constant c_s. Finally, convergence is presented in Figure 3.14 for increasing orders of basis n; the data are generated by $u_I = \exp(x_I)$ and rates of $n+1$ for the L_2 norm in (3.95) and n for the H^1 norm in (3.96) are clearly observed. The rate for linear basis is super-convergent but will approach the theoretical one with further refinement.

For the Galerkin meshfree approximation to second-order partial differential equations (see Chapter 4), the error estimate takes a similar form as (3.94), and for a solution with regularity $u \in \mathrm{H}^r$, the following holds [28, 6]:

$$\left\|u - u^h\right\|_{\mathrm{H}^s} \leq \bar{c}_s a^{\alpha} |u|_{\mathrm{H}^r}, \quad 0 \leq s \leq n, \tag{3.97}$$

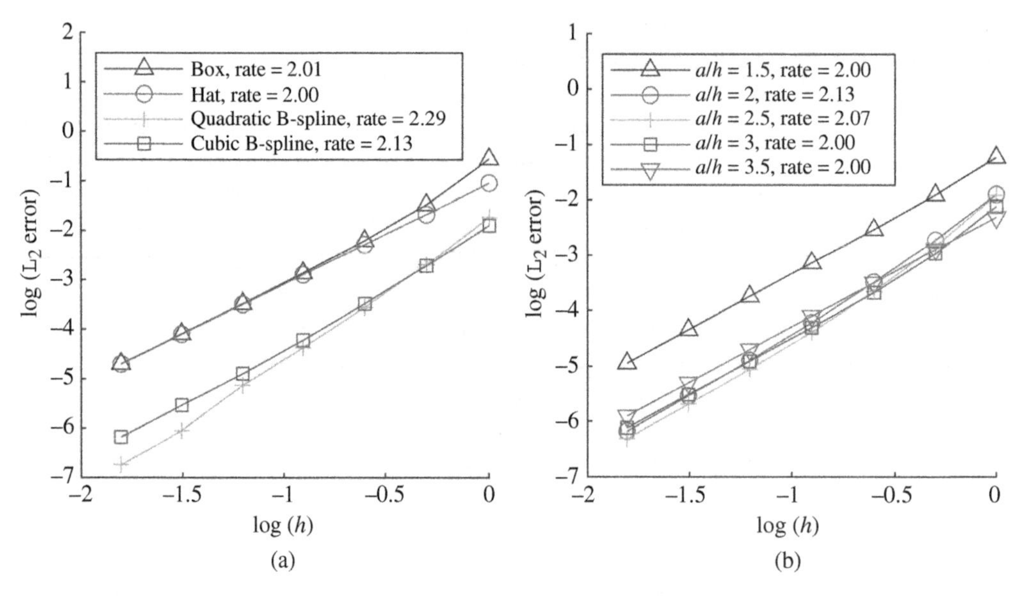

Figure 3.13 Convergence of the RK approximation with linear basis using (a) various kernels and (b) various dilations.

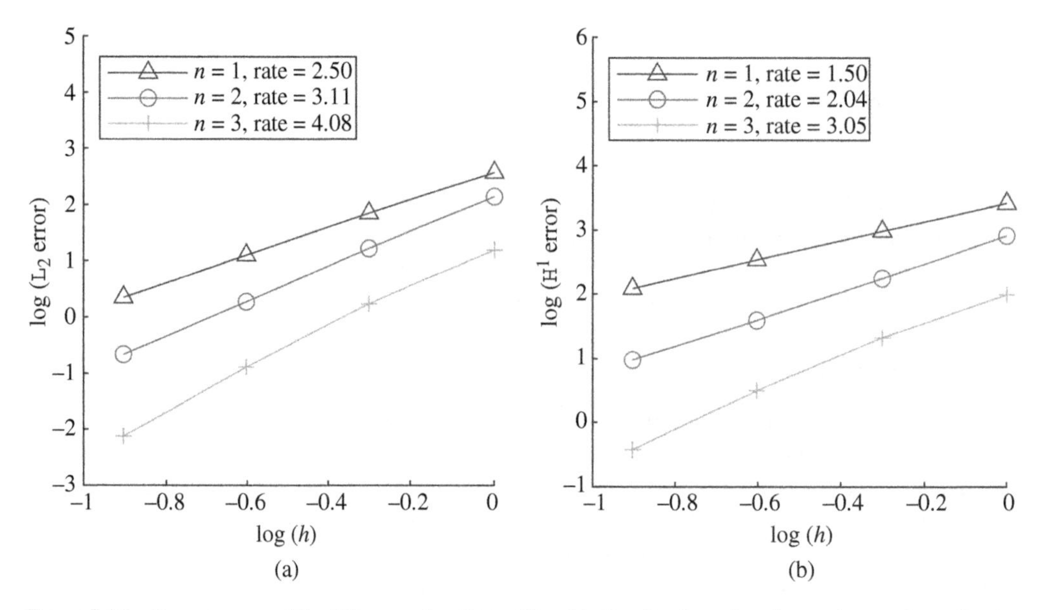

Figure 3.14 Convergence of the RK approximation with cubic B-spline kernels using various orders of basis n: (a) error in L^2 norm and (b) error in H_1 norm.

where $\alpha = \min(n+1-s,\ r-s)$. As in finite elements, the solution must be sufficiently regular ($r \geq n$) for the optimal rate $\alpha = n+1-s$ corresponding to the order n to be obtained. In this case, we have the same rate as the interpolation estimate in (3.94):

$$\left\|u - u^h\right\|_{H^s} \leq \bar{c}_s a^{n+1-s} |u|_{H^r}. \tag{3.98}$$

Again, the rates in the semi-norms are the same as those of the full norms.

Key Takeaways

In summary, the MLS and RK approximations possess the properties:

1) They satisfy the completeness conditions up to order n:

$$\sum_I \mathbf{p}(\mathbf{x}_I)\Psi_I(\mathbf{x}) = \mathbf{p}(\mathbf{x}).$$

2) They have controllable order of continuity k which is *independent* of the polynomial order n in the basis functions by selecting C^k kernels:

$$\{\phi_I(\mathbf{x})\}_{I=1}^{NP} \in C^k \rightarrow \{\Psi_I(\mathbf{x})\}_{I=1}^{NP} \in C^k.$$

The order of basis does not have to be increased to increase the order of continuity as in the finite element method. Global arbitrary-order smoothness can easily be obtained; solving problems such as thin plates and shells requires much less effort than in the finite element method.

3) They do not need a mesh to be constructed and do not have conforming requirements:
   a) h-adaptivity is greatly simplified over the finite element method. Points can be added (or taken away) on the fly as necessary for higher accuracy where needed.
   b) Discretization does not require a high-quality mesh, and a significant amount of time can be saved in constructing a model over FEM.
   c) Extreme deformations are accommodated in simulations of continua (see Chapter 7).
4) The shape functions constitute a partition of unity, which is useful for enriching the solution with special functions, see Chapter 8.
5) The order of completeness n gives the convergence rate of $n+1-s$ in the H^s error norm in the solutions of second-order PDEs, like finite elements, but the error can be made *much lower*, usually by increasing the order of continuity k, or kernel measure a.
6) They are not ensured the Kronecker delta property, and care needs to be taken to enforce boundary conditions (see Chapters 4 and 5):

$$\Psi_I(\mathbf{x}_J) \neq \delta_{IJ}.$$

Exercise 3.3 For the following cases, construct a set of two-dimensional RK shape functions using linear basis with a circular support in a domain $x \in [0, 10]$, $y \in [0, 10]$, where

$$\mathbf{H}^{\mathrm{T}}(\mathbf{x}-\mathbf{x}_I) = [1 \quad x-x_I \quad y-y_I],$$

$$\phi_a(\mathbf{x}-\mathbf{x}_I) = \phi\left(\frac{\|\mathbf{x}-\mathbf{x}_I\|}{a}\right),$$

and ϕ is the cubic B-spline function.

1) Use an equally spaced node discretization with the spacing in x and y selected as $h_x = h_y = 1.0$ (121 nodes in total). Plot shape functions with $a = 2.0 \times h_x = 2.0 \times h_y$ centered at $(x_I, y_I) = (5.0, 5.0)$, $(x_I, y_I) = (5.0, 0.0)$, and $(x_I, y_I) = (0.0, 0.0)$.
2) Now create a nonuniform node distribution by disturbing the uniformly distributed nodal coordinates in (1) by

$$x_I' = x_I + \alpha h_x,$$
$$y_I' = y_I + \beta h_y,$$

where

$$-0.2 \le \alpha, \beta \le 0.2.$$

Let the boundary nodes stay on the boundary after disturbance. Use $a = 2.0 \times h_{\max}$ where $h_{\max}$ is the maximum nodal spacing in each of the x and y directions. Plot the shape functions at $(x_I, y_I) = (5.0 + \alpha h_x, 5.0 + \beta h_y)$, $(x_I, y_I) = (5.0 + \alpha h_x, 0.0)$, and $(x_I, y_I) = (0.0, 0.0)$.

3) Using shape functions constructed in (1) and (2), verify constant and linear consistency:

$$\sum_{I=1}^{121} \Psi_I(\mathbf{x}) = 1 \qquad \left(\text{plot} \sum_{I=1}^{121} \Psi_I(\mathbf{x}) \text{ vs. } (x, y)\right)$$

$$\sum_{I=1}^{121} \Psi_I(\mathbf{x}) x_I = x \qquad \left(\text{plot} \sum_{I=1}^{121} \Psi_I(\mathbf{x}) x_I \text{ vs. } (x, y)\right)$$

$$\sum_{I=1}^{121} \Psi_I(\mathbf{x}) y_I = y \qquad \left(\text{plot} \sum_{I=1}^{121} \Psi_I(\mathbf{x}) y_I \text{ vs. } (x, y)\right)$$

$$\sum_{I=1}^{121} \Psi_{I,x}(\mathbf{x}) = 0 \qquad \left(\text{plot} \sum_{I=1}^{121} \Psi_{I,x}(\mathbf{x}) \text{ vs. } (x, y)\right)$$

$$\sum_{I=1}^{121} \Psi_{I,y}(\mathbf{x}) = 0 \qquad \left(\text{plot} \sum_{I=1}^{121} \Psi_{I,y}(\mathbf{x}) \text{ vs. } (x, y)\right)$$

$$\sum_{I=1}^{121} \Psi_{I,x}(\mathbf{x}) x_I = 1 \qquad \left(\text{plot} \sum_{I=1}^{121} \Psi_{I,x}(\mathbf{x}) x_I \text{ vs. } (x, y)\right)$$

$$\sum_{I=1}^{121} \Psi_{I,y}(\mathbf{x}) y_I = 1 \qquad \left(\text{plot} \sum_{I=1}^{121} \Psi_{I,y}(\mathbf{x}) y_I \text{ vs. } (x, y)\right)$$

Exercise 3.4 Use one-dimensional RK shape functions to approximate a function $f(x)$ by

$$f^h(x) = \sum_{I=1}^{NP} \Psi_I(x) f_I, \qquad 0.0 \le x \le 2\pi, \tag{3.99}$$

where $\Psi_I(x)$ is the RK shape function constructed using the cubic B-spline kernel function, and $f_I = f(x_I)$.

The L^2 norm and $\mathrm{s-H}^1$ semi-norm of the error of the approximation are defined as

$$e_{\mathrm{L}_2} = \left(\int_\Omega \left(f(x) - f^h(x)\right)^2 dx\right)^{1/2}, \tag{3.100}$$

$$e_{\mathrm{s-H}^1} = \left(\int_\Omega \left(f_{,x}(x) - f^h_{,x}(x)\right)^2 dx\right)^{1/2}. \tag{3.101}$$

In this problem, let $f(x) = \sin kx$. In the following, study how the order of basis functions and the kernel function affect the error and rate of convergence in the RK approximation. The rate of convergence can be identified as the slope of $\log(e)$ versus $\log(h)$ lines, where h is the nodal distance and e is an error measure in (3.100)–(3.101). Use at least eight Gauss quadrature points between every two nodes for the evaluation of error norms. Use equally spaced node discretizations with, e.g. 10, 20, 40, 80, and 160 nodes.

1) Study how the order of basis functions affects the convergence rate in the L^2 and $s\text{-}H^1$ semi-norm for the approximation of $f(x) = \sin kx$ with $k = 2$ and $k = 4$. Consider first-order and second-order basis functions, and compare the results to the interpolation estimates in (3.95) and (3.96). Choose your own support sizes.
2) Study the effect of the support size on the L^2 and $s\text{-}H^1$ errors and convergence rates; use at least three support sizes of your choice for the approximation of $f(x) = \sin kx$ with $k = 2$. Use first-order basis functions and a cubic B-spline kernel in the RK approximation. Verify the rates in (3.95) and (3.96), and compare the levels of error achieved for each support size.
3) Study the effect of the kernel function's continuity on the L^2 and $s\text{-}H^1$ error and convergence rates by using the (a) hat function, (b) quadratic B-spline function, and (c) cubic B-spline function, for the approximation of $f(x) = \sin kx$ with $k = 2$. Use first-order basis functions and a support size of twice the nodal distance in the RK approximation. Verify the rates in (3.95) and (3.96), and compare the levels of error achieved for each kernel.

3.5 Derivative Approximations in Meshfree Methods

Several techniques have been employed for approximating derivatives in meshfree methods. Remarkably, though several researchers independently arrived at various approaches that all seem unique, all of the approximations discussed herein are very closely related and usually special cases of one another.

3.5.1 Direct Derivatives

The simplest way to obtain an approximation of derivatives is to directly differentiate shape functions used for the primary variable. In this way, the approximation of all terms in the solution of PDEs is consistent, and derivative reproducing conditions are ensured, as discussed in Section 3.3.

This approach was first introduced in the element-free Galerkin method [26] for solving the PDEs with MLS. If the MLS approximation (3.32) is considered, an approximation for the derivative can be obtained by differentiating the approximation of the primary variable following Section 3.3:

$$\begin{aligned}
\partial^{\boldsymbol{\alpha}} u(\mathbf{x}) &\simeq \partial^{\boldsymbol{\alpha}} u^h(\mathbf{x}) = \sum_{I \in S_{\mathbf{x}}} \partial^{\boldsymbol{\alpha}} \Psi_I(\mathbf{x}) u_I, \\
\partial^{\boldsymbol{\alpha}} \Psi_I(\mathbf{x}) &= \partial^{\boldsymbol{\alpha}} \mathbf{p}^{\mathrm{T}}(\mathbf{x}) \mathbf{A}(\mathbf{x})^{-1} \mathbf{p}(\mathbf{x}_I) w_a(\mathbf{x} - \mathbf{x}_I) \\
&\quad + \mathbf{p}^{\mathrm{T}}(\mathbf{x}) \partial^{\boldsymbol{\alpha}} \mathbf{A}(\mathbf{x})^{-1} \mathbf{p}(\mathbf{x}_I) w_a(\mathbf{x} - \mathbf{x}_I) \\
&\quad + \mathbf{p}^{\mathrm{T}}(\mathbf{x}) \mathbf{A}(\mathbf{x})^{-1} \mathbf{p}(\mathbf{x}_I) \partial^{\boldsymbol{\alpha}} w_a(\mathbf{x} - \mathbf{x}_I).
\end{aligned} \tag{3.102}$$

where $\boldsymbol{\alpha}$ is a d-dimensional multi-index with $\boldsymbol{\alpha} = \alpha_1 \quad \cdots \quad \alpha_d$, $\partial^{\boldsymbol{\alpha}}(\mathbf{x}) \equiv \partial^{|\boldsymbol{\alpha}|} u(\mathbf{x}) / \partial x_1^{\alpha_1} \quad \cdots \quad \partial x_d^{\alpha_d}$, and $|\boldsymbol{\alpha}| = \sum_{i=1}^{d} \alpha_i \leq n$; n is the order of complete monomial basis functions in the basis vector $\mathbf{p}(\mathbf{x})$. The cost of computing the above is relatively high since the MLS/RK shape functions are comprised of matrix operations [29]. Thus, the differentiation of an inverse of a matrix such as $\mathbf{A}(\mathbf{x})^{-1}$, given in (3.89), and the many matrix operations involved in (3.102), makes this type of derivative computationally expensive. On the other hand, one is generally able to obtain higher accuracy in the Galerkin solution of PDEs with the directive derivatives than, say, the more computationally efficient derivatives presented in the next subsections. The direct derivative (3.102) is employed in most Galerkin and collocation methods.

3.5.2 Diffuse Derivatives

In the seminal work of the DEM [5], derivatives in the Galerkin equation are approximated by "diffuse" derivatives of $u^h(\mathbf{x})$. In this method, when differentiating the MLS approximation, the derivatives of the coefficients of the bases in (3.32) are neglected, which actually vary due to the moving nature of the approximation ($\overline{\mathbf{x}} \rightarrow \mathbf{x}$). To make this clear, examining (3.29) and (3.32) one can observe that for the MLS approximation $u^h(\mathbf{x})$ we have

$$u^h(\mathbf{x}) = \mathbf{p}^{\mathrm{T}}(\mathbf{x}) \underbrace{\mathbf{A}(\mathbf{x})^{-1} \sum_{I \in S_{\mathbf{x}}} \mathbf{p}(\mathbf{x}_I) w_a(\mathbf{x} - \mathbf{x}_I) u_I}_{\mathbf{b}(\mathbf{x})}. \tag{3.103}$$

In the diffuse derivatives, the derivatives of $u^h(\mathbf{x})$ are approximated as

$$\begin{aligned} \partial^{\boldsymbol{\alpha}} u(\mathbf{x}) \simeq \partial^{\boldsymbol{\alpha}} u^h(\mathbf{x}) &= \partial^{\boldsymbol{\alpha}} \mathbf{p}^{\mathrm{T}}(\mathbf{x}) \mathbf{b}(\mathbf{x}) \\ &= \partial^{\boldsymbol{\alpha}} \mathbf{p}^{\mathrm{T}}(\mathbf{x}) \mathbf{A}(\mathbf{x})^{-1} \sum_{I \in S_{\mathbf{x}}} \mathbf{p}(\mathbf{x}_I) w_a(\mathbf{x} - \mathbf{x}_I) u_I \\ &\equiv \sum_{I \in S_{\mathbf{x}}} \Psi_I^{(\boldsymbol{\alpha})}(\mathbf{x}) u_I. \end{aligned} \tag{3.104}$$

By employing (3.104), the approximation of $\partial^{\boldsymbol{\alpha}} u(\mathbf{x})$ is just as smooth as the approximation to $u(\mathbf{x})$, and it retains the completeness properties of the true derivative $\partial^{\boldsymbol{\alpha}} u^h(\mathbf{x})$ [5]. One other advantage of this method is that the cost of $\Psi_I^{(\boldsymbol{\alpha})}(\mathbf{x})$ is the exact same as $\Psi_I(\mathbf{x})$ (see Section 3.5.3). Nevertheless, lower accuracy is obtained with diffuse derivatives in the Galerkin solution of PDEs as mentioned previously.

The diffuse derivative can also be derived as follows. Rather than differentiating the approximation of the primary variable to obtain an approximation of derivatives, one can consider constructing an approximation to the derivative directly. First, at a given fixed point $\overline{\mathbf{x}}$, one can construct the weighted least squares approximation in Table 3.1. Then, an approximation to $\partial^{\boldsymbol{\alpha}} u(\mathbf{x})$ at $\overline{\mathbf{x}}$ can be obtained by differentiation of that approximation with respect to the moving variable $\mathbf{x}$:

$$\begin{aligned} &\partial^{\boldsymbol{\alpha}} u^h_{\overline{\mathbf{x}}}(\mathbf{x}) = \sum_{I \in S_{\overline{\mathbf{x}}}} \partial^{\boldsymbol{\alpha}} \Psi_I(\mathbf{x}, \overline{\mathbf{x}}) u_I, \\ &\partial^{\boldsymbol{\alpha}} \Psi_I(\mathbf{x}, \overline{\mathbf{x}}) = \partial^{\boldsymbol{\alpha}} \mathbf{p}^{\mathrm{T}}(\mathbf{x}) \mathbf{A}(\overline{\mathbf{x}})^{-1} \mathbf{p}(\mathbf{x}_I) w_a(\overline{\mathbf{x}} - \mathbf{x}_I), \\ &\mathbf{A}(\overline{\mathbf{x}}) = \sum_{I \in S_{\overline{\mathbf{x}}}} \mathbf{p}(\mathbf{x}_I) \mathbf{p}^{\mathrm{T}}(\mathbf{x}_I) w_a(\overline{\mathbf{x}} - \mathbf{x}_I). \end{aligned} \tag{3.105}$$

The approximation is then made global by taking $\overline{\mathbf{x}} \rightarrow \mathbf{x}$ as in the MLS approximation:

$$\partial^{\boldsymbol{\alpha}} u^h(\mathbf{x}) = \left[\partial^{\boldsymbol{\alpha}} u^h_{\overline{\mathbf{x}}}(\mathbf{x})\right]_{\overline{\mathbf{x}} \rightarrow \mathbf{x}} = \sum_{I \in S_{\mathbf{x}}} \Psi_I^{(\boldsymbol{\alpha})}(\mathbf{x}) u_I, \tag{3.106}$$

where $\Psi_I^{(\boldsymbol{\alpha})}$ is the diffuse derivative shape function in (3.104).

3.5.3 Implicit Gradients and Synchronized Derivatives

The implicit gradient was originally introduced as a regularization in strain localization problems without taking direct derivatives [30] in the spirit of the synchronized RK approximation [31–33]. In the implicit gradient method [34], derivative approximations are constructed by employing the same form as the RK shape function (3.78)–(3.79). For simplicity, one can begin with a nonshifted basis $\mathbf{H}^{\mathrm{T}}(\mathbf{x}_I)$, rather than $\mathbf{H}^{\mathrm{T}}(\mathbf{x}-\mathbf{x}_I)$:

$$\Psi_I^{(\alpha)}(\mathbf{x}) = \mathbf{H}^{\mathrm{T}}(\mathbf{x}_I)\mathbf{b}^{(\alpha)}(\mathbf{x})\phi_a(\mathbf{x}-\mathbf{x}_I). \tag{3.107}$$

The coefficients $\mathbf{b}^{(\alpha)}(\mathbf{x})$ are obtained from the following gradient reproducing conditions [34], analogous to (3.80):

$$\sum_{I\in S_{\mathbf{x}}}\Psi_I^{(\alpha)}(\mathbf{x})\mathbf{H}(\mathbf{x}_I) = \partial^{\alpha}\mathbf{H}(\mathbf{x}). \tag{3.108}$$

By substituting (3.107) into (3.108) and following the same procedures in constructing the RK approximation, the implicit gradient can be obtained as

$$\begin{aligned}
&\partial^{\alpha}u(\mathbf{x}) \simeq \sum_{I\in S_{\mathbf{x}}}\Psi_I^{(\alpha)}(\mathbf{x})u_I,\\
&\Psi_I^{(\alpha)}(\mathbf{x}) = \partial^{\alpha}\mathbf{H}^{\mathrm{T}}(\mathbf{x})\mathbf{M}(\mathbf{x})^{-1}\mathbf{H}(\mathbf{x}_I)\phi_a(\mathbf{x}-\mathbf{x}_I),\\
&\mathbf{M}(\mathbf{x}) = \sum_{I\in S_{\mathbf{x}}}\mathbf{H}(\mathbf{x}_I)\mathbf{H}^{\mathrm{T}}(\mathbf{x}_I)\phi_a(\mathbf{x}-\mathbf{x}_I).
\end{aligned} \tag{3.109}$$

Comparing (3.109) and (3.104), one can see that implicit gradients are indeed diffuse derivatives.

To be consistent with the standard RK approximation with shifted bases $\mathbf{H}^{\mathrm{T}}(\mathbf{x}-\mathbf{x}_I)$, an implicit gradient can also be constructed as follows [30]:

$$\Psi_I^{(\alpha)}(\mathbf{x}) = \mathbf{H}^{\mathrm{T}}(\mathbf{x}-\mathbf{x}_I)\mathbf{b}^{(\alpha)}(\mathbf{x})\phi_a(\mathbf{x}-\mathbf{x}_I). \tag{3.110}$$

For easy illustration, consider a one-dimensional RK approximation with quadratic basis ($n = 2$), where $\mathbf{H}^{\mathrm{T}}(x-x_I) = \begin{bmatrix}1 & x-x_I & (x-x_I)^2\end{bmatrix}$. The first-order implicit gradient, $\alpha = 1$, is expressed as

$$\begin{aligned}
&\partial^{1}u(x) \simeq \sum_{I\in S_x}\Psi_I^{(1)}(x)u_I,\\
&\Psi_I^{(1)}(x) = \mathbf{H}^{\mathrm{T}}(x-x_I)\mathbf{b}^{(1)}(x)\phi_a(x-x_I).
\end{aligned} \tag{3.111}$$

The gradient reproducing conditions for $n = 2$, $\alpha = 1$ are then given as

$$\begin{aligned}
&\sum_{I\in S_{\mathbf{x}}}\Psi_I^{(1)}(x) = 0,\\
&\sum_{I\in S_{\mathbf{x}}}\Psi_I^{(1)}(x)x_I = 1,\\
&\sum_{I\in S_{\mathbf{x}}}\Psi_I^{(1)}(x)x_I^2 = 2x.
\end{aligned} \tag{3.112}$$

These gradient reproducing conditions can be recast with shifted conditions (to be consistent with shifted bases) as follows:

$$\begin{aligned}
&\sum_{I\in S_{\mathbf{x}}}\Psi_I^{(1)}(x) = 0,\\
&\sum_{I\in S_{\mathbf{x}}}\Psi_I^{(1)}(x)(x-x_I) = \left(\sum_{I\in S_{\mathbf{x}}}\Psi_I^{(1)}(x)\right)x - \sum_{I\in S_{\mathbf{x}}}\Psi_I^{(1)}(x)x_I = 0-1 = -1,\\
&\sum_{I\in S_{\mathbf{x}}}\Psi_I^{(1)}(x)(x-x_I)^2 = \left(\sum_{I\in S_{\mathbf{x}}}\Psi_I^{(1)}(x)\right)x^2 - 2\left(\sum_{I\in S_{\mathbf{x}}}\Psi_I^{(1)}(x)x_I\right)x + \sum_{I\in S_{\mathbf{x}}}\Psi_I^{(1)}(x)x_I^2 = 0-2x+2x = 0,
\end{aligned} \tag{3.113}$$

or

$$\sum_{I\in S_x} \Psi_I^{(1)}(x)\mathbf{H}(x-x_I) = \mathbf{H}^{(1)},$$
$$\mathbf{H}^{(1)^{\mathrm{T}}} = [0 \quad -1 \quad 0]. \tag{3.114}$$

The coefficient vector of the first-order implicit gradient $\mathbf{b}^{(1)}(x)$ is obtained by substituting (3.111) into (3.114), and the resulting first-order implicit gradient RK shape function is expressed as

$$\Psi_I^{(1)}(x) = \mathbf{H}^{(1)^{\mathrm{T}}}\mathbf{M}^{-1}(x)\mathbf{H}(x-x_I)\phi_a(x-x_I). \tag{3.115}$$

Similarly, considering the same one-dimensional RK approximation with quadratic basis, the second-order implicit gradient, with $\alpha = 2$, is expressed as

$$\partial^2 u(x) \simeq \sum_{I\in S_x} \Psi_I^{(2)}(x)u_I,$$
$$\Psi_I^{(2)}(x) = \mathbf{H}^{\mathrm{T}}(x-x_I)\mathbf{b}^{(2)}(x)\phi_a(x-x_I). \tag{3.116}$$

The gradient reproducing conditions for $n = 2,\ \alpha = 2$ are expressed as

$$\sum_{I\in S_x} \Psi_I^{(2)}(x) = 0,$$
$$\sum_{I\in S_x} \Psi_I^{(2)}(x)x_I = 0, \tag{3.117}$$
$$\sum_{I\in S_x} \Psi_I^{(2)}(x)x_I^2 = 2.$$

These gradient reproducing conditions can also be recast with the shifted conditions as

$$\sum_{I\in S_x} \Psi_I^{(2)}(x) = 0,$$
$$\sum_{I\in S_x} \Psi_I^{(2)}(x)(x-x_I) = \left(\sum_{I\in S_x} \Psi_I^{(2)}(x)\right)x - \sum_{I\in S_x} \Psi_I^{(2)}(x)x_I = 0-0 = 0,$$
$$\sum_{I\in S_x} \Psi_I^{(2)}(x)(x-x_I)^2 = \left(\sum_{I\in S_x} \Psi_I^{(2)}(x)\right)x^2 - 2\left(\sum_{I\in S_x} \Psi_I^{(2)}(x)x_I\right)x + \sum_{I\in S_x} \Psi_I^{(2)}(x)x_I^2 = 0-0+2 = 2. \tag{3.118}$$

or

$$\sum_{I\in S_x} \Psi_I^{(2)}(x)\mathbf{H}(x-x_I) = \mathbf{H}^{(2)},$$
$$\mathbf{H}^{(2)^{\mathrm{T}}} = [0 \quad 0 \quad 2]. \tag{3.119}$$

Following similar procedures used in the first-order implicit gradient, the second-order implicit gradient RK shape function is obtained as

$$\Psi_I^{(2)}(x) = \mathbf{H}^{(2)^{\mathrm{T}}}\mathbf{M}^{-1}(x)\mathbf{H}(x-x_I)\phi_a(x-x_I). \tag{3.120}$$

For arbitrary αth-order implicit gradient of RK approximation with nth-order bases, $\alpha \leq n$, one can show the following [30, 34]:

$$\begin{aligned} \partial^{\alpha} u(x) &\simeq \sum_{I \in S_{\mathbf{x}}} \Psi_I^{(\alpha)}(x) u_I, \\ \Psi_I^{(\alpha)}(x) &= \mathbf{H}^{(\alpha)^{\mathrm{T}}} \mathbf{M}^{-1}(x) \mathbf{H}(x - x_I) \phi_a(x - x_I). \end{aligned} \tag{3.121}$$

The components of the vector $\mathbf{H}^{(\alpha)^{\mathrm{T}}} = \left[H_0^{(\alpha)}, H_1^{(\alpha)}, \cdots, H_n^{(\alpha)}\right]$ can be shown to be

$$H_i^{(\alpha)} = (-1)^{(\alpha)} \alpha! \delta_{i\alpha}, \quad \alpha \leq n. \tag{3.122}$$

The generalization to three-dimensional implicit gradients is then trivial, with the following notation:

$$\begin{aligned} \partial^{\boldsymbol{\alpha}} u(\mathbf{x}) &\equiv \frac{\partial^{|\boldsymbol{\alpha}|} u(\mathbf{x})}{\partial x_1^{\alpha_1} \partial x_2^{\alpha_2} \partial x_3^{\alpha_3}} \simeq \sum_{I \in S_{\mathbf{x}}} \Psi_I^{(\boldsymbol{\alpha})}(\mathbf{x}) u_I, \\ \Psi_I^{(\boldsymbol{\alpha})}(\mathbf{x}) &= \mathbf{H}^{\mathrm{T}}(\mathbf{x} - \mathbf{x}_I) \mathbf{b}^{(\boldsymbol{\alpha})}(\mathbf{x}) \phi_a(\mathbf{x} - \mathbf{x}_I). \end{aligned} \tag{3.123}$$

The reproducing conditions are expressed as

$$\sum_{I \in S_{\mathbf{x}}} \Psi_I^{(\boldsymbol{\alpha})}(\mathbf{x}) \mathbf{H}(\mathbf{x}_I) = \partial^{\boldsymbol{\alpha}} \mathbf{H}(\mathbf{x}), \tag{3.124}$$

which is equivalent to

$$\sum_{I \in S_{\mathbf{x}}} \Psi_I^{(\boldsymbol{\alpha})}(\mathbf{x}) \mathbf{H}(\mathbf{x} - \mathbf{x}_I) = \mathbf{H}^{(\boldsymbol{\alpha})}, \tag{3.125}$$

where the components of the basis vector $\mathbf{H}^{(\boldsymbol{\alpha})^{\mathrm{T}}} = \left[H_{000}^{(\boldsymbol{\alpha})} \quad H_{100}^{(\boldsymbol{\alpha})} \quad H_{010}^{(\boldsymbol{\alpha})} \quad H_{001}^{(\boldsymbol{\alpha})} \quad \cdots \quad H_{00n}^{(\boldsymbol{\alpha})}\right]$ can be shown to be

$$H_{ijk}^{(\boldsymbol{\alpha})} = (-1)^{|\boldsymbol{\alpha}|} \alpha_1! \alpha_2! \alpha_3! \delta_{i\alpha_1} \delta_{j\alpha_2} \delta_{k\alpha_3}, \quad |\boldsymbol{\alpha}| \leq n. \tag{3.126}$$

Indeed, the following equality can be shown:

$$\mathbf{H}^{(\boldsymbol{\alpha})} = (-1)^{|\boldsymbol{\alpha}|} \partial^{\boldsymbol{\alpha}} \mathbf{H}(\mathbf{x})|_{\mathbf{x} = \mathbf{0}}, \tag{3.127}$$

and in fact

$$\mathbf{H}^{(000)} = \mathbf{H}(\mathbf{0}). \tag{3.128}$$

Solving for $\mathbf{b}^{(\boldsymbol{\alpha})}(\mathbf{x})$ in (3.123) from (3.125), the three-dimensional αth-order implicit gradient RK shape function can be obtained as

$$\Psi_I^{(\boldsymbol{\alpha})}(\mathbf{x}) = \mathbf{H}^{(\boldsymbol{\alpha})^{\mathrm{T}}} \mathbf{M}^{-1}(\mathbf{x}) \mathbf{H}(\mathbf{x} - \mathbf{x}_I) \phi_a(\mathbf{x} - \mathbf{x}_I). \tag{3.129}$$

It can be shown that the implicit gradients have the same expression as the synchronized derivatives [31], scaled by $\alpha_1! \alpha_2! \alpha_3!$, with the difference in sign $(-1)^{|\boldsymbol{\alpha}|}$ emanating from the convention in shifting the bases. In this form, the reproduction of derivative terms can also be seen by examining alternative vanishing moment conditions in (3.56) and (3.57), i.e., matching certain terms in the Taylor expansion other than $u(x)$.

Using this idea, it was shown that by employing certain linear combinations of synchronized derivatives and the RK approximation, synchronized convergence can be obtained in the L_2 norm and H^s norms up to some order s with the proper selection of coefficients $\chi^{\boldsymbol{\alpha}}$ in the following use of the synchronized shape function $\tilde{\Psi}_I$:

$$\tilde{\Psi}_I(\mathbf{x}) = \Psi_I(\mathbf{x}) + \sum_{|\boldsymbol{\alpha}|=1}^{m} \chi^{\boldsymbol{\alpha}} \Psi_I^{(\boldsymbol{\alpha})}(\mathbf{x}). \tag{3.130}$$

Since the additional terms in the above satisfy the partition of nullity, they termed the resulting approximation (3.130) a hierarchical partition of unity [32]. Synchronized derivatives have been developed for improving accuracy in the Helmholtz equation, obtaining high-resolution solutions in localization problems, and stabilization in computational fluid dynamics [33].

Implicit gradients have also been employed for regularization in strain localization problems [30] to avoid the need of ambiguous boundary conditions associated with the standard gradient-type regularization methods. They have also been leveraged to provide highly efficient stabilization of nodal integration (see Section 6.10.3) and to ease the computational cost of meshfree collocation methods (see Section 9.4).

An attractive feature of the implicit gradient is its similarity to the standard RK shape function:

$$\begin{aligned} &\Psi_I^{(\boldsymbol{\alpha})}(\mathbf{x}) = \mathbf{H}^{(\boldsymbol{\alpha})^{\mathrm{T}}} \mathbf{M}^{-1}(\mathbf{x}) \mathbf{H}(\mathbf{x}-\mathbf{x}_I) \phi_a(\mathbf{x}-\mathbf{x}_I), \\ &\Updownarrow \\ &\Psi_I(\mathbf{x}) = \mathbf{H}^{\mathrm{T}}(\mathbf{0}) \mathbf{M}^{-1}(\mathbf{x}) \mathbf{H}(\mathbf{x}-\mathbf{x}_I) \phi_a(\mathbf{x}-\mathbf{x}_I). \end{aligned} \tag{3.131}$$

Together with (3.128), they can be generalized as

$$\begin{aligned} \Psi_I(\mathbf{x}) = \Psi_I^{(000)}(\mathbf{x}) &= \mathbf{H}^{(000)^{\mathrm{T}}} \mathbf{M}^{-1}(\mathbf{x}) \mathbf{H}(\mathbf{x}-\mathbf{x}_I) \phi_a(\mathbf{x}-\mathbf{x}_I) \\ &= \mathbf{H}^{\mathrm{T}}(\mathbf{0}) \mathbf{M}^{-1}(\mathbf{x}) \mathbf{H}(\mathbf{x}-\mathbf{x}_I) \phi_a(\mathbf{x}-\mathbf{x}_I). \end{aligned} \tag{3.132}$$

That is, the RK shape functions and their implicit gradients can be computed together in the same loop in a computer code. For example, in three dimensions, the RK shape functions with linear basis ($n = 1$, $\mathbf{H}^{\mathrm{T}}(\mathbf{x}) = [1 \quad x_1 \quad x_2 \quad x_3]$) and their first-order implicit gradients ($|\boldsymbol{\alpha}| = 1$) are

$$\begin{aligned} &u(\mathbf{x}) \simeq \sum_{I \in S_{\mathbf{x}}} \Psi_I(\mathbf{x}) u_I = \sum_{I \in S_{\mathbf{x}}} \Psi_I^{(000)}(\mathbf{x}) u_I, \\ &\frac{\partial u(\mathbf{x})}{\partial x_1} \simeq \sum_{I \in S_{\mathbf{x}}} \Psi_I^{(100)}(\mathbf{x}) u_I, \\ &\frac{\partial u(\mathbf{x})}{\partial x_2} \simeq \sum_{I \in S_{\mathbf{x}}} \Psi_I^{(010)}(\mathbf{x}) u_I, \\ &\frac{\partial u(\mathbf{x})}{\partial x_3} \simeq \sum_{I \in S_{\mathbf{x}}} \Psi_I^{(001)}(\mathbf{x}) u_I, \end{aligned} \tag{3.133}$$

where

$$\begin{aligned} &\Psi_I(\mathbf{x}) = \Psi_I^{(000)}(\mathbf{x}), \\ &\Psi_I^{(000)}(\mathbf{x}) = \mathbf{H}^{(000)^{\mathrm{T}}} \mathbf{M}^{-1}(\mathbf{x}) \mathbf{H}(\mathbf{x}-\mathbf{x}_I) \phi_a(\mathbf{x}-\mathbf{x}_I), && \mathbf{H}^{(000)^{\mathrm{T}}} = [1 \quad 0 \quad 0 \quad 0], \\ &\Psi_I^{(100)}(\mathbf{x}) = \mathbf{H}^{(100)^{\mathrm{T}}} \mathbf{M}^{-1}(\mathbf{x}) \mathbf{H}(\mathbf{x}-\mathbf{x}_I) \phi_a(\mathbf{x}-\mathbf{x}_I), && \mathbf{H}^{(100)^{\mathrm{T}}} = [0 \quad -1 \quad 0 \quad 0], \\ &\Psi_I^{(010)}(\mathbf{x}) = \mathbf{H}^{(010)^{\mathrm{T}}} \mathbf{M}^{-1}(\mathbf{x}) \mathbf{H}(\mathbf{x}-\mathbf{x}_I) \phi_a(\mathbf{x}-\mathbf{x}_I), && \mathbf{H}^{(010)^{\mathrm{T}}} = [0 \quad 0 \quad -1 \quad 0], \\ &\Psi_I^{(001)}(\mathbf{x}) = \mathbf{H}^{(001)^{\mathrm{T}}} \mathbf{M}^{-1}(\mathbf{x}) \mathbf{H}(\mathbf{x}-\mathbf{x}_I) \phi_a(\mathbf{x}-\mathbf{x}_I), && \mathbf{H}^{(001)^{\mathrm{T}}} = [0 \quad 0 \quad 0 \quad -1]. \end{aligned} \tag{3.134}$$

From Eq. (3.134), the simplicity and computational advantage of employing implicit gradients over direct differentiation in (3.88) should be readily apparent.

3.5.4 Generalized Finite Difference Methods

The method by Liszka and Orkisz [4, 35] generalized previous work in finite differences to arbitrary point distributions and stencils (neighbors). Derivative approximations are constructed directly by the satisfaction of a truncated Taylor expansion. The generalized finite difference (GFD) method starts with the Taylor expansion of function $u(\mathbf{x})$ about at point $u(\mathbf{x}_I)$ truncated to a given order n [4]:

$$u(\mathbf{x}_I) = \sum_{|\alpha|=0}^{n} \frac{(-1)^{|\alpha|}}{\boldsymbol{\alpha}!}(\mathbf{x}-\mathbf{x}_I)^{\boldsymbol{\alpha}}\partial^{\boldsymbol{\alpha}}u(\mathbf{x}) = \mathbf{H}^{\mathrm{T}}(\mathbf{x}-\mathbf{x}_I)\mathbf{J}\mathbf{u}^{\boldsymbol{\alpha}}(\mathbf{x}). \tag{3.135}$$

Here, the following multi-dimensional index notations are used: $(\mathbf{x}-\mathbf{x}_I)^{\boldsymbol{\alpha}} \equiv (x_1-x_{1I})^{\alpha_1}(x_2-x_{2I})^{\alpha_2}(x_3-x_{3I})^{\alpha_3}$, $\boldsymbol{\alpha}! \equiv \alpha_1!\alpha_2!\alpha_3!$; $\mathbf{H}(\mathbf{x}-\mathbf{x}_I)$ is the same vector as in the RK approximation with monomials (3.83), $\mathbf{J}$ is a diagonal matrix of $\left\{(-1)^{|\boldsymbol{\alpha}|}/\boldsymbol{\alpha}!\right\}_{|\boldsymbol{\alpha}|=0}^{n}$, and $\mathbf{u}^{\boldsymbol{\alpha}}(\mathbf{x}) \simeq \{u(\mathbf{x}), \ldots, \partial^{\boldsymbol{\alpha}}u(\mathbf{x}), \ldots, \partial^{\boldsymbol{\alpha}}u(\mathbf{x})|_{|\boldsymbol{\alpha}|=n}\}^{\mathrm{T}}$ is a vector of u and it's derivatives of length N, which are treated as unknowns here (the other variant of GFD does not include u).

To solve for approximations of $u(\mathbf{x})$ and it's derivatives on the right-hand side, (3.135) can be evaluated at M points in a stencil (or "star" in GFD terminology) surrounding $\mathbf{x}$, and one obtains the system

$$\mathbf{U} = \mathbf{P}_{\mathbf{H}}\mathbf{J}\mathbf{u}^{\boldsymbol{\alpha}}(\mathbf{x}), \tag{3.136}$$

where $\mathbf{U} = \{u(\mathbf{x}_1), \ldots, u(\mathbf{x}_I), \ldots, u(\mathbf{x}_M)\}^{\mathrm{T}}$ and

$$\mathbf{P}_{\mathbf{H}} = \begin{bmatrix} \mathbf{H}^{\mathrm{T}}(\mathbf{x}-\mathbf{x}_1) \\ \vdots \\ \mathbf{H}^{\mathrm{T}}(\mathbf{x}-\mathbf{x}_M) \end{bmatrix}. \tag{3.137}$$

In the above, a local point numbering is employed. Now, if the number of points in the "star" (stencil) M is equal to the number of unknowns N, then a solution to the system can be obtained by solving (3.136) directly, which was the method proposed by Jensen [36]. Selecting a suitable star such that the resulting system is not linearly dependent was one of the early troubles in GFD methods, as the number of points in the star was fixed, and each point in the star had to be of sufficient "quality" to avoid linear dependence, leading to a difficult situation. While an effort was made at the time for better a star selection, Liszka and Orkisz [4, 35] greatly improved upon the method by considering that a larger number of points in the star could be used ($M > N$), and the resulting overdetermined system could be solved by least squares, or weighted least squares as follows:

$$\mathbf{P}_{\mathbf{H}}^{\mathrm{T}}\mathbf{B}\mathbf{U} = \mathbf{P}_{\mathbf{H}}^{\mathrm{T}}\mathbf{B}\mathbf{P}_{\mathbf{H}}\mathbf{J}\mathbf{u}^{\boldsymbol{\alpha}}(\mathbf{x}), \tag{3.138}$$

where $\mathbf{B}$ is a matrix of weights. By defining the weight of members of a stencil as a function of nodal distance using a kernel ϕ_a (or equivalently a weight function w_a):

$$\mathbf{B}(\mathbf{x}) = \begin{bmatrix} \phi_a(\mathbf{x}-\mathbf{x}_1) & \cdots & 0 & \cdots & 0 \\ \vdots & \ddots & & & \vdots \\ 0 & & \phi_a(\mathbf{x}-\mathbf{x}_I) & & 0 \\ \vdots & & & \ddots & \vdots \\ 0 & \cdots & 0 & \cdots & \phi_a(\mathbf{x}-\mathbf{x}_M) \end{bmatrix}, \tag{3.139}$$

then $\mathbf{P}_\mathbf{H}^\mathrm{T}\mathbf{B}\mathbf{P}_\mathbf{H} = \mathbf{M}(\mathbf{x})$ is exactly the moment matrix $\mathbf{M}(\mathbf{x})$ in the discrete RK approximation with monomials.

Solving the system (3.138) and using the compact support in the weight matrix $\mathbf{B}$, and nodal indexing previously introduced in the RK and MLS approximations results in

$$\mathbf{u}^{\alpha}(\mathbf{x}) = \sum_{I\in S_\mathbf{x}} \mathbf{J}^{-1}\mathbf{M}^{-1}(\mathbf{x})\mathbf{H}(\mathbf{x}-\mathbf{x}_1)\phi_a(\mathbf{x}-\mathbf{x}_I)u_I. \tag{3.140}$$

where $u_I \equiv u(\mathbf{x}_I)$. Now, let us continue on the premise that the weight matrix to solve the over-determined system is selected as a function of the neighboring points' contribution by distance as in (3.139). Then, to obtain the approximation for $u(\mathbf{x})$, one can pre-multiply both sides of (3.140) by $\mathbf{H}^\mathrm{T}(\mathbf{0})$ to obtain the first row of the vector $\mathbf{u}^{\alpha}$ on the left-hand side of (3.140). As the first entry of $\mathbf{J}^{-1}$ is unity, immediately the RK approximation is obtained. One can also identify a row in the left-hand side of (3.140) corresponding to the approximation of $\partial^{\alpha}u(\mathbf{x})$, as the pre-multiplication of (3.140) by $\mathbf{H}^{(\alpha)^\mathrm{T}}$ in (3.131): this variant of the GFDs is also indeed the implicit gradient (or diffuse derivatives, as shown in Section 3.5.3) when the least-squares system is weighted using ϕ_a and the primary variable is solved for along with its derivatives.

Comparing the GFD and synchronized derivatives to the RK approximation, one can observe that the moment matrix contains information on how to approximate the primary variable as well as derivatives.

3.5.5 Non-ordinary State-based Peridynamics under the Correspondence Principle, and RK Peridynamics

Bessa et al. [37] made a connection between implicit gradients, Savitzky–Golay filters [38], and Peridynamics. They showed that under uniform node distributions in an infinite domain, the deformation gradient in nodally collocated state-based Peridynamics under the correspondence principle [39] is equivalent to employing second-order accurate implicit gradients. More recently, a more general case was established under nonuniform node distributions and finite domains [40], which will be described in the following.

First let a derivative approximation be constructed directly, similar to the implicit gradient approximation (3.109):

$$\partial^{\alpha}u(\mathbf{x}) \simeq \sum_{I\in S_\mathbf{x}} \tilde{\Psi}_I^{(\alpha)}(\mathbf{x})\{u(\mathbf{x}_I) - u(\mathbf{x})\}, \tag{3.141}$$

where

$$\tilde{\Psi}_I^{(\alpha)}(\mathbf{x}) = \tilde{\mathbf{b}}^{(\alpha)}(\mathbf{x})\mathbf{q}(\mathbf{x}-\mathbf{x}_I)\phi_a(\mathbf{x}-\mathbf{x}_I), \tag{3.142}$$

$\tilde{\mathbf{b}}^{(\boldsymbol{\alpha})}$ is a vector of coefficients, and $\mathbf{q}(\mathbf{x})$ is a vector of basis functions containing complete monomials of order ranging from m to n ($m \leq n$):

$$\mathbf{q}^{\mathrm{T}}(\mathbf{x}) = [x_1^m \quad \dots \quad x_3^n]. \tag{3.143}$$

For example, for $m = 0$, the basis function is nothing but the MLS basis $\mathbf{p}(\mathbf{x})$ or the RK monomial basis $\mathbf{H}(\mathbf{x})$. As will be seen, for peridynamics, $m = n = 1$ and $\mathbf{q}^{\mathrm{T}}(\mathbf{x}) = [x_1 \quad x_2 \quad x_3]$.

In [40], a Taylor expansion procedure was given to obtain $\tilde{\mathbf{b}}^{\boldsymbol{\alpha}}$, and the continuous approximations were discretized with quadrature. However, as seen in Section 3.2.2, discrete approximations can be obtained in a very straightforward way by enforcing reproducing conditions directly on (3.141), as in the discrete RK approximation. Therefore here, to simplify the presentation we obtain the coefficients $\tilde{\mathbf{b}}^{(\boldsymbol{\alpha})}$ from the following gradient reproducing conditions, analogous to (3.80):

$$\sum_{I \in S_{\mathbf{x}}} \tilde{\Psi}_I^{(\boldsymbol{\alpha})}(\mathbf{x})\mathbf{q}(\mathbf{x}) = \partial^{\boldsymbol{\alpha}}\mathbf{q}(\mathbf{x}), \tag{3.144}$$

which can be shown to be

$$\sum_{I \in S_{\mathbf{x}}} \Psi_I^{(\boldsymbol{\alpha})}(\mathbf{x})\mathbf{q}(\mathbf{x}_I - \mathbf{x}) = \mathbf{q}^{(\boldsymbol{\alpha})}, \tag{3.145}$$

where $\mathbf{q}^{(\boldsymbol{\alpha})}$ contains the following entries, similar to the implicit gradient (3.126)

$$q_{ijk}^{(\boldsymbol{\alpha})} = \alpha_1!\alpha_2!\alpha_3!\delta_{i\alpha_1}\delta_{j\alpha_2}\delta_{k\alpha_3}, \quad m \leq |\boldsymbol{\alpha}| \leq n. \tag{3.146}$$

In keeping with the conventions of peridynamics, the basis has been shifted as $\mathbf{x}_I - \mathbf{x}$ rather than $\mathbf{x} - \mathbf{x}_I$. This results in the change of $(-1)^{|\boldsymbol{\alpha}|}\alpha_1!\alpha_2!\alpha_3!$ in the implicit gradient vector (3.126) to $\alpha_1!\alpha_2!\alpha_3!$ in (3.146), emanating from Taylor's expansion.

Solving for the coefficients $\tilde{\mathbf{b}}^{(\boldsymbol{\alpha})}$, a unified RK peridynamic approximation is obtained as

$$\begin{aligned} \tilde{\Psi}_I^{(\boldsymbol{\alpha})}(\mathbf{x}) &= \left(\mathbf{q}^{(\boldsymbol{\alpha})}\right)^{\mathrm{T}} \tilde{\mathbf{M}}(\mathbf{x})^{-1}\mathbf{q}(\mathbf{x}_I - \mathbf{x})\phi_a(\mathbf{x}_I - \mathbf{x}), \\ \tilde{\mathbf{M}}(\mathbf{x}) &= \sum_{I \in S_{\mathbf{x}}} \mathbf{q}(\mathbf{x}_I - \mathbf{x})\mathbf{q}^{\mathrm{T}}(\mathbf{x}_I - \mathbf{x})\phi_a(\mathbf{x}_I - \mathbf{x}). \end{aligned} \tag{3.147}$$

For $m = 0$, and letting n be a free variable, it is readily apparent that the above is nothing but the implicit gradient approximation with a simple change of sign in the shifted argument of the basis, since setting $m = 1$ yields $\mathbf{q}(\mathbf{x}) = \mathbf{H}(\mathbf{x})$, $\mathbf{q}^{(\boldsymbol{\alpha})} = \mathbf{H}^{(\boldsymbol{\alpha})}$, $\tilde{\mathbf{M}}(\mathbf{x}) = \mathbf{M}(\mathbf{x})$, and thus $\tilde{\Psi}_I^{(\boldsymbol{\alpha})} = \Psi_I^{(\boldsymbol{\alpha})}$. Finally, using the partition of nullity property inherent $\Psi_I^{(\boldsymbol{\alpha})}$, we obtain for (3.141):

$$\begin{aligned} \partial^{\boldsymbol{\alpha}} u(\mathbf{x}) &\simeq \sum_{I \in S_{\mathbf{x}}} \Psi_I^{(\boldsymbol{\alpha})}(\mathbf{x})\{u(\mathbf{x}_I) - u(\mathbf{x})\} \\ &= \sum_{I \in S_{\mathbf{x}}} \Psi_I^{(\boldsymbol{\alpha})}(\mathbf{x})u(\mathbf{x}_I) - u(\mathbf{x})\underbrace{\sum_{I \in S_{\mathbf{x}}} \Psi_I^{(\boldsymbol{\alpha})}(\mathbf{x})}_{0} \\ &= \sum_{I \in S_{\mathbf{x}}} \Psi_I^{(\boldsymbol{\alpha})}(\mathbf{x})u(\mathbf{x}_I), \end{aligned} \tag{3.148}$$

which is the implicit gradient approximation in (3.109).

While the connection to the deformation gradient approximation in the peridynamics literature may not be readily apparent, consider $m = 1$, and $n = 1$, and the approximation of a first-order derivative ($|\alpha| = 1$) at a point $\mathbf{x}_J$. The displacement gradient is calculated as

$$\begin{aligned} u_{i,j}(\mathbf{x}_J) &\simeq \sum_{I \in S_\mathbf{x}} \tilde{\Psi}^{\nabla}_{Ij}(\mathbf{x}_J)\{u_{iI} - u_{iJ}\}, \\ \tilde{\Psi}^{\nabla}_{Ij} &\equiv \tilde{\Psi}^{\boldsymbol{\alpha}}_{I}\big|_{\alpha=(\delta_{1j},\ \delta_{2j},\ \delta_{3j})}, \\ \tilde{\Psi}^{\nabla}_{Ij}(\mathbf{x}_J) &= \mathbf{q}^{\nabla}_j \tilde{\mathbf{M}}(\mathbf{x}_J)^{-1} \mathbf{q}_j(\mathbf{x}_I)\phi_a(\mathbf{x}_I - \mathbf{x}_J), \\ \mathbf{q}^{\nabla}_j &= [\delta_{1j} \quad \delta_{2j} \quad \delta_{3j}]. \end{aligned} \tag{3.149}$$

Using $\mathbf{q}^{\mathrm{T}}(\mathbf{x}) = [x_1 \quad x_2 \quad x_3] = \mathbf{x}$, and replacing the kernel function ϕ_a with the so-called influence function w_δ with measure δ, we have after re-arranging

$$\begin{aligned} \nabla \mathbf{u}(\mathbf{x}_J) &\simeq \mathbf{I} + \sum_{I \in S_\mathbf{x}} \mathbf{S}(\mathbf{x}_I - \mathbf{x}_J)\tilde{\mathbf{K}}(\mathbf{x}_J)^{-1} w_\delta(\mathbf{x}_I - \mathbf{x}_J), \\ \tilde{\mathbf{K}}(\mathbf{x}_J) &= \tilde{\mathbf{M}}(\mathbf{x}_J)\big|_{n=m=1,\ |\alpha|=1} = \sum_{I \in S_\mathbf{x}} (\mathbf{x}_I - \mathbf{x}_J)^{\mathrm{T}}(\mathbf{x}_I - \mathbf{x}_J) w_\delta(\mathbf{x}_I - \mathbf{x}_J), \\ \mathbf{S}(\mathbf{x}_I - \mathbf{x}_J) &= \sum_{I \in G_\mathbf{x}} ((\mathbf{x}_I + \mathbf{u}_I) - (\mathbf{x}_J + \mathbf{u}_J))^{\mathrm{T}}(\mathbf{x}_I - \mathbf{x}_J) w_\delta(\mathbf{x}_I - \mathbf{x}_J), \end{aligned} \tag{3.150}$$

where in the finite-deformation setting $\mathbf{x}$ represents material coordinates and $\mathbf{x} + \mathbf{u}$ represents the deformed coordinates. Therefore, in terms of the peridynamic terminology, the matrix $\tilde{\mathbf{K}}$ is the undeformed or reference shape tensor (represented as a matrix here), and $\tilde{\mathbf{S}}$ is the deformed shape tensor. Here, quadrature weights in summations have been omitted since we have proceeded directly analogous to the discrete RK approximation. Nevertheless, the exact equivalence can be established easily by including weights [40].

Remark 3.1 *An important recent connection between the peridynamic gradient approximation (3.150) and corrected smoothed particle hydrodynamics (SPH) [41] was given by Ganzenmüller et al. [42]. They showed that these formulations are equivalent in their approximations, which bridges the developments in this section to SPH.*

Remark 3.2 *By employing $m = 1$, the unity term in the "basis" $\mathbf{q}(\mathbf{x})$ of peridyanmics is omitted, and the partition of nullity is not satisfied explicitly as in RK and MLS. However, the use of $u(\mathbf{x}_I) - u(\mathbf{x})$ in (3.141) compensates for this, since when $u =$ constant, derivative approximations yield the correct value of zero. GFDs that do not include the primary variable as an unknown share this feature: the above version of peridynamics is a GFD.*

Since (3.150) is a special case of the more general approximation (3.141), high-order accurate state-based peridynamic derivative approximations for correspondence can be obtained by increasing the order of the monomials in (3.143), that is, leaving $m = 1$, but taking $n > 1$. Using this approach for solving PDEs in the strong form including a non-local divergence of stress was termed reproducing kernel peridynamics (RKPD) in [40], which has subsequently been developed by Behzadinasab and co-workers including fracture of shell structures [43–45]. Note that this approximation is distinct from the so-called peridynamic differential operator [46], which is identical to implicit gradients, see [40].

The late Steve Attaway emphasized the importance of the equivalence of the two methods discussed above: under this unification, state-based peridynamic codes can be extended to use local meshfree approximation methods, while local meshfree codes can be extended to use non-local peridynamics.

Key Takeaways

In the computation of meshfree approximation of derivatives:

1) Diffuse, implicit, and synchronized derivatives all use the same approximation. Generalized finite differences are also equivalent when the least-squares measure is weighted by the distance.
2) Both direct and diffuse derivatives satisfy the nth-order derivative reproducing conditions.
3) Direct differentiation is more expensive, but generally yields lower error.
4) State-based peridynamics with correspondence is distinct from diffuse derivatives. They can be made to be high-order accurate by extending the order of monomials used in the shape tensors.

References

1 Lucy, L.B. (1977). A numerical approach to the testing of the fission hypothesis. *Astron. J.* 82: 1013–1024.

2 Gingold, R.A. and Monaghan, J.J. (1977). Smoothed particle hydrodynamics: theory and application to non-spherical stars. *Mon. Not. R. Astron. Soc.* 181 (3): 375–389.

3 Lancaster, P. and Salkauskas, K. (1981). Surfaces generated by moving least squares methods. *Math. Comput.* 37: 141–158.

4 Liszka, T.J. and Orkisz, J. (1980). The finite difference method at arbitrary irregular grids and its application in applied mechanics. *Comput. Struct.* 11 (1–2): 83–95.

5 Nayroles, B., Touzot, G., and Villon, P. (1992). Generalizing the finite element method: diffuse approximation and diffuse elements. *Comput. Mech.* 10 (5): 307–318.

6 Han, W. and Meng, X. (2001). Error analysis of the reproducing kernel particle method. *Comput. Methods Appl. Mech. Eng.* 190 (46–47): 6157–6181.

7 Monaghan, J.J. (1982). Why particle methods work. *SIAM J. Sci. Stat. Comput.* 3 (4): 422–433.

8 Monaghan, J.J. (1988). An introduction to SPH. *Comput. Phys. Commun.* 48 (1): 89–96.

9 Liu, W.K., Jun, S., and Zhang, Y.F. (1995). Reproducing kernel particle methods. *Int. J. Numer. Methods Fluids.* 20 (8–9): 1081–1106.

10 Chen, J.-S., Wang, H.-P., Yoon, S., and You, Y. (2000). Some recent improvements in meshfree methods for incompressible finite elasticity boundary value problems with contact. *Comput. Mech.* 25: 137–156.

11 Chen, J.-S., Pan, C., Wu, C.-T., and Liu, W.K. (1996). Reproducing kernel particle methods for large deformation analysis of non-linear structures. *Comput. Methods Appl. Mech. Eng.* 139 (1–4): 195–227.

12 Chen, J.-S., Pan, C., Roque, C.M.O.L., and Wang, H.-P. (1998). A Lagrangian reproducing kernel particle method for metal forming analysis. *Comput. Mech.* 22 (3): 289–307.

13 Liu, W.K., Jun, S., Li, S. et al. (1995). Reproducing kernel particle methods for structural dynamics. *Int. J. Numer. Methods Eng.* 38: 1655–1679.

14 Krongauz, Y. and Belytschko, T. (1996). Enforcement of essential boundary conditions in meshless approximations using finite elements. *Comput. Methods Appl. Mech. Eng.* 131 (1–2): 133–145.

15 Fernández-Méndez, S. and Huerta, A. (2004). Imposing essential boundary conditions in mesh-free methods. *Comput. Methods Appl. Mech. Eng.* 193 (12): 1257–1275.

16 Gosz, J. and Liu, W.K. (1996). Admissible approximations for essential boundary conditions in the reproducing kernel particle method. *Comput. Mech.* 19: 120–135.

17 Kaljevic, I. and Saigal, S. (1997). An improved element free Galerkin formulation. *Int. J. Numer. Methods Eng.* 40 (16): 2953–2974.

18 Chen, J.-S., Han, W., You, Y., and Meng, X. (2003). A reproducing kernel method with nodal interpolation property. *Int. J. Numer. Methods Eng.* 56 (7): 935–960.

19 Chen, J.-S. and Wang, H.-P. (2000). New boundary condition treatments in meshfree computation of contact problems. *Comput. Methods Appl. Mech. Eng.* 187 (3–4): 441–468.

20 Atluri, S.N. and Zhu, T.L. (1998). A new meshless local Petrov–Galerkin (MLPG) approach in computational mechanics. *Comput. Mech.* 22 (2): 117–127.

21 Zhu, T.L. and Atluri, S.N. (1998). A modified collocation method and a penalty formulation for enforcing the essential boundary conditions in the element free Galerkin method. *Comput. Mech.* 21 (3): 211–222.

22 Nitsche, J. (1971). Über ein Variationsprinzip zur Lösung von Dirichlet-Problemen bei Verwendung von Teilräumen, die keinen Randbedingungen unterworfen sind. *Abhandlungen Aus Dem Math. Semin. Der Univ. Hambg.* 36: 9–15.

23 Griebel, M. and Schweitzer, M.A. A particle-partition of unity method part V: boundary conditions. In: *Geometric Analysis and Nonlinear Partial Differential Equations* (ed. S. Hildebrandt and H. Karcher). Berlin, Heidelberg: Springer.

24 Babuška, I., Banerjee, U., and Osborn, J.E. (2003). Meshless and generalized finite element methods: a survey of some major results. In: *Meshfree methods for partial differential equations*, 1–20. Berlin, Heidelberg: Springer.

25 Lu, Y.Y., Belytschko, T., and Gu, L. (1994). A new implementation of the element free Galerkin method. *Comput. Methods Appl. Mech. Eng.* 113 (3–4): 397–414.

26 Belytschko, T., Lu, Y.Y., and Gu, L. (1994). Element-free Galerkin methods. *Int. J. Numer. Methods Eng.* 37: 229–256.

27 Belytschko, T., Krongauz, Y., Organ, D. et al. (1996). Meshless methods: an overview and recent developments. *Comput. Methods Appl. Mech. Eng.* 139 (1–4): 3–47.

28 Liu, W.K., Li, S., and Belytschko, T. (1997). Moving least-square reproducing kernel methods (I) Methodology and convergence. *Comput. Methods Appl. Mech. Eng.* 143 (1–2): 113–154.

29 Hu, H.-Y., Lai, C.-K., and Chen, J.-S. (2009). A study on convergence and complexity of reproducing kernel collocation method. *Interact. Multiscale Mech.* 2 (3): 295–319.

30 Chen, J.-S., Zhang, X., and Belytschko, T. (2004). An implicit gradient model by a reproducing kernel strain regularization in strain localization problems. *Comput. Methods Appl. Mech. Eng.* 193 (27–29): 2827–2844.

31 Li, S. and Liu, W.K. (1998). Synchronized reproducing kernel interpolant via multiple wavelet expansion. *Comput. Mech.* 21: 28–47.

32 Li, S. and Liu, W.K. (1999). Reproducing kernel hierarchical partition of unity, part I - formulation and theory. *Int. J. Numer. Methods Eng.* 288: 251–288.

33 Li, S. and Liu, W.K. (1999). Reproducing kernel hierarchical partition of unity, Part II – applications. *Int. J. Numer. Methods Eng.* 45 (3): 289–317.

34 Chi, S.-W., Chen, J.-S., Hu, H.-Y., and Yang, J.P. (2013). A gradient reproducing kernel collocation method for boundary value problems. *Int. J. Numer. Methods Eng.* 93 (13): 1381–1402.

35 Liszka, T.J. (1984). An interpolation method for an irregular net of nodes. *Int. J. Numer. Methods Eng.* 20 (9): 1599–1612.

36 Jensen, P.S. (1972). Finite difference techniques for variable grids. *Comput. Struct.* 2 (1–2): 17–29.

37 Bessa, M.A., Foster, J.T., Belytschko, T., and Liu, W.K. (2014). A meshfree unification: reproducing kernel peridynamics. *Comput. Mech.* 53 (6): 1251–1264.

38 Savitzky, A. and Golay, M.J.E. (1964). Smoothing and differentiation of data by simplified least squares procedures. *Anal. Chem.* 36 (8): 1627–1639.

39 Silling, S.A., Epton, M.A., Weckner, O. et al. (2007). Peridynamic states and constitutive modeling. *J. Elast.* 88 (2): 151–184.

40 Hillman, M., Pasetto, M., and Zhou, G. (2020). Generalized reproducing kernel peridynamics: unification of local and non-local meshfree methods, non-local derivative operations, and an arbitrary-order state-based peridynamic formulation. *Comput. Part. Mech.* 7: 435–469.

41 Randles, P.W. and Libersky, L.D. (1996). Smoothed particle hydrodynamics: some recent improvements and applications. *Comput. Methods Appl. Mech. Eng.* 139 (1–4): 375–408.

42 Ganzenmüller, G.C., Hiermaier, S., and May, M. (2015). On the similarity of meshless discretizations of peridynamics and smooth-particle hydrodynamics. *Comput. & Struct.* 150: 71–78.

43 Behzadinasab, M., Trask, N., and Bazilevs, Y. (2021). A unified, stable and accurate meshfree framework for peridynamic correspondence modeling – Part I: core methods. *J. Peridynamics Nonlocal Model.* 3 (1): 24–45.

44 Behzadinasab, M., Foster, J.T., and Bazilevs, Y. (2021). A unified, stable, and accurate meshfree framework for peridynamic correspondence modeling – Part II: wave propagation and enforcement of stress boundary conditions. *J. Peridynamics Nonlocal Model.* 3 (1): 46–66.

45 Behzadinasab, M., Alaydin, M., Trask, N., and Bazilevs, Y. (2022). A general-purpose, inelastic, rotation-free Kirchhoff-Love shell formulation for peridynamics. *Comput. Methods Appl. Mech. Eng.* 389: 114422.

46 Madenci, E., Barut, A., and Dorduncu, M. (2019). *Peridynamic Differential Operator for Numerical Analysis.* Cham, Switzerland: Springer Nature.

4

Solving PDEs with Galerkin Meshfree Methods

We next consider the discretization of partial differential equations (PDEs) by Galerkin meshfree methods. The domain of the problem is denoted by Ω, and its boundary is denoted by Γ. Distributed in the domain Ω is a set of nodes $S = \{\mathbf{x}_I\}_{I=1}^{NP}$, where NP is the number of points (nodes). In most cases, the distribution of the nodes is quite uniform. The distinct advantages of a meshfree method arise when non-uniformity is introduced by processes such as h-adaptivity or large motions of the nodes. Solving PDEs with the reproducing kernel (RK) approximation in the Galerkin method has been termed the reproducing kernel particle method (RKPM); with moving least squares (MLS), it has been termed the element-free Galerkin method (EFG).

In a meshfree method, the body's boundary must be described in some manner. In finite element methods, the boundaries are described by the positions of the outermost edges. This approach cannot be used in meshfree methods since there are no explicit edges in a point cloud. An approach similar to the finite element description is to label the boundary nodes and then connect them by straight or curved lines. However, describing the boundaries by either computer-aided design ("CAD") data or level sets is preferable. Then there is no need to place any nodes coincident with the boundaries unless strong enforcement of essential boundary conditions at nodal locations is employed (see Chapter 5).

4.1 Linear Diffusion Equation

We will first describe the discretization of the linear diffusion equation. The boundary Γ is subdivided into two parts Γ_u and Γ_q where the primary variable u and its flux q are prescribed, respectively. The two parts of the boundary are complementary in the sense that $\Gamma_u \cap \Gamma_q = \emptyset$ and $\Gamma_u \cup \Gamma_q = \Gamma$. The strong form of the steady-state isotropic diffusion equation is given by (see Section 2.1.1)

$$k\nabla^2 u + s = 0 \quad \text{in} \quad \Omega, \tag{4.1}$$

$$k\nabla u \cdot \mathbf{n} = \overline{q} \quad \text{on} \quad \Gamma_q, \tag{4.2}$$

$$u = \overline{u} \quad \text{on} \quad \Gamma_u, \tag{4.3}$$

where k is the diffusivity, u is the unknown, and s is the source term.

For the boundary conditions on the primary variable, u, the conditions in Eq. (4.3) are called essential boundary conditions in Galerkin methods and Dirichlet conditions in much of the mathematical literature. The conditions on the derivatives of u in (4.2) are called natural boundary

Meshfree and Particle Methods: Fundamentals and Applications, First Edition.
Ted Belytschko, J. S. Chen, and Michael Hillman.

conditions in Galerkin methods since they are satisfied naturally in the weak form; they need not be imposed separately.

A key ingredient in a Galerkin method is a weak form equivalent to the PDE. The weak form for Eqs. (4.1)–(4.3) was developed in Section 2.1.2.1. We repeat it here for convenience:

$$\text{Find } u \in \mathrm{U} \text{ such that } \delta\Pi_{\mathrm{D}} = 0 \,\forall\, \delta u \in \mathrm{U}, \tag{4.4}$$

with

$$\delta\Pi_{\mathrm{D}} = \int_{\Omega} \nabla\delta u \cdot k\nabla u \,\mathrm{d}\Omega - \int_{\Omega} \delta u s \,\mathrm{d}\Omega - \int_{\Gamma_q} \delta u \bar{q} \,\mathrm{d}\Gamma = 0. \tag{4.5}$$

The function u is called the trial function, and the function δu is called the test function; in almost all discussions in this book, the test function is equivalent to the variation of u (see Chapter 2), so we adopt this notation. The set of trial and test functions that meet the conditions in the function spaces $\mathrm{U} = \{u(\mathbf{x}) | u(\mathbf{x}) \in \mathrm{H}^1 \text{ and } u(\mathbf{x}) = \bar{u} \text{ on } \Gamma_u\}$ and $\mathrm{U}_0 = \{\delta u(\mathbf{x}) | \delta u(\mathbf{x}) \in \mathrm{H}^1 \text{ and } \delta u(\mathbf{x}) = 0 \text{ on } \Gamma_u\}$, respectively, are called admissible; these admissibility conditions are given in detail in Chapter 2.

In a Galerkin formulation of the finite element method, the approximations for the trial function $u(\mathbf{x})$ and the test function $\delta u(\mathbf{x})$ are constructed in terms of the element parameters of the approximation. In the case of a meshfree method, they are constructed in terms of the meshfree shape functions $\Psi_I(\mathbf{x})$ in terms of nodal parameters. For purposes of clarity, we will first consider the case where the approximations for the test and trial functions are assumed to satisfy the admissibility conditions, so that we can construct an approximation similar to the finite element method.

With the above assumptions, a set of admissible trial solutions is formulated by the sum of the meshfree shape functions as follows

$$u(\mathbf{x}) \simeq u^h(\mathbf{x}) = \sum_{I \in S_{\mathbf{x}}} \Psi_I(\mathbf{x}) u_I, \tag{4.6}$$

where $S_{\mathbf{x}} = \{I \mid \mathbf{x} \in \operatorname{supp} \Psi_I(\mathbf{x})\}$ is the set of nodes where the corresponding shape function supports cover $\mathbf{x}$. The test functions are constructed with the same set of meshfree shape functions:

$$\delta u(\mathbf{x}) \simeq \delta u^h(\mathbf{x}) = \sum_{I \in S_{\mathbf{x}}} \Psi_I(\mathbf{x}) \delta u_I. \tag{4.7}$$

In the above equations, u_I and δu_I are the coefficients of the trial and test functions, respectively. It is assumed that both of the approximations satisfy the admissibility conditions, that is, $u^h(\mathbf{x}) = \bar{u}$ and $\delta u^h(\mathbf{x}) = 0$, for $\mathbf{x} \in \Gamma_u$, and are sufficiently integrable. The former can be accomplished by using special techniques described in Chapter 5.

In general, the meshfree shape functions are not kinematically admissible, and the imposition of the conditions on the test and trial functions on the essential boundaries is more complicated than in the finite element method, as will be discussed in the following subsections.

We now insert these finite-dimensional approximations for the test and trial functions in (4.6) and (4.7) into the weak form (4.5) to yield:

$$\sum_{I \in S} \delta u_I \left\{ \int_{\Omega} \nabla\Psi_I \cdot \sum_{J \in S} k \nabla\Psi_J u_J \,\mathrm{d}\Omega - \int_{\Omega} \Psi_I s \,\mathrm{d}\Omega - \int_{\Gamma_q} \Psi_I \bar{q} \,\mathrm{d}\Gamma \right\} = 0, \tag{4.8}$$

where S is the set of all nodes. Here, and throughout this chapter, summations are often extended from $I \in S_{\mathbf{x}}$ to all nodes $I \in S$ that discretize the domain and the boundaries; this extension is needed for writing matrix forms and related equations. Since Eq. (4.8) must hold for arbitrary δu_I, it follows that the quantity in the brackets must vanish for each value of I, i.e.,

$$\int_{\Omega} \nabla\Psi_I \cdot \sum_{J\in S} k\nabla\Psi_J u_J \,\mathrm{d}\Omega - \int_{\Omega} \Psi_I s \,\mathrm{d}\Omega - \int_{\Gamma_q} \Psi_I \bar{q} \,\mathrm{d}\Gamma = 0 \quad \forall I \in S. \tag{4.9}$$

The above is a set of linear algebraic equations for the unknowns u_J. In the meshfree literature, these are sometimes referred to as local equilibrium equations; it should be noted that these are the standard equations that result from a Galerkin approach. The above equation also results from what is known as the method of weighted residuals. If the weighted residual is applied with integration by parts, it is identical to a Galerkin method when the problem has a potential (see Chapter 2).

It is customary in meshfree and finite element methods to make the following definitions

$$K_{IJ} = \int_{\Omega} k\nabla\Psi_I \cdot \nabla\Psi_J \,\mathrm{d}\Omega, \tag{4.10}$$

$$f_I = \int_{\Omega} \Psi_I s \,\mathrm{d}\Omega + \int_{\Gamma_q} \Psi_I \bar{q} \,\mathrm{d}\Gamma. \tag{4.11}$$

The above are associated with the three terms in the brackets in (4.8). The conditions that this quantity must vanish for each of the test functions in (4.8), δu_I, $I \in S$ then yields

$$\sum_{J\in S} K_{IJ} u_J = f_I, \quad \forall I \in S, \tag{4.12}$$

or in matrix form

$$\mathbf{Kd} = \mathbf{f}, \tag{4.13}$$

where $\mathbf{K} = [K_{IJ}]$, $\mathbf{f} = \{f_I\}$, $\mathbf{d} = \{u_I\}$; $\mathbf{K}$ is called the conductance matrix in heat conduction and $\mathbf{f}$ is the vector of nodal fluxes and heat source. This is a set of NP linear algebraic equations for the unknowns u_I, $I = 1$ to NP, that is, for $I \in S$. As can be seen from the structure of (4.10), the matrix in (4.13) is symmetric, i.e., $K_{IJ} = K_{JI}$. The matrix $\mathbf{K}$ is also positive semi-definite, and it becomes positive definite after the essential boundary conditions are imposed.

Note that the only nonzero entries in K_{IJ} correspond to the intersection of the supports of nodes I and J. Therefore, the discrete meshfree equations will be sparse, just as the discrete finite element equations. The sparsity diminishes if the size of the support is increased. Consequently, it is advantageous to use a smaller support for computational efficiency, subjected to the support size conditions as discussed in Section 3.4. Large supports can sometimes increase the accuracy of a meshfree approximation, but special solution methods that can handle effectively the resulting larger bandwidth are needed, such as parallel iterative solvers.

Another important related aspect is identifying neighbors for the integration points. Each point will have a list of non-zero contributions to the force vector and stiffness matrix, which must be considered for a code to be of any use (otherwise the CPU time will be immense during assembly and shape function computation). Section 10.3.3 discusses various efficient neighbor search algorithms for arbitrary point distributions and algorithms that consider only the non-zero contributions during assembly and shape function construction. This aspect is also critical for the semi-Lagrangian method as described in Section 7.5, where meshfree shape functions are constructed on the fly, along with their neighbor lists.

Equation (4.9) is sometimes called a local weak form in meshfree methods. It should be pointed out though that any test functions with compact support will result in this form, including the finite element method. The equations in (4.9) are a suitable approximation to the PDE (4.1) only if Eq. (4.9) holds for all δu_I.

It can be noted that the row vector components of the conductance matrix for each shape function (i.e., the Ith row of $\mathbf{K}$) are nonzero only on the support of Ψ_I. Therefore, equations in this form offer some interesting options for integrating the terms in space. Atluri and Zhu [1] and De and Bathe [2] have developed a nice way of dealing with quadrature. Atluri and Zhu called their method the meshless local Petrov–Galerkin method (MLPG), and De and Bathe called their method finite spheres, but they are in most cases Galerkin methods with moving least squares or RK approximations. The essential idea in these methods is to take advantage of the fact that in the discrete equations, each equation (or group of equations for a vector problem) has a nonzero contribution only over the support of the test function. Thus, the integral is evaluated over a subdomain that arises naturally in the meshfree approximation. Note that in these methods subdomain integration is needed for every test function's row of the stiffness matrix and force vector, and this process could be very costly for large discrete systems. Efficient quadrature methods that can avoid a large number of evaluation points in meshfree methods are discussed in Chapter 6.

Since most meshfree approximations are not readily constructed to be admissible, it is often useful to use methods where no "strong" kinematically admissible conditions need to be imposed on the trial or test functions on the essential boundaries. We will describe three methods for imposing the essential boundary conditions in the weak form in this chapter:

1) The penalty method
2) The Lagrange multiplier method
3) Nitsche's method

Nevertheless, it is possible to construct (nodally) kinematically admissible meshfree approximations for a direct imposition of essential boundary conditions similar to the finite element method, which will be discussed in Chapter 5.

4.1.1 Penalty Method for the Diffusion Equation

We next describe how the essential boundary conditions can be applied by the penalty method. The weak form is given in Section 2.1.3.1:

$$\text{Find } u \in H^1, \text{ such that } \delta\Pi_D^P = 0 \ \forall \delta u \in H^1, \tag{4.14}$$

where

$$\Pi_D^P = \Pi_D + \frac{\beta}{2}\int_{\Gamma_u} (u - \bar{u})^2 d\Gamma, \tag{4.15}$$

with

$$\delta\Pi_D^P = \delta\Pi_D + \beta\int_{\Gamma_u} \delta u(u - \bar{u})\, d\Gamma. \tag{4.16}$$

We use the same trial and test functions in (4.6)–(4.7). Substituting these equations into (4.16) gives

$$\sum_{I\in S}\delta u_I\left(\sum_{J\in S}K_{IJ}u_J - f_I\right) + \sum_{I\in S}\delta u_I\left(\sum_{J\in S}A_{IJ}u_J - g_I\right) = 0, \tag{4.17}$$

where

$$\begin{aligned} K_{IJ} &= \int_{\Omega} k\nabla\Psi_I \cdot \nabla\Psi_J \, d\Omega, \\ A_{IJ} &= \beta \int_{\Gamma_u} \Psi_I \Psi_J d\Gamma, \\ f_I &= \int_{\Omega} \Psi_I s \, d\Omega + \int_{\Gamma_q} \Psi_I \overline{q} d\Gamma, \\ g_I &= \beta \int_{\Gamma_u} \Psi_I \overline{u} d\Gamma. \end{aligned} \tag{4.18}$$

Since Eq. (4.17) must hold for arbitrary δu_I, it follows that the quantities in the brackets must vanish for each value of I, i.e.,

$$\sum_{J \in S} (K_{IJ} + A_{IJ}) u_J = f_I + g_I \quad \forall I \in S. \tag{4.19}$$

If we define matrices and vectors $\mathbf{K} = [K_{IJ}]$, $\mathbf{A} = [A_{IJ}]$, $\mathbf{f} = \{f_I\}$, $\mathbf{g} = \{g_I\}$, and $\mathbf{d} = \{u_I\}$, then the above can be written as

$$(\mathbf{K} + \mathbf{A})\mathbf{d} = \mathbf{f} + \mathbf{g}. \tag{4.20}$$

It can be seen that no additional degrees of freedom are needed in the penalty method, as compared to the Lagrange multiplier method, where discretization of the multiplier is needed, as shown in the next section. However, in the penalty method, the parameter β has to be sufficiently large with respect to k for essential boundary conditions $u = \overline{u}$ to be imposed (with a slight discrepancy due to the term $\frac{1}{\beta} k u_{,n}$ in (2.42) as discussed in Section 2.1.3.1). An excessively large penalty parameter β, on the other hand, could yield ill conditioning of the discrete equations in (4.20) and consequently, an unstable solution. Therefore, an appropriate selection of β is essential to obtaining a reliable solution in the penalty method. In [3], numerical testing showed that accurate and convergent solutions could be obtained if $\beta \sim h^{-2}$ with h the nodal spacing, scaled by a sufficiently large constant, but also that the condition number of the left-hand side matrix scales logarithmically with the constant. This illustrates the essential difficulty with the penalty method. In a general setting, β should also be proportional to the constants in the PDE at hand [4].

Key Takeaways

For the penalty method:

1) There are no extra degrees of freedom (compared to the Lagrange multiplier approach), and the method is straightforward.
2) A sufficiently large parameter needs to be used to properly enforce essential boundary conditions and avoid excessive solution error.
3) A parameter too large leads to ill conditioning of the system matrix.
4) Numerical experiments show the parameter should be selected with $\beta \sim h^{-2}$ for convergence.

4.1.2 The Lagrange Multiplier Method for the Diffusion Equation

The weak form for the diffusion equation using the Lagrange multiplier method is developed in Section 2.1.3.2. The trial and test functions need not be admissible except for the integrability conditions. For convenience, we recall the weak form here:

$$\text{Find } u \in \mathrm{H}^1,\ \lambda \in \mathrm{L}_2, \text{ such that } \delta\Pi_{\mathrm{D}}^{\mathrm{L}} = 0\ \forall\ \delta u \in \mathrm{H}^1,\ \delta\lambda \in \mathrm{L}_2, \tag{4.21}$$

where

$$\Pi_{\mathrm{D}}^{\mathrm{L}} = \Pi_{\mathrm{D}} + \int_{\Gamma_u} \lambda(u - \overline{u})\, d\Gamma, \tag{4.22}$$

with

$$\delta\Pi_{\mathrm{D}}^{\mathrm{L}} = \delta\Pi_{\mathrm{D}} + \int_{\Gamma_u} \delta\lambda(u - \overline{u})\, d\Gamma + \int_{\Gamma_u} \delta u\, \lambda\, d\Gamma. \tag{4.23}$$

As can be seen from (4.21), no admissibility conditions are imposed on the test or trial functions other than integrability.

We also need to construct approximations for the Lagrange multiplier test and trial functions. Since these functions only appear in boundary integrals, we can use several approaches to construct these approximations. In most early meshfree work, the problems were two-dimensional, and Lagrange interpolants approximated the Lagrange multipliers along the boundary. Extension of this technique to three dimensions would require the construction of a surface mesh on the body. An alternative technique is to use either the same meshfree shape functions, or a different set of meshfree shape functions constructed using only the nodes on the boundary.

Here, we write these approximations in terms of an alternate generic set of shape functions $\psi_\alpha(\boldsymbol{\xi})$, where $\boldsymbol{\xi}$ are the coordinates that parameterize the edge $\boldsymbol{\xi} = [\xi]$ in 2D and the surface $\boldsymbol{\xi} = [\xi_1, \xi_2]$ in 3D. The approximation of the Lagrange multiplier field is then given by

$$\lambda(\boldsymbol{\xi}) \simeq \lambda^h(\boldsymbol{\xi}) = \sum_{\alpha \in S_u} \psi_\alpha(\boldsymbol{\xi})\lambda_\alpha, \tag{4.24}$$

where S_u is the set of nodes associated with the essential boundary where $u(\mathbf{x})$ is prescribed. Since the shape functions typically have compact support, the approximation can be expressed more concisely as

$$\lambda^h(\boldsymbol{\xi}) = \sum_{\alpha \in S_{\boldsymbol{\xi}} \cap S_u} \psi_\alpha(\boldsymbol{\xi})\lambda_\alpha, \tag{4.25}$$

where $S_{\boldsymbol{\xi}} = \{\alpha \mid \psi_\alpha(\boldsymbol{\xi}) \neq 0\}$, and the approximation $\lambda^h(\boldsymbol{\xi})$ in (4.25) involves only nodes on the essential boundary where the supports cover the evaluation location $\boldsymbol{\xi}$, hence nodes where $\alpha \in S_{\boldsymbol{\xi}} \cap S_u$. This field must satisfy the Babuska–Brezzi (BB) condition [5, 6] if the approximation is to be stable and convergent.

The test function for the Lagrange multiplier (or the variation) is similarly approximated by

$$\delta\lambda(\boldsymbol{\xi}) \simeq \delta\lambda^h(\boldsymbol{\xi}) = \sum_{\alpha \in S_{\boldsymbol{\xi}} \cap S_u} \psi_\alpha(\boldsymbol{\xi})\delta\lambda_\alpha. \tag{4.26}$$

To obtain the discrete equation, test and trial functions for u in (4.6)–(4.7) and the Lagrange multiplier λ in (4.24)–(4.26) are substituted into (4.23). This gives

$$\delta\Pi_{\mathrm{D}}^{\mathrm{L}} = \delta\Pi_{\mathrm{D}} + \int_{\Gamma_u} \sum_{\alpha \in S_u} \delta\lambda_\alpha\psi_\alpha \left(\sum_{I \in S} \Psi_I u_I - \overline{u} \right) d\Gamma + \int_{\Gamma_u} \sum_{I \in S} \delta u_I \Psi_I \left(\sum_{\alpha \in S_u} \psi_\alpha \lambda_\alpha \right) d\Gamma = 0, \tag{4.27}$$

where the sets for the approximations have been expanded to the entire node sets. Equation (4.27) can then be written as

$$\begin{aligned}\delta\Pi_{\mathrm{D}}^{\mathrm{L}} = &\sum_{I\in S}\delta u_I\sum_{J\in S}K_{IJ}u_J - \sum_{I\in S}\delta u_I f_I + \sum_{\alpha\in S_u}\delta\lambda_\alpha\sum_{I\in S}G_{I\alpha}u_I - \sum_{\alpha\in S_u}\delta\lambda_\alpha\gamma_\alpha \\ &+ \sum_{I\in S}\delta u_I\sum_{\alpha\in S_u}G_{I\alpha}\lambda_\alpha = 0,\end{aligned} \tag{4.28}$$

where

$$\begin{aligned} K_{IJ} &= \int_\Omega k\nabla\Psi_I\cdot\nabla\Psi_J\,\mathrm{d}\Omega, \\ f_I &= \int_\Omega \Psi_I s\,\mathrm{d}\Omega + \int_{\Gamma_q}\Psi_I\bar{q}\,d\Gamma, \\ G_{I\alpha} &= \int_{\Gamma_u}\Psi_I\psi_\alpha\,d\Gamma, \\ \gamma_\alpha &= \int_{\Gamma_u}\psi_\alpha\bar{u}\,d\Gamma. \end{aligned} \tag{4.29}$$

Invoking the arbitrariness of δu_I and $\delta\lambda_\alpha$, we obtain two sets of equations

$$\sum_{J\in S}K_{IJ}u_J + \sum_{\alpha\in S_u}G_{I\alpha}\lambda_\alpha - f_I = 0 \quad \forall I\in S, \tag{4.30}$$

$$\sum_{I\in S}G_{I\alpha}u_I - \gamma_\alpha = 0 \quad \forall\alpha\in S_u. \tag{4.31}$$

If we define matrices and vectors $\mathbf{K} = [K_{IJ}]$, $\mathbf{G} = [G_{I\alpha}]$, $\mathbf{f} = \{f_I\}$, $\boldsymbol{\gamma} = \{\gamma_\alpha\}$, $\mathbf{d} = \{u_I\}$, and $\boldsymbol{\ell} = \{\lambda_\alpha\}$, then the above can be written as

$$\begin{bmatrix}\mathbf{K} & \mathbf{G}\\ \mathbf{G}^{\mathrm{T}} & \mathbf{0}\end{bmatrix}\begin{Bmatrix}\mathbf{d}\\ \boldsymbol{\ell}\end{Bmatrix} = \begin{Bmatrix}\mathbf{f}\\ \boldsymbol{\gamma}\end{Bmatrix}. \tag{4.32}$$

This is the standard form of a Lagrange multiplier problem. Note that the imposition of the essential boundary conditions is consistently enforced by this method, but it results in an increase in the number of unknowns and number of equations. Furthermore, the equations are no longer positive definite, and this limits solver selection. And, as previously mentioned, Lagrange multiplier methods also require that the BB conditions be met [5, 6], which in practice may cause stability issues.

Figure 4.1 shows the solution of the diffusion equation by RKPM using the Lagrange multiplier approach to enforce essential boundary conditions. The approximation of u in the domain is performed using RK shape functions with order n, while the multiplier λ on the boundary is approximated with RK shape functions with order $n-1$. Optimal rates are achieved ($n+1$ in the L_2 norm and n in the semi-H^1 norm, see Section 3.4) when the nodal spacing for the Lagrange multiplier is equal to the meshfree node spacing. However, when the density is increased by two-fold (2× finer), the solution is no longer stable and subsequently, convergence is not achieved, demonstrating the potential issue with this method.

In practice, we have observed that this pair of approximations with order n and $n-1$ for u and λ, respectively, using equal nodal spacings, can often be employed without issue (as observed in Figure 4.1). Nevertheless, BB conditions are required to ensure stability in the Lagrange multiplier method, and this motivates the introduction of Nitsche's method for enforcing boundary conditions where stability can be ensured.

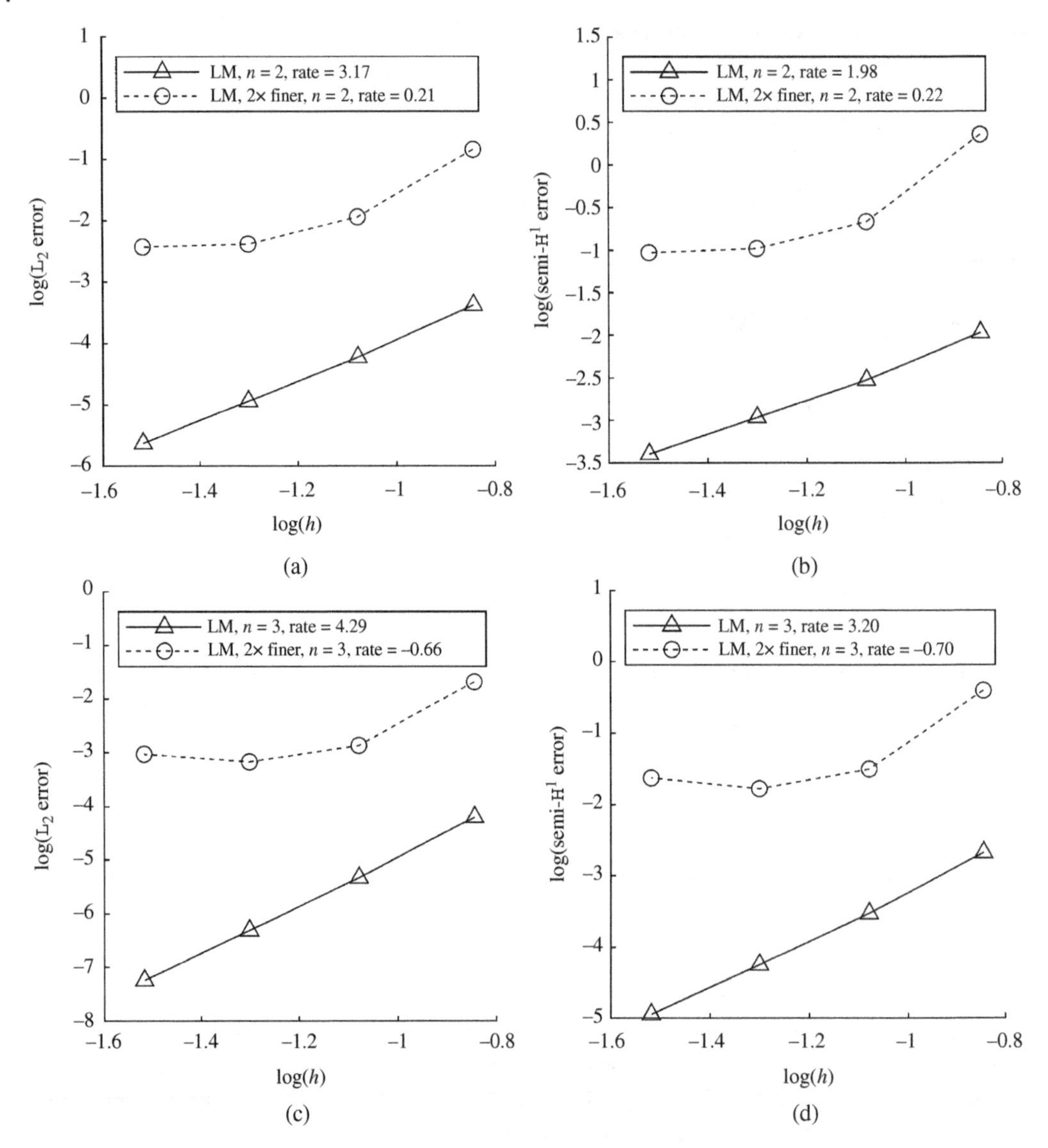

Figure 4.1 Covergence using Lagrange multipliers for essential boundary conditions: (a) $n = 2$, L_2 norm, (b) $n = 2$, H^1 semi-norm, (c) $n = 3$, L_2 norm, and (d) $n = 3$, H^1 semi-norm.

Key Takeaways

For the Lagrange multiplier method:

1) There are no parameters to select (compared to the penalty and Nitsche approaches).
2) The multipliers must be approximated which leads to extra degrees of freedom.
3) The method is subject to the BB condition; if the pair of spaces for the variables is not chosen properly, the solution will not be stable or convergent.
4) The system matrix is positive semi-definite and the solver needs to be selected accordingly.

4.1.3 Nitsche's Method for the Diffusion Equation

In Nitsche's method [3, 7], several terms are added to the weak form so that the essential boundary conditions are exactly enforced in the weak form, as shown in Chapter 2. Neither test functions nor trial functions need to satisfy the usual kinematically admissible conditions on the essential boundaries.

We first discuss Nitsche's method for the diffusion equation, which was developed in Section 2.1.3.3. The Nitsche weak form is

$$\text{Find } u \in \mathrm{H}^1 \text{ such that } \delta\Pi_{\mathrm{D}}^{\mathrm{N}} = 0 \ \forall \delta u \in \mathrm{H}^1, \tag{4.33}$$

where

$$\Pi_{\mathrm{D}}^{\mathrm{N}} = \Pi_{\mathrm{D}} \underbrace{- \int_{\Gamma_u} u_{,n} k(u - \overline{u}) d\Gamma + \frac{\beta}{2} \int_{\Gamma_u} (u - \overline{u})^2 d\Gamma}_{\Pi_{\mathrm{D}}^{\mathrm{Nit}}}, \tag{4.34}$$

with

$$\delta\Pi_{\mathrm{D}}^{\mathrm{N}} = \delta\Pi_{\mathrm{D}} + \delta\Pi_{\mathrm{D}}^{\mathrm{Nit}} = \delta\Pi_{\mathrm{D}} - \int_{\Gamma_u} \delta u k u_{,n} d\Gamma - \int_{\Gamma_u} \delta u_{,n} k(u - \overline{u}) d\Gamma + \beta \int_{\Gamma_u} \delta u (u - \overline{u}) d\Gamma. \tag{4.35}$$

The second and third terms on the right-hand side of (4.35) symmetrize the discrete equations; the fourth term is a penalty with a parameter β. Note that without the second and the third integrals on the right-hand side of (3.32), this reduces to a penalty method. However, the parameter in Nitsche's method can be much smaller than in the standard penalty methods, so the discrete equations will be better conditioned. This term is added to stabilize the discrete equations, and it actually improves conditioning.

For purposes of clarity, we segregate the Nitsche terms from those in the standard variational principle. The variation on $\Pi_{\mathrm{D}}^{\mathrm{Nit}}$ can be expressed as

$$\delta\Pi_{\mathrm{D}}^{\mathrm{Nit}} = \int_{\Gamma_u} (\delta u_{,n} k \overline{u} - \beta \delta u \overline{u}) d\Gamma - \int_{\Gamma_u} (\delta u_{,n} k u + \delta u (k u_{,n} - \beta u)) d\Gamma. \tag{4.36}$$

Substituting the test and trial functions of u in (4.6) and (4.7) into (4.36) yields

$$\begin{aligned} \delta\Pi_{\mathrm{D}}^{\mathrm{Nit}} &= \sum_{I \in S} \delta u_I \int_{\Gamma_u} (k \Psi_{I,n} \overline{u} - \beta \Psi_I \overline{u}) d\Gamma - \sum_{I \in S} \delta u_I \int_{\Gamma_u} \sum_{J \in S} (\Psi_{I,n} k \Psi_J + \Psi_I (k \Psi_{J,n} - \beta \Psi_J)) u_J d\Gamma \\ &= \sum_{I \in S} \delta u_I \left(\sum_{J \in S} K_{IJ}^{\mathrm{N}} u_J - f_I^{\mathrm{N}} \right), \end{aligned} \tag{4.37}$$

where

$$\begin{aligned} K_{IJ}^{\mathrm{N}} &= - \int_{\Gamma_u} (\Psi_{I,n} k \Psi_J + \Psi_I k \Psi_{J,n} - \beta \Psi_I \Psi_J) d\Gamma, \\ f_I^{\mathrm{N}} &= \int_{\Gamma_u} (- k \Psi_{I,n} \overline{u} + \beta \Psi_I \overline{u}) d\Gamma. \end{aligned} \tag{4.38}$$

The terms in Eq. (4.38) are additional contributions to the conductance matrix and the nodal fluxes that arise from the Nitsche implementation, and hence the superscript "N." Substituting (4.5), (4.8), and (4.37) into (4.34) and invoking the arbitrariness of δu_I gives

$$\sum_{J\in S}\left(K_{IJ} + K_{IJ}^{N}\right)u_J = f_I + f_I^{N} \quad \forall I \in S. \tag{4.39}$$

If we define the matrices and vectors $\mathbf{K} = [K_{IJ}]$, $\mathbf{K}^{N} = \left[K_{IJ}^{N}\right]$, $\mathbf{f} = \{f_I\}$, $\mathbf{f}^{N} = \left\{f_I^{N}\right\}$, and $\mathbf{d} = \{u_I\}$, then the above can be written in a matrix form as

$$\left(\mathbf{K} + \mathbf{K}^{N}\right)\mathbf{d} = \mathbf{f} + \mathbf{f}^{N}. \tag{4.40}$$

The additional terms, $\mathbf{K}^{N}$ and $\mathbf{f}^{N}$, are both boundary integrals. Note that the augmented conductance matrix $\mathbf{K} + \mathbf{K}^{N}$ is symmetric. This can be verified by interchanging I and J in (4.38), which gives the same result, and recalling that the conductance matrix $\mathbf{K}$ in (4.10) is symmetric. However, $\mathbf{K}^{N}$ is not necessarily positive definite, since the integrands $\Psi_{I,n}\Psi_J$ may be negative, and these terms have negative signs as shown in (4.38). Nevertheless, by making β large enough, positive definiteness can be recovered. Thus, the penalty serves as a stabilizing factor, but it need not be as large as the regular penalty parameters.

As noted in [3], Nitsche's method can be viewed as a consistent improvement of the penalty method. The penalty method is not consistent with the strong form since solutions will not satisfy the essential boundary conditions. The consistency of a penalty method is improved by increasing the penalty parameter, i.e., the essential boundary conditions are met more closely as the penalty parameter is increased (see Section 2.1.3.1), but this comes at the cost of poorer conditioning of the equations.

A simple way to properly select the parameter in Nitsche's method can be obtained via dimensional analysis of each term. The parameter should be on the order of $c_{\text{max, PDE}} \times h^{-1}$ where $c_{\text{max, PDE}}$ is the maximum of the problem constants in the equation at hand (e.g. diffusivity, Lamé's parameters) and h is the nodal spacing [3, 4], so that the left-hand side system matrix remains well-conditioned. Thus, the parameter employed should be selected as $\beta \sim c_{\text{max, PDE}} \times h^{-1}$, usually multiplied by another independent constant to ensure that the boundary conditions are sufficiently satisfied. In Nitsche's method, the independent constant can be much smaller than in the penalty method.

Figure 4.2 shows a comparison between Nitsche's method with and without scaling by h^{-1}. The diffusion constant is unity, and the results using $\beta = 50$ (fixed) are not stable or convergent. On the other hand, using $\beta = 50h^{-1}$, the solution is well-behaved and optimally convergent for the linear approximation employed (rates of two and one in L_2 and semi-H^1, see Section 3.4).

In [8], Griebel and Schweitzer showed that an eigenvalue analysis related to the discretization can yield a sharper estimate of this parameter if needed. However, in our experience, this procedure is only necessary to ensure good convergence behavior of the solution when higher-order bases are used in the meshfree approximation ($n > 1$); Figure 4.3 shows optimal convergence ($n + 1$ in the L_2 norm and n in the semi-H^1 norm, see Section 3.4) for order of approximations $n = 1, \ldots, 4$ using this estimate.

Key Takeaways

For Nitsche's method:

1) There are no extra degrees of freedom (compared to the Lagrange multiplier approach).
2) The penalty parameter is used to ensure stability rather than enforce boundary conditions, and it can be much smaller than the one used in the penalty approach.
3) The parameter should be scaled with the constants in the PDE and as $\beta \sim h^{-1}$.
4) A reliable way to obtain a parameter that ensures stability is to use the eigenvalue estimate in [8].

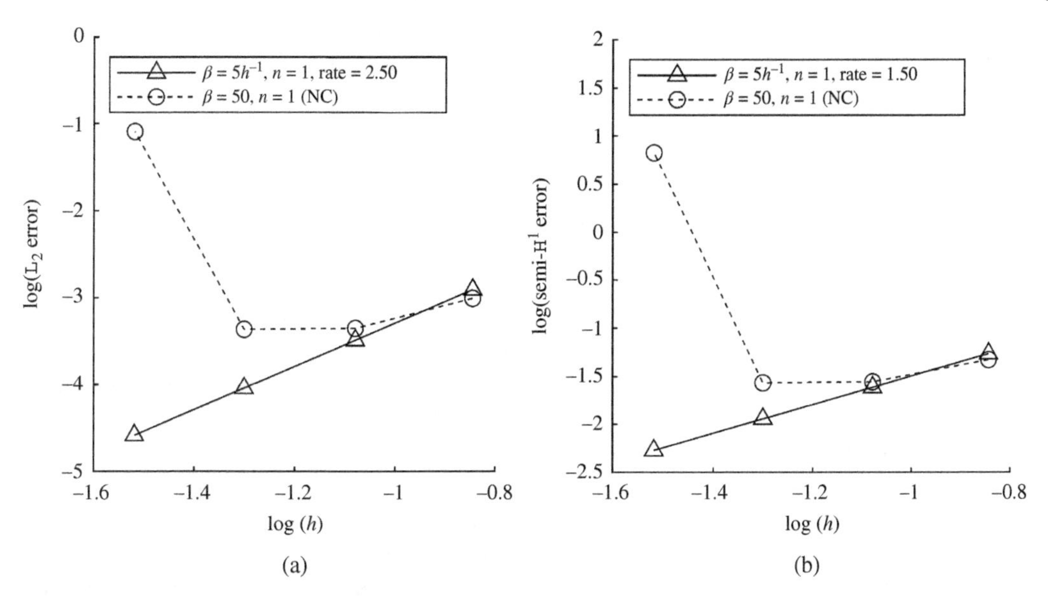

Figure 4.2 Convergence using Nitsche's method with and without scaling with h^{-1}: (a) L_2 error norm, and (b) semi-H^1 error norm. NC denotes no convergence.

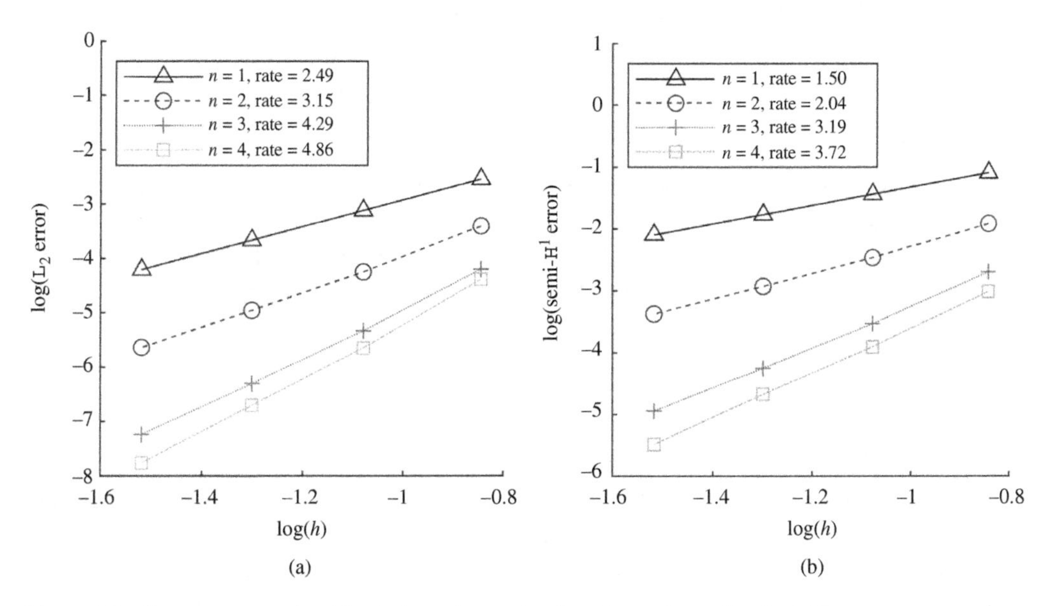

Figure 4.3 Convergence using Nitsche's method using the eigenvalue estimate: (a) L_2 error norm, and (b) semi-H^1 error norm. NC denotes no convergence.

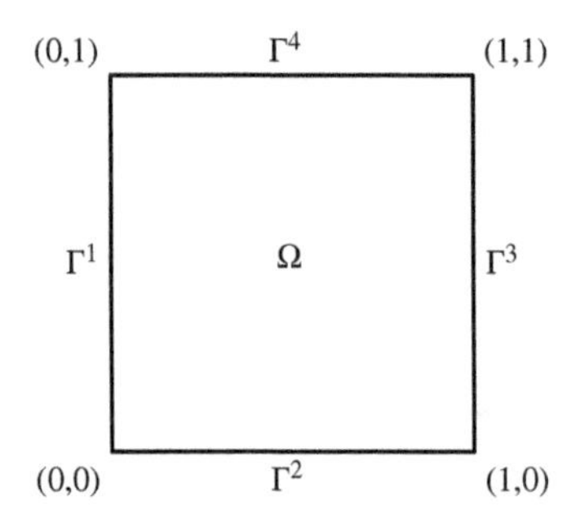

Figure 4.4 Domain and boundary definitions for Exercise 4.1.

Exercise 4.1 Consider the following two-dimensional boundary value problem with reference to Figure 4.4:

$$\begin{aligned} u_{,xx} + u_{,yy} &= -2\pi^2 \sin \pi x \sin \pi y \quad \text{in} \quad \Omega, \\ u &= 0 \quad \text{on} \quad \Gamma^1 \cup \Gamma^2, \\ u_{,n} &= -\pi \sin \pi y \quad \text{on} \quad \Gamma^3, \\ u_{,n} &= -\pi \sin \pi x \quad \text{on} \quad \Gamma^4. \end{aligned} \tag{4.41}$$

The exact solution of this problem is $u(x, y) = \sin \pi x \sin \pi y$. Use RKPM to solve this problem with linear basis and cubic B-spline kernels and impose the essential boundary conditions using Nitsche's method. Integrate the weak form by Gauss integration with 5×5 quadrature for the domain and 5-point quadrature for the boundary terms (see Sections 4.3 and 6.2).

Obtain the RKPM solution using (i) direct derivatives and (ii) implicit gradients (see Section 3.5.3), and compare the results in the following studies.

1) Obtain the numerical solution using 11×11, 21×21, and 41×41 nodes. Use a normalized support size $a/h = 2.0$, where a is the kernel support size and h is the nodal distance. Compare numerical and exact solutions along lines $x = 0.25$, $x = 0.50$, and $x = 0.75$.
2) Compute the L_2 error norm of u, denoted by e_{L^2}, and obtain rate of convergence by plotting $\log_{10}(e_{\mathrm{L}^2})$ vs. $\log_{10}(h)$ and finding the slope, where e_{L^2} is defined as

$$e_{\mathrm{L}^2} = \left[\int_{\Omega} \left(u^h(x,y) - u(x,y)\right)^2 d\Omega \right]^{1/2}.$$

Use at least eight-point Gauss quadrature to integrate the above error measure. Compare the rate to the theoretical one given in Chapter 3, Eq. (3.95).

3) Compute the H^1 error semi-norm of u, denoted by $e_{\mathrm{s\text{-}H}^1}$, and obtain rate of convergence by plotting $\log_{10}(e_{\mathrm{s\text{-}H}^1})$ vs. $\log_{10}(h)$ and finding the slope, where $e_{\mathrm{s\text{-}H}^1}$ is defined as

$$e_{\mathrm{s\text{-}H}^1} = \left[\int_{\Omega} \left(u^h_{,x}(x,y) - u_{,x}(x,y)\right)^2 + \left(u^h_{,y}(x,y) - u_{,y}(x,y)\right)^2 d\Omega \right]^{1/2}.$$

Use at least eight-point Gauss quadrature to integrate the above error measure. Compare the rate to the theoretical one given in Chapter 3, Eq. (3.96).

4.2 Elasticity

The strong form for linear elasticity consists of the equilibrium equations and the essential and natural boundary conditions, which are given, respectively, by (see Section 2.2.1)

$$\begin{aligned} &\nabla \cdot \boldsymbol{\sigma} + \mathbf{b} = \mathbf{0} \quad \text{in } \Omega, \\ &\text{or} \\ &\sigma_{ij,j} + b_i = 0 \quad \text{in } \Omega, \end{aligned} \tag{4.42}$$

$$\begin{aligned} &\mathbf{u} = \bar{\mathbf{u}} \quad \text{on } \Gamma_u, \\ &\text{or} \\ &u_i = \bar{u}_i \quad \text{on } \Gamma_u, \end{aligned} \tag{4.43}$$

$$\begin{aligned} &\boldsymbol{\sigma} \cdot \mathbf{n} = \bar{\mathbf{t}} \quad \text{on } \Gamma_t, \\ &\text{or} \\ &\sigma_{ij} n_j = \bar{t}_i \quad \text{on } \Gamma_t, \end{aligned} \tag{4.44}$$

where b_i is the body force, and $\bar{\mathbf{u}}$ and $\bar{\mathbf{t}}$ are the prescribed boundary displacements and traction, respectively.

Since the penalty method is a special case of Nitsche's method, for elasticity, we discuss only the Lagrange multiplier method and Nitsche's method for meshfree approximations that are not kinematically admissible with respect to the essential boundary conditions.

We will develop the discrete equations corresponding to the variational principle of elasticity (see Section 2.2.2); this principle is called the principle of minimum potential energy and states that the solution $u_i \in \mathrm{U}$ is the minimizer of

$$\Pi_{\mathrm{E}}(\mathbf{u}) = W^{\mathrm{int}}(\mathbf{u}) - W^{\mathrm{ext}}(\mathbf{u}), \tag{4.45}$$

where

$$W^{\mathrm{int}}(\mathbf{u}) = \frac{1}{2}\int_{\Omega} u_{(i,j)} C_{ijkl} u_{(k,l)} d\Omega = \frac{1}{2}\int_{\Omega} \mathcal{E}(\mathbf{u})^{\mathrm{T}} \mathbf{D} \mathcal{E}(\mathbf{u}) d\Omega, \tag{4.46}$$

$$W^{\mathrm{ext}}(\mathbf{u}) = \int_{\Omega} u_i b_i d\Omega + \int_{\Gamma_t} u_i \bar{t}_i d\Gamma = \int_{\Omega} \mathbf{u}^{\mathrm{T}} b_i d\Omega + \int_{\Gamma_t} \mathbf{u}^{\mathrm{T}} \bar{\mathbf{t}} d\Gamma, \tag{4.47}$$

and $\mathrm{U} = \{u_i(\mathbf{x}) | u_i(\mathbf{x}) \in \mathrm{H}^1 \text{ and } u_i(\mathbf{x}) = \bar{u}_i \text{ on } \Gamma_u\}$; here, $\mathbf{D}$ is the matrix form of C_{ijkl}, and $\mathcal{E}(\mathbf{u})$ is the vector form of $u_{(i,j)}$. We will adopt Voigt notation whenever possible for tensors, for instance, in 2D: $\mathcal{E} = [\varepsilon_{11} \ \ \varepsilon_{22} \ \ 2\varepsilon_{12}]^{\mathrm{T}}$ for kinematic quantities (describing deformation), and later, for kinetic (stress-like) quantities we use, e.g. $\boldsymbol{\Sigma} = [\sigma_{11} \ \ \sigma_{22} \ \ \sigma_{12}]^{\mathrm{T}}$. Fourth-order tensors with major and minor symmetries like C_{ijkl} follow the same numbering convention:

$$\mathbf{D} = \begin{bmatrix} C_{1111} & C_{1122} & C_{1112} \\ C_{2211} & C_{2222} & C_{2212} \\ C_{1211} & C_{1222} & C_{1212} \end{bmatrix}. \tag{4.48}$$

The weak form using the standard variational principle with kinematically admissible fields has been developed in Section 2.2.2 and is stated as

$$\text{Find } u_i \in \mathrm{U} \text{ such that } \delta\Pi_{\mathrm{E}} = 0 \, \forall \, \delta u_i \in \mathrm{U}_0, \tag{4.49}$$

with

$$\delta\Pi_{\mathrm{E}} = \int_{\Omega} \delta u_{(i,j)} C_{ijkl} u_{(k,l)} d\Omega - \int_{\Omega} \delta u_i b_i d\Omega - \int_{\Gamma_t} \delta u_i \bar{t}_i d\Gamma = 0, \tag{4.50}$$

and $\mathrm{U}_0 = \{\delta u_i(\mathbf{x}) | \delta u_i(\mathbf{x}) \in \mathrm{H}^1 \text{ and } \delta u_i(\mathbf{x}) = 0 \text{ on } \Gamma_u\}$.

To proceed with the discretization, we then invoke the approximation for the displacement in the form

$$\begin{aligned}\mathbf{u}(\mathbf{x}) &\simeq \mathbf{u}^h(\mathbf{x}) = \sum_{I\in S_{\mathbf{x}}} \Psi_I(\mathbf{x})\mathbf{u}_I,\\ \delta\mathbf{u}(\mathbf{x}) &\simeq \delta\mathbf{u}^h(\mathbf{x}) = \sum_{I\in S_{\mathbf{x}}} \Psi_I(\mathbf{x})\delta\mathbf{u}_I,\end{aligned} \tag{4.51}$$

where $S_{\mathbf{x}}$ is the set of nodes with non-zero kernel at a point $\mathbf{x}$. Consequently, the strain vector $\mathcal{E}(\mathbf{u})$ is approximated as

$$\mathcal{E}(\mathbf{u}(\mathbf{x})) \simeq \mathcal{E}\left(\mathbf{u}^h(\mathbf{x})\right) = \sum_{I\in S_{\mathbf{x}}} \mathbf{B}_I(\mathbf{x})\mathbf{u}_I, \tag{4.52}$$

where $\mathbf{B}_I$ is the strain matrix associated with Ψ_I as follows (in 2D):

$$\mathbf{B}_I = \begin{bmatrix} \Psi_{I,1} & 0 \\ 0 & \Psi_{I,2} \\ \Psi_{I,2} & \Psi_{I,1} \end{bmatrix}. \tag{4.53}$$

Substituting (4.51) and (4.52) into (4.50) following the diffusion equation discretization, we have

$$\sum_{I\in S} \delta\mathbf{u}_I^{\mathrm{T}} \left(\int_\Omega \mathbf{B}_I^{\mathrm{T}}\mathbf{D}\sum_{J\in S}\mathbf{B}_J\mathbf{u}_J d\Omega - \int_\Omega \Psi_I b_i d\Omega - \int_{\Gamma_t} \Psi_I \bar{\mathbf{t}} d\Gamma \right) = 0, \tag{4.54}$$

where the node sets for the domain integration terms have again been extended to include all nodes. Since $\delta\mathbf{u}_I$ is arbitrary, (4.54) can be expressed as the condition

$$\int_\Omega \mathbf{B}_I^{\mathrm{T}}\mathbf{D}\sum_{J\in S}\mathbf{B}_J\mathbf{u}_J d\Omega - \int_\Omega \Psi_I b_i d\Omega - \int_{\Gamma_t} \Psi_I \bar{\mathbf{t}} d\Gamma = 0 \ \forall I \in S. \tag{4.55}$$

Now, if we define the following matrices and vectors:

$$\begin{aligned}\mathbf{K}_{IJ} &= \int_\Omega \mathbf{B}_I^{\mathrm{T}}\mathbf{D}\mathbf{B}_J d\Omega,\\ \mathbf{f}_I &= \int_\Omega \Psi_I b_i d\Omega + \int_{\Gamma_t} \Psi_I \bar{\mathbf{t}} d\Gamma,\end{aligned} \tag{4.56}$$

then (4.55) reduces to

$$\sum_{J\in S} \mathbf{K}_{IJ}\mathbf{u}_J - \mathbf{f}_I = 0 \ \forall I \in S. \tag{4.57}$$

Finally, if we let $\mathbf{d} = \{\mathbf{u}_I\}$, $\mathbf{K} = [\mathbf{K}_{IJ}]$, and $\mathbf{f} = \{\mathbf{f}_I\}$ then we have the familiar matrix system of equations:

$$\mathbf{Kd} = \mathbf{f}. \tag{4.58}$$

Again, most meshfree approximations are not kinematically admissible unless they are specially constructed. Chapter 5 will discuss how to accomplish this. Here, we revisit the augmented variational principles to enforce essential boundary conditions for elasticity.

4.2.1 The Lagrange Multiplier Method for Elasticity

To modify this variational principle so that it can be applied to approximations that are not kinematically admissible, we first append a Lagrange multiplier to the total potential Π_E as follows:

$$\Pi_E^L(\mathbf{u}, \boldsymbol{\lambda}) = \Pi_E(\mathbf{u}) + \int_{\Gamma_u} \boldsymbol{\lambda} \cdot (\mathbf{u} - \bar{\mathbf{u}}) d\Gamma. \tag{4.59}$$

The weak form for the elasticity equation for the Lagrange multiplier method was developed in Section 2.2.3.2:

$$\text{Find } u_i \in \mathrm{H}^1,\ \lambda_i \in \mathrm{L}_2,\ \text{such that } \delta\Pi_E^L = 0\ \forall\, \delta u_i \in \mathrm{H}^1,\ \delta\lambda_i \in \mathrm{L}_2, \tag{4.60}$$

with

$$\delta\Pi_E^L = \delta\Pi_E + \int_{\Gamma_u} (\delta\lambda_i(u_i - \bar{u}_i) + \lambda_i \delta u_i) d\Gamma = 0. \tag{4.61}$$

The stationary point of $\Pi_E^L(\mathbf{u}, \boldsymbol{\lambda})$ is equivalent to the strong form (4.42)–(4.44) as discussed in Section 2.2.3.2.

We also need to discretize the Lagrange multiplier field on the prescribed displacement surface (an edge in 2D). The approximation of the Lagrange multiplier field is then given by

$$\begin{aligned} \boldsymbol{\lambda}(\boldsymbol{\xi}) &\simeq \boldsymbol{\lambda}^h(\boldsymbol{\xi}) = \sum_{\alpha \in S_\xi \cap S_u} \psi_\alpha(\boldsymbol{\xi}) \boldsymbol{\lambda}_\alpha, \\ \delta\boldsymbol{\lambda}(\boldsymbol{\xi}) &\simeq \delta\boldsymbol{\lambda}^h(\boldsymbol{\xi}) = \sum_{\alpha \in S_\xi \cap S_u} \psi_\alpha(\boldsymbol{\xi}) \delta\boldsymbol{\lambda}_\alpha. \end{aligned} \tag{4.62}$$

Again, these fields must satisfy the BB conditions for stability and thus also convergence.

Substituting (4.62) and (4.51) into (4.61) gives

$$\sum_{I \in S} \delta\mathbf{u}_I^T \left(\sum_{J \in S} (\mathbf{K}_{IJ} \mathbf{u}_J) - \mathbf{f}_I \right) + \sum_{I \in S} \delta\mathbf{u}_I^T \sum_{\alpha \in S_u} (\mathbf{G}_{I\alpha} \boldsymbol{\lambda}_\alpha) + \sum_{\alpha \in S_u} \delta\boldsymbol{\lambda}_\alpha^T \left(\sum_{I \in S} \mathbf{G}_{I\alpha} \mathbf{u}_I - \boldsymbol{\gamma}_\alpha \right) = 0, \tag{4.63}$$

where

$$\begin{aligned} \mathbf{K}_{IJ} &= \int_\Omega \mathbf{B}_I^T \mathbf{D} \mathbf{B}_J d\Omega, \\ \mathbf{f}_I &= \int_\Omega \Psi_I b_i d\Omega + \int_{\Gamma_t} \Psi_I \bar{\mathbf{t}} d\Gamma, \\ \mathbf{G}_{I\alpha} &= \int_{\Gamma_u} \Psi_I \psi_\alpha \mathbf{I} d\Gamma, \\ \boldsymbol{\gamma}_\alpha &= \int_{\Gamma_u} \psi_\alpha(\boldsymbol{\xi}) \bar{\mathbf{u}} d\Gamma. \end{aligned} \tag{4.64}$$

Note that the node sets for the domain integration terms and essential boundary integration terms in (4.63) have been extended to the entire domain and essential boundary node sets S and S_u, respectively, as before. Invoking the arbitrariness of $\delta\mathbf{u}_I$ and $\delta\boldsymbol{\lambda}_\alpha$, we obtain two sets of equations

$$\sum_{J\in S}\mathbf{K}_{IJ}\mathbf{u}_J-\mathbf{f}_I+\sum_{\alpha\in S_u}\mathbf{G}_{I\alpha}\boldsymbol{\lambda}_\alpha=0\quad\forall I\in S, \tag{4.65}$$

$$\sum_{I\in S}\mathbf{G}_{I\alpha}\mathbf{u}_I-\boldsymbol{\gamma}_\alpha=0\quad\forall\alpha\in S_u. \tag{4.66}$$

If we now define a column matrix of nodal displacements and Lagrange multipliers for the entire model by $\mathbf{d}^{\mathrm{T}}=\left[\mathbf{u}_1^{\mathrm{T}}\,\mathbf{u}_2^{\mathrm{T}}...\mathbf{u}_{NP}^{\mathrm{T}}\right]$ and $\boldsymbol{\ell}^{\mathrm{T}}=\left[\boldsymbol{\lambda}_1^{\mathrm{T}}\,\boldsymbol{\lambda}_2^{\mathrm{T}}...\boldsymbol{\lambda}_{Neb}^{\mathrm{T}}\right]$ with *Neb* the number of essential boundary points, and define $\mathbf{K}=[\mathbf{K}_{IJ}]$, $\mathbf{G}=[\mathbf{G}_{I\alpha}]$, and $\boldsymbol{\gamma}=\{\boldsymbol{\gamma}_\alpha\}$, we can write the above equations in matrix form as

$$\begin{bmatrix}\mathbf{K} & \mathbf{G}\\ \mathbf{G}^{\mathrm{T}} & \mathbf{0}\end{bmatrix}\begin{Bmatrix}\mathbf{d}\\ \boldsymbol{\ell}\end{Bmatrix}=\begin{Bmatrix}\mathbf{f}\\ \boldsymbol{\gamma}\end{Bmatrix}. \tag{4.67}$$

This is the standard Lagrange multiplier form. As can be seen from the fact that the lower right matrix is a null matrix, the system matrix is not positive definite. The augmented Lagrangian approach to enforcing the displacement boundary conditions consists of adding a quadratic penalty term to the Lagrange potential (4.59) which becomes

$$\Pi_{\mathrm{E}}^{\mathrm{L}}=\Pi_{\mathrm{E}}+\int_{\Gamma_u}\boldsymbol{\lambda}\cdot\left(\mathbf{u}-\overline{\mathbf{u}}-\frac{1}{2\beta}\boldsymbol{\lambda}\right)d\Gamma, \tag{4.68}$$

where β is a penalty parameter. Following similar procedures discussed in this chapter, the discretization of weak form is

$$\begin{aligned}&\sum_{I\in S}\delta\mathbf{u}_I^{\mathrm{T}}\left(\sum_{J\in S}(\mathbf{K}_{IJ}\mathbf{u}_J)-\mathbf{f}_I\right)+\sum_{I\in S}\delta\mathbf{u}_I^{\mathrm{T}}\left(\sum_{\alpha\in S_u}\mathbf{G}_{I\alpha}\boldsymbol{\lambda}_\alpha\right)+\sum_{\alpha\in S_u}\delta\boldsymbol{\lambda}_\alpha^{\mathrm{T}}\left(\sum_{J\in S}\mathbf{G}_{J\alpha}\mathbf{u}_J-\boldsymbol{\gamma}_\alpha\right)\\ &\quad-\sum_{\alpha\in S_u}\delta\boldsymbol{\lambda}_\alpha^{\mathrm{T}}\left(\sum_{J\in S}\mathbf{M}_{\alpha\beta}\boldsymbol{\lambda}_\beta\right)=0,\end{aligned} \tag{4.69}$$

where

$$\mathbf{M}_{\alpha\beta}=\frac{1}{\beta}\int_{\Gamma_u}\psi_\alpha\psi_\beta\mathbf{I}d\Gamma. \tag{4.70}$$

The corresponding matrix form of this augmented Lagrangian formulation is

$$\begin{bmatrix}\mathbf{K} & \mathbf{G}\\ \mathbf{G}^{T} & -\mathbf{M}\end{bmatrix}\begin{Bmatrix}\mathbf{d}\\ \boldsymbol{\ell}\end{Bmatrix}=\begin{Bmatrix}\mathbf{f}\\ \boldsymbol{\gamma}\end{Bmatrix}, \tag{4.71}$$

and $\mathbf{M}=[\mathbf{M}_{\alpha\beta}]$. This approach leads to a positive definite matrix system:

$$\begin{aligned}&\left(\mathbf{K}+\mathbf{G}\mathbf{M}^{-1}\mathbf{G}^{\mathrm{T}}\right)\mathbf{d}=\mathbf{f}+\mathbf{G}\mathbf{M}^{-1}\boldsymbol{\gamma},\\ &\boldsymbol{\ell}=\mathbf{M}^{-1}\left(\mathbf{G}^{\mathrm{T}}\mathbf{d}-\boldsymbol{\gamma}\right).\end{aligned} \tag{4.72}$$

4.2.2 Nitsche's Method for Elasticity

The Nitsche weak form for elasticity is analogous to that for the diffusion equation. We can replace u by u_i, $u_{,n}$ by $u_{i,n}$ and obtain the weak form. However, it is customary to work with the traction t_i, so we will develop it in that way. Following Nitsche's method for elasticity discussed in Section 2.2.3.3, the weak form is

$$\text{Find } u_i \in \mathrm{H}^1 \text{ such that } \delta\Pi_{\mathrm{E}}^{\mathrm{N}} = 0 \ \forall \delta u_i \in \mathrm{H}^1, \tag{4.73}$$

with

$$\Pi_{\mathrm{E}}^{\mathrm{N}} = \Pi_{\mathrm{E}} + \Pi_{\mathrm{E}}^{\mathrm{Nit}}, \tag{4.74}$$

$$\begin{aligned} \Pi_{\mathrm{E}}^{\mathrm{Nit}} &= -\int_{\Gamma_u} t_i(u_i - \overline{u}_i)d\Gamma + \frac{\beta}{2}\int_{\Gamma_u}(u_i - \overline{u}_i)(u_i - \overline{u}_i)d\Gamma \\ &= -\int_{\Gamma_u} C_{ijkl}u_{(k,l)}n_j(u_i - \overline{u}_i)d\Gamma + \frac{\beta}{2}\int_{\Gamma_u}(u_i - \overline{u}_i)(u_i - \overline{u}_i)d\Gamma, \end{aligned} \tag{4.75}$$

where we have used $t_i = \sigma_{ij}n_j = C_{ijkl}u_{(k,l)}n_j$, i.e., Cauchy's law and the stress–strain law. By following the same steps as in the proof for the diffusion equation in Section 2.1.3.3, it is straightforward to show that the Eq. (4.73) implies the equilibrium equation, the traction boundary conditions, and the essential boundary conditions.

The discrete equations are obtained by constructing the meshfree approximations as in (4.51):

$$\begin{aligned} \mathbf{u}(\mathbf{x}) &\simeq \mathbf{u}^h(\mathbf{x}) = \sum_{I \in S_{\mathbf{x}}} \Psi_I(\mathbf{x})\mathbf{u}_I, \\ \delta\mathbf{u}(\mathbf{x}) &\simeq \delta\mathbf{u}^h(\mathbf{x}) = \sum_{I \in S_{\mathbf{x}}} \Psi_I(\mathbf{x})\delta\mathbf{u}_I. \end{aligned}$$

The form of $\delta\Pi_{\mathrm{E}}$ is identical to that in (4.50), so we have

$$\delta\Pi_{\mathrm{E}} \approx \sum_{I \in S} \delta\mathbf{u}_I^{\mathrm{T}} \left(\sum_{J \in S} (\mathbf{K}_{IJ}\mathbf{u}_J) - \mathbf{f}_I \right). \tag{4.76}$$

The additional terms in (4.75) can be rearranged and expressed as

$$\begin{aligned} \Pi_{\mathrm{E}}^{\mathrm{Nit}} &= -\int_{\Gamma_u} u_i n_j C_{ijkl}u_{(k,l)}d\Gamma + \int_{\Gamma_u} \overline{u}_i n_j C_{ijkl}u_{(k,l)}d\Gamma + \frac{\beta}{2}\int_{\Gamma_u}(u_i - \overline{u}_i)(u_i - \overline{u}_i)d\Gamma \\ &= -\int_{\Gamma_u} a_{ij}C_{ijkl}u_{(k,l)}d\Gamma + \int_{\Gamma_u} \overline{a}_{ij}C_{ijkl}u_{(k,l)}d\Gamma + \frac{\beta}{2}\int_{\Gamma_u}(u_i - \overline{u}_i)(u_i - \overline{u}_i)d\Gamma, \end{aligned} \tag{4.77}$$

where $a_{ij} = u_i n_j$ and $\overline{a}_{ij} = \overline{u}_i n_j$. For the construction of the discrete equation, it is easier to first transform (4.77) into a vector/matrix form:

$$\Pi_{\mathrm{E}}^{\mathrm{Nit}} = -\int_{\Gamma_u} \mathbf{a}^{\mathrm{T}}\mathbf{D}\mathcal{E}d\Gamma + \int_{\Gamma_u} \overline{\mathbf{a}}^{\mathrm{T}}\mathbf{D}\mathcal{E}d\Gamma + \frac{\beta}{2}\int_{\Gamma_u}(\mathbf{u} - \overline{\mathbf{u}})^{\mathrm{T}}(\mathbf{u} - \overline{\mathbf{u}})d\Gamma, \tag{4.78}$$

where $\mathbf{D}$ is the matrix form of C_{ijkl}, $\mathcal{E}$ is the vector form of $u_{(k,l)}$, and $\mathbf{a}$ and $\overline{\mathbf{a}}$ are vector forms of $u_i n_j$ and $\overline{u}_i n_j$, respectively, and can be expressed as (in 2D):

$$\mathbf{a} = \begin{bmatrix} u_1 n_1 \\ u_2 n_2 \\ u_1 n_2 + u_2 n_1 \end{bmatrix} = \underbrace{\begin{bmatrix} n_1 & 0 \\ 0 & n_2 \\ n_2 & n_1 \end{bmatrix}}_{\boldsymbol{\eta}} \underbrace{\begin{bmatrix} u_1 \\ u_2 \end{bmatrix}}_{\mathbf{u}} \equiv \boldsymbol{\eta}\mathbf{u}, \tag{4.79}$$

$$\bar{\mathbf{a}} = \begin{bmatrix} \bar{u}_1 n_1 \\ \bar{u}_2 n_2 \\ \bar{u}_1 n_2 + \bar{u}_2 n_1 \end{bmatrix} = \underbrace{\begin{bmatrix} n_1 & 0 \\ 0 & n_2 \\ n_2 & n_1 \end{bmatrix}}_{\boldsymbol{\eta}} \underbrace{\begin{bmatrix} \bar{u}_1 \\ \bar{u}_2 \end{bmatrix}}_{\bar{\mathbf{u}}} \equiv \boldsymbol{\eta}\bar{\mathbf{u}}. \tag{4.80}$$

By introducing meshfree approximations on test and trial functions in (4.51), we have for the Nitsche term in (4.79)

$$\mathbf{a}(\mathbf{u}(\mathbf{x})) \simeq \mathbf{a}(\mathbf{u}^h(\mathbf{x})) = \boldsymbol{\eta}\left(\sum_{I\in S_{\mathbf{x}}} \Psi_I(\mathbf{x})\mathbf{u}_I\right). \tag{4.81}$$

Introducing (4.51), (4.52), and (4.81) into the variation of the Nitsche constraint terms in (4.78), one obtains

$$\begin{aligned}
\delta\Pi_{\mathrm{E}}^{\mathrm{Nit}} &= -\int_{\Gamma_u} \delta\mathbf{a}^{\mathrm{T}}\mathbf{D}\mathcal{E}d\Gamma - \int_{\Gamma_u} \delta\mathcal{E}^{\mathrm{T}}\mathbf{D}\mathbf{a}d\Gamma + \int_{\Gamma_u} \delta\mathcal{E}^{\mathrm{T}}\mathbf{D}\bar{\mathbf{a}}d\Gamma + \beta\int_{\Gamma_u} \delta\mathbf{u}^{\mathrm{T}}(\mathbf{u}-\bar{\mathbf{u}})d\Gamma \\
&= -\int_{\Gamma_u} \delta\mathbf{u}^{\mathrm{T}}\boldsymbol{\eta}^{\mathrm{T}}\mathbf{D}\mathcal{E}d\Gamma - \int_{\Gamma_u} \delta\mathcal{E}^{\mathrm{T}}\mathbf{D}\boldsymbol{\eta}\mathbf{u}d\Gamma + \int_{\Gamma_u} \delta\mathcal{E}^{\mathrm{T}}\mathbf{D}\boldsymbol{\eta}\bar{\mathbf{u}}d\Gamma + \beta\int_{\Gamma_u} \delta\mathbf{u}^{\mathrm{T}}(\mathbf{u}-\bar{\mathbf{u}})d\Gamma \\
&\approx -\sum_{I\in S} \delta\mathbf{u}_I^{\mathrm{T}}\left(\int_{\Gamma_u} \Psi_I\boldsymbol{\eta}^{\mathrm{T}}\mathbf{D}\sum_{J\in S}\mathbf{B}_J d\Gamma + \int_{\Gamma_u} \mathbf{B}_I^{\mathrm{T}}\mathbf{D}\boldsymbol{\eta}\sum_{J\in S}\Psi_J d\Gamma\right)\mathbf{u}_J + \sum_{I\in S}\delta\mathbf{u}_I^{\mathrm{T}}\int_{\Gamma_u} \mathbf{B}_I^{\mathrm{T}}\mathbf{D}\boldsymbol{\eta}\bar{\mathbf{u}}d\Gamma \\
&\quad + \sum_{I\in S}\delta\mathbf{u}_I^{\mathrm{T}}\left(\beta\int_{\Gamma_u} \Psi_I\sum_{J\in S}\Psi_J d\Gamma\right)\mathbf{u}_J - \sum_{I\in S}\delta\mathbf{u}_I^{\mathrm{T}}\beta\int_{\Gamma_u} \Psi_I\bar{\mathbf{u}}d\Gamma \\
&= \sum_{I\in S}\delta\mathbf{u}_I^{\mathrm{T}}\sum_{J\in S}(-\mathbf{H}_{IJ} - \mathbf{H}_{IJ}^{\mathrm{T}} + \mathbf{A}_{IJ})\mathbf{u}_J + \sum_{I\in S}\delta\mathbf{u}_I^{\mathrm{T}}(\mathbf{h}_I - \mathbf{g}_I),
\end{aligned} \tag{4.82}$$

where

$$\mathbf{H}_{IJ} = \int_{\Gamma_u} \Psi_I\boldsymbol{\eta}^{\mathrm{T}}\mathbf{D}\mathbf{B}_J d\Gamma, \tag{4.83}$$

$$\mathbf{A}_{IJ} = \beta\int_{\Gamma_u} \Psi_I\Psi_J\mathbf{I}d\Gamma, \tag{4.84}$$

$$\mathbf{h}_I = \int_{\Gamma_u} \mathbf{B}_I^{\mathrm{T}}\mathbf{D}\boldsymbol{\eta}\bar{\mathbf{u}}d\Gamma, \tag{4.85}$$

$$\mathbf{g}_I = \beta\int_{\Gamma_u} \Psi_I\bar{\mathbf{u}}d\Gamma. \tag{4.86}$$

Setting the variation on (4.74) equal to zero, using (4.82) and (4.76), and invoking the arbitrariness of $\delta\mathbf{u}_I$, we have

$$\sum_{J\in S}(\mathbf{K}_{IJ} + \mathbf{K}_{IJ}^{\mathrm{N}})\mathbf{u}_J = \mathbf{f}_I + \mathbf{f}_I^{\mathrm{N}} \quad \forall I \in S, \tag{4.87}$$

where

$$\begin{aligned} \mathbf{K}_{IJ}^{\mathrm{N}} &= -\mathbf{H}_{IJ} - \mathbf{H}_{IJ}^{\mathrm{T}} + \mathbf{A}_{IJ}, \\ \mathbf{f}_{I}^{\mathrm{N}} &= -\mathbf{h}_{I} + \mathbf{g}_{I}. \end{aligned} \tag{4.88}$$

By defining $\mathbf{K} = [\mathbf{K}_{IJ}]$, $\mathbf{K}^{\mathrm{N}} = \left[\mathbf{K}_{IJ}^{\mathrm{N}}\right]$, $\mathbf{f} = \{\mathbf{f}_I\}$, $\mathbf{f}^{\mathrm{N}} = \left\{\mathbf{f}_I^{\mathrm{N}}\right\}$, and $\mathbf{d}^{\mathrm{T}} = \left[\mathbf{u}_1^{\mathrm{T}}\,\mathbf{u}_2^{\mathrm{T}}...\mathbf{u}_{NP}^{\mathrm{T}}\right]$, (4.87) can be expressed in the following matrix equation:

$$\left(\mathbf{K} + \mathbf{K}^{\mathrm{N}}\right)\mathbf{d} = \mathbf{f} + \mathbf{f}^{\mathrm{N}}. \tag{4.89}$$

As in the diffusion equation, a symmetric matrix is added to the stiffness matrix associated with the standard potential functional of the elasticity problem. Additional terms are also added to the right-hand side. These terms are nonzero only for nodes whose support covers the essential boundary. It is easily seen that by removing the surface traction terms in $\Pi_{\mathrm{E}}^{\mathrm{Nit}}$, that $\mathbf{H}_{IJ} = \mathbf{0}$ and $\mathbf{h}_I = \mathbf{0}$, and the method reduces to the penalty method for imposing the essential boundary conditions in elasticity. Nitsche's method is employed in the RKPM2D code described in Chapter 10, with the associated computer implementation detailed in Section 10.1.

As discussed in Section 4.1.3, the stability parameter in Nitsche's method should either be obtained with an eigenvalue estimate or scaled as $\beta \sim c_{\mathrm{max,\ PDE}} \times h^{-1}$. For elasticity, the first and second Lame parameters λ and μ can be employed with $\beta \sim h^{-1} \times \max(\lambda,\ \mu)$. In RKPM2D, the parameter is normalized following this scaling.

4.3 Numerical Integration

So far, we have constructed several meshfree discretizations for boundary value problems, and in practice, the domain integrals in the weak form must be carried out numerically. One of the major issues in the implementation of Galerkin meshfree methods is the evaluation of these integrals. In finite element methods, the integrals are nearly polynomials, so Gauss quadrature is very effective. Furthermore, finite element methods naturally provide a procedure for partitioning the domain into subdomains over which the integrals of the Galerkin weak form can be easily evaluated: the elements of the mesh. In meshfree methods, the shape functions, as described in Chapter 3, are rational functions, and their supports overlap in an arbitrary fashion in general.

The earliest Galerkin meshfree methods employed a "background" mesh of elements for quadrature. One example of this idea is illustrated in Figure 4.5. As seen here, one can use the nodes of the background mesh for simplicity, yet this is not necessary: the background mesh and meshfree nodes can be completely independent (see Chapters 6 and 10). A nice advantage of these procedures over finite elements is that the elements do not have to be arranged in any special way so long as they cover the entire domain; they are only used for quadrature. In the finite element method, the elements are also used to construct the approximation, so their geometry requires special care to maintain accuracy, and they cannot be too distorted.

In any case, the integrals using the above procedure are then evaluated in each element by Gauss quadrature. The standard formulas for Gauss quadrature are used. For example, for the diffusion equation in two dimensions, the quadrature for the conductance matrix is 3×3 in Figure 4.5.

Yet Gauss quadrature is not as effective in meshfree methods as it is in finite elements, and in most cases, introduces quadrature error due to the non-matching domain integration cells and shape function supports, which manifests as error in the numerical solution. Methods have been

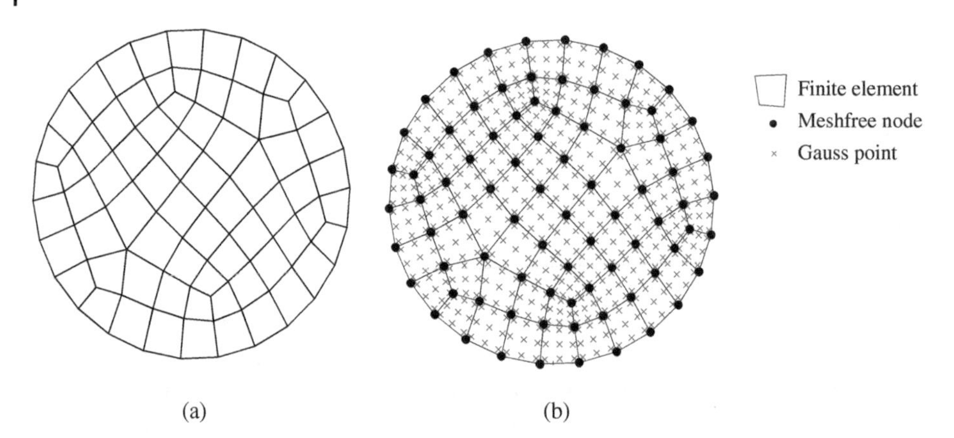

Figure 4.5 Using finite elements to generate nodes and perform quadrature for a meshfree discretization: (a) finite element mesh and (b) meshfree discretization with Gauss integration based on the mesh and finite element nodal locations.

introduced to match supports and integration cells [9], including the particle partition of unity method by Griebel and Schweitzer [10], although this introduces some complexity in the final meshfree formulation.

Methods have also been developed to avoid a mesh for integration completely and yield a method that is meshfree in totality. The partition of unity property of the meshfree approximation (see Section 3.1.3) has been exploited to avoid a background mesh [11, 12]. The MLPG method [1] and method of finite spheres [2] integrate the stiffness matrix over the support of the test function for each of the row equations. In the end however, the quadrature needed to achieve the desired accuracy is expensive. Nodal integration methods can also avoid a mesh altogether in the solution of the discrete equations, but can lead to rank deficiency and stabilization is needed.

A stabilized conforming nodal integration (SCNI) [13] has been introduced to ensure passing the linear patch test and to remedy rank deficiency of direct nodal integration, which is similar to stabilization of the hour-glass modes in finite elements. Extensions of SCNI for higher-order accuracy have been proposed [14, 15].

Other stabilized nodal integrations have been developed, including adding a residual of the equilibrium equation to the nodally integrated potential energy functional [16], the stress point method by taking derivatives away from the nodal points [17], least-squares type stabilization [18], and gradient approaches [19, 20] which add nonzero contributions to nodal integration to increase coercivity.

It should also be pointed out that the MLS and RK shape functions have a rather high degree of complexity. This can be seen in Figure 3.5 for a one-dimensional shape function and Figure 3.6 for the two-dimensional case (see Chapter 3). This is not as bad as it looks for the two-dimensional isotropic heat conduction equation, for the terms in the conductance matrix are the squares of the derivatives of the shape functions, $\Psi_{I,x}$ and $\Psi_{I,y}$. However, in the stiffness matrix for two-dimensional and three-dimensional elasticity, terms like $\Psi_{I,x}\Psi_{I,y}$ appear, and they are quite complex. As a result of the non-matching supports and integration, and complexity of shape functions, domain integration of the weak form poses considerable difficulty in the Galerkin meshfree method. The error in quadrature can prevent optimal convergence [21, 22], and special techniques must be introduced to circumvent this problem. SCNI can ensure passing the patch test (linear exactness), and convergence consistent with the optimal rates for linear approximations. A generalized arbitrary-order method has also been introduced [23], where a correction of any given quadrature rule can be obtained to achieve optimal rates of convergence.

Stabilized and corrected methods for integration of the weak form will be discussed in Chapter 6, where rank instability in nodal integration, quadrature requirements for exactness in the Galerkin approximation, and several stabilization methods will be discussed.

4.4 Further Discussions on Essential Boundary Conditions

For meshfree approximations that are not interpolants, imposition of the boundary conditions on the primary variables, such as the temperature in heat conduction and the displacement in elasticity, requires special procedures, as discussed in this chapter. However, there are other methods for the imposition of the essential boundary conditions, where the test function and the trial functions are kinematically admissible, and they will be discussed in detail in Chapter 5. This chapter will focus on the particular technique of coupling with finite elements on the essential boundaries.

In this method, the essential boundary conditions are imposed by coupling the meshfree approximation with a finite element approximation. The meshfree approximation is used for most of the domain, except in the vicinity of essential (Dirichlet) boundaries. A finite element approximation is used along the essential boundaries. Any node connected to an element is called an element node; the other nodes are called meshfree nodes. The essential boundary conditions are then applied to the element nodes that lie on the essential boundary.

An example of this type of model is shown in Figure 4.6. Here, a single layer of finite elements has been placed along the essential boundary; elements have also been placed on the natural (Neumann) boundaries adjacent to the edge of the essential boundaries. This is done because the shape functions for meshfree nodes in close proximity to the essential boundary may overlap the natural boundary and cause violation of the essential boundary condition. Usually, a single layer of elements is used as shown in Figure 4.6. However, more layers can be used to mitigate any dispersion error due to the transition in approximation in dynamic analysis.

To apply this method, it is necessary to couple the meshfree approximation to the finite element approximation. Such coupling methods are not only useful for applying the essential boundary conditions, but also have more general areas of application. For example, because of the greater cost of meshfree methods, it is often worthwhile to mesh the part of the domain with finite elements where the capabilities of meshfree shape functions are not needed. It is then also necessary to have a method for coupling finite element and meshfree approximations.

Two types of methods have been most successful:

1) Blending methods, where the approximation passes from an MLS approximation within the interior to a finite element approximation along the displacement boundaries, as proposed in [24]

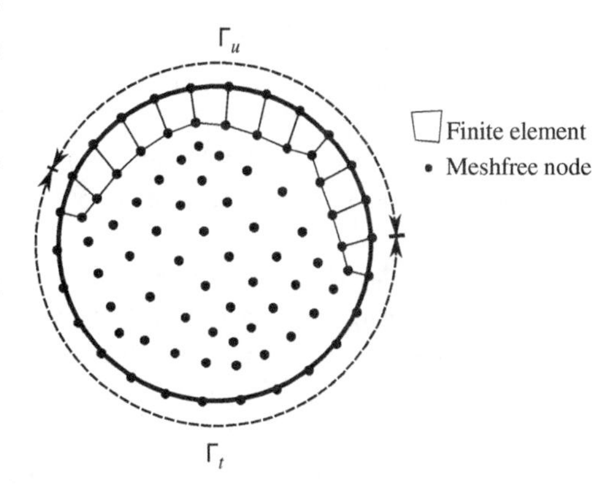

Figure 4.6 Meshfree and finite element coupling for enforcing essential boundary conditions.

2) Constructing new approximation functions that are linear combinations of the finite element and meshfree shape functions and satisfy the reproducing conditions [3, 25].

In the blending methods, a ramp function is defined in a subdomain that includes some of the meshfree nodes and finite element nodes. This ramp function can easily be constructed by using a distance function from the boundary or *r*-methods. A ramp function is then constructed in that subdomain [24]. It is best to use a distance function for the surface that defines the exterior when the method is used for imposition of displacement boundary conditions, and the interface when the method is used for coupling finite elements with meshfree approximations. An implementation of this method is given in Chapter 5.

Methods that recover nodal values in the approximation have also been proposed. Lancaster and Salkauskas [26] introduced a singularity to the weight function to yield interpolants in the MLS approximation, and this concept has been introduced for the imposition of essential boundary conditions in meshfree methods [27]. This approach, however, yields oscillations in the approximation since the MLS and RK shape functions with singular kernels pass through points and become wavy. As such, reduced convergence rates have been observed in solving PDEs. Introducing singular kernels only at the nodes on the essential boundary has been proposed to improve the rate of convergence [28]. Another approach is to construct a RK approximation with the interpolation property without singular weight functions [29], and a similar approach for MLS was also given in [30].

Several researchers [25, 31–34] have also introduced a transformation method to recover interpolation properties of an RK or MLS approximation, apparently independently proposing this approach around the same time. To reduce the computational burden in the coordinate transformation, a transformation for only the nodes on the essential boundaries can be employed, which has been termed a mixed transformation [28]. The formulation and implementation details of these types of methods will be discussed in Chapter 5.

References

1 Atluri, S.N. and Zhu, T.L. (1998). A new meshless local Petrov–Galerkin (MLPG) approach in computational mechanics. *Comput. Mech.* 22 (2): 117–127.

2 De, S. and Bathe, K.J. (2000). The method of finite spheres. *Comput. Mech.* 25 (4): 329–345.

3 Fernández-Méndez, S. and Huerta, A. (2004). Imposing essential boundary conditions in mesh-free methods. *Comput. Methods Appl. Mech. Eng.* 193 (12): 1257–1275.

4 Wang, J., Zhou, G., Hillman, M. et al. (2021). Consistent immersed volumetric Nitsche methods for composite analysis. *Comput. Methods Appl. Mech. Eng. (under Minor Revis.)* 385: 114042.

5 Babuška, I. (1973). The finite element method with Lagrangian multipliers. *Numer. Math.* 20 (3): 179–192.

6 Brezzi, F. (1974). On the existence, uniqueness and approximation of saddle-point problems arising from Lagrangian multipliers. *Publ. Mathématiques Inform. Rennes.* S4: 1–26.

7 Nitsche, J. (1971). Über ein Variationsprinzip zur Lösung von Dirichlet-Problemen bei Verwendung von Teilräumen, die keinen Randbedingungen unterworfen sind. *Abhandlungen Aus Dem Math. Semin. Der Univ. Hambg.* 36: 9–15.

8 Griebel, M. and Schweitzer, M.A. (2003). A particle-partition of unity method part V: boundary conditions. In: *Geometric Analysis and Nonlinear Partial Differential Equations* (ed. S. Hildebrandt and H. Karcher). Berlin, Heidelberg: Springer.

9 Dolbow, J. and Belytschko, T. (1999). Numerical integration of the Galerkin weak form in meshfree methods. *Comput. Mech.* 23 (3): 219–230.

10 Griebel, M. and Schweitzer, M.A. (2002). A particle-partition of unity method – Part II: efficient cover construction and reliable integration. *SIAM J. Sci. Comput.* 23 (5): 1655–1682.

11 Duflot, M. and Nguyen-Dang, H. (2002). A truly meshless Galerkin method based on a moving least squares quadrature. *Commun. Numer. Methods Eng.* 18 (6): 441–449.

12 Romero, I. and Armero, F. (2002). The partition of unity quadrature in meshless methods. *Int. J. Numer. Methods Eng.* 54 (7): 987–1006.

13 Chen, J.-S., Wu, C.-T., and Yoon, S. (2001). A stabilized conforming nodal integration for Galerkin mesh-free methods. *Int. J. Numer. Methods Eng.* 50 (2): 435–466.

14 Duan, Q., Li, X., Zhang, H., and Belytschko, T. (2012). Second-order accurate derivatives and integration schemes for meshfree methods. *Int. J. Numer. Methods Eng.* 92 (4): 399–424.

15 Wang, D. and Wu, J. (2019). An inherently consistent reproducing kernel gradient smoothing framework toward efficient Galerkin meshfree formulation with explicit quadrature. *Comput. Methods Appl. Mech. Eng.* 349: 628–672.

16 Beissel, S.R. and Belytschko, T. (1996). Nodal integration of the element-free Galerkin method. *Comput. Methods Appl. Mech. Eng.* 139: 49–74.

17 Randles, P.W. and Libersky, L.D. (2000). Normalized SPH with stress points. *Int. J. Numer. Meth. Engng.* 48: 1445–1462.

18 Bonet, J. and Kulasegaram, S. (2000). Correction and stabilization of smooth particle hydrodynamics methods with applications in metal forming simulations. *Int. J. Numer. Methods Eng.* 47: 1189–1214.

19 Wu, C.-T., Koishi, M., and Hu, W. (2015). A displacement smoothing induced strain gradient stabilization for the meshfree Galerkin nodal integration method. *Comput. Mech.*.

20 Hillman, M. and Chen, J.-S. (2016). An accelerated, convergent, and stable nodal integration in Galerkin meshfree methods for linear and nonlinear mechanics. *Int. J. Numer. Methods Eng.* 107: 603–630.

21 Babuška, I., Banerjee, U., Osborn, J.E., and Li, Q. (2008). Quadrature for meshless methods. *Int. J. Numer. Methods Eng.* 76 (9): 1434–1470.

22 Wu, J. and Wang, D. (2021). An accuracy analysis of Galerkin meshfree methods accounting for numerical integration. *Comput. Methods Appl. Mech. Eng.* 375: 113631.

23 Chen, J.-S., Hillman, M., and Rüter, M. (2013). An arbitrary order variationally consistent integration for Galerkin meshfree methods. *Int. J. Numer. Methods Eng.* 95 (5): 387–418.

24 Krongauz, Y. and Belytschko, T. (1996). Enforcement of essential boundary conditions in meshless approximations using finite elements. *Comput. Methods Appl. Mech. Eng.* 131 (1–2): 133–145.

25 Wagner, G.J. and Liu, W.K. (2000). Application of essential boundary conditions in mesh-free methods: a corrected collocation method. *Int. J. Numer. Methods Eng.* 47 (8): 1367–1379.

26 Lancaster, P. and Salkauskas, K. (1981). Surfaces generated by moving least squares methods. *Math. Comput.* 37: 141–158.

27 Kaljevic, I. and Saigal, S. (1997). An improved element free Galerkin formulation. *Int. J. Numer. Methods Eng.* 40 (16): 2953–2974.

28 Chen, J.-S. and Wang, H.-P. (2000). New boundary condition treatments in meshfree computation of contact problems. *Comput. Methods Appl. Mech. Eng.* 187 (3–4): 441–468.

29 Chen, J.-S., Han, W., You, Y., and Meng, X. (2003). A reproducing kernel method with nodal interpolation property. *Int. J. Numer. Methods Eng.* 56 (7): 935–960.

30 Zhuang, X., Zhu, H., and Augarde, C. (2014). An improved meshless Shepard and least squares method possessing the delta property and requiring no singular weight function. *Comput. Mech.* 53 (2): 343–357.

31 Chen, J.-S., Pan, C., Wu, C.-T., and Liu, W.K. (1996). Reproducing kernel particle methods for large deformation analysis of non-linear structures. *Comput. Methods Appl. Mech. Eng.* 139 (1–4): 195–227.

32 Zhu, T.L. and Atluri, S.N. (1998). A modified collocation method and a penalty formulation for enforcing the essential boundary conditions in the element free Galerkin method. *Comput. Mech.* 21 (3): 211–222.

33 Günther, F.C. and Liu, W.K. (1998). Implementation of boundary conditions for meshless methods. *Comput. Methods Appl. Mech. Eng.* 163 (1–4): 205–230.

34 Atluri, S.N., Kim, H.G., and Cho, J.Y. (1999). Critical assessment of the truly Meshless Local Petrov-Galerkin (MLPG), and Local Boundary Integral Equation (LBIE) methods. *Comput. Mech.* 24 (5): 348–372.

5

Construction of Kinematically Admissible Shape Functions

Previously, we have seen weak formulations considering meshfree approximations are not directly applicable (called *kinematically admissible*) to the standard variational formulation. However, it is possible to construct meshfree approximations that are. In this chapter, several kinematically admissible meshfree approximations are given. We will use two-dimensional elastostatics as a model problem; the application to other boundary-value problems is straightforward. The spirit of this approach to boundary conditions is, instead of modifying the weak formulation itself, one constructs modified meshfree approximations that can be used in the standard weak formulation. Then, enforcement of essential boundary conditions is performed just as in the finite element method. This has the further advantage of easily implementing meshfree methods into an existing finite element code, e.g. a production code.

5.1 Strong Enforcement of Essential Boundary Conditions

Here, we recall the requirements of the standard weak formulation: we must enforce $\mathbf{u}^h = \overline{\mathbf{u}}$ on Γ_u and $\delta\mathbf{u}^h = \mathbf{0}$ on Γ_u *strongly*, or exactly, since these approximations must be in the subspaces of $\mathbb{U}$ and $\mathbb{U}_0$, respectively. As such, the conditions for $\mathbf{u} = \overline{\mathbf{u}}$ on Γ_u are therefore called *essential boundary conditions* (also called Dirichlet boundary conditions). The enforcement of these conditions in the traditional finite element method is straightforward due to the property that nodal coefficients in the approximation are exactly the values of displacements (or other field variables in the governing equation) at nodal locations. Therefore, these conditions can be imposed directly on the nodal coefficients by simply setting them equal to the desired values in the matrix equation.

This feature of numerical methods is generally referred to as the Kronecker delta property, although weaker requirements can also yield this ability (see Section 5.2.2). Nevertheless, for a shape function N_I evaluated at a nodal location $\mathbf{x}_J$, the delta property is

$$N_I(\mathbf{x}_J) = \delta_{IJ}, \tag{5.1}$$

where

$$\delta_{IJ} = \begin{cases} 1 & \text{if } I = J \\ 0 & \text{if } I \neq J \end{cases}, \tag{5.2}$$

is the Kronecker delta. When (5.1) holds, an approximation to displacement $\mathbf{u}^h$ evaluated at any given nodal location $\mathbf{x}_J$ is obtained as

Meshfree and Particle Methods: Fundamentals and Applications, First Edition.
Ted Belytschko, J. S. Chen, and Michael Hillman.

$$\mathbf{u}^h(\mathbf{x}_J) = \sum_{I=1}^{NP} N_I(\mathbf{x}_J)\mathbf{u}_I = \sum_{I=1}^{NP} \delta_{IJ}\mathbf{u}_I = \mathbf{u}_J, \tag{5.3}$$

where the last equivalence is the essential property of the Kronecker delta function (δ_{IJ} allows one to switch I for any J in a summation, and one can show this with (5.2)). Using the above, it is easy to see that one can achieve $\mathbf{u}^h(\mathbf{x}_J) = \overline{\mathbf{u}}(\mathbf{x}_J)$ on Γ_u and $\delta\mathbf{u}^h(\mathbf{x}_J) = \mathbf{0}$ on Γ_u for all nodes J on the essential boundary. While the values may deviate between the nodes, the proper subsets of test and trial spaces can still be approximately constructed.

On the other hand, most meshfree methods do not enjoy the Kronecker Delta property, i.e., $\Psi_I(\mathbf{x}_J) \neq \delta_{IJ}$ and thus $\mathbf{u}^h(\mathbf{x}_J) \neq \mathbf{u}_J$. A displacement coefficient $\mathbf{u}_J$ is therefore not the true displacement, and instead it is called a *generalized displacement*. The value $\mathbf{u}^h(\mathbf{x}_J)$ is termed a *physical displacement*, which we will differentiate from the generalized one with a "hat," i.e., $\hat{\mathbf{u}}_J \equiv \mathbf{u}^h(\mathbf{x}_J)$. It is incorrect to impose constraints on generalized quantities and doing so will result in spurious solutions; for instance, the impenetration condition will no longer hold in contact problems. Significant research has therefore gone into special techniques for properly enforcing essential boundary conditions.

Key Takeaway
Observing a meshfree solution that does not satisfy imposed kinematic conditions (e.g. essential boundary conditions, impenetration in contact) is commonly caused by improperly imposing them on generalized quantities.

5.2 Basic Ideas, Notation, and Formal Requirements

In this section, we introduce the basic ideas and the formal requirements for enforcing essential boundary conditions at nodal locations, as well as some remarks and notations.

5.2.1 Basic Ideas

The approach of strong enforcement of boundary conditions consists of two steps. The meshfree approximation is modified such that the nodal coefficients for essential boundary nodes coincide with their field variables. Then, essential boundary conditions can be imposed directly. The set of shape functions $\{\Psi_I\}_{I=1}^{NP}$ is replaced by a set of new shape functions, $\{\hat{\Psi}_I\}_{I=1}^{NP}$ which satisfies $\hat{\Psi}_I(\mathbf{x}_J) = \delta_{IJ}$ for nodes on the essential boundary. Then $\mathbf{u}^h(\mathbf{x}_J) = \overline{\mathbf{u}}(\mathbf{x}_J)$ ($\overline{\mathbf{u}}(\mathbf{x})$ being the prescribed boundary displacements) and $\delta\mathbf{u}^h(\mathbf{x}_J) = \mathbf{0}$ can be achieved for all nodes J on the essential boundary.

5.2.2 Formal Requirements

Unlike weak methods, constrained and unconstrained degrees of freedom must be identified and treated seperately in the strong enforcement methods. For the complete collection of nodes S, we call the subset of nodes S_C the set of nodes that are constrained. Therefore, the compliment $S_\mathrm{F} = S \backslash S_\mathrm{C}$ defines the nodes that are free from constraints. Formally, we construct the approximations for the trial functions using $\{\hat{\Psi}_I\}_{I=1}^{NP}$ as

$$u_i^h(\mathbf{x}) = \sum_{I \in S_\mathrm{F}} \hat{\Psi}_I(\mathbf{x}) u_{iI} + \sum_{I \in S_\mathrm{C}} \hat{\Psi}_I(\mathbf{x}) \overline{u}_{iI}, \tag{5.4}$$

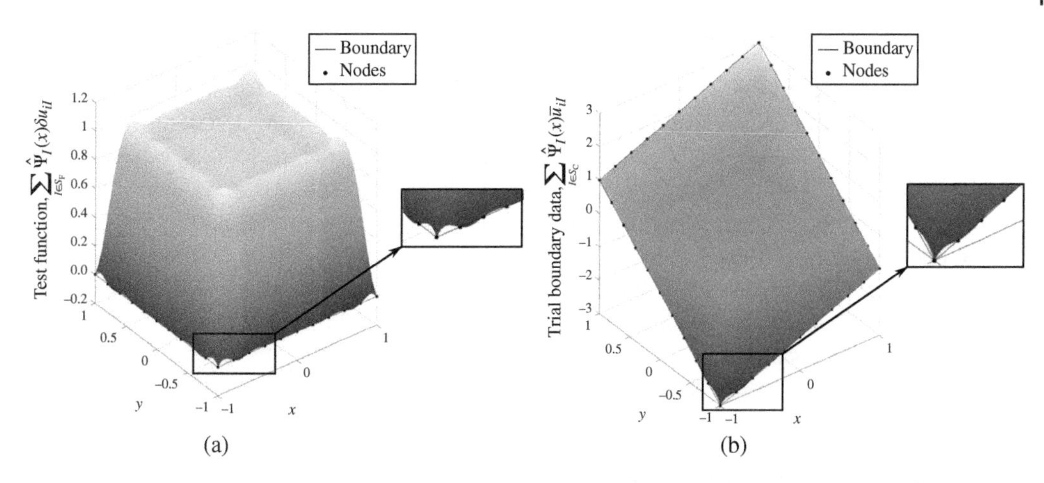

Figure 5.1 Strong enforcement of boundary conditions at nodal locations with $\Gamma = \Gamma_u$: (a) test function resulting from the transformation method, and (b) approximated boundary displacement using transformation method. The arbitrary coefficients of the test function in S_F are set to unity, and the essential boundary condition used for S_C is $\overline{u} = x + 2y$. *Source:* Hillman and Lin [1], figure 2, p. 7 and figure 3, p. 9 / With permission of Elsevier.

with $\overline{u}_{iI} \equiv \overline{u}_i(\mathbf{x}_I)$, and the set of test functions is constructed as

$$\delta u_i^h(\mathbf{x}) = \sum_{I \in S_F} \hat{\Psi}_I(\mathbf{x}) \delta u_{iI}. \tag{5.5}$$

An example of this approach is shown in Figure 5.1, where it is seen that the desired conditions are satisfied at nodes.

For $u_i^h(\mathbf{x})$ and $\delta u_i^h(\mathbf{x})$ to be nodally kinematically admissible, $u_i^h(\mathbf{x}_J) = \overline{u}_i(\mathbf{x}_J)$ and $\delta u_i^h(\mathbf{x}_J) = 0$, for all $\mathbf{x}_J \in S_C$, and the specific properties that our new shape functions $\{\hat{\Psi}_I\}_{I=1}^{NP}$ must satisfy are

$$\begin{aligned} \hat{\Psi}_I(\mathbf{x}_J) &= 0 \quad \forall \quad I \in S_F, \quad J \in S_C, \\ \hat{\Psi}_I(\mathbf{x}_J) &= \delta_{IJ} \quad \forall \quad I \in S_C, \quad J \in S_C. \end{aligned} \tag{5.6}$$

The above conditions mean that free nodes should not contribute to the approximation at nodal locations on the essential boundary, while all constrained nodes need to verify the delta property at nodal locations on the essential boundary. The conditions (5.6) can be utilized in (5.4) and (5.5) in order to verify $u_i^h(\mathbf{x}_J) = \overline{u}_i(\mathbf{x}_J)$ and $\delta u_i^h(\mathbf{x}_J) = 0$ for all constrained nodes $J \in S_C$.

Here, we remark that the condition (5.6) is distinct from both the delta property in (5.1) and the so-called *weak delta property*

$$N_I(\mathbf{x}) = 0 \quad \forall \quad \mathbf{x} \in \Gamma_u, \quad I \in S_F, \tag{5.7}$$

and is a comparatively less stringent requirement. In the literature, meshfree methods are described as lacking the delta property and therefore require additional effort in enforcing boundary conditions, but this is not concise. The methods discussed here satisfy (5.6) and in fact the exact delta property is undesirable since this can cause oscillations in the solution by forcing interpolation of the displacement coefficients through nodes.

Exercise 5.1 Show that $u_i^h(\mathbf{x}_J) = \overline{u}_i(\mathbf{x}_J)$ and $\delta u_i^h(\mathbf{x}_J) = 0$ for all constrained nodes $J \in S_C$ using the properties of the modified shape function (5.6) and definitions in (5.4) and (5.5).

5.2.3 Comment on Procedures

Due to the nature of meshfree approximations, rather than employing the formal definitions of trial and test approximations in (5.4) and (5.5), it is easier to explain the imposition of most boundary condition methods using the full statically uncondensed systems for all nodes in S, with boundary conditions applied to the nodes in S_C. It is still desirable however, to move the boundary conditions to the right-hand side of the system during assembly, provided the proper bookkeeping is in place (cf. the book by Hughes [2]).

Next, we shall discuss these procedures for the uncondensed matrix systems $\mathbf{Kd} = \mathbf{f}$ that result from employing the discrete variational forms introduced for the diffusion equation in Section 4.1 and for elasticity in Section 4.2.

5.3 Transformation Methods

5.3.1 Full Transformation Method: Matrix Implementation

A simple idea to impose values on the displacements at nodal locations is to utilize the relationship between the *generalized* and *physical* displacements [3–8]. This has been called the *collocation method* or *transformation method* [3, 8]. However, as one can imagine, the straightforward application of this idea results in the need for inverting a matrix the size of the total number of degrees of freedom and is somewhat impractical for real applications. Therefore, alternative approaches, which are both more practical and easier to implement, have been proposed, as will be seen in the next section. Nevertheless, the first method, which we will term the "matrix implementation" of the transformation method, serves as an illustrative example of enforcing essential boundary conditions strongly in meshfree methods.

To obtain the relationship between generalized and physical displacements, we first recall the meshfree approximation of displacement $\mathbf{u}$ at a location $\mathbf{x}$:

$$\mathbf{u}^h(\mathbf{x}) = \sum_{I \in S} \Psi_I(\mathbf{x})\mathbf{u}_I = \boldsymbol{\Psi}(\mathbf{x})\mathbf{d}, \tag{5.8}$$

where for two-dimensional elasticity, the matrix forms for the above are

$$\begin{aligned}
\boldsymbol{\Psi}(\mathbf{x}) &= [\boldsymbol{\Psi}_1(\mathbf{x}) \;\; \boldsymbol{\Psi}_2(\mathbf{x}) \ldots \boldsymbol{\Psi}_{NP}(\mathbf{x})], \\
\boldsymbol{\Psi}_I(\mathbf{x}) &= \begin{bmatrix} \Psi_I(\mathbf{x}) & 0 \\ 0 & \Psi_I(\mathbf{x}) \end{bmatrix}, \\
\mathbf{d} &= [u_{1I} \;\; u_{2I} \ldots u_{2NP}]^{\mathrm{T}}.
\end{aligned} \tag{5.9}$$

Thus, the physical displacement $\mathbf{u}^h$ at any node $\mathbf{x}_J$ is obtained as

$$\mathbf{u}^h(\mathbf{x}_J) = \boldsymbol{\Psi}(\mathbf{x}_J)\mathbf{d}. \tag{5.10}$$

Proceeding to express the physical displacements for all nodes, we have several equations:

$$\begin{aligned}
\mathbf{u}^h(\mathbf{x}_1) &= \boldsymbol{\Psi}(\mathbf{x}_1)\mathbf{d}, \\
\mathbf{u}^h(\mathbf{x}_2) &= \boldsymbol{\Psi}(\mathbf{x}_2)\mathbf{d}, \\
&\vdots \\
\mathbf{u}^h(\mathbf{x}_{NP}) &= \boldsymbol{\Psi}(\mathbf{x}_{NP})\mathbf{d}.
\end{aligned} \tag{5.11}$$

Since all of the equations involve the coefficients $\mathbf{d}$, it can be recognized that the above series of equations is a determined system of size $2 \times NP$, representing the transformation of generalized displacements to physical displacements. If we express the left-hand side of physical displacements as a column vector $\hat{\mathbf{d}}$, we obtain the system $\hat{\mathbf{d}} = \boldsymbol{\Lambda}\mathbf{d}$:

$$\begin{bmatrix} \mathbf{u}^h(\mathbf{x}_1) \\ \mathbf{u}^h(\mathbf{x}_2) \\ \vdots \\ \mathbf{u}^h(\mathbf{x}_{NP}) \end{bmatrix} \equiv \hat{\mathbf{d}} = \underbrace{\begin{bmatrix} \boldsymbol{\Psi}(\mathbf{x}_1) \\ \boldsymbol{\Psi}(\mathbf{x}_2) \\ \vdots \\ \boldsymbol{\Psi}(\mathbf{x}_{NP}) \end{bmatrix}}_{\boldsymbol{\Lambda}} \mathbf{d}, \tag{5.12}$$

where the entries of $\boldsymbol{\Lambda}$ are

$$\boldsymbol{\Lambda} = \begin{bmatrix} \Psi_1(\mathbf{x}_1) & \Psi_2(\mathbf{x}_1) & \cdots & \Psi_{NP}(\mathbf{x}_1) \\ \Psi_1(\mathbf{x}_2) & \Psi_2(\mathbf{x}_2) & & \Psi_{NP}(\mathbf{x}_2) \\ \vdots & & \ddots & \vdots \\ \Psi_1(\mathbf{x}_{NP}) & \Psi_2(\mathbf{x}_{NP}) & \cdots & \Psi_{NP}(\mathbf{x}_{NP}) \end{bmatrix}. \tag{5.13}$$

Therefore, given any physical displacements, we can obtain the associated generalized displacements as $\mathbf{d} = \boldsymbol{\Lambda}^{-1}\hat{\mathbf{d}}$. In order to utilize this to enforce boundary conditions, we substitute this relation into the matrix system of equations $\mathbf{Kd} = \mathbf{f}$ as

$$\mathbf{K}\boldsymbol{\Lambda}^{-1}\hat{\mathbf{d}} = \mathbf{f}. \tag{5.14}$$

Thus, we have a new system of equations with the physical displacements at the nodes as unknowns, just as in the finite element method. Boundary conditions are then directly imposed on $\hat{\mathbf{d}}$ when solving the system.

Finally, one can also pre-multiply both sides by $\boldsymbol{\Lambda}^{-\mathrm{T}}$ for symmetry:

$$\underbrace{\boldsymbol{\Lambda}^{-\mathrm{T}}\mathbf{K}\boldsymbol{\Lambda}^{-1}}_{\hat{\mathbf{K}}}\hat{\mathbf{d}} = \underbrace{\boldsymbol{\Lambda}^{-\mathrm{T}}\mathbf{f}}_{\hat{\mathbf{f}}} \rightarrow \hat{\mathbf{K}}\hat{\mathbf{d}} = \hat{\mathbf{f}}. \tag{5.15}$$

A standard static condensation procedure to impose the boundary conditions is given here for convenience. First, the node order is rearranged such that the physical displacements can be expressed as

$$\hat{\mathbf{d}} \rightarrow \begin{bmatrix} \hat{\mathbf{d}}_{\mathrm{C}} \\ \hat{\mathbf{d}}_{\mathrm{F}} \end{bmatrix}, \tag{5.16}$$

where $\hat{\mathbf{d}}_{\mathrm{F}}$ is the set of free physical displacements, which are unknown, and $\hat{\mathbf{d}}_{\mathrm{C}}$ is the set of constrained physical displacements which are prescribed in the problem. Similarly, the stiffness matrix and force vector are re-ordered:

$$\begin{aligned} \hat{\mathbf{K}} &\rightarrow \begin{bmatrix} \hat{\mathbf{K}}_{\mathrm{CC}} & \hat{\mathbf{K}}_{\mathrm{CF}} \\ \hat{\mathbf{K}}_{\mathrm{FC}} & \hat{\mathbf{K}}_{\mathrm{FF}} \end{bmatrix}, \\ \hat{\mathbf{f}} &\rightarrow \begin{bmatrix} \hat{\mathbf{f}}_{\mathrm{C}} \\ \hat{\mathbf{f}}_{\mathrm{F}} \end{bmatrix}. \end{aligned} \tag{5.17}$$

Thus, the system reads after the partition:

$$\begin{bmatrix} \hat{\mathbf{K}}_{\mathrm{CC}} & \hat{\mathbf{K}}_{\mathrm{CF}} \\ \hat{\mathbf{K}}_{\mathrm{FC}} & \hat{\mathbf{K}}_{\mathrm{FF}} \end{bmatrix} \begin{bmatrix} \hat{\mathbf{d}}_{\mathrm{C}} \\ \hat{\mathbf{d}}_{\mathrm{F}} \end{bmatrix} = \begin{bmatrix} \hat{\mathbf{f}}_{\mathrm{C}} \\ \hat{\mathbf{f}}_{\mathrm{F}} \end{bmatrix}. \tag{5.18}$$

The unknown displacements $\hat{\mathbf{d}}_{\mathrm{F}}$ are obtained by solving the system in the second set of row equations associated with the "free" nodes and using the known displacements on the boundary $\hat{\mathbf{d}}_{\mathrm{C}} = \{\overline{u}_i(\mathbf{x}_I)\}_{I \in S_{\mathrm{C}}}$:

$$\hat{\mathbf{K}}_{\mathrm{FF}}\hat{\mathbf{d}}_{\mathrm{F}} = \hat{\mathbf{f}}_{\mathrm{F}} - \hat{\mathbf{K}}_{\mathrm{FC}}\hat{\mathbf{d}}_{\mathrm{C}}. \tag{5.19}$$

Note that $\hat{\mathbf{d}}_{\mathrm{F}}$ is the set of physical displacements at the free nodes, not the generalized ones, which are needed to obtain the field variables such as stress. Therefore, for post-processing, the generalized displacements are obtained by the transformation back to the generalized displacements $\mathbf{d} = \mathbf{\Lambda}^{-1}\hat{\mathbf{d}}$.

The forces $\hat{\mathbf{f}}_{\mathrm{C}}$ are reaction forces and are treated as unknowns, and can be obtained if desired by examining the first set of equations once the displacements $\hat{\mathbf{d}}_{\mathrm{F}}$ are obtained:

$$\hat{\mathbf{f}}_{\mathrm{C}} = \hat{\mathbf{K}}_{\mathrm{CC}}\hat{\mathbf{d}}_{\mathrm{C}} + \hat{\mathbf{K}}_{\mathrm{CF}}\hat{\mathbf{d}}_{\mathrm{F}}. \tag{5.20}$$

The matrix version of the transformation method is given in **Box 5.1**.

Now, one may wonder, have we really modified the approximation space such that the boundary conditions can be employed directly now? The answer lies in examining the new system in the following exercise.

Box 5.1 Matrix Transformation Method

Given that one has constructed the full uncondensed $\mathbf{K}$ and $\mathbf{f}$ (of size $2NP$):

Step 1. Generate the transformation matrix $\mathbf{\Lambda}$: Loop over nodal locations and generate the shape functions at those locations. Fill in the matrix in the loop.

Step 2. Pre- and post-multiply $\mathbf{K}$ and $\mathbf{f}$ to form $\hat{\mathbf{K}} = \mathbf{\Lambda}^{-\mathrm{T}}\mathbf{K}\mathbf{\Lambda}^{-1}$ and $\hat{\mathbf{f}} = \mathbf{\Lambda}^{-\mathrm{T}}\mathbf{f}$.

Step 3. Reorder the numbering on the degrees of freedom, and partition the stiffness matrix and force vector such that the following system is obtained for the constrained nodes S_{C} and free nodes S_{F}:

$$\hat{\mathbf{K}}_{\mathrm{CC}}\hat{\mathbf{d}}_{\mathrm{C}} + \hat{\mathbf{K}}_{\mathrm{CF}}\hat{\mathbf{d}}_{\mathrm{F}} = \hat{\mathbf{f}}_{\mathrm{C}},$$
$$\hat{\mathbf{K}}_{\mathrm{FC}}\hat{\mathbf{d}}_{\mathrm{C}} + \hat{\mathbf{K}}_{\mathrm{FF}}\hat{\mathbf{d}}_{\mathrm{F}} = \hat{\mathbf{f}}_{\mathrm{F}}.$$

Step 4. Obtain the unknown interior nodal displacements by solving the system $\hat{\mathbf{K}}_{\mathrm{FF}}\hat{\mathbf{d}}_{\mathrm{F}} = \hat{\mathbf{f}}_{\mathrm{F}} - \hat{\mathbf{K}}_{\mathrm{FC}}\hat{\mathbf{d}}_{\mathrm{C}}$ where $\hat{\mathbf{d}}_{\mathrm{C}}$ contains the known essential boundary conditions.

Step 5. For post-processing, reorder the displacements and obtain the generalized displacements by $\mathbf{d} = \mathbf{\Lambda}^{-1}\hat{\mathbf{d}}$. Obtain the reaction forces as $\hat{\mathbf{f}}_{\mathrm{C}} = \hat{\mathbf{K}}_{\mathrm{CC}}\hat{\mathbf{d}}_{\mathrm{C}} + \hat{\mathbf{K}}_{\mathrm{CF}}\hat{\mathbf{d}}_{\mathrm{F}}$ if desired.

Exercise 5.2 Show the new system at hand using $\hat{\mathbf{K}}$ and $\hat{\mathbf{f}}$ employs a strain–displacement matrix $\hat{\mathbf{B}}$ and shape function vector $\hat{\boldsymbol{\Psi}}$ constructed with the transformed shape functions $\{\hat{\Psi}_I^{\mathrm{TM}}\}_{I=1}^{NP}$ from the operation $\hat{\boldsymbol{\Psi}} = \boldsymbol{\Psi}\boldsymbol{\Lambda}^{-1}$ and $\hat{\mathbf{B}} = \mathbf{B}\boldsymbol{\Lambda}^{-1}$.

Solution

Recalling the definition of the matrix $\hat{\mathbf{K}}$ and vector $\hat{\mathbf{f}}$ in (5.15) and the conventional discrete formulation for elasticity (see Section 4.2), we have

$$\begin{aligned}
\hat{\mathbf{K}} &= \boldsymbol{\Lambda}^{-\mathrm{T}} \int_{\Omega} \mathbf{B}^{\mathrm{T}}\mathbf{D}\mathbf{B}\mathrm{d}\Omega\boldsymbol{\Lambda}^{-1} = \int_{\Omega} \left(\mathbf{B}\boldsymbol{\Lambda}^{-1}\right)^{\mathrm{T}}\mathbf{D}\left(\mathbf{B}\boldsymbol{\Lambda}^{-1}\right)\mathrm{d}\Omega \equiv \int_{\Omega} \hat{\mathbf{B}}^{\mathrm{T}}\mathbf{D}\hat{\mathbf{B}}\mathrm{d}\Omega, \\
\hat{\mathbf{f}} &= \boldsymbol{\Lambda}^{-\mathrm{T}} \int_{\Omega} \boldsymbol{\Psi}^{\mathrm{T}}\mathbf{b}\mathrm{d}\Omega + \boldsymbol{\Lambda}^{-\mathrm{T}} \int_{\partial\Omega_h} \boldsymbol{\Psi}^{\mathrm{T}}\mathbf{h}\mathrm{d}\Gamma = \int_{\Omega} \left(\boldsymbol{\Psi}\boldsymbol{\Lambda}^{-1}\right)^{\mathrm{T}}\mathbf{b}\mathrm{d}\Omega + \int_{\partial\Omega_h} \left(\boldsymbol{\Psi}\boldsymbol{\Lambda}^{-1}\right)^{\mathrm{T}}\mathbf{h}\mathrm{d}\Gamma \\
&\equiv \int_{\Omega} \hat{\boldsymbol{\Psi}}^{\mathrm{T}}\mathbf{b}\mathrm{d}\Omega + \int_{\partial\Omega_h} \hat{\boldsymbol{\Psi}}^{\mathrm{T}}\mathbf{h}\mathrm{d}\Gamma.
\end{aligned} \tag{5.21}$$

Exercise 5.3 Show that the set of transformed shape functions $\{\hat{\Psi}_I^{\mathrm{TM}}\}_{I=1}^{NP}$ from Exercise 5.2 satisfies $\hat{\Psi}_I^{\mathrm{TM}}(\mathbf{x}_J) = \delta_{IJ}$.

From the exercises, it should be clear that $\{\hat{\Psi}_I^{\mathrm{TM}}\}_{I=1}^{NP}$ satisfies the requirements in (5.6). Indeed, $\{\hat{\Psi}_I^{\mathrm{TM}}\}_{I=1}^{NP}$ satisfies strong delta properties with the boundary condition imposition shown in Figure 5.1 for the transformation method.

5.3.2 Full Transformation Method: Row-Swap Implementation

A less cumbersome transformation procedure [8] involves swapping rows of the system matrix and force vector. The derivation leverages the fact that collocation of the boundary conditions is equivalent to employing Lagrange multipliers to enforce them as constraints *pointwise*, rather than continuously.

Consider the constrained variational principle for a single constraint at a node J in the ith direction $u_{iJ}^h = \bar{u}_i(\mathbf{x}_J) \equiv \bar{u}_{iJ}$, enforced with a scalar Lagrange multiplier λ:

$$\Pi_{\mathrm{E}}^{\mathrm{L}}\left(\mathbf{u}^h, \lambda\right) = \Pi_{\mathrm{E}}\left(\mathbf{u}^h\right) + \lambda\left(u_{iJ}^h - \bar{u}_{iJ}\right), \tag{5.22}$$

where

$$\Pi_{\mathrm{E}}\left(\mathbf{u}^h\right) = \frac{1}{2}\int_{\Omega} u_{(i,j)}^h C_{ijkl} u_{(k,l)}^h d\Omega - \int_{\Omega} u_i^h b_i d\Omega - \int_{\Gamma_t} u_i^h \bar{t}_i d\Gamma. \tag{5.23}$$

Here, we will use the notation $m = 2(J-1) + i$ to denote the mth degree of freedom, corresponding to the mth row of the stiffness matrix and force vector in 2-dimensional elasticity. Setting the variation of (5.22) equal to zero gives

$$\delta\Pi_{\mathrm{E}}^{\mathrm{L}} = \delta\Pi_{\mathrm{E}} + \delta\lambda\left(u_{iJ}^h - \bar{u}_{iJ}\right) + \lambda\left(\delta u_{iJ}^h\right) = 0. \tag{5.24}$$

The displacement and its variation can be expressed as $u_{iJ}^h = \mathbf{A}_{iJ}^{\mathrm{T}}\mathbf{d}$ and $\delta\mathbf{u}_{iJ}^h = \mathbf{A}_{iJ}^{\mathrm{T}}\delta\mathbf{d}$ where $\mathbf{d} = \{u_I\}$, $\delta\mathbf{d} = \{\delta u_I\}$, and

$$\begin{aligned} \mathbf{A}_{1J}^{\mathrm{T}} &= [\Psi_1(\mathbf{x}_J) \quad 0 \quad \Psi_2(\mathbf{x}_J) \quad 0 \; \ldots \; \Psi_{NP}(\mathbf{x}_J) \quad 0], \\ \mathbf{A}_{2J}^{\mathrm{T}} &= [0 \quad \Psi_1(\mathbf{x}_J) \quad 0 \quad \Psi_2(\mathbf{x}_J) \; \ldots \; 0 \quad \Psi_{NP}(\mathbf{x}_J)]. \end{aligned} \tag{5.25}$$

Comparing (5.25) and (5.13), it can be seen that $\mathbf{A}_{iJ}$ is simply the mth row of the transformation matrix $\mathbf{\Lambda}$ from the previous section.

It can be shown using (5.25), that (5.24) results in the following system of equations

$$\begin{bmatrix} \mathbf{K} & \mathbf{A}_{iJ} \\ \mathbf{A}_{iJ}^{\mathrm{T}} & 0 \end{bmatrix} \begin{bmatrix} \mathbf{d} \\ \lambda \end{bmatrix} = \begin{bmatrix} \mathbf{f} \\ \overline{u}_{iJ} \end{bmatrix}. \tag{5.26}$$

A row-swap procedure to enforce boundary conditions can be obtained by pre-multiplying the first row of block equations by $\mathbf{\Lambda}^{-\mathrm{T}}$:

$$\begin{bmatrix} \tilde{\mathbf{K}} & \mathbf{L}_{iJ} \\ \mathbf{A}_{iJ}^{\mathrm{T}} & 0 \end{bmatrix} \begin{bmatrix} \mathbf{d} \\ \lambda \end{bmatrix} = \begin{bmatrix} \tilde{\mathbf{f}} \\ \overline{u}_{iJ} \end{bmatrix}, \tag{5.27}$$

where $\tilde{\mathbf{K}} = \mathbf{\Lambda}^{-\mathrm{T}}\mathbf{K}$, $\tilde{\mathbf{f}} = \mathbf{\Lambda}^{-\mathrm{T}}\mathbf{f}$, and

$$\mathbf{L}_{iJ} = \begin{bmatrix} 0 \\ \vdots \\ 0 \\ 1 \\ 0 \\ \vdots \\ 0 \end{bmatrix} \begin{matrix} \\ \\ \\ \leftarrow \text{row } m, \\ \\ \\ \\ \end{matrix} \tag{5.28}$$

and we have used the fact that $\mathbf{\Lambda}^{-\mathrm{T}}\mathbf{A}_{iJ} = \mathbf{L}_{iJ}$.

The last row of (5.27) can be swapped with the mth row of $\tilde{\mathbf{K}}$ and $\tilde{\mathbf{f}}$ to obtain a convenient way to implement the boundary conditions. First, we express (5.27) explicitly:

$$\begin{bmatrix} \tilde{\mathbf{k}}_1 & 0 \\ \vdots & \vdots \\ \tilde{\mathbf{k}}_m & 1 \\ \vdots & \vdots \\ \tilde{\mathbf{k}}_{2NP} & 0 \\ \mathbf{A}_{iJ}^{\mathrm{T}} & 0 \end{bmatrix} \begin{bmatrix} u_1 \\ \vdots \\ u_m \\ \vdots \\ u_{2NP} \\ \lambda \end{bmatrix} = \begin{bmatrix} \tilde{f}_1 \\ \vdots \\ \tilde{f}_m \\ \vdots \\ \tilde{f}_{2NP} \\ \overline{u}_{iJ} \end{bmatrix}, \tag{5.29}$$

where $\tilde{\mathbf{k}}_m, \tilde{f}_m$, and u_m are the mth row of the matrix $\tilde{\mathbf{K}}$, vector $\tilde{\mathbf{f}}$, and vector $\mathbf{d}$, respectively. Swapping row m with the last row, we have

$$\begin{bmatrix} \tilde{\mathbf{k}}_1 & 0 \\ \vdots & \vdots \\ \mathbf{A}_{iJ}^{\mathrm{T}} & 0 \\ \vdots & \vdots \\ \tilde{\mathbf{k}}_{2NP} & 0 \\ \tilde{\mathbf{k}}_m & 1 \end{bmatrix} \begin{bmatrix} u_{11} \\ \vdots \\ u_m \\ \vdots \\ u_{2NP} \\ \lambda \end{bmatrix} = \begin{bmatrix} \tilde{f}_{11} \\ \vdots \\ \overline{u}_{iJ} \\ \vdots \\ \tilde{f}_{2NP} \\ \tilde{f}_m \end{bmatrix}. \tag{5.30}$$

This operation on the matrix $\tilde{\mathbf{K}}$ and vector $\tilde{\mathbf{f}}$ can be expressed as $\boldsymbol{K} = \mathcal{L}_{iJ}(\tilde{\mathbf{K}})$ and $\boldsymbol{f} = l_{iJ}(\tilde{\mathbf{f}})$, respectively, where $\mathcal{L}_{iJ}$ is the matrix operator switching the mth ($m = 2(J-1)+i$) row of $\tilde{\mathbf{K}}$ with $\mathbf{A}_{iJ}^{\mathrm{T}}$, and l_{iJ} is the vector operator switching the mth row of $\tilde{\mathbf{f}}$ with $\overline{u}_{iJ}$, so that the following equations for the generalized displacement $\mathbf{d}$ and the reaction force λ are obtained:

$$\begin{aligned} \boldsymbol{K}\mathbf{d} &= \boldsymbol{f}, \\ \lambda &= -\tilde{\mathbf{K}}_m\mathbf{d} + \tilde{f}_m. \end{aligned} \tag{5.31}$$

For multiple constraints, the operation is simply performed for all the prescribed degrees of freedom:

$$\begin{aligned} \boldsymbol{K} &= \mathcal{L}_{iJ}\mathcal{L}_{jK}...\mathcal{L}_{kL}(\tilde{\mathbf{K}}), \\ \boldsymbol{f} &= l_{iJ}l_{jK}...l_{kL}(\tilde{\mathbf{f}}). \end{aligned} \tag{5.32}$$

Compared to the matrix version, static condensation and reordering of degrees of freedom are not necessary, and the procedure is simple, as summarized in **Box 5.2**. The generalized displacements are also conveniently obtained directly from the system (5.31), and no transformation is necessary for postprocessing.

Exercise 5.4 Derive the system of equations in (5.26).

Box 5.2 Row-Swap Transformation Method

Given that one has constructed the full $\mathbf{K}$ and $\mathbf{f}$ (of size $2NP$):

Step 1. Generate the transformation matrix $\boldsymbol{\Lambda}$: Loop over nodal locations and generate the shape functions at those locations. Fill in the matrix in the loop.
Step 2. Pre-multiply $\mathbf{K}$ and $\mathbf{f}$ by $\boldsymbol{\Lambda}^{-\mathrm{T}}$ to form $\tilde{\mathbf{K}} = \boldsymbol{\Lambda}^{-\mathrm{T}}\mathbf{K}$, $\tilde{\mathbf{f}} = \boldsymbol{\Lambda}^{-\mathrm{T}}\mathbf{f}$.
Step 3. Loop over the constrained degrees of freedom and replace the mth row ($m = 2(J-1)+i$) of $\tilde{\mathbf{K}}$ with the mth row of $\boldsymbol{\Lambda}$ (that is, $\mathbf{A}_i(\mathbf{x}_J)$), and the mth entry of $\tilde{\mathbf{f}}$ with $\overline{u}_{iJ}$.
Step 4. Solve the system $\boldsymbol{K}\mathbf{d} = \boldsymbol{f}$ for the generalized displacements $\mathbf{d}$.

5.3.3 Mixed Transformation Method

A block-inverse method can be used to avoid inverting the matrix $\mathbf{\Lambda}$ (whose size is on the order of $NP \times NP$) for the transformation method [4, 7, 8], which has been termed the *mixed transformation method* [8]. Instead, inversion of a matrix only related to the constrained degrees of freedom is necessary, on the order of $NPC \times NPC$ where NPC is the number of constrained nodes; generally $NPC \ll NP$ which greatly eases the computational time since the operation count for an inversion is cubic. In addition, the resulting final left-hand side system is less dense (see Section 5.3.4), requiring less storage.

First, the nodes are separated into displacement vectors for nodes with degrees of freedom constrained $\mathbf{d}_\mathrm{C}$ and free $\mathbf{d}_\mathrm{F}$, along with the associated matrices:

$$\mathbf{d} \to \begin{bmatrix} \mathbf{d}_\mathrm{C} \\ \mathbf{d}_\mathrm{F} \end{bmatrix}, \quad \mathbf{K} \to \begin{bmatrix} \mathbf{K}_\mathrm{CC} & \mathbf{K}_\mathrm{CF} \\ \mathbf{K}_\mathrm{FC} & \mathbf{K}_\mathrm{FF} \end{bmatrix}, \quad \mathbf{f} \to \begin{bmatrix} \mathbf{f}_\mathrm{C} \\ \mathbf{f}_\mathrm{F} \end{bmatrix}. \tag{5.33}$$

The relationship between the physical displacements is

$$\hat{\mathbf{d}} = \begin{bmatrix} \hat{\mathbf{d}}_\mathrm{C} \\ \hat{\mathbf{d}}_\mathrm{F} \end{bmatrix} = \begin{bmatrix} \mathbf{\Lambda}_\mathrm{CC} & \mathbf{\Lambda}_\mathrm{CF} \\ \mathbf{\Lambda}_\mathrm{FC} & \mathbf{\Lambda}_\mathrm{FF} \end{bmatrix} \begin{bmatrix} \mathbf{d}_\mathrm{C} \\ \mathbf{d}_\mathrm{F} \end{bmatrix}. \tag{5.34}$$

A mixed displacement vector $\mathbf{d}^*$ can also be obtained and defined as

$$\mathbf{d}^* \equiv \begin{bmatrix} \hat{\mathbf{d}}_\mathrm{C} \\ \mathbf{d}_\mathrm{F} \end{bmatrix}. \tag{5.35}$$

The relationship between the mixed displacement $\mathbf{d}^*$ and the generalized displacement $\mathbf{d}$ is then

$$\mathbf{d}^* = \begin{bmatrix} \hat{\mathbf{d}}_\mathrm{C} \\ \mathbf{d}_\mathrm{F} \end{bmatrix} = \begin{bmatrix} \mathbf{\Lambda}_\mathrm{CC} & \mathbf{\Lambda}_\mathrm{CF} \\ \mathbf{0} & \mathbf{I} \end{bmatrix} \begin{bmatrix} \mathbf{d}_\mathrm{C} \\ \mathbf{d}_\mathrm{F} \end{bmatrix} \equiv \mathbf{\Lambda}^* \mathbf{d}. \tag{5.36}$$

As will be seen, only the inversion of $\mathbf{\Lambda}^*$ is needed rather than the matrix $\mathbf{\Lambda}$, which only requires inverting a much smaller matrix $\mathbf{\Lambda}_\mathrm{CC}$ (note again that performing an inverse of a matrix scales cubically and this is a significant improvement):

$$(\mathbf{\Lambda}^*)^{-1} = \begin{bmatrix} (\mathbf{\Lambda}_\mathrm{CC})^{-1} & -(\mathbf{\Lambda}_\mathrm{CC})^{-1}\mathbf{\Lambda}_\mathrm{CF} \\ \mathbf{0} & \mathbf{I} \end{bmatrix}. \tag{5.37}$$

Pre-multiplying the first row of blocks by $(\mathbf{\Lambda}^*)^{-\mathrm{T}}$ in $\mathbf{K}\mathbf{d} = \mathbf{f}$ using (5.33), and considering a single degree of freedom as in (5.26), the following system is obtained:

$$\begin{bmatrix} \mathbf{K}^* & \mathbf{I}_{iJ} \\ \mathbf{A}_{iJ}^\mathrm{T} & 0 \end{bmatrix} \begin{bmatrix} \mathbf{d} \\ \lambda \end{bmatrix} = \begin{bmatrix} \mathbf{f}^* \\ \bar{u}_{iJ} \end{bmatrix}. \tag{5.38}$$

where $\mathbf{K}^* = (\mathbf{\Lambda}^*)^{-\mathrm{T}}\mathbf{K}$ and $\mathbf{f}^* = (\mathbf{\Lambda}^*)^{-\mathrm{T}}\mathbf{f}$.

As before, using the row swap procedure, the following system can be solved for the displacement $\mathbf{d}$:

$$\boldsymbol{K}\mathbf{d} = \boldsymbol{f}, \tag{5.39}$$

Box 5.3 Mixed Transformation Method

Given that one has constructed the full $\mathbf{K}$ and $\mathbf{f}$ (of size $2NP$):

Step 1. Generate the transformation matrix $\mathbf{\Lambda}$: Loop over nodal locations and generate the shape functions at those locations. Fill in the matrix in the loop.
Step 2. Reorder the degrees of freedom and form $\mathbf{\Lambda}^*$.
Step 3. Pre-multiply $\mathbf{K}$ and $\mathbf{f}$ by $(\mathbf{\Lambda}^*)^{-\mathrm{T}}$ to form $\mathbf{K}^* = (\mathbf{\Lambda}^*)^{-\mathrm{T}}\mathbf{K}$ and $\mathbf{f}^* = (\mathbf{\Lambda}^*)^{-\mathrm{T}}\mathbf{f}$.
Step 4. Loop over the constrained degrees of freedom and replace the mth row ($m = 2(J-1)+i$) of $\mathbf{K}^*$ with the mth row of $\mathbf{\Lambda}$ (that is, $\mathbf{A}_i(\mathbf{x}_J)$), and the mth entry in $\mathbf{f}^*$ with $\bar{u}_{iJ}$.
Step 4. Solve the system $\boldsymbol{K}\mathbf{d} = \boldsymbol{f}$ for the generalized displacements $\mathbf{d}$, and reorder degrees of freedom for postprocessing as necessary.

where, for multiple constraints, the operation is again performed for all of the prescribed degrees of freedom:

$$\begin{aligned} \boldsymbol{K} &= \mathcal{L}_{iJ}\mathcal{L}_{jK}...\mathcal{L}_{kL}(\mathbf{K}^*), \\ \boldsymbol{f} &= l_{iJ}l_{jK}...l_{kL}(\mathbf{f}^*). \end{aligned} \tag{5.40}$$

where $\mathcal{L}_{iJ}$ and l_{iJ} are the matrix switching operator and the vector switching operator described in Section 5.3.2. A summary of the procedure is given in **Box 5.3**.

5.3.4 The Sparsity of Transformation Methods

One key consideration in the selection of these methods is the effort involved and the properties of the system matrices at hand. For full and mixed transformation methods, the matrices significantly differ in sparsity. Figure 5.2 shows the sparsity patterns for mixed and full transformation methods at various stages of computation. Since the inverse of the full transformation matrix $\mathbf{\Lambda}$ is not sparse, neither is the product with the stiffness matrix; recall the system matrix for the full transformation method $\hat{\mathbf{K}}$ in (5.15) and $\tilde{\mathbf{K}}$ (5.27) involve the multiplication by $\mathbf{\Lambda}^{-\mathrm{T}}$. For the mixed transformation on the other hand, the block inverse in (5.37) retains significant sparsity, and subsequently, the system matrix $\mathbf{K}^*$ in (5.38) does as well.

Nevertheless, it is possible to *ignore* entries in the system matrices with a small contribution to the overall solution, provided a proper tolerance is chosen; here, these are denoted as "reduced" cases; reduced refers to only retaining entries whose absolute value is greater than a certain tolerance. As an example, for the convergence of the reproducing kernel particle method (RKPM) in the Poisson equation shown in Figure 5.3, a tolerance of only retaining stiffness matrix entries with $|K_{IJ}| > 10^{-5}$ maintains optimal convergence for the linear basis chosen. For other partial differential equations, this tolerance should depend on the problem constants.

5.3.5 Preconditioners in Transformation Methods

Nonstationary iterative solvers rely on the condition number and spectral properties of the system matrix at hand for convergence. As a result, the number of iterations for convergence in these solvers is significantly lower for better-conditioned matrices. For the solvers and preconditioners we have tested, LU factorization with no fill and incomplete Cholesky factorization with no fill preconditioning for the full transformation method can converge in *one* iteration for many solvers. Omitting entries in the system matrix, as previously discussed, increases the number of iterations

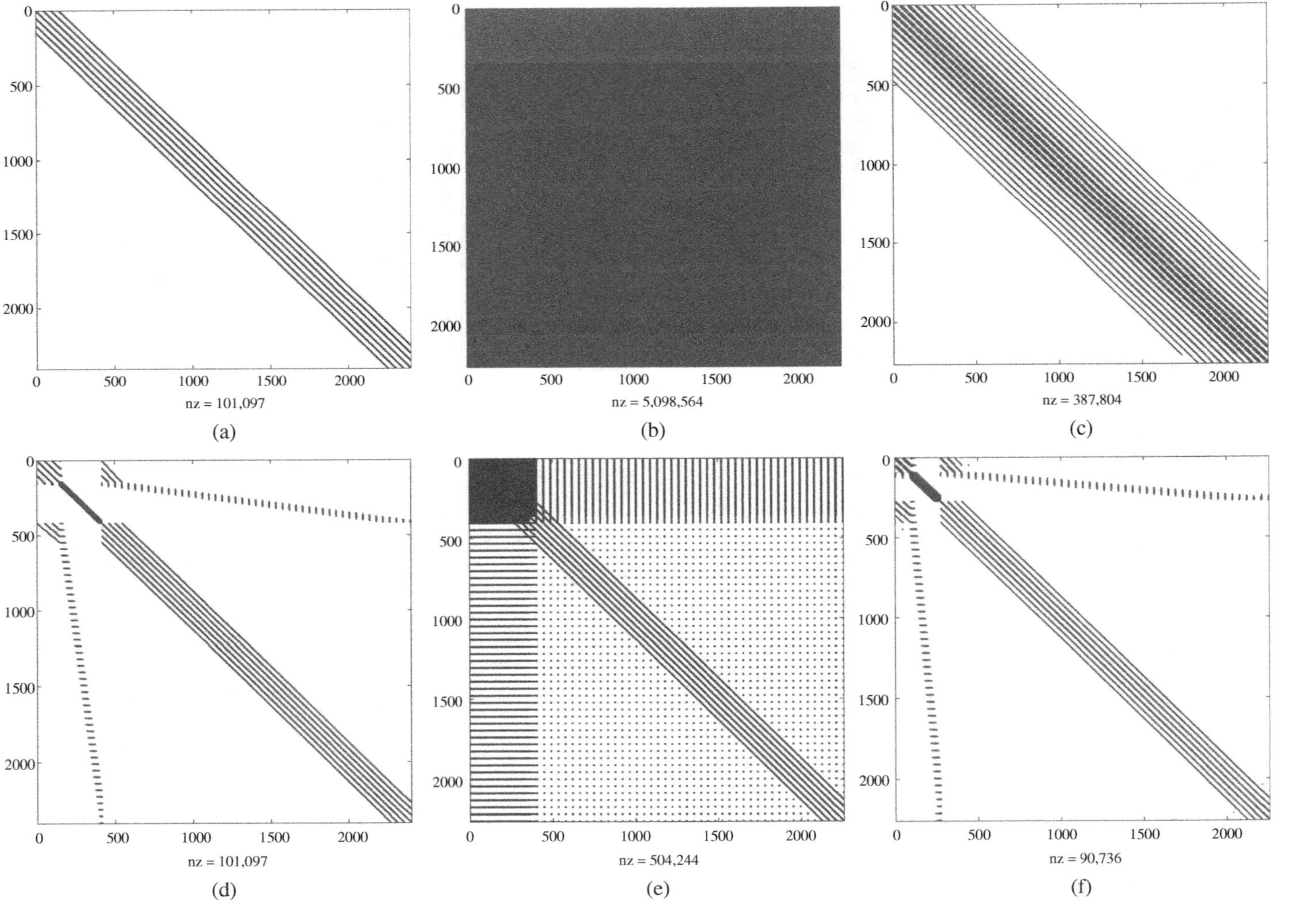

Figure 5.2 Sparsity patterns (nz = number of nonzero elements): (a) original stiffness matrix without modification; (b) fully transformed stiffness matrix for constrained nodes (5.18); (c) fully transformed stiffness matrix for constrained nodes reduced to only contain entries with significant contribution to the system of equations; (d) original stiffness matrix with reordered nodes for mixed method (5.33); (e) transformed mixed stiffness matrix (5.38) for constrained nodes; and (f) transformed mixed stiffness matrix for constrained nodes, reduced to only contain entries with significant contribution to the system of equations.

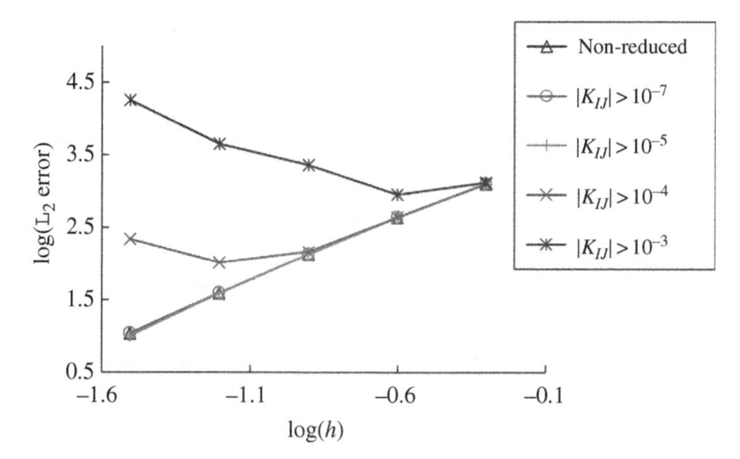

Figure 5.3 Convergence of reproducing kernel particle method in the Poisson equation with a reduced full transformation method with various tolerances for the reduction.

to converge. Clearly, there are tradeoffs that are worth investigating. Elaboration of the proper selection of solvers based on other system characteristics (e.g. positive definite, symmetric, etc.) can be found in [9].

5.4 Boundary Singular Kernel Method

The boundary singular kernel method [8] modifies the kernel of the reproducing kernel (RK) approximation (or weight function in the moving least squares (MLS) approximation) for nodes located on the essential boundary such that the generalized displacements are the physical displacements. Nonetheless, while boundary conditions can be directly imposed, this method does not recover the Kronecker delta property of the shape functions unless singular kernels are used for all nodes. However, it is important to note that this is not necessary (see Section 5.2.2) and is in fact undesirable since it would force the solution to pass through nodal locations, leading to oscillations.

Singular kernels were first suggested by Shepard [10] and Lancaster Salkauskas [11], and subsequently rediscovered in the advent of meshfree methods for the Galerkin solution to PDEs [15, 8]. The essential idea is that when a singular kernel is employed in an MLS or RK approximation, the generalized displacement becomes the physical displacement. The only task then is to develop a singular kernel for use in the approximation and employ it for nodes located on the essential boundary. A relatively simple singular kernel was proposed in [8], where the order of the singularity can be controlled. For a node $\tilde{\mathbf{x}}_I$ located on the essential boundary, the following singular kernel $\tilde{\phi}_a$ can be employed:

$$\tilde{\phi}_a(\mathbf{x}-\tilde{\mathbf{x}}_I) = \frac{\phi_a(\mathbf{x}-\tilde{\mathbf{x}}_I)}{f(\mathbf{x}-\tilde{\mathbf{x}}_I)}, \tag{5.41}$$

where the function f satisfies $f(\mathbf{0}) = 0$, resulting in a singularity when $\mathbf{x} = \tilde{\mathbf{x}}_I$. The function f can be constructed as

$$f(\mathbf{x}-\tilde{\mathbf{x}}_I) = \left[\left(\frac{x-\tilde{x}_I}{a_{Ix}}\right)^2 + \left(\frac{y-\tilde{y}_I}{a_{Iy}}\right)^2 + \left(\frac{z-\tilde{z}_I}{a_{Iz}}\right)^2\right]^p, \quad p > 0, \tag{5.42}$$

where $2p$ is the order of the singularity (order one is acceptable [8]). Singular kernels are introduced for all nodes on the essential boundary; thus, the modified approximation is constructed as

$$\hat{\Psi}_I^{\mathrm{BSK}}(\mathbf{x}) = \begin{cases} \mathbf{H}^{\mathrm{T}}(\mathbf{0})\tilde{\mathbf{M}}^{-1}(\mathbf{x})\mathbf{H}(\mathbf{x}-\mathbf{x}_I)\phi_a(\mathbf{x}-\mathbf{x}_I) & I \in S_{\mathrm{F}} \\ \mathbf{H}^{\mathrm{T}}(\mathbf{0})\tilde{\mathbf{M}}^{-1}(\mathbf{x})\mathbf{H}(\mathbf{x}-\mathbf{x}_I)\tilde{\phi}_a(\mathbf{x}-\mathbf{x}_I) & I \in S_{\mathrm{C}} \end{cases}, \tag{5.43}$$

and since the moment matrix also involves the kernel function, it is computed as

$$\tilde{\mathbf{M}}(\mathbf{x}) = \sum_{I \in S_{\mathrm{F}}} \mathbf{H}(\mathbf{x}-\mathbf{x}_I)\mathbf{H}^{\mathrm{T}}(\mathbf{x}-\mathbf{x}_I)\phi_a(\mathbf{x}-\mathbf{x}_I) + \sum_{I \in S_{\mathrm{C}}} \mathbf{H}(\mathbf{x}-\mathbf{x}_I)\mathbf{H}^{\mathrm{T}}(\mathbf{x}-\mathbf{x}_I)\tilde{\phi}_a(\mathbf{x}-\mathbf{x}_I). \tag{5.44}$$

It can then be shown that the RK and MLS approximations with a singular kernel shape function has the following properties (see [8] for the proof):

$$\begin{aligned} &\lim_{\mathbf{x} \longrightarrow \mathbf{x}_J} \hat{\Psi}_I^{\mathrm{BSK}}(\mathbf{x}) = 0 \quad \forall \;\; I \in S_{\mathrm{F}}, \;\; J \in S_{\mathrm{C}}, \\ &\lim_{\mathbf{x} \longrightarrow \mathbf{x}_J} \hat{\Psi}_I^{\mathrm{BSK}}(\mathbf{x}) = \delta_{IJ} \quad \forall \;\; I \in S_{\mathrm{C}}, \;\; J \in S_{\mathrm{C}}. \end{aligned} \tag{5.45}$$

Thus, the function does not satisfy the Kronecker delta property, but does satisfy the necessary requirements to impose boundary conditions (5.6) in the limit. A summary of the construction of shape functions with boundary singular kernels is given in **Box 5.4** (it is useful to compare this to **Box 3.1** to see the modifications to the standard RK shape functions).

Note that in practice, one cannot evaluate the function at the nodal location due to the singularity. Therefore, either logic can be considered in the code to set the limit values, or for ease of implementation, a small perturbation can be introduced into the distance measures in the kernel function:

$$\tilde{\phi}_a(\mathbf{x}-\mathbf{x}_I) = \tilde{\phi}_a\left(\frac{\|\mathbf{x}-\mathbf{x}_I\| + \varepsilon}{a}\right), \tag{5.46}$$

Box 5.4 Boundary Singular Kernel Method: Construction of Shape Functions

Given: The set of indices S_{C} with constrained degrees of freedom, $\mathbf{x}$, the set of nodes $\{\mathbf{x}\}_{I=1}^{NP}$, indices $S_{\mathbf{x}}$, kernel function, set of nodal supports $\{a_I\}_{I=1}^{NP}$, and order of basis n.

Step 1. Moment matrix. Loop over the indices of the nodes in $S_{\mathbf{x}}$ with index I:

- Compute $\mathbf{H}(\mathbf{x}-\mathbf{x}_I)$ and kernel function and store in memory. If this index I is in the set S_{C}, use $\tilde{\phi}_{a_I}(\mathbf{x}-\mathbf{x}_I) = \frac{\phi_{a_I}(\mathbf{x}-\mathbf{x}_I)}{f(\mathbf{x}-\mathbf{x}_I)}$, else use $\phi_{a_I}(\mathbf{x}-\mathbf{x}_I)$.
- Add the contributions to the moment matrix $\tilde{\mathbf{M}}(\mathbf{x})$.

Step 2. Get correction function. Solve $\tilde{\mathbf{M}}(\mathbf{x})\tilde{\mathbf{b}}(\mathbf{x}) = \mathbf{H}(\mathbf{0})$.

Step 3. Shape function construction. Loop over the indices of the nodes in $S_{\mathbf{x}}$ with index I to obtain the shape functions:

$$\hat{\Psi}_I^{\mathrm{BSK}}(\mathbf{x}) = \begin{cases} \tilde{\mathbf{b}}^{\mathrm{T}}(\mathbf{x})\mathbf{H}(\mathbf{x}-\mathbf{x}_I)\phi_{a_I}(\mathbf{x}-\mathbf{x}_I) & I \in S_{\mathrm{F}} \\ \tilde{\mathbf{b}}^{\mathrm{T}}(\mathbf{x})\mathbf{H}(\mathbf{x}-\mathbf{x}_I)\tilde{\phi}_{a_I}(\mathbf{x}-\mathbf{x}_I) & I \in S_{\mathrm{C}} \end{cases}.$$

where ε is a perturbation that can be taken on the order of $a \times 10^{-6}$. We have found that significant error can be introduced into the approximation if this number is too close to machine precision (usually 10^{-16}).

5.5 RK with Nodal Interpolation

Another way to obtain a meshfree shape function with the necessary properties for direct enforcement of boundary conditions is to utilize a *primitive function* as a type of enrichment [12]. For example, the RK shape function Ψ_I supplemented with a function $\overline{\Psi}_I(\mathbf{x})$ in the approximation. The function is only introduced for constrained nodes in the set S_C:

$$\mathbf{u}^h(\mathbf{x}) = \sum_{I \in S_F} \Psi_I(\mathbf{x})\mathbf{u}_I + \sum_{I \in S_C} \{\Psi_I(\mathbf{x}) + \overline{\Psi}_I(\mathbf{x})\}\mathbf{u}_I \equiv \sum_{I \in S} \hat{\Psi}_I^{\text{RKNI}}(\mathbf{x})\mathbf{u}_I, \tag{5.47}$$

where $\overline{\Psi}_I(\mathbf{x})$ satisfies the Kronecker delta property $\overline{\Psi}_I(\mathbf{x}_J) = \delta_{IJ}$. Enforcing the reproducing conditions on the shape functions $\{\hat{\Psi}_I^{\text{RKNI}}\}_{I=1}^{NP}$ will result in $\hat{\Psi}_I^{\text{RKNI}}(\mathbf{x}_J) = \delta_{IJ}, \forall\, I \in S_C, J \in S_C$. The construction of $\overline{\Psi}_I(\mathbf{x})$ can be easily accomplished by using a compact support $\overline{a}$ for the function, small enough such that it does not cover any other node (in a uniform discretization with equal nodal spacing h, $\overline{a} < h$), with $\overline{\Psi}_I(\mathbf{x}_I) = 1$. This can be achieved with the following kernel approximation:

$$\overline{\Psi}_I(\mathbf{x}) = \frac{\phi_{\overline{a}}(\mathbf{x} - \mathbf{x}_I)}{\phi_{\overline{a}}(\mathbf{0})}, \tag{5.48}$$

since if $\mathbf{x} = \mathbf{x}_I$ then $\overline{\Psi}_I(\mathbf{x}) = \phi_{\overline{a}}(\mathbf{0})/\phi_{\overline{a}}(\mathbf{0}) = 1$, and if $\mathbf{x} = \mathbf{x}_J$ with $I \neq J$ then $\overline{\Psi}_I(\mathbf{x}) = 0$. Proceeding to enforce the reproducing conditions on $\{\hat{\Psi}_I^{\text{RKNI}}\}_{I=1}^{NP}$, the following equations for the unknown coefficients are obtained:

$$\begin{aligned} &\mathbf{M}(\mathbf{x})\overline{\mathbf{b}}(\mathbf{x}) = \mathbf{H}(\mathbf{0}) - \overline{\mathbf{f}}(\mathbf{x}), \\ &\overline{\mathbf{f}}(\mathbf{x}) = \sum_{I \in S_C} \overline{\Psi}_I(\mathbf{x})\mathbf{H}(\mathbf{x} - \mathbf{x}_I), \end{aligned} \tag{5.49}$$

where all the terms are the same as in the previous definitions in Chapter 3, and the shape functions are thus constructed as

$$\hat{\Psi}_I^{\text{RKNI}}(\mathbf{x}) = \begin{cases} \Psi_I(\mathbf{x}) & I \in S_F \\ \Psi_I(\mathbf{x}) + \overline{\Psi}_I(\mathbf{x}) & I \in S_C \end{cases}, \tag{5.50}$$

where

$$\Psi_I(\mathbf{x}) = \overline{\mathbf{b}}^{\mathrm{T}}(\mathbf{x})\mathbf{H}(\mathbf{x} - \mathbf{x}_I)\phi_a(\mathbf{x} - \mathbf{x}_I). \tag{5.51}$$

It can be shown that the shape functions (5.50) satisfy similar properties as the singular kernel shape function, but no longer in the limit sense:

$$\begin{aligned} &\hat{\Psi}_I^{\text{RKNI}}(\mathbf{x}_J) = 0 \quad \forall\; I \in S_F,\; J \in S_C, \\ &\hat{\Psi}_I^{\text{RKNI}}(\mathbf{x}_J) = \delta_{IJ} \quad \forall\; I \in S_C,\; J \in S_C. \end{aligned} \tag{5.52}$$

The construction of the set of shape functions with nodal interpolation is summarized in **Box 5.5** (again, it is useful to compare this to **Box 3.1** to see the modifications to the standard RK shape functions).

Exercise 5.5 Prove that the RK-interpolation shape function satisfies (5.52).

Box 5.5 RK with Interpolation: Construction of Shape Functions

Given: The set of indices S_C with constrained degrees of freedom, $\mathbf{x}$, the set of nodes $\{\mathbf{x}\}_{I=1}^{NP}$, indices $S_{\mathbf{x}}$, kernel function, set of nodal supports $\{a_I\}_{I=1}^{NP}$, and order of basis n.

Step 1. Moment matrix. Loop over the indices of the nodes in $S_{\mathbf{x}}$ with index I:

- Compute $\mathbf{H}(\mathbf{x}-\mathbf{x}_I)$ and $\phi_{a_I}(\mathbf{x}-\mathbf{x}_I)$ and store in memory.
- Add the contributions to the moment matrix:

$$\mathbf{M}(\mathbf{x}) = \sum_{I\in S_{\mathbf{x}}} \mathbf{H}(\mathbf{x}-\mathbf{x}_I)\mathbf{H}^{\mathrm{T}}(\mathbf{x}-\mathbf{x}_I)\phi_{a_I}(\mathbf{x}-\mathbf{x}_I).$$

- Also add the contributions to $\bar{\mathbf{f}}(\mathbf{x})$:

$$\bar{\mathbf{f}}(\mathbf{x}) = \sum_{I\in S_C} \overline{\Psi}_I(\mathbf{x})\mathbf{H}(\mathbf{x}-\mathbf{x}_I).$$

Step 2. Get correction function. Solve $\mathbf{M}(\mathbf{x})\bar{\mathbf{b}}(\mathbf{x}) = \mathbf{H}(\mathbf{0}) - \bar{\mathbf{f}}(\mathbf{x})$.

Step 3. Shape function construction. Loop over the indices of the nodes in $S_{\mathbf{x}}$ with index I to obtain the shape functions:

$$\hat{\Psi}_I^{\mathrm{RKNI}}(\mathbf{x}) = \begin{cases} \Psi_I(\mathbf{x}) & I\in S_F \\ \Psi_I(\mathbf{x}) + \overline{\Psi}_I(\mathbf{x}) & I\in S_C \end{cases}, \quad \Psi_I(\mathbf{x}) = \bar{\mathbf{b}}^{\mathrm{T}}(\mathbf{x})\mathbf{H}(\mathbf{x}-\mathbf{x}_I)\phi_{a_I}(\mathbf{x}-\mathbf{x}_I).$$

5.6 Coupling with Finite Elements on the Boundary

Since finite elements satisfy the necessary conditions to impose essential boundary conditions directly, we now describe a simple way to enforce boundary conditions, which is to *blend* or *couple* meshfree and finite element approximations. For example, an averaged approximation [13, 14] can be introduced to define the coupling between the two sets of shape functions:

$$u_i^h(\mathbf{x}) = \sum_{I=1}^{NP} [r(\mathbf{x})N_I(\mathbf{x}) + (1-r(\mathbf{x}))\Psi_I(\mathbf{x})]u_{iI} \equiv \sum_{I=1}^{NP} \hat{\Psi}_I^{\mathrm{CFEM}}(\mathbf{x})u_{iI}, \tag{5.53}$$

where N_I is a finite element (FE) shape function, and $r(\mathbf{x})$ is a ramping or coupling function. The ramping function $r(\mathbf{x})$ can be defined in various ways but needs to satisfy $r(\mathbf{x}) = 1$ on Γ_u and should decrease to zero away from the boundary. One possible way to accomplish this is to use the FEM shape functions *themselves*. Overall, the coupling width should be several finite elements wide to avoid spurious wave reflection due to the mismatch of the two discretizations in dynamic problems.

The convex linear combination in (5.53) guarantees that the constant and linear completeness conditions hold

$$\sum_{I=1}^{NP} \hat{\Psi}_I^{\mathrm{CFEM}}(\mathbf{x}) = 1, \quad \sum_{I=1}^{NP} \hat{\Psi}_I^{\mathrm{CFEM}}(\mathbf{x})\mathbf{x}_I = \mathbf{x}. \tag{5.54}$$

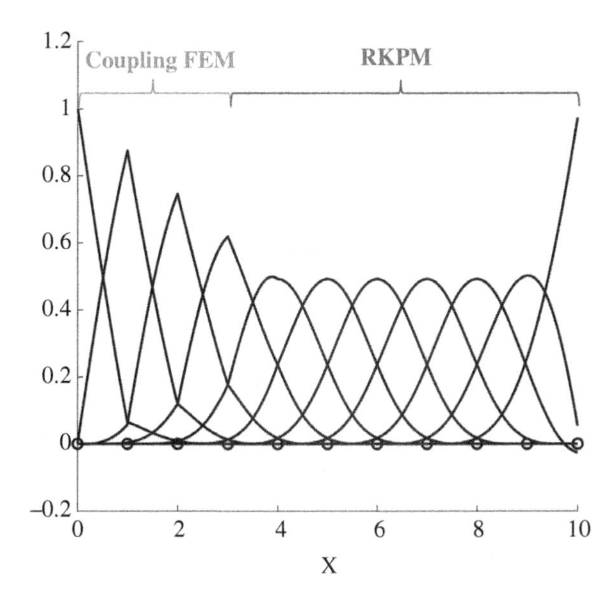

Figure 5.4 Coupled meshfree and finite element shape functions in one dimension.

Higher-order consistency conditions can also be obtained in a straightforward manner by enforcing them directly on $\{\hat{\Psi}_I^{\mathrm{CFEM}}\}_{I=1}^{NP}$.

Exercise 5.6 Prove that the coupled shape functions satisfy (5.54).

A distance function $d(\mathbf{x})$ can be utilized for $r(\mathbf{x})$

$$r(\mathbf{x}) = \begin{cases} \dfrac{w_c - d(\mathbf{x})}{w_c} & \text{for } d(\mathbf{x}) \leq w_c \\ 0 & \text{otherwise} \end{cases}, \tag{5.55}$$

where w_c is the width of the coupling region and

$$d(\mathbf{x}) = \|\mathbf{x} - \mathbf{x}_c\|, \tag{5.56}$$

where $\mathbf{x}_c$ is the closest point on the essential boundary. The width w_c should be sufficiently large to yield a smooth transition of shape functions. A width w_c several times the nodal spacing is sufficient, so several layers of finite elements should be utilized in practice. Figure 5.4 shows the averaged approximation scheme for coupling FEM and RK shape functions in one dimension.

5.7 Comparison of Strong Methods

As we have seen, the implementation of the various methods so far varies considerably; a comparison between the methods discussed is presented in Table 5.1. For the transformation methods, the transformation matrices must be constructed which is trivial, but need inversion which can be

Table 5.1 Comparison of operation counts and other considerations in strong methods.

Method	Operation count considerations	Sparsity of system	Implementation
Full transformation with matrix	Inverse: total operations $O(NP^3)$	Fully dense	Vector/matrix partitioning for free and constrained nodes.
Full transformation with row-swap	Inverse: total operations $O(NP^3)$	Fully dense	Standard bookkeeping as in FEM.
Mixed transformation	Inverse: total operations $O(NPC^3)$	Partially Dense	Vector/matrix partitioning for free and constrained nodes.
Singular kernel	None	Sparse	Modification to meshfree approximation; standard bookkeeping as in FEM.
RK with interpolation	None	Sparse	Modification to meshfree approximation; standard bookkeeping as in FEM.
Coupling with finite elements	None	Sparse	Implementation of element coupling; standard bookkeeping as in FEM.

computationally intensive depending on the method. From Table 5.1, it can be seen that the mixed transformation method is far more efficient in this regard.

On the other hand, RK with interpolation, singular kernels, and coupling with FEM can avoid an inverse altogether. The additional effort for these methods is the modification of meshfree approximations themselves. Forming the system and solving it is overall more straightforward after this step and follows FEM. Because of this feature, if a meshfree method is implemented into an existing finite element code (e.g. a production code), these methods are preferred. These methods are also more favorable in explicit dynamic codes as essential boundary conditions can be easily enforced during time integration.

In any case, these methods generally all perform equally well. The optimal convergence in a Poisson equation using RKPM with linear basis and transformation (mixed and full have the same solution and only differ in implementation), singular kernel, and RK with interpolation is shown in Figure 5.5, where nearly identical results are obtained. Singular kernels and interpolation functions are only implemented on boundary nodes. Further discussion and comparisons can be found in [8].

For the following exercises, consider the linear RK or MLS approximation with a normalized support size $a/h = 2.0$ for a Galerkin meshfree solution in conjunction with any of the methods presented. It is suggested to use the row-swap version of the full transformation method to avoid excessive bookkeeping. Use at least eight-point Gauss integration (between each node in 1D, 8×8 integration using finite elements coincident with nodal positions in 2D) to compute the stiffness matrix, force vectors, and norms. If it is unclear how to perform the numerical integration, consult Section 6.2 on Gaussian integration for meshfree methods.

Exercise 5.7

1) For the following two-point boundary value problem for diffusion with $k = 1$: find u such that

$$\begin{aligned} &u_{,xx} = 0 \quad \text{for} \quad 0 < x < 1, \\ &u(0) = 0, \\ &u(1) = 1. \end{aligned} \tag{5.57}$$

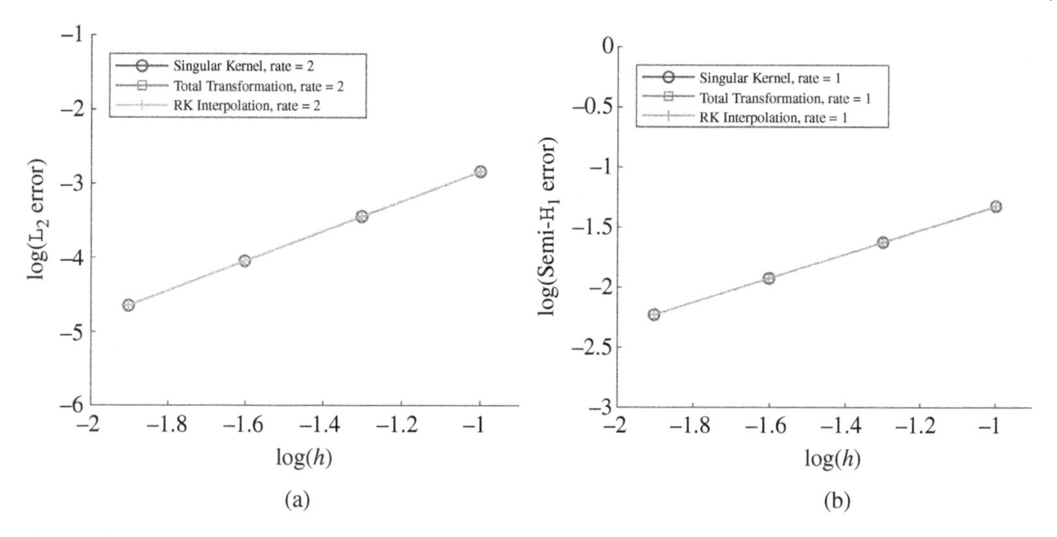

Figure 5.5 Optimal convergence for the singular kernel, transformation, and RK approximation with interpolation using linear basis in a Poisson equation.

The exact solution is $u = x$. Solve the problem with linear RK or MLS approximations using the Galerkin formulation and a set of eleven ($NP = 11$) *evenly* distributed nodes $\{x_I\}_{I=1}^{NP}$ over the domain $\overline{\Omega} = [0, 1]$ (the nodal spacing h is 1/10). Plot the exact solution and the RK or MLS solution in one plot and confirm the agreement between the solutions. Obtain the L_2 norm of the error e_{L_2} defined as

$$e_{\mathrm{L}_2} = \left(\int_0^{10} \left(u(x) - u^h(x)\right)^2 dx \right)^{1/2}. \tag{5.58}$$

Keep in mind that you will have to use numerical quadrature to integrate both the weak form and the norm.

2) Now consider the following two-point boundary value problem for diffusion with $k = 1$: find u such that

$$\begin{aligned} u_{,xx} &= e^x \quad \text{for} \quad 0 < x < 1, \\ u(0) &= e^0 = 1, \\ u(1) &= e^1 = e. \end{aligned} \tag{5.59}$$

The exact solution is $u = e^x$. Solve the problem with linear RK or MLS approximations using the Galerkin formulation. Use a set of eleven ($NP = 11$) evenly distributed nodes $\{x_I\}_{I=1}^{NP}$ over a domain $\overline{\Omega} = [0, 1]$ (the nodal spacing h is 1/10), then refine the solution by letting $h = 1/10, 1/20, 1/40$. Use Gaussian quadrature to evaluate the integrals. Plot the exact solution and the RK/MLS solution for the first refinement, compute the L_2 error norm of the error denoted by e_{L_2}, and obtain the rate of convergence (slope of the data) by plotting $\log_{10}(e_{\mathrm{L}_2})$ vs. $\log_{10}(h)$, where h is the nodal distance. Compare to the optimal rate of two (see Chapter 3, Eq. (3.95)).

Exercise 5.8 Consider the multi-dimensional RK approximation for a Galerkin solution. Solve the following boundary value problem for diffusion with $k = 1$, for the scalar u on $\overline{\Omega} = [0, 1] \times [0, 1]$:

$$
\begin{aligned}
u_{,ii} &= 0 \text{ on } \Omega, \\
u &= x + 2y \text{ on } \Gamma_u,
\end{aligned} \tag{5.60}
$$

where $u_{,ii} = u_{,xx} + u_{,yy}$ and $\Gamma_u = \Gamma$, $\Gamma_h = \emptyset$. The exact solution is $u = x + 2y$. Plot the exact and Galerkin solutions, along with the error, and confirm the agreement between the two solutions. Note that in general, meshfree methods will not recover the exact solution despite using linear basis; Chapter 6 introduces special techniques to correct this error.

Exercise 5.9 Solve the following boundary value problem for diffusion with $k = 1$ for the scalar u on $\overline{\Omega} = [0, 1] \times [0, 1]$:

$$
\begin{aligned}
u_{,ii} &= \left(x^2 + y^2\right)e^{xy} \quad \text{on} \quad \Omega, \\
u &= e^{xy} \quad \text{on} \quad \Gamma_u,
\end{aligned} \tag{5.61}
$$

where $u_{,ii} = u_{,xx} + u_{,yy}$, and $\Gamma_u = \Gamma, \Gamma_h = \emptyset$. The exact solution is $u = e^{xy}$. Obtain numerical solution using 11×11, 21×21, and 41×41 nodes.

1) Plot the solution error for 11×11 nodes.
2) Compare numerical and exact solutions along lines $x = 0.25$, $x = 0.5$, and $x = 0.75$ for 11×11 nodes.
3) Compute the L_2 error norm of the error denoted by e_{L_2} and obtain the rate of convergence by plotting $\log_{10}(e_{L_2})$ vs. $\log_{10}(h)$, where h is the nodal distance, and e_{L_2} is defined as

$$
e_{L_2} = \left(\int_\Omega \left(u(x,y) - u^h(x,y)\right)^2 dxdy\right)^{1/2}. \tag{5.62}
$$

Compare to the optimal rate of two (see Chapter 3, Eq. (3.95)).

5.8 Higher-Order Accuracy and Convergence in Strong Methods

So far, this chapter has established meshfree approximations that can be used in the standard weak formulations. The next step will be to consider some more appropriate weak formulations which accommodate the inconsistencies in enforcing boundary conditions at nodal positions alone. Since the following weak equations may or may not correspond to variational forms, we denote the arbitrary test function as $\mathbf{v}$ with discrete counterpart $\mathbf{v}^h$, and nodal coefficients $\mathbf{v}_I$, rather than $\delta\mathbf{u}$ with discrete counterpart $\delta\mathbf{u}^h$ and coefficients $\delta\mathbf{u}_I$.

5.8.1 Standard Weak Form

For the standard weak formulation, recall from Chapter 2 we wish to find $\mathbf{u}(\mathbf{x}) \in \mathrm{U}$ such that for all $\mathbf{v} \in \mathrm{U}_0$ (now using $\mathbf{v}$ instead of $\delta\mathbf{u}$):

$$
\int_\Omega v_{(i,j)} C_{ijkl} u_{(k,l)} d\Omega = \int_\Omega v_i b_i d\Omega + \int_{\Gamma_t} v_i \bar{t}_i d\Gamma, \tag{5.63}
$$

where

$$\mathrm{U} = \{\mathbf{u}(\mathbf{x}) | \mathbf{u}(\mathbf{x}) \in \mathrm{H}^1 \text{ and } \mathbf{u}(\mathbf{x}) = \bar{\mathbf{u}} \text{ on } \Gamma_u \},$$

and

$$\mathrm{U}_0 = \{\mathbf{v}(\mathbf{x}) | \mathbf{v}(\mathbf{x}) \in \mathrm{H}^1 \text{ and } \mathbf{v}(\mathbf{x}) = 0 \text{ on } \Gamma_u \}.$$

As previously discussed, it is possible to satisfy the above conditions at nodal locations, yet meshfree approximations generally cannot strictly meet the above requirements along the entire essential boundary. In [1], it was shown that this inconsistency manifests itself as an $\mathcal{O}(h)$ error (on the order of the nodal spacing h) in the energy norm of the problem at hand, and an $\mathcal{O}(h^2)$ error in the displacements, independent of the order of basis used in the meshfree shape functions. Therefore, this is only consistent with *linear basis* in the meshfree approximation and thus places a barrier on the order of accuracy. More details can be found in [1]. Here, we will introduce two weak formulations to remedy this situation. And, to avoid any confusion: these weak forms *are to be used in conjunction with the strong methods* in this chapter to obtain higher-order accuracy. It is however important to note that if linear basis is employed, the weak form in (5.63) is perfectly acceptable from an accuracy standpoint. Nevertheless, the adaptation will yield counter-intuitive results like not passing the patch test in methods that are designed to pass it (see Chapter 6).

5.8.2 Consistent Weak Formulation One (CWF I)

A weak formulation consistent with the fact that test functions may deviate between nodal locations on the essential boundary (see Figure 5.1) can be derived by considering the possibility of $\mathbf{v}(\mathbf{x}) \neq 0$ on Γ_u in between nodes. Recall the strong form of elasticity (see Section 2.2.1):

$$\begin{aligned} &\nabla\cdot\boldsymbol{\sigma} + \mathbf{b} = \mathbf{0} \quad \text{in } \Omega, \\ &\text{or} \\ &\sigma_{ij,j} + b_i = 0 \quad \text{in } \Omega, \end{aligned} \tag{5.64}$$

$$\mathbf{u} = \bar{\mathbf{u}} \quad \text{on } \Gamma_u, \tag{5.65}$$

$$\boldsymbol{\sigma}\cdot\mathbf{n} = \bar{\mathbf{t}} \quad \text{on } \Gamma_t. \tag{5.66}$$

The weighted residual method (see Chapter 2) starts by integrating the residual of (5.64) over the domain of interest with a test function v_i:

$$\int_\Omega v_i\left(\sigma_{ij,j} + b_i\right) d\Omega = 0. \tag{5.67}$$

Integrating by parts, invoking the stress–strain and strain–displacement relationships, and the symmetry of the stress one has

$$\int_\Omega v_{(i,j)} C_{ijkl} u_{(k,l)} d\Omega = \int_\Omega v_i b_i d\Omega + \int_{\Gamma_t} v_i \sigma_{ij} n_j d\Gamma + \int_{\Gamma_u} v_i \sigma_{ij} n_j d\Gamma. \tag{5.68}$$

Now, the second term on the right-hand side of the equation can be recognized as the prescribed traction, while the third term is typically zero applying the definition of the test function. However, again regarding Figure 5.1, we must consider that $\mathbf{v}(\mathbf{x}) \neq 0$ on Γ_u. Therefore, this term is kept

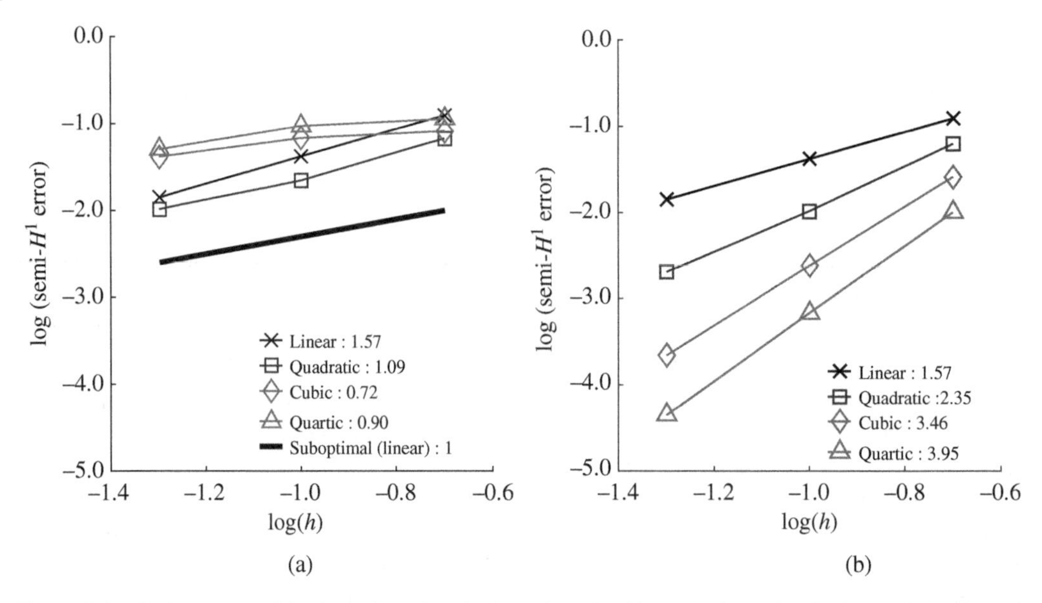

Figure 5.6 Convergence of the boundary singular kernel method (rates in legend): (a) with traditional weak formulation, linear rates (suboptimal), despite increasing the basis order, and (b) with consistent weak formulation I, optimal rates (rate = order of basis n). *Source:* Hillman and Lin [1], figure 22, p. 25 / With permission of Elsevier.

rather than dropped and becomes an additional stiffness term in the left-hand side matrix. The resulting CWF I can be stated as

$$\int_{\Omega} v_{(i,j)} C_{ijkl} u_{(k,l)} d\Omega - \int_{\Gamma_u} v_i \left(C_{ijkl} u_{(k,l)} \right) n_j d\Gamma = \int_{\Omega} v_i b_i d\Omega + \int_{\Gamma_t} v_i \bar{t}_i d\Gamma. \tag{5.69}$$

The above is sufficient to rectify the deficiencies in the conventional weak formulation with strong methods of manifesting errors that limit the order of accuracy in the Galerkin solution (see [1] for further discussion), and higher-order convergence for $n > 1$ can be obtained. Figure 5.6 shows the convergence behavior for the standard weak formulation, and CWF I: optimal rates of n in the H^1 error norm (see Chapter 3, Eq. (3.95)) are only obtained by introducing the additional term on the left-hand side in (5.69).

Exercise 5.10 Galerkin orthogonality is an important property of the Galerkin solution u_i^h. This states the solution u_i is the closest point projection onto the space of trial functions spanning U. Orthogonality refers to the fact that the error $u_i^h - u_i$ is orthogonal to the test space, and this minimizes the error in the energy norm. Show that unless $\mathbf{u}^h(\mathbf{x}) = \bar{\mathbf{u}}(\mathbf{x})$ on all of Γ_u (not just nodal locations), that $a(\mathbf{v}^h, \mathbf{u}^h - \mathbf{u})_\Omega \neq 0$ where $a(\mathbf{v}, \mathbf{u}) = \int_\Omega v_{(i,j)} C_{ijkl} u_{(k,l)} d\Omega$ is the so-called bilinear form using the standard weak formulation (5.63), i.e., Galerkin orthogonality ($a(\mathbf{v}^h, \mathbf{u}^h - \mathbf{u})_\Omega = 0$) is violated.

Solution

Using the standard weak form of the problem, u_i^h satisfies, by definition:

$$\int_{\Omega} v_{(i,j)}^h C_{ijkl} u_{(k,l)}^h d\Omega = \int_{\Omega} v_i^h b_i d\Omega + \int_{\Gamma_t} v_i^h \bar{t} d\Gamma. \tag{5.70}$$

Now consider the following integration by parts identity:

$$\begin{aligned}\int_{\Omega} v^h_{(i,j)} C_{ijkl} u_{(k,l)} d\Omega &= -\int_{\Omega} v^h_i \left(C_{ijkl} u_{(k,l)}\right)_{,j} d\Omega + \int_{\Gamma} v^h_i C_{ijkl} u_{(k,l)} n_j d\Gamma \\ &= -\int_{\Omega} v^h_i \left(C_{ijkl} u_{(k,l)}\right)_{,j} d\Omega + \int_{\Gamma_t} v^h_i C_{ijkl} u_{(k,l)} n_j d\Gamma + \int_{\Gamma_u} v^h_i C_{ijkl} u_{(k,l)} n_j d\Gamma,\end{aligned} \tag{5.71}$$

where the symmetry of C_{ijkl} and the decomposition $\Gamma = \Gamma_t \cup \Gamma_u$, $\Gamma_t \cap \Gamma_u = \emptyset$ was utilized. The last term in the right-hand side of the second row of (5.71) is usually dropped using $v^h_i = 0$ on Γ_u, but this may not be true in meshfree methods (see Figure 5.1), so it must be retained in this case.

Now, using the strain–displacement, stress–strain law, and the data from the strong form (5.64) and (5.66), we have $C_{ijkl} u_{(k,l)} n_j = \sigma_{ij} n_j = \bar{t}_i$ on Γ_t and $(C_{ijkl} u_{(k,l)})_{,j} = \sigma_{ij,j} = -b_i$ on Ω, and (5.71) yields

$$\int_{\Omega} v^h_{(i,j)} C_{ijkl} u_{(k,l)} d\Omega = \int_{\Omega} v^h_i b_i d\Omega + \int_{\Gamma_t} v^h_i \bar{t} d\Gamma + \int_{\Gamma_u} v^h_i C_{ijkl} u_{(k,l)} n_j d\Gamma. \tag{5.72}$$

Subtracting (5.70) from (5.72), one obtains

$$\int_{\Omega} v^h_{(i,j)} C_{ijkl} \left(u_{(k,l)} - u^h_{(k,l)}\right) d\Omega = \int_{\Gamma_u} v^h_i C_{ijkl} u_{(k,l)} n_j d\Gamma \neq 0, \tag{5.73}$$

or,

$$a\left(\mathbf{v}^h, \mathbf{u}^h - \mathbf{u}\right)_{\Omega} = \int_{\Gamma_u} \mathbf{v}^h \cdot \mathbf{C} : \boldsymbol{\varepsilon} \cdot \mathbf{n} d\Gamma \neq 0. \tag{5.74}$$

Therefore unless $v^h_i = 0$ on Γ_u, which may not be satisfied by the *nodally* kinematically admissible shape functions (e.g. transformed, RK with interpolation), the employment of the traditional weak formulation violates Galerkin orthogonality.

Exercise 5.11 Show that Galerkin orthogonality is restored using CWF I, i.e., $a(\mathbf{v}^h, \mathbf{u}^h - \mathbf{u})_{\Omega} = 0$.

Solution

The proof follows Exercise 5.10. Rearranging (5.69), u^h_i satisfies the following equation using CWF I:

$$\int_{\Omega} v^h_{(i,j)} C_{ijkl} u^h_{(k,l)} d\Omega = \int_{\Omega} v^h_i b_i d\Omega + \int_{\Gamma_t} v^h_i \bar{t} d\Gamma + \int_{\Gamma_u} v^h_i C_{ijkl} u^h_{(k,l)} n_j d\Gamma. \tag{5.75}$$

Subtracting (5.75) from the result (5.72) gives

$$\int_{\Omega} v^h_{(i,j)} C_{ijkl} \left(u_{(k,l)} - u^h_{(k,l)}\right) d\Omega = 0, \tag{5.76}$$

or

$$a\left(\mathbf{v}^h, \mathbf{u}^h - \mathbf{u}\right)_{\Omega} = 0, \tag{5.77}$$

and Galerkin orthognality is restored (compare to Exercise 5.10).

5.8.3 Consistent Weak Formulation Two (CWF II)

The employment of (5.69) yields a nonsymmetric stiffness matrix, and the asymmetry is often undesirable from the standpoint of solver selection. In addition, unless trial functions can satisfy the essential boundary conditions exactly, this is still not consistent with a meshfree discretization. To address these two issues, consider a general form of the weighted residual formulation which considers $\mathbf{u} \neq \bar{\mathbf{u}}$ on Γ_u and must therefore weigh the residual of this equation over Γ_u:

$$\int_{\Omega} v_i \left(\sigma_{ij,j} + b_i\right) d\Omega + \int_{\Gamma_u} \widehat{v}_i (u_i - \bar{u}_i) d\Gamma = 0, \tag{5.78}$$

where $\widehat{v}_i = C_{ijkl} v_{(k,l)} n_j$ is the weight on the essential boundary. As will be seen, this weight will yield a symmetric stiffness matrix and is also consistent with the fact that the weight must necessarily have the units of flux to be consistent with all other terms in the formulation. Now, following the previous procedure in deriving the weak formulation, we have the following CWF II:

$$\begin{aligned} &\int_{\Omega} v_{(i,j)} C_{ijkl} u_{(k,l)} d\Omega - \int_{\Gamma_u} v_i \left(C_{ijkl} u_{(k,l)}\right) n_j d\Gamma - \int_{\Gamma_u} u_i \left(C_{ijkl} v_{(k,l)}\right) n_j d\Gamma = \\ &\int_{\Omega} v_i b_i d\Omega + \int_{\Gamma_t} v_i \bar{t}_i d\Gamma - \int_{\Gamma_u} \bar{u}_i \left(C_{ijkl} v_{(k,l)}\right) n_j d\Gamma, \end{aligned} \tag{5.79}$$

where both test and trial function subspaces allow the fact that the solution can deviate between nodes. It is noted that CWF II is in fact the form of the modified variational principle [16], which is closely related to Nitsche's method without the penalty term.

Both weak formulations restore the ability to attain higher-order ($n > 1$) accuracy in the Galerkin solution as well as other important properties, as shown in [1].

Exercise 5.12 Show that CWF II can be derived using the following variational principle that considers the work done by the error on the boundary:

$$\Pi_{\mathrm{E}}\left(\mathbf{u}^h\right) = \Pi_{\mathrm{E}}\left(\mathbf{u}^h\right) - \int_{\Gamma_u} \left(u_i^h - \bar{u}_i\right) t_i d\Gamma, \tag{5.80}$$

where $t_i = \left(C_{ijkl} u^h_{(k,l)}\right) n_j$ and

$$\Pi_{\mathrm{E}}\left(\mathbf{u}^h\right) = \frac{1}{2} \int_{\Omega} u^h_{(i,j)} C_{ijkl} u^h_{(k,l)} d\Omega - \int_{\Omega} u_i^h b_i d\Omega - \int_{\Gamma_t} u_i^h \bar{t}_i d\Gamma. \tag{5.81}$$

Exercise 5.13 Show that for the Galerkin discretization of CWF II, the following orthogonality relationship is satisfied for all $\mathbf{v}^h$:

$$a\left(\mathbf{v}^h, \mathbf{u}^h - \mathbf{u}\right)_{\Omega} - \int_{\Gamma_u} \left(C_{ijkl} v_{(k,l)}\right) n_j (u_i - \bar{u}_i) d\Gamma + \int_{\Gamma_u} \left(C_{ijkl} v_{(k,l)} - C_{ijkl} \bar{u}_{(k,l)}\right) n_j v_i d\Gamma = 0. \tag{5.82}$$

5.9 Comparison Between Weak Methods and Strong Methods

As we have seen in the previous chapters, *constrained variational principles*, or more generally, *weighted residual methods* can also be employed. In these methods, approximations do not need to satisfy any particular requirement related to the essential boundary, and instead impose boundary conditions weakly (see Chapter 2 for more details).

In summary, there are two types of enforcement:

1) Strong enforcement: The meshfree approximation is modified such that the nodal coefficients for boundary nodes coincide with their field variables. Then, boundary conditions can be imposed directly. The set of shape functions $\{\Psi_I\}_{I=1}^{NP}$ is replaced by a set of new shape functions $\{\hat{\Psi}_I\}_{I=1}^{NP}$, which satisfies $\hat{\Psi}_I(\mathbf{x}_J) = \delta_{IJ}$ on the essential boundary. Then $u_i^h(\mathbf{x}_J) = \overline{u}_i(\mathbf{x}_J)$ on Γ_u and $v_i^h(\mathbf{x}_J) = 0$ on Γ_u can be achieved for all nodes J on the boundary. Several methods to achieve this have been described in this chapter.
2) Weak enforcement of conditions along the essential boundary: This requires modification of the weak form to relax the admissibility requirement (refer to Chapter 2) but generally still maintains the equivalence to the strong form, including the essential boundary conditions. No modification to the set of shape functions is needed.

These methods are compared by the various aspects involved in Table 5.2.

Table 5.2 Comparisons of the different aspects of boundary condition enforcement methods.

Method	Degrees of freedom	Parameters	Stability condition	System characteristics	Implementation effort
Lagrange multiplier	Order *NP*+*Neb* (*Neb* = number of Lagrange multiplier nodes)	No	On space of $(\mathbf{u}^h, \boldsymbol{\lambda}^h)$	Positive semi-definite	Essential boundary approximation and integration.
Penalty	Order *NP*	Penalty parameter β	Penalty parameter β sufficiently large	Positive definite with sufficient β, possibly ill-conditioned with large β	Essential boundary integration.
Nitsche	Order *NP*	Stability parameter β	Stability parameter β sufficiently large	Positive definite with sufficient β	Essential boundary integration.
Strong	Order *NP*	No	No	Positive definite	Modification of shape functions or matrix operations, bookkeeping.

References

1 Hillman, M. and Lin, K.-C. (2021). Consistent weak forms for meshfree methods: full realization of h-refinement, p-refinement, and a-refinement in strong-type essential boundary condition enforcement. *Comput. Methods Appl. Mech. Eng.* 373: 113448.

2 Hughes, T.J.R. (2012). *The Finite Element Method: Linear Static and Dynamic Finite Element Analysis*. Courier Corporation.

3 Chen, J.-S., Pan, C., Wu, C.-T., and Liu, W.K. (1996). Reproducing kernel particle methods for large deformation analysis of non-linear structures. *Comput. Methods Appl. Mech. Eng.* 139: 195–227.

4 Zhu, T.L. and Atluri, S.N. (1998). A modified collocation method and a penalty formulation for enforcing the essential boundary conditions in the element free Galerkin method. *Comput. Mech.* 21: 211–222.

5 Günther, F.C. and Liu, W.K. (1998). Implementation of boundary conditions for meshless methods. *Comput. Methods Appl. Mech. Eng.* 163: 205–230.

6 Atluri, S.N., Kim, H.G., and Cho, J.Y. (1999). Critical assessment of the truly Meshless Local Petrov-Galerkin (MLPG), and Local Boundary Integral Equation (LBIE) methods. *Comput. Mech.* 24: 348–372.

7 Wagner, G.J. and Liu, W.K. (2000). Application of essential boundary conditions in mesh-free methods: a corrected collocation method. *Int. J. Numer. Methods Eng.* 47: 1367–1379.

8 Chen, J.-S. and Wang, H.-P. (2000). New boundary condition treatments in meshfree computation of contact problems. *Comput. Methods Appl. Mech. Eng.* 187: 441–468.

9 Barrett, R., Berry, M., Chan, T.F. et al. (1994). *Templates for the Solution of Linear Systems: Building Blocks for Iterative Methods*. SIAM.

10 Shepard, D. (1968). A two-dimensional interpolation function for irregularly-spaced data. *Proceedings of the 1968 23rd ACM National Conference* (27–29 August 1968), 517–524. New York, NY: Computing Machinery.

11 Lancaster, P. and Salkauskas, K. (1981). Surfaces generated by moving least squares methods. *Math. Comput.* 141–158.

12 Chen, J.-S., Han, W., You, Y., and Meng, X. (2003). A reproducing kernel method with nodal interpolation property. *Int. J. Numer. Methods Eng.* 56: 935–960.

13 Krongauz, Y. and Belytschko, T. (1996). Enforcement of essential boundary conditions in meshless approximations using finite elements. *Comput. Methods Appl. Mech. Eng.* 131: 133–145.

14 Fernández-Méndez, S. and Huerta, A. (2004). Imposing essential boundary conditions in mesh-free methods. *Comput. Methods Appl. Mech. Eng.* 193: 1257–1275.

15 Kaljevic, I. and Saigal, S. (1997). An improved element free Galerkin formulation. *Int. J. Numer. Meth. Eng.* 40: 2953–2974.

16 Lu, Y.Y., Belytschko, T., and Gu, L. (1994). A new implementation of the element free Galerkin method. *Comput. Meth. Appl. Mech. Eng.* 113: 397–414.

6

Quadrature in Meshfree Methods

In weak form-based Galerkin methods, the evaluation of integrals must be carried out numerically, with quadrature rules that approximate true integration. Meshfree methods are unique in the space of numerical methods in that the selection of quadrature rules is particularly nontrivial. To illustrate why, it is helpful to examine why this is not a problem in the finite element method (FEM). Based on the element topology, Gauss rules are directly applied to the domain integrals in the weak form at the element level in the parent domain. The approximation itself, defined at the element level in natural coordinates in terms of polynomials, is thus very amenable to Gaussian quadrature which can integrate these functions exactly (disregarding the rational nature of the Jacobian which has a negligible effect unless the elements are highly distorted). Thus, in finite elements, quadrature is unified with the approximation itself and everything is done "neatly": the supports of the functions are perfectly aligned with the quadrature scheme, and we know what must be done to integrate the FEM shape functions accurately and efficiently. The analysis by Strang and Fix [1] informs us exactly what order of Gauss quadrature is needed to maintain optimal convergence.

In meshfree methods, the situation is quite the opposite: the approximation is only associated with nodes, and no elements are needed. Numerical integration is also performed on rational approximations, with no intrinsic tie of their support to integration cells. The early approach was to borrow the ideas from FEM and use a "background" mesh independent of the approximation, using Gauss rules to generate quadrature weights and locations. These conventional techniques, however, necessitate high-order integration rules to ensure accurate and stable meshfree solutions, yet the associated computational cost makes this intractable for applications. Low-order integration is much faster of course, but this cannot yield the accuracy and stability of the former, presenting an apparent impasse. As will be seen in this chapter, the key to resolving this issue is to employ a low-cost quadrature rule in conjunction with efficient corrections for accuracy and stability.

6.1 Nomenclature and Acronyms

The development of corrected and stabilized quadrature for meshfree methods over the past several decades has resulted in somewhat of an "alphabet soup" due to the increasing variety and therefore increasing permutations of combined and enhanced methods. For instance, SCNI refers to stabilized *conforming* nodal integration, SNNI refers to stabilized *non-conforming* nodal integration, VC-SNNI refers to *variationally consistent* SNNI, and so on. Therefore, Table 6.1 is provided to list the acronyms used in this chapter and in the literature for reference.

Meshfree and Particle Methods: Fundamentals and Applications, First Edition.
Ted Belytschko, J. S. Chen, and Michael Hillman.

Table 6.1 Common acronyms used in numerical integration for meshfree methods.

Acronym	Quadrature method
GI	Gauss integration
DNI	Direct nodal integration
SCNI	Stabilized conforming nodal integration
SNNI	Stabilized non-conforming nodal integration
MSCNI	Modified stabilized conforming nodal integration
MSNNI	Modified stabilized non-conforming nodal integration
VC-	Prefix denotes a method corrected with the variationally consistent (VC) approach
NSNI	Naturally stabilized nodal integration
NSCNI	Naturally stabilized conforming nodal integration

6.2 Gauss Integration: An Introduction to Quadrature in Meshfree Methods

A straightforward way (and therefore an introductory technique) to perform numerical integration in meshfree methods is to utilize the same concepts in classical FEM. That is, the global integral weak form is carried out element-by-element. The meshfree nodes themselves need not be tied to the elements, but rather can live on top of a *background mesh* which is purely for purposes of carrying out numerical integration. This was the first type of numerical integration used in the Galerkin class of meshfree methods. Figure 6.1 shows the concept of an independent mesh used only for quadrature, with a 5×5 Guass rule employed.

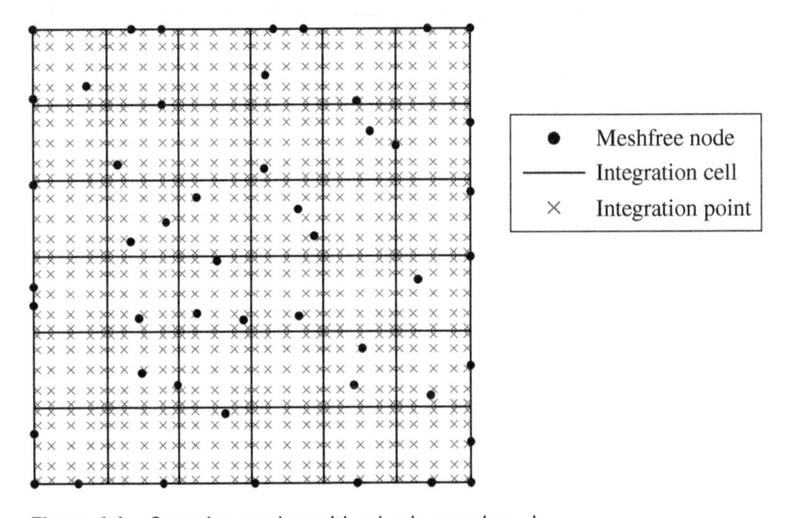

Figure 6.1 Gauss integration with a background mesh.

To illustrate how this works, we first show two-dimensional domain integrals of a function $f(x, y)$ in the weak formulation decomposed into element domains:

$$\begin{aligned}
&\int_{\Omega} f(x, y)\mathrm{d}\Omega \\
&= \sum_{e}\int_{\Omega_e} f(x, y)\mathrm{d}\Omega \\
&= \sum_{e}\int_{\square} f(x(\xi, \eta), y(\xi, \eta))J(\xi, \eta)d\xi d\eta \\
&\approx \sum_{e}\sum_{i=1}^{NQ}\sum_{j=1}^{NQ} f\left(x\left(\xi_i, \eta_j\right), y\left(\xi_i, \eta_j\right)\right)J\left(\xi_i, \eta_j\right)W_iW_j,
\end{aligned} \tag{6.1}$$

where e denotes an element, $\square \equiv [-1, +1] \times [-1, +1]$ denotes the domain of the parent element, J is the Jacobian, (ξ, η) are the natural coordinates, (ξ_i, η_j) is the location of a quadrature point, and W_iW_j is associated tensor-product quadrature weight. Unlike FEM, the integrand in (6.1) always contains the meshfree shape functions, which are built from the global coordinates. Thus, it is important to find the associated global coordinates at the evaluation points by properly defining $\mathbf{x} = \mathbf{x}(\boldsymbol{\xi})$. The easiest approach of defining this map is to use finite element shape functions $N_I(\xi, \eta)$:

$$\begin{aligned}
x(\xi, \eta) &= \sum_{I=1}^{4} N_I(\xi, \eta)x_I, \\
y(\xi, \eta) &= \sum_{I=1}^{4} N_I(\xi, \eta)y_I,
\end{aligned} \tag{6.2}$$

where x_I and y_I are the physical coordinates of background element nodes. The Jacobian is defined using standard finite element procedures:

$$\begin{aligned}
J &= \det\left(\mathbf{x}_{,\boldsymbol{\xi}}\right), & \mathbf{x}_{,\boldsymbol{\xi}} &= \begin{bmatrix} x_{,\xi} & x_{,\eta} \\ y_{,\xi} & y_{,\eta} \end{bmatrix}, \\
x_{,\xi}(\xi, \eta) &= \sum_{I=1}^{4} N_{I,\xi}(\xi, \eta)x_I, & y_{,\xi}(\xi, \eta) &= \sum_{I=1}^{4} N_{I,\xi}(\xi, \eta)y_I, \\
x_{,\eta}(\xi, \eta) &= \sum_{I=1}^{4} N_{I,\eta}(\xi, \eta)x_I, & y_{,\eta}(\xi, \eta) &= \sum_{I=1}^{4} N_{I,\eta}(\xi, \eta)y_I.
\end{aligned} \tag{6.3}$$

Therefore, once cells are defined the procedure is quite simple, as shown in **Box 6.1**. Interestingly, Step 5 defines a natural connectivity in meshfree methods rather than the FEM connectivity where the current element is tied to the four nodes (in 2D). This idea was first illustrated in Chapter 3 (see Figure 3.2). An elaboration on efficiently constructing neighbor lists for meshfree approximations is given in Sections 7.5 and 10.3.3.

Box 6.1 Gaussian Integration in Meshfree Methods
Given: the background cells, number of quadrature points per cell, and nodal locations. **Loop over cells and quadrature points inside cells:** **Step 1.** Compute or recall the finite element shape functions and their derivatives.

(Continued)

Box 6.1 (Continued)

Step 2. Compute the Cartesian coordinate associated with the location of the Gauss quadrature point in the natural coordinate using (6.2).

Step 3. Evaluate the meshfree shape functions and other quantities that are defined in the Cartesian coordinate, needed for the integrals in the weak form.

Step 4. Compute the Jacobian using (6.3).

Step 5. Carry out the numerical integration and assemble the result using the neighbors (the list of nodes with non-zero support over this point, $S_{\mathbf{x}}$, see Chapter 3).

6.3 Issues with Quadrature in Meshfree Methods

As it turns out, Gauss integration using background cells is a very ineffective approach in meshfree methods. To demonstrate this, consider various orders of Gauss rules in solving the following two-dimensional isotropic diffusion problem with $\Gamma_u = \Gamma$ and $\Omega = [0, 1] \times [0, 1]$:

$$\begin{aligned} k\nabla^2 u &= \sin(\pi x)\sin(\pi y) \quad &&\text{in } \Omega, \\ u &= 0 &&\text{on } \Gamma. \end{aligned} \tag{6.4}$$

Linear basis and cubic B-spline kernels are employed in a reproducing kernel particle method (RKPM) solution (see Chapters 3 and 4), with non-uniform discretizations shown in Figure 6.2, and $k = 1$; a uniform nodal distribution with spacing $h = 1/3$ in each direction is perturbed to generate the first refinement, and the normalized dilation (normalized support size) is two. Nitsche's method is selected with $\beta = 1{,}000/h$. It is important to note that meshfree solutions under uniform discretizations generally do not suffer from integration issues, while the solutions under non-uniform ones are sensitive to the choice of domain integration.[1] For this smooth problem, linear accuracy should yield rates of two and one in the L_2 norm and H^1 semi-norm, respectively (see Chapter 3, Eq. (3.94)). Figure 6.3 shows that five-point Gauss integration is needed to attain these optimal convergence rates. In three dimensions, this would yield 125 quadrature points per cell, which makes this method computationally intractable. Recently, it has been shown in [2] that as the refinement keeps increasing, it is theoretically impossible to attain optimal rates of convergence using these traditional types of quadrature schemes, due to their lack of *variational consistency* (see Section 6.7).

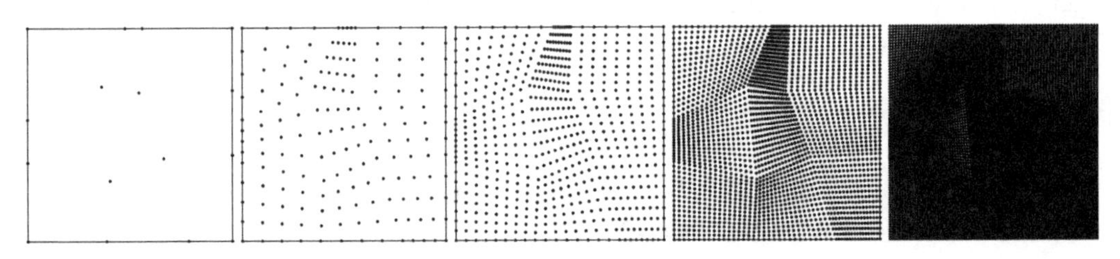

Figure 6.2 Refinements for convergence of RKPM with linear basis.

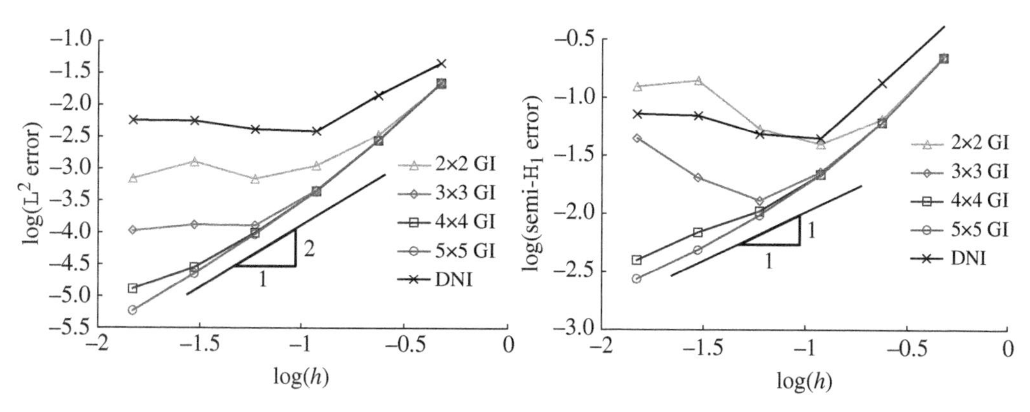

Figure 6.3 Convergence of RKPM with linear basis with increasing orders NQ of Gauss quadrature (denoted $NQ \times NQ$ GI) and direct nodal integration (denoted DNI).

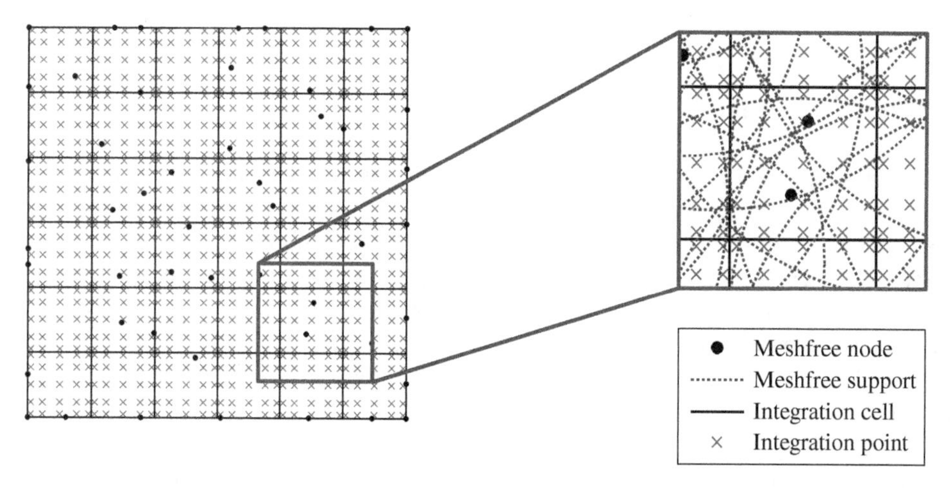

Figure 6.4 Gauss integration with a zoom-in showing nodal supports are not coincident with cells.

Why does this occur? As previously discussed, in Gaussian integration, the cells may not necessarily align with the supports of the shape functions, as nodes can be arranged in an arbitrary fashion, which causes a mismatch as shown in Figure 6.4. This mismatch introduces a significant amount of error in numerical integration of the weak form, and subsequently, error in the Galerkin solution itself. In addition, Gauss quadrature has difficulty in accurately integrating meshfree approximations, which are rational functions.

The more critical of these two issues is the former (see [3] for an elaboration); therefore, several schemes have been developed to align quadrature cells with supports [3, 4]. Some meshfree methods such as the meshless local Petrov–Galerkin method (MLPG) [5] and the method of finite spheres [6] share similar features. However, due to the number of quadrature points still needed per node for accuracy in these methods, this chapter will focus on methods with quadrature at nodes, where the nodal point is coincident with the integration point. This is referred to as nodal integration, which is depicted in Figure 6.5.

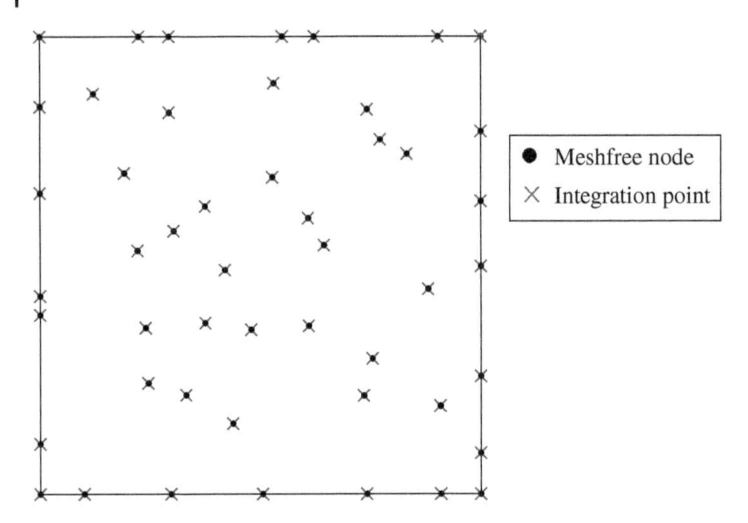

Figure 6.5 Nodal integration: nodal locations are the locations of the quadrature points.

6.4 Introduction to Nodal integration

Nodally integrating the Galerkin equation is a natural choice for meshfree methods. This avoids using a mesh altogether and keeps the meshfree method truly "meshfree": the solution is obtained only with a set of scattered points. Associated techniques to effectively do so have largely been the focus of research in quadrature for meshfree methods. Weights can be determined in a variety of ways, generally carried out during the generation of the computational grid. For instance, for a set of scattered notes, as shown in Figure 6.6a, Voronoi diagrams (Figure 6.6b) or triangulation (Figure 6.6c) can be employed to obtain weights. Or, if meshing software is used for preprocessing, the mesh itself can be used to generate tributary areas or volumes for each node (Figure 6.6d), with the mesh discarded after.

Not only does nodal integration keep meshfree methods "truly meshfree," it offers great simplicity, ease of implementation, and is very fast. The downside, of course, is that the numerical integration is crude. "Doing nothing" and carrying out the nodal integration in naivety has been termed *direct nodal integration* (DNI) in the literature [7]. The accuracy in evaluating integrals in the weak form is quite low in general, and this approach encounters the same difficulties as Gauss quadrature (misalignment of supports and integration domains, rational functions) due to so few integration points. This in turn causes low accuracy of the numerical solution and low rates of convergence (and possibly non-convergent solutions), comparable to other low-order quadrature rules. Figure 6.3 shows that DNI performs similar to two-point Gauss integration, and the solution eventually diverges with refinement. This situation is obviously undesirable, and clearly "doing nothing" while performing nodal integration is not satisfactory in terms of solution accuracy.

Nodal integration must also confront another major obstacle: the instability of the solution, analogous to the hourglass instability in FEM. In meshfree methods, under-integration results in node-to-node oscillations, as shown in Figure 6.7. This point will be revisited in Section 6.9. We will first discuss how to address the issue of accuracy.

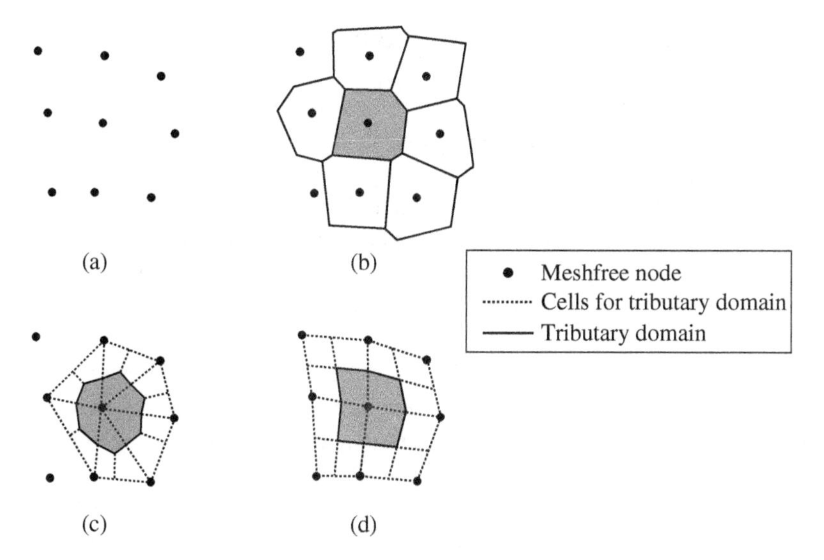

Figure 6.6 (a) scattered nodes; weights in nodal integration (shaded areas): (b) Voronoi diagram; (c) triangulation; and (d) FEM hex mesh.

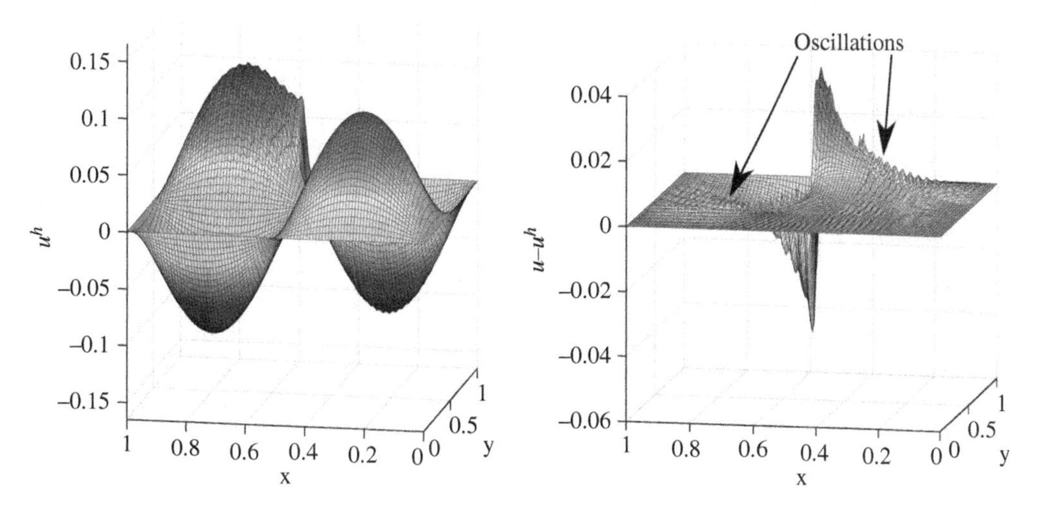

Figure 6.7 Instability due to nodal integration in the Poisson equation with DNI. The solution is from the last refinment shown in Figure 6.2.

Key Takeaway

Observing nonconvergence of the meshfree numerical solution where it is expected to converge is usually the result of using an inappropriate domain integration technique. When the quadrature is low order, the solution will not converge properly, and this is most often observed in nonuniform discretizations.

6.5 Integration Constraints and the Linear Patch Test

The issue of quadrature in Galerkin meshfree methods presents somewhat of a conundrum. High-order integration ensures accuracy but is costly. On the other end of the spectrum, low-order integration is much faster but yields inaccurate results.

Clearly, something must be done to remedy the situation, as high-order quadrature is of course undesirable. That *something* is what researchers started to arrive at around the same time in the mid 1990s to the early 2000s [7–9]: the requirements to pass the *linear patch test*, eventually termed the *integration constraints* [7] and first-order *variational consistency conditions* [10], and methods that satisfy them.

In a linear patch test, one seeks to recover the exact solution in the Galerkin equation when the solution is linear. In the context of solid mechanics, if displacements are linear, strains are constant, and the stresses throughout the entire domain are constant if the body is homogeneous: the test is then to see if the numerical method can exactly represent a state of constant stress.

Historically this test has been used to assess various formulations of FEM, although the requirements to actually pass the patch test appear not to have been closely examined until meshfree methods required something to be done about quadrature. As we shall see, from a mechanics point of view, this requirement is equivalent to equilibrium of the discrete form of the problem.

For illustration, we shall consider the two-dimensional case of elasticity (plane strain or plane stress) for the remainder of this chapter.

So, one seeks to recover an exact solution of the form $u_i(x, y) = u_i^{\mathrm{P}}(x, y) = a_i + b_i x + c_i y$, with a_i, b_i, and c_i arbitrary constants, independent in each direction i, with "P" denoting a polynomial solution. The specific boundary value problem associated with this solutions states[2]:

$$\begin{aligned}
&\sigma_{ij,j} = 0 \text{ on } \Omega, \\
&\sigma_{ij} n_j = \bar{t}_i^{\mathrm{P}} \text{ on } \Gamma_t, \\
&u_i = \bar{u}_i^{\mathrm{P}}(x, y) \text{ on } \Gamma_u,
\end{aligned} \tag{6.5}$$

where $\bar{t}_i^{\mathrm{P}} = \sigma_{ij}^{\mathrm{P}} n_j$ with $\sigma_{ij}^{\mathrm{P}} \equiv C_{ijkl} u_{(k,l)}^{\mathrm{P}}$ the constant stress associated with $u_i^{\mathrm{P}}(x, y)$, and $\bar{u}_i^{\mathrm{P}}(x, y) = u_i^{\mathrm{P}}(x, y)$ is the manufactured polynomial solution. The body force is zero, since any body force will cause the displacement solution to deviate from linear, e.g. a constant body force is associated with a quadratic solution since the above represents a second-order differential equation. Or explained differently, the body force is calculated from the equilibrium equation with the constant stress σ_{ij}^{P} as $b_i = -\sigma_{ij,j}^{\mathrm{P}} = 0$.

As a first requirement, the approximation space should be able to represent this displacement field:

$$u_i^h(x, y) = \sum_{J=1}^{NP} \Psi_I(x, y) u_i^{\mathrm{P}}(x_J, y_J) = u_i^{\mathrm{P}}(x, y), \tag{6.6}$$

which is satisfied if the shape functions are linearly complete:

$$\begin{aligned}
\sum_{J=1}^{NP} \Psi_J(x, y)(a_i + b_i x_J + c_i y_J) &= a_i \left(\sum_{J=1}^{NP} \Psi_J(x, y) \right) + b_i \left(\sum_{J=1}^{NP} \Psi_J(x, y) x_J \right) \\
&\quad + c_i \left(\sum_{J=1}^{NP} \Psi_I(x, y) y_J \right) = u_i^{\mathrm{P}}(x, y),
\end{aligned} \tag{6.7}$$

That is, it should be clear from Chapter 3 that the requirement (6.6) is nothing but the linear completeness requirement.

Using the nodal coefficients corresponding to $u_i^{\mathrm{P}}(x, y)$:

$$\mathbf{u}_J = \mathbf{u}_J^{\mathrm{P}} \equiv \begin{bmatrix} a_1 + b_1 x_J + c_1 y_J \\ a_2 + b_2 x_J + c_2 y_J \end{bmatrix}, \tag{6.8}$$

the associated left-hand side of the Galerkin equation is calculated using the consistent weak form one (CWF I, see Chapter 5) as[3]:

$$\begin{aligned} \sum_{J=1}^{NP} (\mathbf{K}_{IJ} - \mathbf{H}_{IJ})\mathbf{u}_J^{\mathrm{P}} &= \sum_{J=1}^{NP} \left\{ \int_{\Omega} \mathbf{B}_I^{\mathrm{T}} \mathbf{D} \mathbf{B}_J d\Omega - \int_{\Gamma_u} \Psi_I \boldsymbol{\eta}^{\mathrm{T}} \mathbf{D} \mathbf{B}_J d\Gamma \right\} \mathbf{u}_J^{\mathrm{P}} \\ &= \int_{\Omega} \mathbf{B}_I^{\mathrm{T}} \underbrace{\sum_{J=1}^{NP} \mathbf{D} \mathbf{B}_J \mathbf{u}_J^{\mathrm{P}}}_{\boldsymbol{\sigma}^{\mathrm{P}}} d\Omega - \int_{\Gamma_u} \Psi_I \boldsymbol{\eta}^{\mathrm{T}} \underbrace{\sum_{J=1}^{NP} \mathbf{D} \mathbf{B}_J \mathbf{u}_J^{\mathrm{P}}}_{\boldsymbol{\sigma}^{\mathrm{P}}} d\Gamma = \left(\int_{\Omega} \mathbf{B}_I^{\mathrm{T}} d\Omega - \int_{\Gamma_u} \Psi_I \boldsymbol{\eta}^{\mathrm{T}} d\Gamma \right) \boldsymbol{\sigma}^{\mathrm{P}}, \end{aligned} \tag{6.9}$$

where $\boldsymbol{\sigma}^{\mathrm{P}} = \sum_{J=1}^{NP} \mathbf{D} \mathbf{B}_J \mathbf{u}_J^{\mathrm{P}}$ is the vector of constant stress σ_{ij}^{P}, which has been moved outside of the integrals, $\mathbf{B}_I$ is the strain matrix:

$$\mathbf{B}_I = \begin{bmatrix} \Psi_{I,1} & 0 \\ 0 & \Psi_{I,2} \\ \Psi_{I,2} & \Psi_{I,1} \end{bmatrix}, \tag{6.10}$$

and $\boldsymbol{\eta}$ is the matrix of outward unit normals to Γ:

$$\boldsymbol{\eta} = \begin{bmatrix} n_1 & 0 \\ 0 & n_2 \\ n_2 & n_1 \end{bmatrix}. \tag{6.11}$$

Here, the matrix forms for all terms are invoked (rather than indicial notation) to simplify the presentation.

The external force for the solution $u_i^{\mathrm{P}}(x, y)$ is obtained using the prescribed traction $\bar{\mathbf{t}}^{\mathrm{P}}$ associated with the constant stress $\boldsymbol{\sigma}^{\mathrm{P}}$ (recall the body force is zero):

$$\mathbf{f}_I = \int_{\Gamma_t} \Psi_I \underbrace{\bar{\mathbf{t}}^{\mathrm{P}}}_{\boldsymbol{\eta}^{\mathrm{T}} \boldsymbol{\sigma}^{\mathrm{P}}} d\Gamma = \int_{\Gamma_t} \Psi_I \boldsymbol{\eta}^{\mathrm{T}} \boldsymbol{\sigma}^{\mathrm{P}} d\Gamma = \int_{\Gamma_t} \Psi_I \boldsymbol{\eta}^{\mathrm{T}} d\Gamma \boldsymbol{\sigma}^{\mathrm{P}}, \tag{6.12}$$

where again the constant stress has been moved outside the integral.

For satisfaction of discrete equilibrium, (6.9) and (6.12) should be in balance, which requires

$$\left(\int_{\Omega} \mathbf{B}_I^{\mathrm{T}} d\Omega - \int_{\Gamma_u} \Psi_I \boldsymbol{\eta}^{\mathrm{T}} d\Gamma \right) \boldsymbol{\sigma}^{\mathrm{P}} = \int_{\Gamma_t} \Psi_I \boldsymbol{\eta}^{\mathrm{T}} d\Gamma \boldsymbol{\sigma}^{\mathrm{P}}. \tag{6.13}$$

Since $\boldsymbol{\sigma}^{\mathrm{P}}$ contains arbitrary coefficients, the above should hold for any $\boldsymbol{\sigma}^{\mathrm{P}}$, and using the fact that $\Gamma = \Gamma_u \cup \Gamma_t$, the resulting condition considering quadrature is

$$\hat{\int_{\Omega}} \mathbf{B}_I^{\mathrm{T}} d\Omega = \hat{\int_{\Gamma}} \Psi_I \boldsymbol{\eta}^{\mathrm{T}} d\Gamma,$$

or

$$\hat{\int_{\Omega}} \begin{bmatrix} \Psi_{I,1} & 0 \\ 0 & \Psi_{I,2} \\ \Psi_{I,2} & \Psi_{I,1} \end{bmatrix} d\Omega = \hat{\int_{\Gamma}} \Psi_I \begin{bmatrix} n_1 & 0 \\ 0 & n_2 \\ n_2 & n_1 \end{bmatrix} d\Gamma, \tag{6.14}$$

or

$$\hat{\int_{\Omega}} \Psi_{I,i} d\Omega = \hat{\int_{\Gamma}} \Psi_I n_i d\Gamma,$$

where $\hat{\int}$ denotes the quadrature version of an integral. Here, it is easy to see that this is simply a divergence condition: if exact integration is employed, it is automatically satisfied. However, since numerical integration is employed, the condition (6.14) is in general not satisfied with quadrature due to the nature of meshfree approximations. The above condition has been termed the *integration constraint* [7] and first-order *variational consistency condition* [10].

This condition for linear exactness in the Galerkin solution is a relationship between the test function (here recall Ψ_I in (6.14) came from the test function approximation), and the numerical integration, and can be expressed as follows in terms of quadrature sums:

$$\sum_{L=1}^{NINT} \mathbf{B}_I^{\mathrm{T}}(\mathbf{x}_L) W_L = \sum_{L=1}^{NINTb} \Psi_I \boldsymbol{\eta}^{\mathrm{T}}(\mathbf{x}_L) S_L \quad \forall I,$$

or (6.15)

$$\sum_{L=1}^{NINT} \Psi_{I,i}(\mathbf{x}_L) W_L = \sum_{L=1}^{NINTb} \Psi_I n_i(\mathbf{x}_L) S_L \quad \forall I,$$

where *NINT* and *NINTb* are the number of integration points used in integrating the domain terms and boundary terms in the Galerkin equation, respectively, $\mathbf{x}_L$ are the spatial coordinates of the quadrature points, and W_L and S_L are the weights of the domain and contour quadrature, respectively.

Now, for illustration, consider the following problem with a linear solution:

$$\begin{aligned} \nabla^2 u &= 0 && \text{in } \Omega, \\ u &= .1x + .3y && \text{on } \Gamma_u, \end{aligned} \tag{6.16}$$

with $\Gamma_u = \Gamma$ and $\Omega = [0, 1] \times [0, 1]$. The solution is $u = .1x + .3y$. Linear basis is employed, which can exactly represent the solution. The domain is discretized with the irregular node distribution shown in Figure 6.8. The violation of the integration constraint (6.15) is shown in Figure 6.9 for five-point Gaussian quadrature and DNI, where it is seen that both methods fail to satisfy the constraint. Since the condition is simply the divergence theorem with quadrature, the Gauss quadrature scheme chosen has lower residual than DNI, since it better approximates the true integrals. Nevertheless, the linear solution is not recovered using either method as shown in Figure 6.10, with large solution errors for DNI. These errors eventually lead to nonconvergence of the numerical solution, as shown in Figure 6.3.

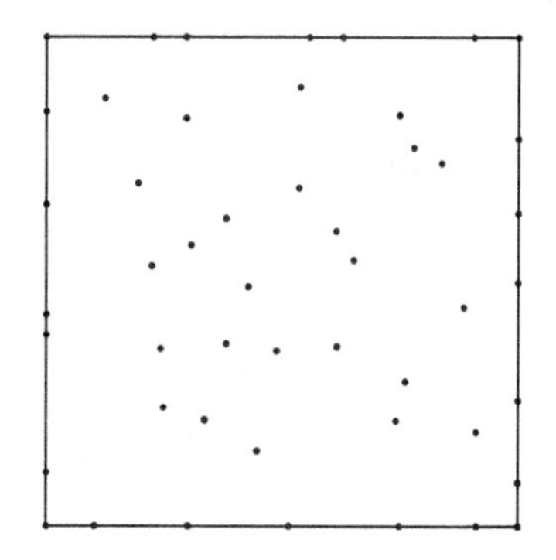

Figure 6.8 Irregular meshfree discretization for the patch test.

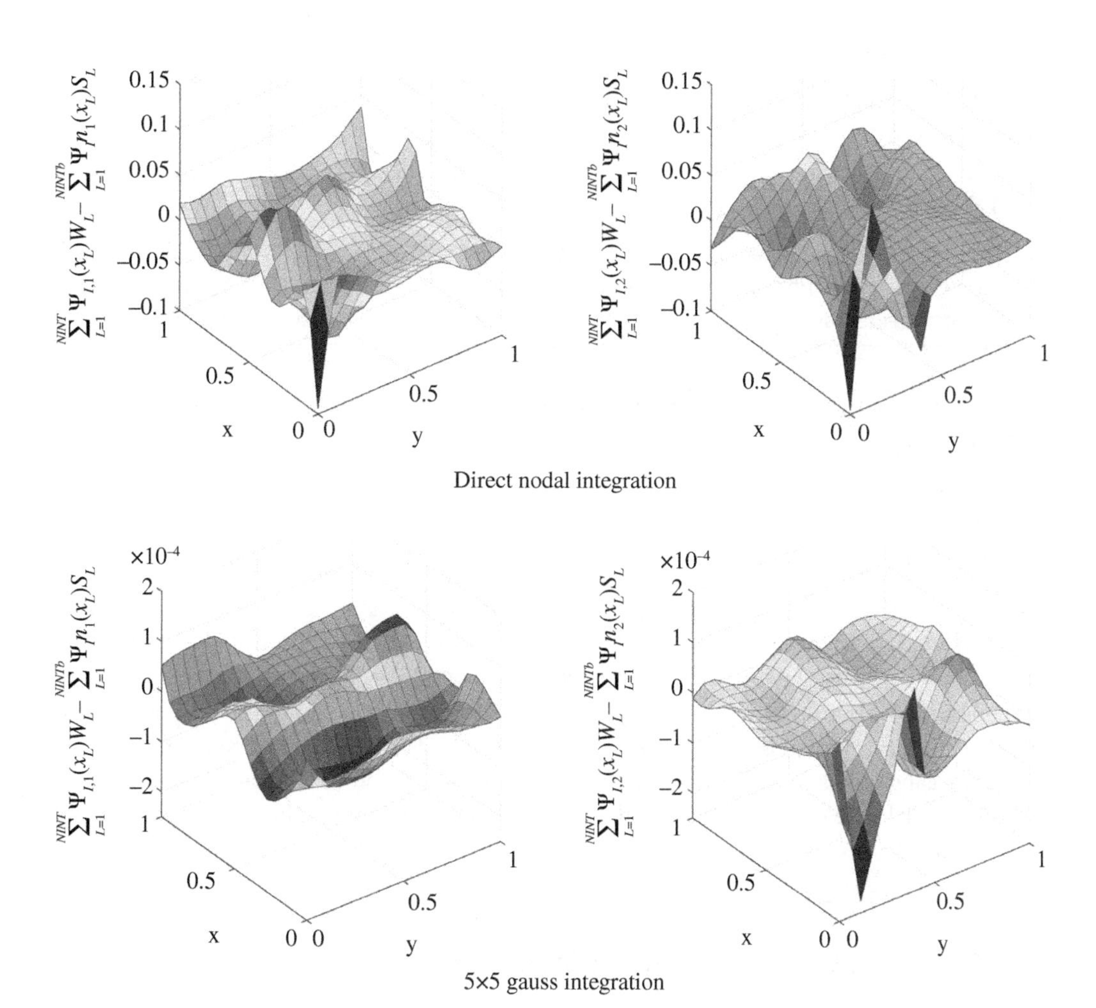

Figure 6.9 Violation of integration constraint in direct nodal integration and Gauss integration.

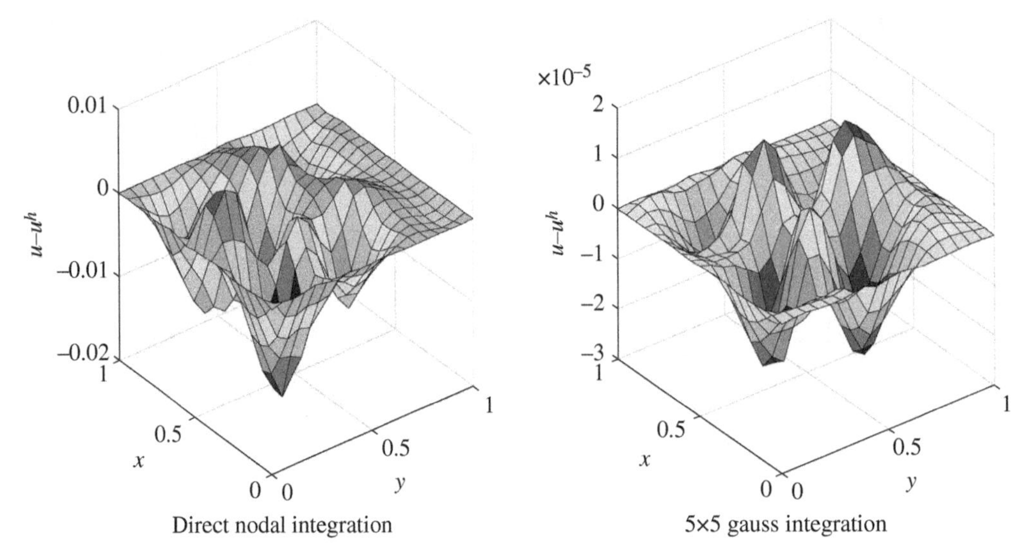

Figure 6.10 Errors in linear patch test using linear basis.

Key Takeaway

Passing a patch test is not guaranteed for meshfree methods and is the result of using conventional quadrature which does not satisfy the integration constraints. Special integration techniques such as SCNI are needed for effective meshfree analysis; these are presented later in this chapter.

Exercise 6.1 Consider the following two-point boundary value problem for the linear patch test: find u such that

$$\begin{aligned} &u_{,xx} = 0 \quad \text{for} \quad 0 < x < 1, \\ &u(0) = 0, \\ &u(1) = 1. \end{aligned} \tag{6.17}$$

The exact solution is $u = x$. Solve the problem with linear basis in the reproducing kernel approximation and the Galerkin formulation, with any of the methods presented in Chapters 4 or 5 for enforcing boundary conditions. Use a set of eleven ($NP = 11$) nonuniformly distributed nodes perturbed from uniform points $\{x_I\}_{I=1}^{NP}$ over the domain $\overline{\Omega} = [0, 1]$ (nodal spacing h is 1/10, with final nodal positions $x_I' = x_I + \alpha h, -0.05 \leq \alpha \leq 0.05$). Use a normalized support $a/h = 2.0$.

Use increasing orders of Gauss quadrature (one to five points) and DNI. Verify that none of the methods can yield the exact solution and that the error decreases with an increase in the order of Gauss integration.

6.6 Stabilized Conforming Nodal Integration

Constructing a method that satisfies the integration constraint (6.14) is not immediately straightforward, given the rational nature of meshfree approximations. However, the strain smoothing technique introduced in [7] can be employed to satisfy the conditions.

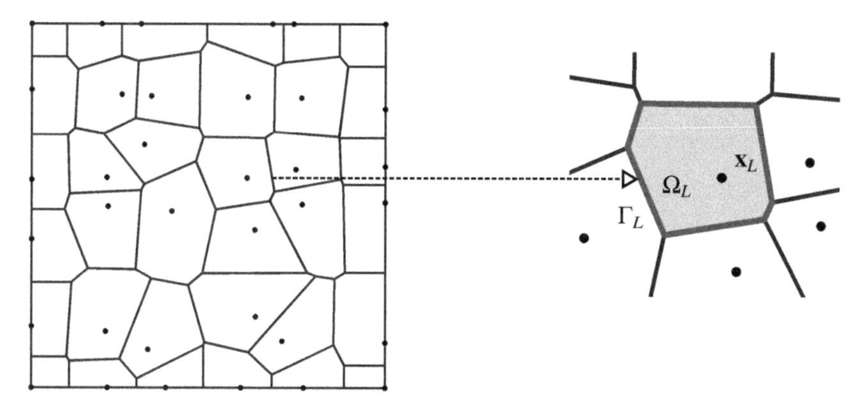

Figure 6.11 Stabilized conforming nodal integration: conforming nodal subdivision of domain and individual smoothing cell for node L.

First, consider the decomposition of the domain Ω into conforming nodal representative domains Ω_L for each node at $\mathbf{x}_L$, as shown in Figure 6.11. This subdivision can be accomplished using Voronoi diagrams for instance (see Section 8.3.1). The domains will be used for both nodal integration and strain smoothing as will be seen.

Now consider the strain at a node smoothed using a distribution function Φ:

$$\tilde{\varepsilon}_{ij}(\mathbf{x}_L) = \int_{\Omega} \varepsilon_{ij}(\mathbf{x})\Phi(\mathbf{x};\mathbf{x}-\mathbf{x}_L)d\Omega. \tag{6.18}$$

The following function can be chosen to facilitate the satisfaction of the integration constraints:

$$\Phi(\mathbf{x};\mathbf{x}-\mathbf{x}_L) = \begin{cases} 1/W_L, & \mathbf{x}\in\Omega_L \\ 0, & \mathbf{x}\notin\Omega_L \end{cases}, \tag{6.19}$$

where $W_L = \int_{\Omega_L} d\Omega$ is the nodal weight, which will also be used in the nodal quadrature of domain integrals. The strain smoothing operation in (6.18) can be manipulated using the function (6.19) and the divergence theorem:

$$\begin{aligned}
\tilde{\varepsilon}_{ij}(\mathbf{x}_L) &= \int_{\Omega} \varepsilon_{ij}(\mathbf{x})\Phi(\mathbf{x};\mathbf{x}-\mathbf{x}_L)d\Omega \\
&= \frac{1}{W_L}\int_{\Omega_L} \varepsilon_{ij}(\mathbf{x})d\Omega \\
&= \frac{1}{2W_L}\int_{\Omega_L} \left(u^h_{i,j}(\mathbf{x}) + u^h_{j,i}(\mathbf{x})\right)d\Omega \\
&= \frac{1}{2W_L}\int_{\Gamma_L} \left(u^h_i(\mathbf{x})n_j(\mathbf{x}) + u^h_j(\mathbf{x})n_i(\mathbf{x})\right)d\Gamma,
\end{aligned} \tag{6.20}$$

where Γ_L is the boundary of the nodal domain Ω_L. Introducing the shape functions u^h_i, we have the following approximation for the smoothed strain:

$$\tilde{\mathcal{E}}(\mathbf{x}_L) = \sum_{I=1}^{NP} \tilde{\mathbf{B}}_I(\mathbf{x}_L)\mathbf{u}_I, \tag{6.21}$$

were $\tilde{\mathcal{E}}(\mathbf{x}_L)$ and $\tilde{\mathbf{B}}_I(\mathbf{x}_L)$ are the smoothed strain vector and smoothed strain gradient matrix, respectively, for instance, in 2-D:

$$\tilde{\mathcal{E}}(\mathbf{x}_L) = \begin{bmatrix} \tilde{\varepsilon}_{11} \\ \tilde{\varepsilon}_{22} \\ \tilde{\varepsilon}_{12} \end{bmatrix}, \qquad \tilde{\mathbf{B}}_I(\mathbf{x}_L) = \begin{bmatrix} \tilde{b}_{1I}^L & 0 \\ 0 & \tilde{b}_{2I}^L \\ \tilde{b}_{2I}^L & \tilde{b}_{1I}^L \end{bmatrix}, \qquad \tilde{b}_{iI}^L = \frac{1}{A_L} \hat{\int}_{\Gamma_L} \Psi_I(\mathbf{x}) n_i(\mathbf{x}) d\Gamma, \tag{6.22}$$

where $A_L = W_L$ is the nodal weight (area). Note that quadrature is carried out to evaluate the smoothed gradient terms; the nature of this quadrature will be revisited. The integration constraints for the strain smoothing method, using (6.15) can be stated as

$$\begin{gathered} \sum_{L=1}^{NINT} \tilde{\mathbf{B}}_I^{\mathrm{T}}(\mathbf{x}_L)\, W_L = \sum_{L=1}^{NINTb} \boldsymbol{\Psi}_I(\mathbf{x}_L)\boldsymbol{\eta}^{\mathrm{T}}(\mathbf{x}_L) S_L \quad \forall I, \\ or \\ \sum_{L=1}^{NINT} \tilde{b}_{Ii}^L(\mathbf{x}_L)\, W_L = \sum_{L=1}^{NINTb} \Psi_I(\mathbf{x}_L) n_i(\mathbf{x}_L) S_L \quad \forall I, \end{gathered} \tag{6.23}$$

which can be shown to be satisfied when (6.19) is introduced in conjunction with the divergence theorem (6.20) resulting in (6.22):

$$\begin{aligned} \sum_{L=1}^{NINT} \tilde{b}_{Ii}^L(\mathbf{x}_L) W_L &= \sum_{L=1}^{NINT} \left(\frac{1}{W_L} \hat{\int}_{\Gamma_L} \Psi_I(\mathbf{x}) n_i(\mathbf{x}) d\Gamma \right) W_L \\ &= \underbrace{\sum_{L=1}^{NINT} \hat{\int}_{\Gamma_L} \Psi_I(\mathbf{x}) n_i(\mathbf{x}) d\Gamma = \hat{\int}_{\Gamma} \Psi_I(\mathbf{x}) n_i(\mathbf{x}) d\Gamma}_{\text{conforming condition}} \\ &= \sum_{L=1}^{NINTb} \Psi_I n_i(\mathbf{x}_L) S_L \quad \forall I. \end{aligned} \tag{6.24}$$

Since the nodal representative domains are conforming, if the cells share integration points along interior faces, the terms containing normals will cancel for the integrals along the internal shared edges of Γ_L (third equivalence), and so long as the contour integration performed on Γ coincides with the contour integration on Γ_L (fourth equivalence), the constraint is satisfied. Therefore, it is important to keep all contour integration consistent in order to pass the patch test.

Nodal integration of the weak form using the smoothed strains in (6.21) is termed stabilized conforming nodal integration (SCNI) [7]. The term *conforming* refers to the fact that all nodal domains conform to one another. *Stabilized* refers to the fact that it eliminates spurious unstable zero-energy modes when using the strains obtained with direct differentiation (see Section 6.9). The smoothing of strain in the RKPM2D code described in Chapter 10 is performed using the function `getSmoothedDerivative`, see Section 10.3.3.

It is important to note that SCNI is a type of nodal integration, so it is highly efficient. Further, because of the employment of the divergence theorem in (6.20), the order of differentiation of the shape functions needed in the weak form is reduced to zero, which also increases the efficiency (see Section 3.5). It both passes the linear patch test and converges at optimal rates consistent with a linear approximation.

At this point, one may ask, what is the order of quadrature needed on the boundary of each nodal cell? In [10], it was shown that one-point quadrature at the midpoint of each cell boundary is sufficient to maintain linear exactness. Further increasing the quadrature order has little to no effect since the constraints are already met, and consistency is already satisfied.

The SCNI method can be summarized in the following discrete equations for elasticity:

$$\sum_{J \in S} \mathbf{K}_{IJ}\mathbf{u}_J - \mathbf{f}_I = 0 \ \forall I \in S, \tag{6.25}$$

where $\mathbf{K}$ and $\mathbf{f}$ are the stiffness matrix and force vector respectfully, carried out by nodal integration along with the smoothed nodal gradients (6.21):

$$\begin{aligned} \mathbf{K}_{IJ} &= \sum_{L=1}^{NP} \tilde{\mathbf{B}}_I^{\mathrm{T}}(\mathbf{x}_L)\mathbf{D}(\mathbf{x}_L)\tilde{\mathbf{B}}_J(\mathbf{x}_L)W_L, \\ \mathbf{f}_I &= \sum_{L=1}^{NP} \Psi_I(\mathbf{x}_L)\mathbf{b}(\mathbf{x}_L)W_L + \sum_{L=1}^{NINTb} \Psi_I(\mathbf{x}_L)\bar{\mathbf{t}}(\mathbf{x}_L)S_L. \end{aligned} \tag{6.26}$$

We should point out that employing either the consistent weak forms (see Chapter 5), Lagrange multipliers, or Nitsche method (see Chapter 4) can only truly pass the patch test; see [10, 11] for a discussion. So here, we give the consistent weak formulation one for convenience:

$$\sum_{J \in S} (\mathbf{K}_{IJ} - \mathbf{H}_{IJ})\mathbf{u}_J - \mathbf{f}_I = 0 \ \forall I \in S, \tag{6.27}$$

where $\mathbf{K}_{IJ}$ is the same as in (6.26), and

$$\mathbf{H}_{IJ} = \sum_{L=1}^{NPb} \Psi_I(\mathbf{x}_L)\boldsymbol{\eta}^{\mathrm{T}}(\mathbf{x}_L)\mathbf{D}(\mathbf{x}_L)\tilde{\mathbf{B}}_J(\mathbf{x}_L)S_L. \tag{6.28}$$

The employment of this form is necessary to obtain higher-order convergence (for $n > 1$) as discussed in Section 5.8, and it is also strictly needed for passing a patch test, which is useful for code verification.

Now, when solving the Poisson problem in (6.4), one can observe that the solution converges at optimal rates as shown in Figure 6.12, in contrast to DNI. A relaxed version of SCNI, stabilized *non-conforming* nodal integration (SNNI) [12, 13], is also shown, where the nodal domains no longer conform (see Figure 6.13) and the method violates the integration constraints due to the inability to satisfy the equivalences in (6.24). Here, it is seen that optimal convergence is not achieved–demonstrating the importance of the conforming condition.

The concept of strain smoothing in SCNI has also been applied to FEM. The PhD thesis of Guan [14] first introduced the strain smoothing in SCNI to finite elements as the basis for coupling FEM and RKPM with unified discretization and domain integration. Various nodal and element strain smoothing techniques have been proposed by Liu et al. [15–17] for finite elements and are termed smoothed FEMs. The SCNI method has been shown to be grounded in variational principles [18].

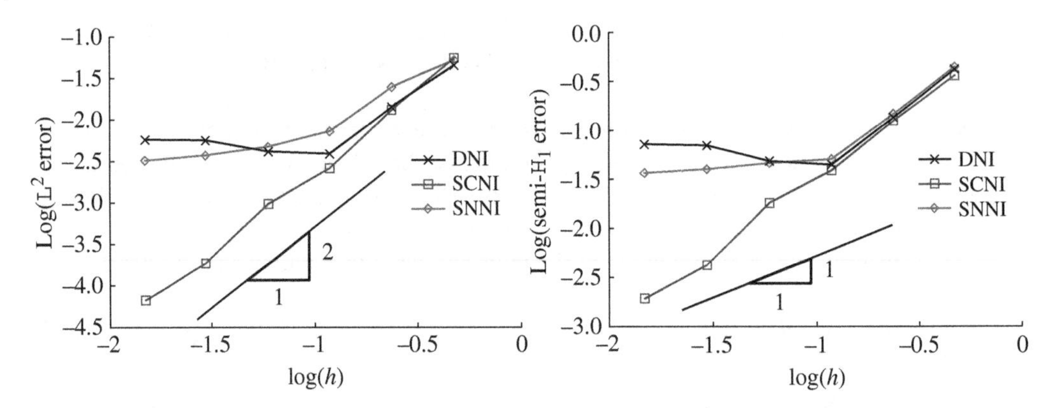

Figure 6.12 Convergence of nodal integration in the Poisson problem (diffusion). Optimal rates are achieved for SCNI, but not for SNNI due to relaxation of the conforming condition. DNI shown for reference, which provides a solution similar to SNNI.

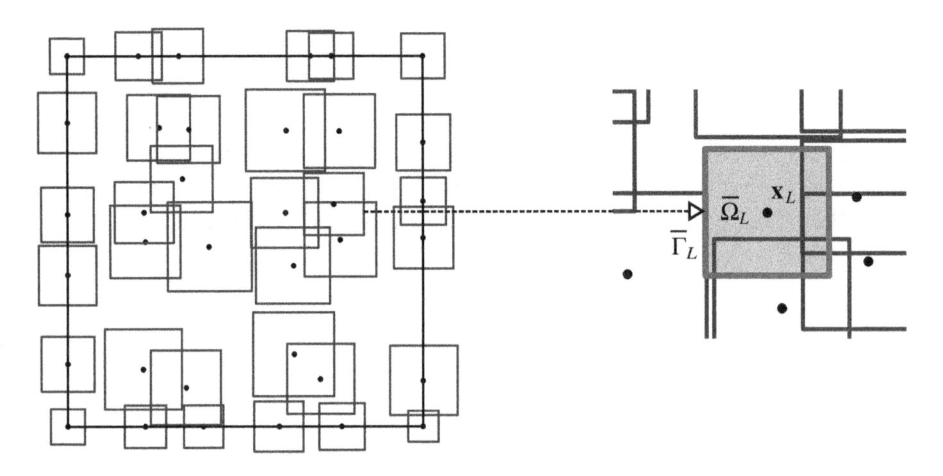

Figure 6.13 Stabilized non-conforming nodal integration (SNNI) with domains $\overline{\Omega}_L$ and boundary $\overline{\Gamma}_L$.

Key Takeaway
SCNI can be introduced to attain linear exactness in the Galerkin solution (passing the linear patch test) and restore optimal convergence rates when a linear basis is used in the meshfree approximation. The method provides simplicity being node-based and is highly efficient.

Exercise 6.2 Consider the problem in Exercise 6.1. Solve the problem with SCNI and compare it to the solution by DNI and Gauss quadrature. Verify that SCNI can pass the patch test. Note that for SCNI in 1-D, we have (divergence is just the first fundamental theorem of calculus in 1-D):

$$u^h_{,x}(x_L) = \sum_{I=1}^{NP} \tilde{\Psi}_{I,x}(x_L) u_I \equiv \tilde{\mathbf{B}}(x_L)\mathbf{d}, \tag{6.29}$$

with

$$\tilde{\Psi}_{I,x}(x_L) = \frac{1}{S_L}\int_{x_L^-}^{x_L^+} \Psi_{I,x}(x_L)dx = \frac{1}{S_L}\{\Psi_I(x_L^+) - \Psi_I(x_L^-)\}, \tag{6.30}$$

where $\tilde{\Psi}_{I,x}(x_L)$ is the smoothed derivative of the shape function, $S_L = x_L^+ - x_L^-$ is the weight associated with node L, x_L^+ is the location of the conforming smoothing point to the right of the node, and on x_L^- is the location of the conforming smoothing point to the left of the node, the collection of which for all nodes form the conforming cells over the domain.

Exercise 6.3 Consider the two-dimensional reproducing kernel (RK) approximation (see Chapter 3) in the Galerkin method (RKPM—see Chapter 4) with SCNI for the diffusion equation with $k = 1$, where the entries of the stiffness are evaluated as

$$\mathbf{K}_{IJ} = \sum_{L=1}^{NP} \tilde{\mathbf{B}}_I^{\mathrm{T}}(\mathbf{x}_L)\tilde{\mathbf{B}}_J(\mathbf{x}_L)A_L, \tag{6.31}$$

where

$$\tilde{\mathbf{B}}_I(\mathbf{x}_L) = \begin{bmatrix} \tilde{b}_{I1}^L \\ \tilde{b}_{I2}^L \end{bmatrix}, \tag{6.32}$$

with

$$\tilde{b}_{Ii}^L = \frac{1}{A_L}\int_{\Gamma_L} \Psi_I(\mathbf{x})n_i(\mathbf{x})d\Gamma, \tag{6.33}$$

and $A_L = \int_{\Omega_L} d\Omega$ is the nodal area. Note that the flux vector for the source term is integrated with standard nodal integration:

$$f_I = \sum_{N=1}^{NP} \Psi_I(\mathbf{x}_N)s(\mathbf{x}_N)A_N. \tag{6.34}$$

1) Solve the following diffusion equation with $k = 1$ and $s = 0$ for the scalar u on $\Omega = [0, 1] \times [0, 1]$:

$$\begin{aligned} u_{,ii} &= 0 \quad \text{on } \Omega, \\ u &= x + 2y \quad \text{on } \Gamma_u, \end{aligned} \tag{6.35}$$

where $u_{,ii} = u_{,xx} + u_{,yy}$, and $\Gamma_u = \Gamma$, $\Gamma_q = \emptyset$. The exact solution is $x + 2y$. Plot the solution error for two cases of uniform and nonuniform nodes. Note that SCNI won't pass the patch test exactly (to machine precision) unless the consistent weak forms are used with the transformation method (due to inconsistencies described in Chapter 5), but it will come close with the standard weak form (standard variational principle). If desired, use a consistent weak form from Section 5.8 to verify machine precision error is obtained.

2) Solve the diffusion equation with $k = 1$ and $s = -(x^2+y^2)e^{xy}$ for the scalar u on $\Omega = [0, 1] \times [0, 1]$:

$$\begin{aligned} u_{,ii} &= (x^2 + y^2)e^{xy} \quad \text{on} \quad \Omega, \\ u &= e^{xy} \quad \text{on} \quad \Gamma_u, \end{aligned} \tag{6.36}$$

where $u_{,ii} = u_{,xx} + u_{,yy}$, and $\Gamma_u = \Gamma$, $\Gamma_q = \emptyset$. The exact solution is e^{xy}. Use the RKPM with the following formulation: linear basis, cubic B-spline kernel, and stabilized conforming nodal integration. Obtain the numerical solution using 11×11, 21×21, and 41×41 nodes. Use a normalized support size $a/h = 2.0$. Compare numerical and exact solutions along lines $x = 0.25$, $x = 0.5$, and $x = 0.75$. Also compute the L_2 error norm of the error denoted by e_{L_2} and obtain the rate of convergence by plotting $\log_{10}(e_{L_2})$ vs. $\log_{10}(h)$, where h is the nodal distance, and e_{L_2} is defined as

$$e_{L_2} = \left(\int_\Omega \left(u(x) - u^h(x)\right)^2 d\Omega \right)^{1/2}. \tag{6.37}$$

Use at least eight-point Gauss quadrature to carry out the above integral. Confirm the optimal rate of convergence of two (see Chapter 3, Eq. (3.95)).

6.7 Variationally Consistent Integration

Meshfree methods show their true strength and advantage when deformations are extreme: the absence of a mesh and any related mapping circumvents issues with mesh distortion and mesh entanglement. However, while SCNI is an effective nodal integration, it still employs some mesh-like features: conforming domains are needed, whereas in extreme deformations, the mapping between the current and reference configuration may become ill-conditioned (the mapping may become extremely distorted) or not exist (the topology of the domain may change). Therefore, in this class of problems, conforming cells would need to be updated rather than be defined in the material configuration. The updates would become cumbersome and also involve the clipping of Voronoi cells where the topology of the domain changes in the formation or closure of free surfaces.

Therefore, a non-conforming integration is desirable. However, as we have seen, satisfaction of linear exactness by SCNI relies on the conforming nature of the cells, and alternative methods must be adopted to maintain accuracy.

6.7.1 Variational Consistency Conditions

First, it is useful to generalize the concept of the integration constraint to nth-order solutions, or in other words, the case of a high-order patch test. This concept was first introduced in [10] and was termed *variational consistency*. It can be utilized to obtain a unified correction of various types of quadrature, conforming or nonconforming, to be seen in the next section.

Consider an nth-order solution of the form:

$$u_i(x, y) = u_i^{\mathrm{P}}(x, y) \equiv c_{00i} + c_{10i}x + c_{01i}y + c_{20i}x^2 + c_{11i}xy + \ldots + c_{\alpha\beta i}x^\alpha y^\beta + \ldots + c_{0ni}y^n, \tag{6.38}$$

where $c_{\alpha\beta i}$ are arbitrary constants independent in each direction i. The two-dimensional elasticity boundary value problem associated with this high-order solution is

$$\begin{aligned} \sigma_{ij,j} &= \overline{b}_i^{\mathrm{P}} \text{ on } \Omega, \\ \sigma_{ij} n_j &= \overline{t}_i^{\mathrm{P}} \text{ on } \Gamma_t, \\ u_i &= \overline{u}_i^{\mathrm{P}}(x,y) \text{ on } \Gamma_u, \end{aligned} \tag{6.39}$$

where again $\overline{t}_i^{\mathrm{P}} = \sigma_{ij}^{\mathrm{P}} n_j$ with $\sigma_{ij}^{\mathrm{P}} \equiv C_{ijkl} u_{(k,l)}^{\mathrm{P}}(x,y)$ the stress associated with $u_i^{\mathrm{P}}(x,y)$, and $\overline{u}_i(x,y) = u_i^{\mathrm{P}}(x,y)$ is the polynomial solution. Note that since the stress is no longer constant, the body force in the problem at hand is now nonzero and is formed as $\overline{b}_i^{\mathrm{P}} = -\sigma_{ij,j}^{\mathrm{P}}$. Similar to the linear case, the approximation space should be able to represent this nth order displacement field, which is satisfied if the shape functions are nth order complete:

$$\sum_{J=1}^{NP} \Psi_J(x,y) u_i^{\mathrm{P}}(x_J, y_J) = u_i^{\mathrm{P}}(x,y). \tag{6.40}$$

Exercise 6.4

Show that the solution $u_i^{\mathrm{P}}(x,y)$ satisfies (6.39) for arbitrary n.

Hint

Use $\overline{t}_i^{\mathrm{P}} = \sigma_{ij}^{\mathrm{P}} n_j$ and $\overline{b}_i^{\mathrm{P}} = -\sigma_{ij,j}^{\mathrm{P}}$ with $\sigma_{ij}^{\mathrm{P}} \equiv C_{ijkl} u_{(k,l)}^{\mathrm{P}}(x,y)$.

Exercise 6.5

Show that (6.40) is satisfied using the completeness conditions of the RK shape function from Chapter 3.

Solution

The proof follows the linear case:

$$\begin{aligned} &\sum_{J=1}^{NP} \Psi_I \left(c_{00i} + c_{10i} x_J + c_{01i} y_J + c_{20i} x_J^2 + c_{11i} x_J y_J + \ldots + c_{0ni} y_J^n \right) \\ &= c_{00i} \left(\sum_{J=1}^{NP} \Psi_J \right) + c_{10i} \left(\sum_{J=1}^{NP} \Psi_J x_J \right) + c_{01i} \left(\sum_{J=1}^{NP} \Psi_J y_J \right) + c_{20i} \left(\sum_{J=1}^{NP} \Psi_J x_J^2 \right) \\ &\quad + c_{11i} \left(\sum_{J=1}^{NP} \Psi_J x_J y_J \right) + \ldots + c_{0ni} \left(\sum_{J=1}^{NP} \Psi_J y_J^n \right) \\ &= c_{00i} + c_{10i} x + c_{01i} y + c_{20i} x^2 + c_{11i} xy + \ldots + c_{0ni} y^n. \end{aligned} \tag{6.41}$$

Employing the nodal coefficients corresponding to $u_i^{\mathrm{P}}(x,y)$ and following the linear case using the consistency conditions to obtain $\sum_{J=1}^{NP} \mathbf{D}\mathbf{B}_J \mathbf{u}_J^{\mathrm{P}} = \boldsymbol{\sigma}^{\mathrm{P}}$, the associated left-hand side of the Galerkin equations using CWF I is calculated as

$$\sum_{J=1}^{NP}(\mathbf{K}_{IJ}-\mathbf{H}_{IJ})\mathbf{u}_J^{\mathrm{P}} = \sum_{J=1}^{NP}\left\{\int_{\Omega}\mathbf{B}_I^{\mathrm{T}}\mathbf{D}\mathbf{B}_J d\Omega - \int_{\Gamma_u}\Psi_I\boldsymbol{\eta}^{\mathrm{T}}\mathbf{D}\mathbf{B}_J d\Gamma\right\}\mathbf{u}_J^{\mathrm{P}} = \int_{\Omega}\mathbf{B}_I^{\mathrm{T}}\boldsymbol{\sigma}^{\mathrm{P}}d\Omega - \int_{\Gamma_u}\Psi_I\boldsymbol{\eta}^{\mathrm{T}}\boldsymbol{\sigma}^{\mathrm{P}}d\Gamma. \tag{6.42}$$

The external force for the solution $u_i^{\mathrm{P}}(x,y)$ is again obtained using the prescribed values of the vector traction $\bar{\mathbf{t}}^{\mathrm{P}}$, but now also considering the vector body force $\bar{\mathbf{b}}^{\mathrm{P}}$:

$$\mathbf{f}_I = \int_{\Gamma_t}\Psi_I\underbrace{\bar{\mathbf{t}}^{\mathrm{P}}}_{\boldsymbol{\eta}^{\mathrm{T}}\boldsymbol{\sigma}^{\mathrm{P}}}d\Gamma + \int_{\Omega}\Psi_I\underbrace{\bar{\mathbf{b}}^{\mathrm{P}}}_{-\mathbf{L}^{\mathrm{T}}\boldsymbol{\sigma}^{\mathrm{P}}}d\Omega = \int_{\Gamma_t}\Psi_I\boldsymbol{\eta}^{\mathrm{T}}\boldsymbol{\sigma}^{\mathrm{P}}d\Gamma - \int_{\Omega}\Psi_I\mathbf{L}^{\mathrm{T}}\boldsymbol{\sigma}^{\mathrm{P}}d\Omega, \tag{6.43}$$

where $\mathbf{L}$ is a matrix of differential operators:

$$\mathbf{L} = \begin{bmatrix}\partial/\partial x_1 & 0 \\ 0 & \partial/\partial x_2 \\ \partial/\partial x_2 & \partial/\partial x_1\end{bmatrix}. \tag{6.44}$$

For satisfaction of discrete equilibrium, (6.42) and (6.43), should be in balance, which requires

$$\int_{\Omega}\mathbf{B}_I^{\mathrm{T}}\boldsymbol{\sigma}^{\mathrm{P}}d\Omega - \int_{\Gamma_u}\Psi_I\boldsymbol{\eta}^{\mathrm{T}}\boldsymbol{\sigma}^{\mathrm{P}}d\Gamma = \int_{\Gamma_t}\Psi_I\boldsymbol{\eta}^{\mathrm{T}}\boldsymbol{\sigma}^{\mathrm{P}}d\Gamma - \int_{\Omega}\Psi_I\mathbf{L}^{\mathrm{T}}\boldsymbol{\sigma}^{\mathrm{P}}d\Omega. \tag{6.45}$$

This should hold for any $\boldsymbol{\sigma}^{\mathrm{P}}$. From the above, it is immediately apparent that when $\boldsymbol{\sigma}^{\mathrm{P}}$ = constant, the requirement reduces to the original integration constraint (6.14). However, since $\boldsymbol{\sigma}^{\mathrm{P}}$ is no longer constant, the constraints become more complex, but are still tractable. Using the fact that $\Gamma = \Gamma_u \cup \Gamma_t$, the resulting condition is

$$\int_{\Omega}(\mathbf{L}\Psi_I)^{\mathrm{T}}\boldsymbol{\sigma}^{\mathrm{P}}d\Omega = \int_{\Gamma}\Psi_I\boldsymbol{\eta}^{\mathrm{T}}\boldsymbol{\sigma}^{\mathrm{P}}d\Gamma - \int_{\Omega}\Psi_I\mathbf{L}^{\mathrm{T}}\boldsymbol{\sigma}^{\mathrm{P}}d\Omega, \tag{6.46}$$

where $\mathbf{B}_I = \mathbf{L}\Psi_I$ has been employed. Here one can observe that the above is nothing but integration by parts: transferring the differential operator $\mathbf{L}$ from Ψ_I to $\boldsymbol{\sigma}^{\mathrm{P}}$, one obtains the negative product $\mathbf{L}^{\mathrm{T}}\boldsymbol{\sigma}^{\mathrm{P}}$ and the contour integral involving $\boldsymbol{\eta}^{\mathrm{T}}\boldsymbol{\sigma}^{\mathrm{P}}$.

Finally, introducing numerical integration, and using the fact that $\boldsymbol{\sigma}^{\mathrm{P}}$ is arbitrary ($\mathbf{D}$ contains arbitrary constants, as does $\mathbf{u}^{\mathrm{P}}$), it can be shown that (6.46) reduces to [10]

$$\begin{aligned}\hat{\int_{\Omega}}\Psi_{I,j}u_{k,i}^{\mathrm{P}}(x,y)d\Omega &= \hat{\int_{\Gamma}}\Psi_I u_{k,i}^{\mathrm{P}}(x,y)n_j d\Gamma - \hat{\int_{\Omega}}\Psi_I u_{k,ij}^{\mathrm{P}}(x,y)d\Omega, \\ &\text{or} \\ \hat{\int_{\Omega}}\Psi_{I,j}\mathbf{u}_{,i}^{\mathrm{P}}(x,y)d\Omega &= \hat{\int_{\Gamma}}\Psi_I\mathbf{u}_{,i}^{\mathrm{P}}(x,y)n_j d\Gamma - \hat{\int_{\Omega}}\Psi_I\mathbf{u}_{,ij}^{\mathrm{P}}(x,y)d\Omega.\end{aligned} \tag{6.47}$$

Approximations with compatible numerical integration in the form of the above have been termed *variationally consistent integration* (VCI) methods. Using this technique, optimal convergence can be attained using far lower-order quadrature than would otherwise be required. For VC conditions for other partial differential equations, consult [19].

6.7.2 Petrov–Galerkin Correction: VCI

As before, due to the rational nature of the shape functions involved in the constraints (6.47), designing methods that satisfy these conditions does not seem straightforward. However, noticing that the constraints contain the test functions rather than the trial functions which need only be nth order complete, a correction can be obtained by designing a separate test function. The idea is akin to the satisfaction of the reproducing conditions in Chapter 3: for m conditions, one can introduce m functions with m unknowns and simply solve the equation. For the equation in (6.47), this correction can be performed on a node-by-node basis since the constraint is for each test function Ψ_I individually. Since nothing is said of the quadrature at hand, this method can be employed to correct nodal integration methods which do not require conforming cells (that are cumbersome for extreme deformation problems).

First, as noted above, the constraint is only on the test function, which we shall now distinguish from the set of trial functions using the notation $\overline{\Psi}_I$. Also, only the gradient of $\mathbf{u}^{\mathrm{P}}$ shows up in the constraint; therefore, a reduced column vector of (n-1)th order polynomials $\mathcal{P}$ of length m can be used to represent the conditions:

$$\mathcal{P} \equiv \left[1 \quad x \quad y \quad x^2 \quad xy \quad y^2 \;\dots\quad x^{\alpha}y^{\beta} \;\dots\; y^{(n-1)}\right]^{\mathrm{T}}. \tag{6.48}$$

Interestingly, this implies that constant exactness is automatic for second-order partial differential equations such as elasticity, so long as the partition of unity holds.

The constraint can then be simplified as

$$\hat{\int_{\Omega}} \overline{\Psi}_{I,j} \mathcal{P} d\Omega \;=\; \hat{\int_{\Gamma}} \overline{\Psi}_I \mathcal{P} n_j d\Gamma - \hat{\int_{\Omega}} \overline{\Psi}_I \mathcal{P}_{,j} d\Omega. \tag{6.49}$$

Now, the number of constraints is m, the size of $\mathcal{P}$. Noticing that the left-hand side only contains derivatives of $\overline{\Psi}_I$, an assumed strain technique is introduced for $\overline{\Psi}_{I,j}$ with m unknowns:

$$\begin{aligned} u_i^h &= \sum_{I=1}^{NP} \Psi_I u_{iI}, \qquad u_{i,j}^h = \sum_{I=1}^{NP} \Psi_{I,j} u_{iI}, \\ v_i^h &= \sum_{I=1}^{NP} \Psi_I v_{iI}, \qquad v_{i,j}^h = \sum_{I=1}^{NP} \overline{\Psi}_{I,j} v_{iI}, \end{aligned} \tag{6.50}$$

where

$$\overline{\Psi}_{I,j}(x,y) = \Psi_{I,j}(x,y) + \Theta_I(x,y)\mathcal{P}^{\mathrm{T}}(x,y)\zeta_{jI}. \tag{6.51}$$

The term ζ_{jI} is a column vector of constant unknowns for each j and I, and Θ_I is a kernel-type localizing function which for simplicity can be taken as

$$\Theta_I = \begin{cases} 1 & \operatorname{supp}(\Psi_I) \neq 0 \\ 0 & \text{else} \end{cases}. \tag{6.52}$$

Substituting (6.51) into (6.49), one obtains the equation

$$\zeta_{jI} = \mathcal{M}_I^{-1}\mathcal{R}_{jI}, \tag{6.53}$$

where

$$\mathcal{M}_I = \hat{\int_\Omega} \Theta_I \mathcal{P}\mathcal{P}^{\mathrm{T}} d\Omega, \tag{6.54}$$

and $\mathcal{R}_{jJ}$ is the residual of the integration constraint (6.49):

$$\mathcal{R}_{jI} = \hat{\int_\Gamma} \Psi_I \mathcal{P} n_j d\Gamma - \hat{\int_\Omega} \Psi_I \mathcal{P}_{,j} d\Omega - \hat{\int_\Omega} \Psi_{I,j} \mathcal{P} d\Omega. \tag{6.55}$$

As can be seen from (6.51), (6.53), and (6.55), the method is driven by the residual of integration constraints. Therefore, a correction is only introduced as needed. For SCNI, no first-order correction is performed, although it can made to be higher order.

The correction in (6.51)–(6.55) is also both simple and efficient. First, it involves systems that can be solved on a local, node-by-node basis, and no global system is involved. The matrices in (6.54) are also symmetric and small (e.g. for a linear correction, they are scalars, and for a quadratic correction, the matrices are size $d + 1$ for dimension d). The loop needed for (6.54) and (6.55) in coding is also essentially the same as the one for integrating a force vector in the Galerkin equation and can easily be made parallel using the same logic.

Figure 6.14 shows that optimal convergence rates can be attained using variationally consistent methods with various low-order quadrature schemes, including DNI and the nonconforming strain smoothing of SNNI. Therefore, this method can be used to obtain high-order (linear or higher) accuracy in extreme deformations problems where reconstruction of conforming integration cells would be impractical. However, it is important to note that since the trial function remains unmodified, unstable modes can still be admitted in the solution, and low-order quadrature such as nodal integration still needs stabilization (see Section 6.10).

Several other schemes to attain variational consistency using Bubnov–Galerkin methods (using the same test and trial functions) have been proposed for quadratic accuracy [21, 22] and arbitrary-order accuracy [23]. In general, these methods can be considered extensions of SCNI to higher-order cases. The method in [23] employs a minimum number of global quadrature points and presents a quite

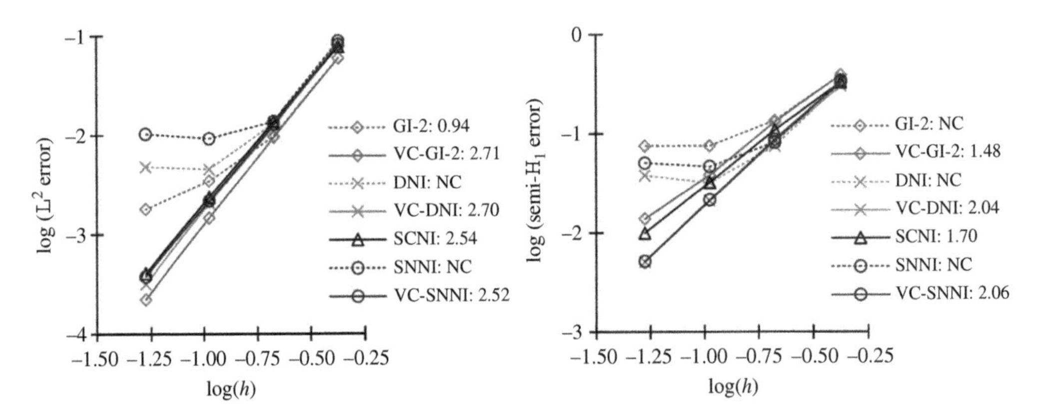

Figure 6.14 Convergence of RKPM with linear basis using various integration methods with and without variational consistency. Variationally consistent methods are denoted with the prefix "VC-"; 2nd order of Gauss integration (GI) is denoted with the suffix "-2". Convergence rates are indicated in the legend. *Source:* Chen et al. [20], figure 14, p. 17 / ASCE.

favorable option. Nevertheless, these methods all rely on conforming cells to achieve both gradient consistency (needed for the trial functions), and variational consistency (needed for the test functions), and limits their use in extreme-deformation problems. Conforming integration appears to be an inherent prerequisite to achieve a variationally consistent Bubnov–Galerkin method.

6.8 Quasi-Conforming SNNI for Extreme Deformations: Adaptive Cells

As previously discussed, SCNI would require the reconstruction of smoothing cells in extreme-deformation problems. Therefore, in practice, it is much more convenient to simply use nonconforming cells throughout the simulation which are "floated" as boxes or spheres around the cells (called SNNI, see Figure 6.13). However, the relaxation of the conforming condition results in sub-optimal or non-convergent solutions, as we have discussed. While the variationally consistent correction in Section 6.7 can rectify this situation, the total boundary of the domain Γ, needed in (6.55), is not readily available in problems where the topology of the domain is evolving. Nevertheless, the geometry strain smoothing cells employed for SNNI can be adjusted to maintain a high degree of conformity without any additional computational cost as follows (in 2D):

$$\begin{aligned} \dot{\theta} &= \dot{u}_{y',x'}, \\ \dot{D}_{x'} &= D^0_{x'}\dot{u}_x, \\ \dot{D}_{y'} &= D^0_{y'}\dot{u}_y, \end{aligned} \tag{6.56}$$

where $\dot{\theta}$ is the time rate of change of the orientation of the smoothing cell, $\dot{u}_x$ and $\dot{u}_y$ are the velocities in the x and y directions, respectively, $D_{x'}$ and $D_{y'}$ are the updated dimensions of a smoothing cell in the x' and y' directions, respectively, and the superscript "0" denote values at the initial configuration. Figure 6.15 schematically illustrates the effect of using the updated smoothing cells

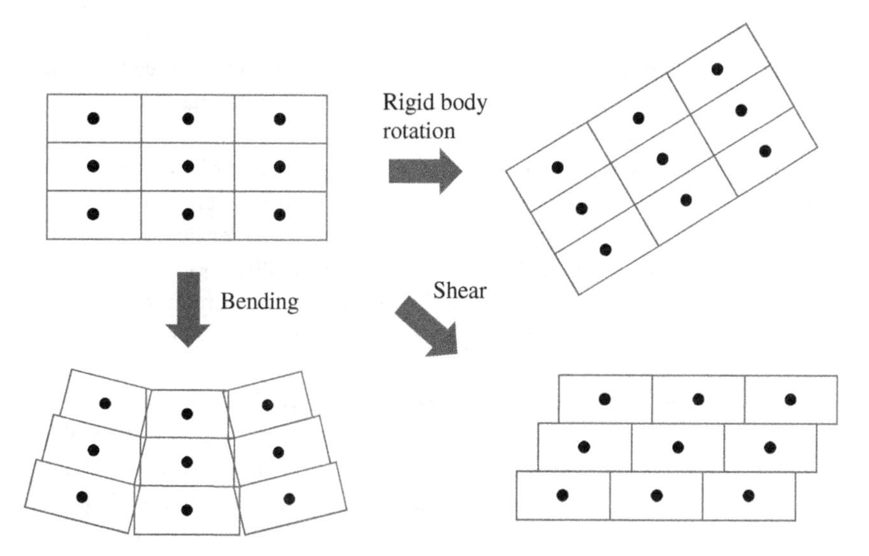

Figure 6.15 Schematic illustration of updated smoothing cells. *Source:* Baek et al. [24], figure 8, p. 11 / Springer Nature / Public Domain CC BY.

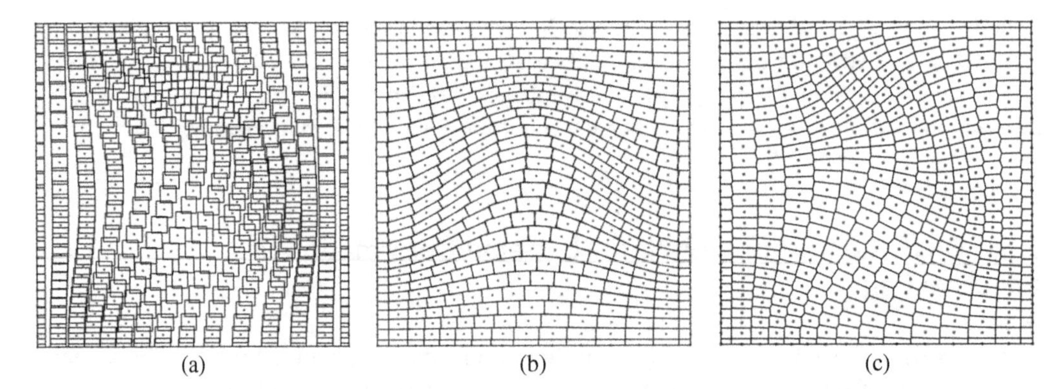

Figure 6.16 Smoothing domains using (a) non-conforming cells, (b) adaptive rectangular cells, and (c) Voronoi diagram cells. *Source:* Adapted from Baek et al. [24], figure 10, p. 13.

under rigid body motion, bending, and shear deformation. Although the cells are still non-conforming under arbitrary deformations, they maintain a high level of conformity, as shown in Figure 6.16 for a manufactured deformation. The studies presented in [24] show that the quasi-conforming cell update (as in Figure 6.16b) is an effective approach to maintain optimal convergence rates. Nevertheless, when the cells are not highly conforming due to extreme distortions, the residual of the integration constraint (6.55) can be used as a quadrature error indicator to locally correct the approximation with VCI.

6.9 Instability in Nodal Integration

In addition to providing low accuracy of the solution due to a relatively crude evaluation of integrals in the weak formulation, employing nodes as integration points results in energy underestimation of oscillatory modes. The sampling points are essentially just "no good," and the mode contributes virtually no strain energy to the solution and can thus grow essentially unbounded. An important focus of research in meshfree methods over the past several decades has been to resolve this issue.

To illustrate the essential difficulty, consider the one-dimensional problem of a bar discretized by RKPM with a mode of oscillating displacement as shown in Figure 6.17. The associated strain energy density for this mode is shown, which is zero at nodal locations in the interior of the domain. The choice of nodal integration will thus severely underestimate the true energy of this mode (the area under the curve shown), and the stiffness of the formulation will offer little to no resistance. The clustering of nodes shown in Figure 6.17 is the pattern of instability observed in simulations and serves as a red flag that the chosen nodal integration is not providing stability. Figure 6.18a–c shows the typical manifestation of this pattern in an engineering analysis of penetration.

This mode is also easily observed in the eigenvalue analysis of a nodally integrated stiffness matrix for elasticity: the lowest non-zero energy mode is almost always a spurious oscillatory mode as shown in Figure 6.19 for a meshfree discretization of a plane-strain solid (compare to the fully-integrated FEM mode). In the Galerkin method, because of the low energy, the modal participation in the solution will usually be large and pollute the result. In dynamics, this mode tends to grow virtually unbounded, and in many solid mechanics problems of interest for meshfree methods, which involve failure, this results in over prediction of damage and spurious numerical fracture seen in Figure 6.18b and c.

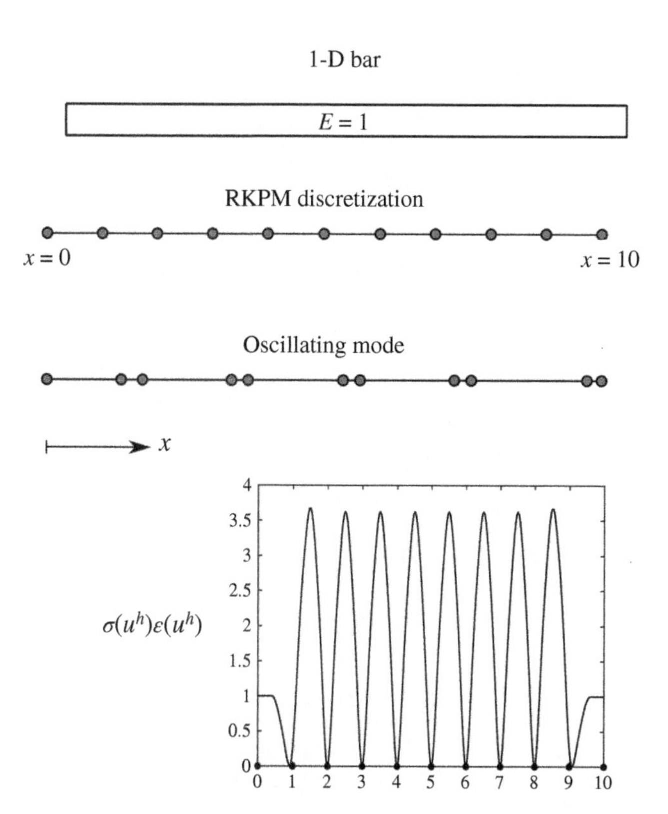

Figure 6.17 RKPM discretization of a bar, alternating displacement mode, and associated strain energy density: the strain energy density is zero at nodal locations in the interior of the domain.

Key Takeaway

Observing a node-to-node oscillating pattern in nodally integrated meshfree solutions is *almost always* the result of using an insufficiently stable nodal integration technique.

6.10 Stabilization of Nodal Integration

The instability in nodal integration can be attributed to vanishing derivatives of short-wavelength (two times the nodal spacing) modes, and thus have little or no energy and can grow unbounded in the solution [7, 25, 26]. As shown in Figure 6.17, the strain energy density is severely underestimated for this mode when using nodal integration.

One way to circumvent this instability is to employ techniques which minimize a proportion of residual in the least-squares sense [9, 25, 27]. These methods, however, involve computationally intensive second-order derivatives of approximation functions and typically a stabilization parameter. In addition, it is not clear how to effectively compute the residual of the divergence of stress in a nonlinear setting.

Another way to circumvent this difficulty is to calculate gradients in locations other than the nodes, often called the stress point method [28–30], which gives a more reasonable estimate of the total strain energy (Figure 6.17 demonstrates how this method works). This method does, however, require additional stabilization in order to ensure convergent solutions in all situations [31].

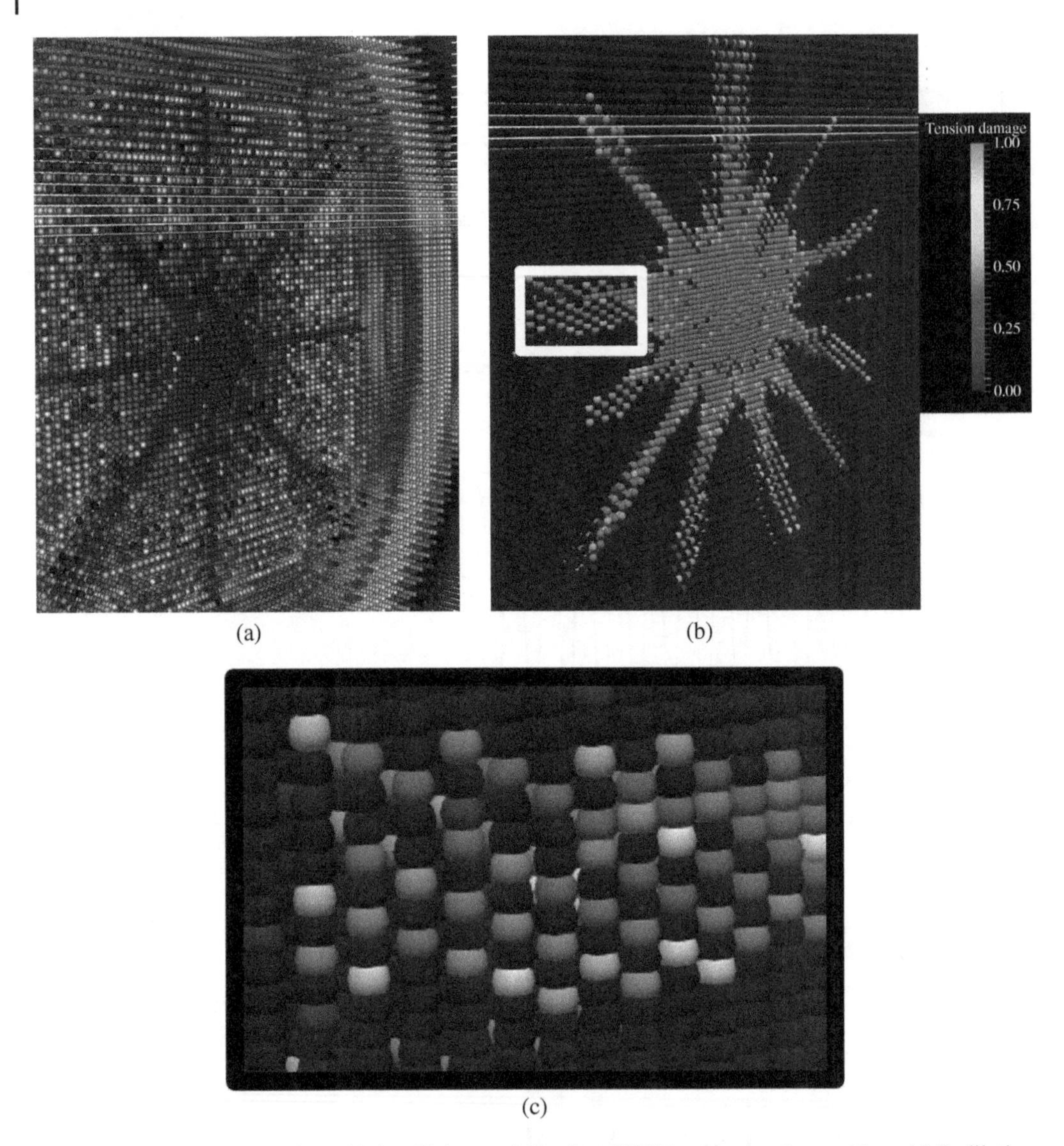

Figure 6.18 Nodal integration with insufficient stabilization (SNNI) in a penetration problem: (a) Oscillations in von-Mises stress, (b) spurious damage pattern induced by oscillations, and (c) zoom-in of damage pattern. *Source:* Adapted from Hillman et al. [42] figure 5, p. 13.

The addition of sampling points can also carry a significant computational burden, akin to higher-order integration.

Utilizing Taylor series expansions allows one to obtain "extra" information around the integration points, and thus stabilize modes which have zero or near-zero energy associated with them. The origin of this method is the unification of stabilization in finite elements [32]. The technique has been subsequently utilized to stabilize nodal integration in meshfree methods [33, 34]. The drawback is that it requires the calculation of high-order derivatives (which are computationally expensive for meshfree approximations, see Chapter 3).

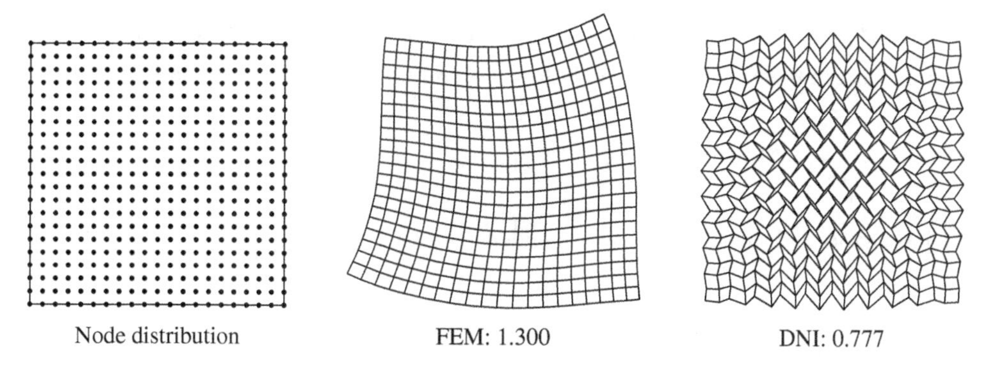

Figure 6.19 First non-zero eigenvalue modes with eigenvalues for fully-integrated FEM, and DNI for meshfree.

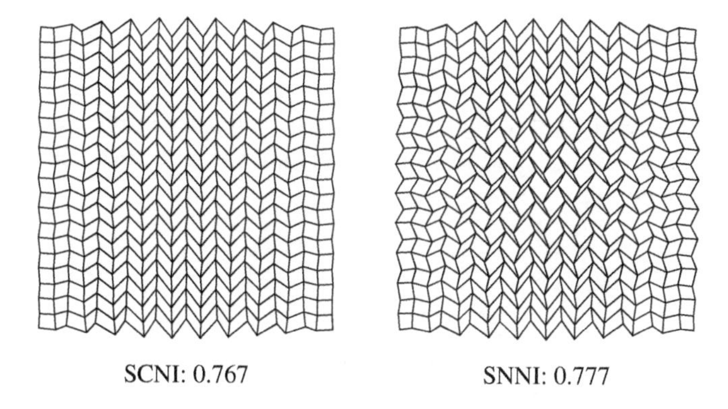

Figure 6.20 First non-zero eigenvalue modes with eigenvalues for SCNI and SNNI.

Part of the dual motivation for the SCNI technique was that it also stabilizes zero-energy modes in DNI by avoiding evaluating derivatives directly at nodal points [35]. In this method, the order of differentiation for constructing a stable method is reduced, also leading to enhanced efficiency. However, the smoothed integration in SCNI and SNNI are subject to spurious, oscillatory low-energy modes which can show up when the surface to volume ratio is sufficiently small, or in sufficiently fine discretizations [36]. An example of this mode obtained by an eigenvalue analysis of the stiffness matrix for SCNI and SNNI in a finer discretization is shown in Figure 6.20.

6.10.1 Notation for Stabilized Nodal Integration

Before describing stabilized nodal integration methods, it is useful to define some notations to make the presentation concise.

Let the variation on approximated strain energy be denoted using the so-called *bilinear form* $a(\cdot, \cdot)$ as follows:

$$a\left(\delta \mathbf{u}^h, \mathbf{u}^h\right) = \int_\Omega \delta u^h_{(i,j)} C_{ijkl} u^h_{(k,l)} d\Omega, \tag{6.57}$$

where $u^h_{(i,j)} = \left(u^h_{i,j} + u^h_{j,i}\right)/2$ and likewise for $\delta u^h_{(i,j)}$. The term *bilinearity* means $a(\cdot, \cdot)$ is linear in each argument, as can be seen in the above. In terms of strains, using bold-face notation, we have

$$a(\delta\mathbf{u}, \mathbf{u}) = \int_\Omega \boldsymbol{\varepsilon}\left(\delta\mathbf{u}^h\right) : \mathbf{C} : \boldsymbol{\varepsilon}\left(\mathbf{u}^h\right) d\Omega, \tag{6.58}$$

where we have expressed the strain with a dependence on an argument, that is,

$$\begin{aligned} \varepsilon_{ij}\left(\mathbf{u}^h\right) &= u^h_{(i,j)}, \\ \varepsilon_{ij}\left(\delta\mathbf{u}^h\right) &= \delta u^h_{(i,j)}. \end{aligned} \tag{6.59}$$

Numerical integration of (6.58) yields a quadrature version of the nodally integrated discrete strain energy which we denote $a\langle\cdot, \cdot\rangle$. For DNI, we have

$$a_{\mathrm{D}}\left\langle\delta\mathbf{u}^h, \mathbf{u}^h\right\rangle = \sum_{L=1}^{NP} \boldsymbol{\varepsilon}_L\left(\delta\mathbf{u}^h\right) : \mathbf{C} : \boldsymbol{\varepsilon}_L\left(\mathbf{u}^h\right) W_L, \tag{6.60}$$

where the subscript "D" denotes that it is evaluated using DNI, and we have further introduced $\boldsymbol{\varepsilon}_L(\mathbf{u}^h) \equiv \boldsymbol{\varepsilon}(\mathbf{u}^h(\mathbf{x}_L))$, which is the nodal strain evaluated at $\mathbf{x}_L$ using direct derivatives. To make all of this clear, the explicit expression of (6.60) is

$$a_{\mathrm{D}}\left\langle\delta\mathbf{u}^h, \mathbf{u}^h\right\rangle = \sum_{L=1}^{NP} \delta u^h_{(i,j)}(\mathbf{x}_L) C_{ijkl} u^h_{(k,l)}(\mathbf{x}_L) W_L, \tag{6.61}$$

although this form is not as useful as (6.60) for the subsequent developments.

For SCNI, one can express the quadrature version of (6.58) as

$$a_{\mathrm{S}}\left\langle\delta\mathbf{u}^h, \mathbf{u}^h\right\rangle = \sum_{L=1}^{NP} \tilde{\boldsymbol{\varepsilon}}_L\left(\delta\mathbf{u}^h\right) : \mathbf{C} : \tilde{\boldsymbol{\varepsilon}}_L\left(\mathbf{u}^h\right) W_L, \tag{6.62}$$

where $a_{\mathrm{S}}\langle\cdot, \cdot\rangle$ is the variation on strain energy with *strain smoothing* (hence the subscript "S") at nodal locations $\mathbf{x}_L$ using smoothed nodal strains $\tilde{\boldsymbol{\varepsilon}}_L$ from (6.20), with $\tilde{\boldsymbol{\varepsilon}}_L\left(\mathbf{u}^h\right) \equiv \tilde{\boldsymbol{\varepsilon}}\left(\mathbf{u}^h(\mathbf{x}_L)\right)$. SNNI, the nonconforming version of strain smoothing, can be expressed similarly. In this section, we will not make the distinction between which type of cells are employed.

As discussed in Section 6.9, both (6.60) and (6.62) can yield severe oscillations in the solution. Stabilization is most conveniently expressed using these types of bilinear expressions, since it necessarily must add coercivity for a stable solution. The practical implementation of these types of forms in terms of the strain-displacement (gradient) and stiffness matrices is detailed in Chapter 10.

6.10.2 Modified Strain Smoothing

Stabilization of SCNI and SNNI has been proposed in [36, 37] where these modes are penalized throughout the smoothing domain, resulting in modified SCNI (MSCNI) and modified SNNI (MSNNI) in the strain energy:

$$a_{\mathrm{SM}}\left\langle\delta\mathbf{u}^h, \mathbf{u}^h\right\rangle = a_{\mathrm{S}}\left\langle\delta\mathbf{u}^h, \mathbf{u}^h\right\rangle + a_{\mathrm{M}}\left\langle\delta\mathbf{u}^h, \mathbf{u}^h\right\rangle, \tag{6.63}$$

where $a_S\langle\cdot,\cdot\rangle$ is the nodally integrated bilinear form using strain smoothing (6.62), and $a_M\langle\cdot,\cdot\rangle$ is the stabilization term (a "modified" type, denoted with "M"):

$$a_M\langle\delta\mathbf{u}^h,\mathbf{u}^h\rangle = \sum_{L=1}^{NP}\sum_{K=1}^{NS} c_S\left[\left(\tilde{\boldsymbol{\varepsilon}}_L(\delta\mathbf{u}^h)-\boldsymbol{\varepsilon}_L^K(\delta\mathbf{u}^h)\right):\mathbf{C}:\left(\tilde{\boldsymbol{\varepsilon}}_L(\mathbf{u}^h)-\boldsymbol{\varepsilon}_L^K(\mathbf{u}^h)\right)W_L^K\right], \tag{6.64}$$

where $0.0 \leq c_S \leq 1.0$ is a stabilization parameter, $\boldsymbol{\varepsilon}_L^K(\mathbf{u}^h) \equiv \boldsymbol{\varepsilon}(\mathbf{u}^h(\hat{\mathbf{x}}_L^K))$ is the strain evaluated at the centroid of subcells $\hat{\mathbf{x}}_L^K$, W_L^K is the weight of the subcell calculated from the weight W_L, and NS is the number of subdomain quadrature points. Stabilization schemes for MSCNI and MSNNI are shown in Figure 6.21 including the cell divisions. Note that the nodal strains $\boldsymbol{\varepsilon}_L$ and the subcell strains $\boldsymbol{\varepsilon}_L^K$ can be calculated using either direct differentiation, or smoothing over the individual nodal domains or subdomains, as in SCNI. Both approaches are implemented into RKPM2D in the function `Pre_GenerateShapeFunction`, with `getSmoothedDerivative` for the smoothed strains, see Chapter 10.

It should be clear from (6.64) that for $c_S > 0$ additional coercivity is added to the solution ($a_M\langle\cdot,\cdot\rangle > 0$ for $\boldsymbol{\varepsilon} \neq \mathbf{0}$, since $\mathbf{C}$ is positive definite). The key to the success of this method in SCNI is that the additional stabilization maintains the linear exactness of SCNI. If strains are constant, as in the linear patch test, the method reduces to SCNI since $a_M\langle\delta\mathbf{u}^h,\mathbf{u}^h\rangle = 0$ in this case. The modes stabilized by using MSCNI and VC-MSNNI (MSNNI with VCI) are shown in Figure 6.22; good agreement with fully-integrated FEM (see Figure 6.19) is obtained with stabilization and satisfaction of variational consistency (SCNI is inherently consistent).

An interesting property of this method is that when $c_S = 1$, (6.63) can facilitate high-order integration in SCNI (termed high-order SCNI in [39]). That is, as the number of subdomain quadrature

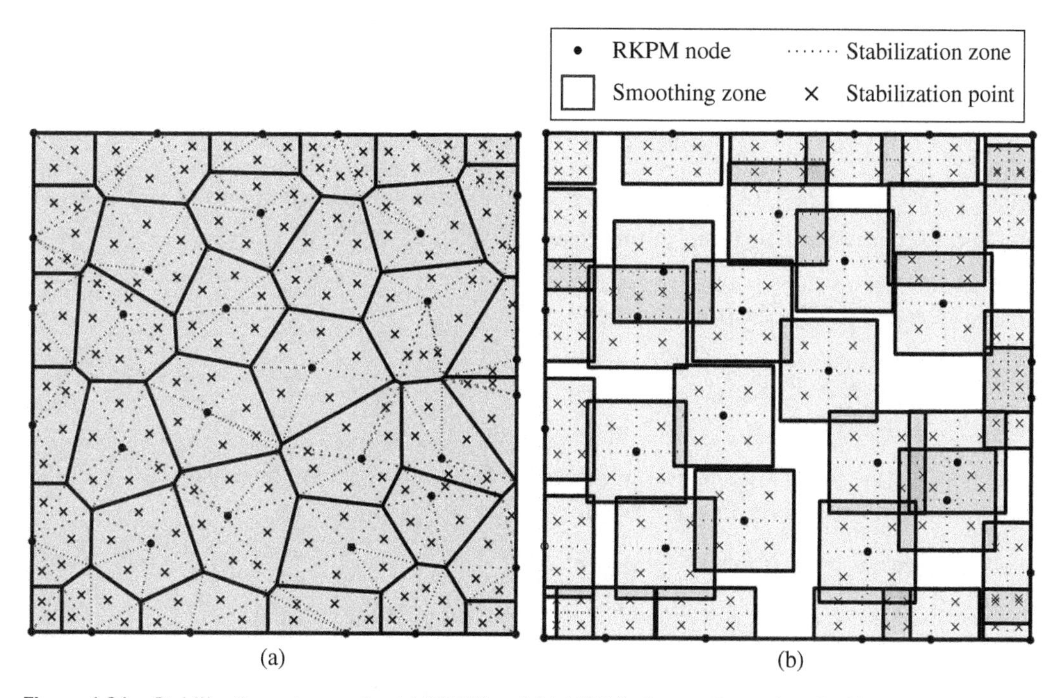

Figure 6.21 Stabilization schemes for (a) MSCNI and (b) MSNNI. *Source:* Reproduced with permission from Hillman and Chen [38], Figure 4 p. 8 / John Wiley & Sons.

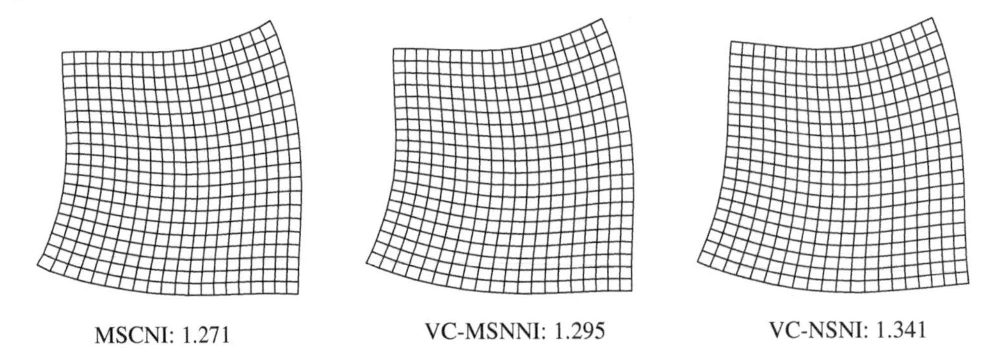

Figure 6.22 Lowest energy eigenvector modes with eigenvalues of stabilized and corrected integration methods.

points *NS* increases, the accuracy of evaluating integrals is increased. And, since it is formulated under the strain-smoothing framework, it always retains linear exactness. In [39], it was used to accurately integrate enrichment functions in RKPM, and in [40], it was used to obtain additional coercivity in nodal integration of convection-dominated problems.

6.10.3 Naturally Stabilized Nodal Integration

A stabilization has been proposed in [38] that employs implicit gradients (see Section 3.5.3) in a Taylor series expansion to yield stabilization without explicit computation of higher-order derivatives. It also avoids adding additional sampling points in the stress point method and sub-domain type stabilization methods such as modified SCNI and SNNI, which can be costly.

In this approach, the implicit gradient expansion of the strain around a node $\mathbf{x}_L$ in two dimensions is defined as

$$\boldsymbol{\varepsilon}\left(\mathbf{u}^h(\mathbf{x})\right) \approx \boldsymbol{\varepsilon}_L\left(\mathbf{u}^h\right) + (x - x_L)\boldsymbol{\varepsilon}_L\left(\mathbf{u}_x^h\right) + (y - y_L)\boldsymbol{\varepsilon}_L\left(\mathbf{u}_y^h\right), \tag{6.65}$$

where $\boldsymbol{\varepsilon}_L\left(\mathbf{u}_x^h\right) \equiv \boldsymbol{\varepsilon}\left(\mathbf{u}_x^h(\mathbf{x}_L)\right)$, $\boldsymbol{\varepsilon}_L\left(\mathbf{u}_y^h\right) \equiv \boldsymbol{\varepsilon}\left(\mathbf{u}_y^h(\mathbf{x}_L)\right)$, and first-order implicit gradients $\left\{\Psi_I^{\boldsymbol{\alpha}}(\mathbf{x})\right\}_{|\alpha|=1}$ are used to approximate the terms in the expansion:

$$\begin{aligned}
\mathbf{u}_x^h(\mathbf{x}_L) &\equiv \sum_{I=1}^{NP} \Psi_I^{(10)}(\mathbf{x}_L)\mathbf{u}_I, \\
\mathbf{u}_y^h(\mathbf{x}_L) &\equiv \sum_{I=1}^{NP} \Psi_I^{(01)}(\mathbf{x}_L)\mathbf{u}_I,
\end{aligned} \tag{6.66}$$

where $\Psi_I^{(\boldsymbol{\alpha})}$ is the implicit gradient shape function (see Section 3.5.3). Employing (6.65) for the strains near each node, one obtains a stabilized strain energy $a_{\mathrm{DN}}\langle\cdot,\cdot\rangle$:

$$a_{\mathrm{DN}}\left\langle\delta\mathbf{u}^h, \mathbf{u}^h\right\rangle = a_{\mathrm{D}}\left\langle\delta\mathbf{u}^h, \mathbf{u}^h\right\rangle + a_{\mathrm{N}}\left\langle\delta\mathbf{u}^h, \mathbf{u}^h\right\rangle, \tag{6.67}$$

where $a_{\mathrm{D}}\langle\cdot,\cdot\rangle$ is the nodally integrated bilinear form using DNI (6.60), and

$$a_{\mathrm{N}}\left\langle\delta\mathbf{u}^h, \mathbf{u}^h\right\rangle = \sum_{L=1}^{NP}\left\{\boldsymbol{\varepsilon}_L\left(\delta\mathbf{u}_x^h\right) : \mathbf{C} : \boldsymbol{\varepsilon}_L\left(\mathbf{u}_x^h\right)M_{Lx} + \boldsymbol{\varepsilon}_L\left(\delta\mathbf{u}_y^h\right) : \mathbf{C} : \boldsymbol{\varepsilon}_L\left(\mathbf{u}_y^h\right)M_{Ly}\right\}, \tag{6.68}$$

is the additional stabilization resulting from (6.65) (natural-type, denoted with "N"), where M_{Lx} and M_{Ly}, are the second moments of inertia of the nodal domains about node L:

$$M_{Lx_i} = \int_{\Omega_L} (x_i - x_{iL})^2 d\Omega. \tag{6.69}$$

Clearly, for coercive forms of $a(\cdot, \cdot)$, (6.68) adds additional stabilization for the strain energy. This stabilized nodal integration has been termed naturally stabilized nodal integration (NSNI) since the constants in the stabilization come naturally from the discretization. In [38], it was demonstrated through complexity analysis that speed-up factors of up to 20 times could be achieved over methods such as MSCNI and MSNNI that involve additional sampling points. Since the method employs DNI for strains, the variationally consistent correction should be applied in conjunction with this stabilization. Note that since $a_N\langle\cdot, \cdot\rangle$ vanishes for linear solutions, the introduction of a linear VCI correction (see Section 6.7.2) is trivial; consult [38] for further discussions. Alternatively, conforming cells can be employed for linear variational consistency, as described in the following section.

Performing an eigenvalue analysis of the associated VC-NSNI (NSNI with VCI) stiffness matrix, the results in Figure 6.22 show good agreement with fully-integrated FEM (see Figure 6.19). Note that here (as in Figure 6.21) we did not bother to show the non-VC versions of each method, as they do not agree with FEM; a demonstration of the role of VCI in stabilization is given in reference [41].

The convergence of several nodal integration methods with and without stabilization and variational consistency is shown in Figure 6.23. Here, it can be confirmed that both are necessary to obtain optimal rates (the only methods in this study that satisfy both are VC-NSNI and VC-MSNNI, but MSCNI also performs equally well).

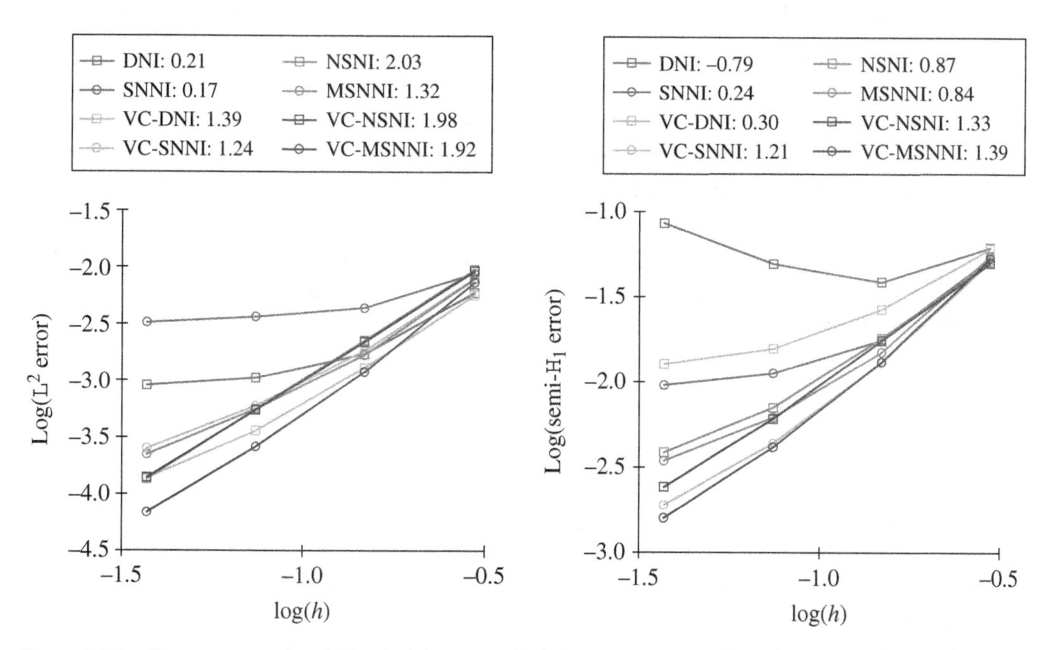

Figure 6.23 Convergence of stabilized and corrected nodal integration methods (rates indicated in legend). *Source:* Reproduced with permission from Hillman and Chen [38], figure 10 p. 16 / John Wiley & Sons.

6.10.4 Naturally Stabilized Conforming Nodal Integration

As an alternative to correcting NSNI with VCI for variational consistency, one can employ the smoothing cells in SCNI as a basis for nodal integration, denoted as NSCNI (naturally stabilized conforming nodal integration), as follows:

$$a_{\mathrm{SN}}\left\langle \delta \mathbf{u}^h, \mathbf{u}^h \right\rangle = a_{\mathrm{S}}\left\langle \delta \mathbf{u}^h, \mathbf{u}^h \right\rangle + a_{\mathrm{N}}\left\langle \delta \mathbf{u}^h, \mathbf{u}^h \right\rangle, \tag{6.70}$$

where $a_{\mathrm{S}}\langle\cdot,\cdot\rangle$ is the quadrature version of the nodally integrated strain energy using conforming smoothing (6.62), and $a_N\langle\cdot,\cdot\rangle$ is the natural stabilization term in (6.68). This form maintains the linear exactness of SCNI since the derivatives of a constant strain vanish in $a_N\langle\cdot,\cdot\rangle$.

It can be noted that (6.68) still contains direct derivatives of the shape functions. Therefore, a *smoothed* and *implicit* stabilization term $\tilde{a}_{\mathrm{N}}\langle\cdot,\cdot\rangle$ can be introduced in place of $a_{\mathrm{N}}\langle\cdot,\cdot\rangle$:

$$\tilde{a}_{\mathrm{SN}}\left\langle \delta \mathbf{u}^h, \mathbf{u}^h \right\rangle = a_{\mathrm{S}}\left\langle \delta \mathbf{u}^h, \mathbf{u}^h \right\rangle + \tilde{a}_{\mathrm{N}}\left\langle \delta \mathbf{u}^h, \mathbf{u}^h \right\rangle, \tag{6.71}$$

where

$$\tilde{a}_{\mathrm{N}}\left\langle \delta \mathbf{u}^h, \mathbf{u}^h \right\rangle = \sum_{L=1}^{NP} \left\{ \tilde{\boldsymbol{\varepsilon}}_L\left(\delta \mathbf{u}_x^h\right) : \mathbf{C} : \tilde{\boldsymbol{\varepsilon}}_L\left(\mathbf{u}_x^h\right) M_{Lx} + \tilde{\boldsymbol{\varepsilon}}_L\left(\delta \mathbf{u}_y^h\right) : \mathbf{C} : \tilde{\boldsymbol{\varepsilon}}_L\left(\mathbf{u}_y^h\right) M_{Ly} \right\}, \tag{6.72}$$

with

$$\begin{aligned} \tilde{\varepsilon}_{ijL}\left(\mathbf{u}_x^h\right) &= \frac{1}{2W_L} \int_{\Gamma_L} u_{ix}^h(\mathbf{x}) n_j(\mathbf{x}) + u_{jx}^h(\mathbf{x}) n_i(\mathbf{x}) d\Gamma, \\ \tilde{\varepsilon}_{ijL}\left(\mathbf{u}_y^h\right) &= \frac{1}{2W_L} \int_{\Gamma_L} u_{iy}^h(\mathbf{x}) n_j(\mathbf{x}) + u_{jy}^h(\mathbf{x}) n_i(\mathbf{x}) d\Gamma, \end{aligned} \tag{6.73}$$

where $u_{ix}^h \equiv \left(\mathbf{u}_x^h\right)_i$ and $u_{iy}^h \equiv \left(\mathbf{u}_y^h\right)_i$ are the ith components of the implicit gradients in (6.66). The employment of both smoothing in (6.73) and implicit gradient approximations in (6.66) reduces the order of differentiation in the formulation to zero—no expensive meshfree derivatives are needed whatsoever. It also unifies computation of nodal strains and stabilization terms, as they are all evaluated on the contour integration points of cells. Efficiency is also gained since the implicit gradient can be obtained easily when computing the shape functions at these locations by replacing $\mathbf{H}(\mathbf{0})$ with $\mathbf{H}^{(\alpha)}$ in the same loop (see Section 3.5.3).

For both formulations (6.70) and (6.71), no correction for optimal convergence is required when the smoothing cells are conforming. The direct nodal stabilization approach (6.70) is implemented into RKPM2D for both DNI and SCNI, see Chapter 10.

Notes

1 This is somewhat intuitive as the misalignment of cells and supports is more severe in nonuniform discretizations, and the rational nature of the meshfree shape functions also increases; little or no correction is needed to the kernel to attain reproducing conditions in uniform discretizations, see Chapter 3.

2 To understand the concept of a linear patch test, it is useful to consider the pure Dirichlet case of the problem ($\Gamma = \Gamma_u$):

$$\sigma_{ij,j} = 0 \quad \text{on} \quad \Omega,$$
$$u_i = u_i^{\mathrm{P}} = a_i + b_i x + c_i y \quad \text{on} \quad \Gamma_u.$$

In the above, it can be easily seen that the exact solution is $u_i = u_i^{\mathrm{P}}$, since this displacement satisfies both $\sigma_{ij,j} = \left(C_{ijkl}u_{(k,l)}^{\mathrm{P}}\right)_{,j} = C_{ijkl}u_{k,lj}^{\mathrm{P}} = C_{ijkl}(a_k + b_k x + c_k y)_{,lj} = 0$ and $u_i = u_i^{\mathrm{P}}$ on Γ_u.

3 A weak form fully consistent with the strong form is strictly necessary to pass a patch test. This includes the Lagrange multiplier method, Nitsche's method, and the consistent weak forms introduced in Section 5.8. It generally does not include strong-type methods, which impose essential boundary conditions point-wise (see Section 5.3.2). For further discussion, see [10] and [11].

References

1 Strang, G. and Fix, G.J. (1973). *An Analysis of the Finite Element Method*. Englewood Cliffs, NJ: Prentice-Hall.

2 Wu, J. and Wang, D. (2021). An accuracy analysis of Galerkin meshfree methods accounting for numerical integration. *Comput. Methods Appl. Mech. Eng.* 375: 113631.

3 Dolbow, J. and Belytschko, T. (1999). Numerical integration of the Galerkin weak form in meshfree methods. *Comput. Mech.* 23 (3): 219–230.

4 Griebel, M. and Schweitzer, M.A. (2002). A particle-partition of unity method—Part II: efficient cover construction and reliable integration. *SIAM J. Sci. Comput.* 23 (5): 1655–1682.

5 Atluri, S.N. and Shen, S. (2002). *The Meshless Local Petrov–Galerkin (MLPG) Method*. Los Angeles, CA: Tech Science Press.

6 De, S. and Bathe, K.J. (2000). The method of finite spheres. *Comput. Mech.* 25 (4): 329–345.

7 Chen, J.-S., Wu, C.-T., and Yoon, S. (2001). A stabilized conforming nodal integration for Galerkin mesh-free methods. *Int. J. Numer. Methods Eng.* 50 (2): 435–466.

8 Krongauz, Y. and Belytschko, T. (1997). Consistent pseudo-derivatives in meshless methods. *Comput. Methods Appl. Mech. Eng.* 146 (3–4): 371–386.

9 Bonet, J. and Kulasegaram, S. (2000). Correction and stabilization of smooth particle hydrodynamics methods with applications in metal forming simulations. *Int. J. Numer. Methods Eng.* 47 (6): 1189–1214.

10 Chen, J.-S., Hillman, M., and Rüter, M. (2013). An arbitrary order variationally consistent integration for Galerkin meshfree methods. *Int. J. Numer. Methods Eng.* 95 (5): 387–418.

11 Hillman, M. and Lin, K.-C. (2021). Consistent weak forms for meshfree methods: full realization of *h*-refinement, *p*-refinement, and *a*-refinement in strong-type essential boundary condition enforcement. *Comput. Methods Appl. Mech. Eng.* 373: 113448.

12 Guan, P.C., Chen, J.-S., Wu, Y. et al. (2009). Semi-Lagrangian reproducing kernel formulation and application to modeling earth moving operations. *Mech. Mater.* 41 (6): 670–683.

13 Guan, P.C., Chi, S.W., Chen, J.-S. et al. (2011). Semi-Lagrangian reproducing kernel particle method for fragment-impact problems. *Int. J. Impact Eng.* 38 (12): 1033–1047.

14 Guan, P.-C. (2009). Adaptive coupling of FEM and RKPM formulations for contact and impact problems. Ph.D. Dissertation, University of California, Los Angeles.

15 Liu, G.-R., Dai, K.Y., and Nguyen, T.T. (2007). A smoothed finite element method for mechanics problems. *Comput. Mech.* 39 (6): 859–877.

16 Liu, G.-R. (2008). A generalized gradient smoothing technique and the smoothed bilinear form for Galerkin formulation of a wide class of computational methods. *Int. J. Comput. Methods.* 5 (2): 199–236.

17 Liu, G.-R. and Nguyen-Thoi, T. (2010). *Smoothed Finite Element Methods*. Boca Raton, FL: CRC Press.

18 Sze, K.Y., Chen, J.-S., Sheng, N., and Liu, X.H. (2004). Stabilized conforming nodal integration: exactness and variational justification. *Finite Elem. Anal. Des.* 41 (2): 147–171.

19 Hillman, M.C. (2013). An arbitrary order variationally consistent integration method for Galerkin meshfree methods. Ph.D. Dissertation, University of California, Los Angeles.

20 Chen, J.-S., Hillman, M., and Chi, S.W. (2017. 04017001). Meshfree methods: progress made after 20 years. *J. Eng. Mech.* 143 (4): 38pp.

21 Duan, Q., Li, X., Zhang, H. et al. (2012). Quadratically consistent one-point (QC1) quadrature for meshfree Galerkin methods. *Comput. Methods Appl. Mech. Eng.* 245–246: 256–272.

22 Wang, D. and Wu, J. (2016). An efficient nesting sub-domain gradient smoothing integration algorithm with quadratic exactness for Galerkin meshfree methods. *Comput. Methods Appl. Mech. Eng.* 298: 485–519.

23 Wang, D. and Wu, J. (2019). An inherently consistent reproducing kernel gradient smoothing framework toward efficient Galerkin meshfree formulation with explicit quadrature. *Comput. Methods Appl. Mech. Eng.* 349: 628–672.

24 Baek, J., Chen, J.-S., Zhou, G., Arnett, K.P., Hillman, M.C., Hegemier, G., and Hardesty, S. (2021). A semi-Lagrangian reproducing kernel particle method with particle-based shock algorithm for explosive welding simulation. *Comput. Mech.* 67 (6): 1601–1627.

25 Beissel, S.R. and Belytschko, T. (1996). Nodal integration of the element-free Galerkin method. *Comput. Methods Appl. Mech. Eng.* 139: 49–74.

26 Belytschko, T., Guo, Y., Liu, W.K., and Xiao, S.P. (2000). A unified stability analysis of meshless particle methods. *Int. J. Numer. Methods Eng.* 48 (9): 1359–1400.

27 Duan, Q. and Belytschko, T. (2009). Gradient and dilatational stabilizations for stress-point integration in the element-free Galerkin method. *Int. J. Numer. Methods Eng.* 77 (6): 776–798.

28 Randles, P.W. and Libersky, L.D. (2000). Normalized SPH with stress points. *Int. J. Numer. Meth. Engng.* 48: 1445–1462.

29 Dyka, C.T., Randles, P.W., and Ingel, R.P. (1997). Stress points for tension instability in SPH. *Int. J. Numer. Methods Eng.* 40 (13): 2325–2341.

30 Dyka, C.T. and Ingel, R.P. (1995). An approach for tension instability in smoothed particle hydrodynamics (SPH). *Comput. Struct.* 57 (4): 573–580.

31 Fries, T.-P. and Belytschko, T. (2008). Convergence and stabilization of stress-point integration in mesh-free and particle methods. *Int. J. Numer. Methods Eng.* 74 (7): 1067–1087.

32 Liu, W.K., Ong, J.S.-J., and Uras, R.A. (1985). Finite element stabilization matrices—a unification approach. *Comput. Methods Appl. Mech. Eng.* 53 (1): 13–46.

33 Nagashima, T. (1999). Node-by-node meshless approach and its applications to structural analyses. *Int. J. Numer. Methods Eng.* 46 (3): 341–385.

34 Liu, G.-R., Zhang, G.Y., Wang, Y.Y. et al. (2007). A nodal integration technique for meshfree radial point interpolation method (NI-RPIM). *Int. J. Solids Struct.* 44 (11–12): 3840–3860.

35 Chen, J.-S., Yoon, S., and Wu, C.-T. (2002). Non-linear version of stabilized conforming nodal integration for Galerkin mesh-free methods. *Int. J. Numer. Methods Eng.* 53 (12): 2587–2615.

36 Puso, M.A., Chen, J.-S., Zywicz, E., and Elmer, W. (2008). Meshfree and finite element nodal integration methods. *Int. J. Numer. Methods Eng.* 74 (3): 416–446.

37 Chen, J.-S., Hu, W., Puso, M.A., Wu, Y., and Zhang, X. (2007). Strain smoothing for stabilization and regularization of galerkin meshfree methods. *Lect. Notes Comput. Sci. Eng.* 57: 57–75.

38 Hillman, M. and Chen, J.-S. (2016). An accelerated, convergent, and stable nodal integration in Galerkin meshfree methods for linear and nonlinear mechanics. *Int. J. Numer. Methods Eng.* 107: 603–630.

39 Chen, J.-S., Hu, W., and Puso, M.A. (2007). Orbital HP-clouds for solving Schrödinger equation in quantum mechanics. *Comput. Methods Appl. Mech. Eng.* 196 (37): 3693–3705.

40 Hillman, M. and Chen, J.-S. (2016). Nodally integrated implicit gradient reproducing kernel particle method for convection dominated problems. *Comput. Methods Appl. Mech. Eng.* 299: 381–400.

41 Hillman, M., Chen, J.-S., and Chi, S.-W. (2014). Stabilized and variationally consistent nodal integration for meshfree modeling of impact problems. *Comp. Part. Mech.* 1: 245–256.

42 Hillman, M., Chen, J.S., and Roth, M.J. (2016). Advanced computational methods to understand & mitigate extreme events. *IACM Expressions* 36: 12–16.

7

Nonlinear Meshfree Methods

Meshfree methods have proven particularly effective for solving large deformation problems. Since approximations such as the reproducing kernel (RK) and moving least squares (MLS) are constructed entirely based on a set of discrete points, they have been successfully applied to model extreme deformations without the burden of dealing with mesh distortion or entanglement present in the conventional finite element method. The Lagrangian reproducing kernel particle method (RKPM) [1], where the kernel functions maintain the same coverage of neighboring points at all stages of the simulation, as shown in Figure 7.1a–d, has been introduced to model hyperelastic (path-independent) materials [3, 4] and plastic (path-dependent) materials [5–7], among other applications. However, for problems involving severe material damage and separation, the deformation gradient is no longer positive definite at all material points, and Lagrangian RKPM is neither applicable to the total nor updated Lagrangian formulations. For such cases, semi-Lagrangian RKPM [8, 9] has been proposed, where the nodal points are attached to the material (and thus are Lagrangian), while the supports of kernel functions do not necessarily deform with the material (and thus the combined formulation is "semi-Lagrangian"), as shown in Figure 7.1a, b, e, and f. In this approach, to be discussed in Section 7.3, the neighbor points are redefined during the deformation process to account for large flows of material motion and the formation of free surfaces, and the approximation is constructed in the current configuration.

While no "explicit" mesh is necessary for constructing the meshfree approximation functions, the temporal stability for time integration in transient problems is still influenced by the temporal discretization, the node-based spatial discretization, and the associated meshfree shape functions. The stability conditions for meshfree methods with explicit time integration have been studied [9] and will be discussed in Section 7.4. Neighbor search algorithms for the semi-Lagrangian discretization are discussed in Section 7.5.

Contact problems are nonlinear in nature, with kinematic constraints and frictional conditions considered in the interaction between bodies. Proper smoothness in the representation of associated surfaces strongly influences the computational accuracy and effectiveness, particularly in the iteration processes of sliding contact. Meshfree approximation functions such as MLS and RK are particularly useful for surface representation as they offer the flexibility of specifying arbitrary-order smoothness in space. Section 7.6 presents how this unique property can be applied to formulate a continuum-based contact formulation with an exact tangent containing up to third order derivatives. A natural kernel contact algorithm [10] has been developed under the semi-Lagrangian RKPM framework for multibody problems with evolving contact surfaces that utilizes the interaction of kernel functions to satisfy the impenetration conditions automatically. This approach has been applied to model fragment impact and penetration processes [10, 11], among other challenging applications, and is presented in Section 7.7.

Meshfree and Particle Methods: Fundamentals and Applications, First Edition.
Ted Belytschko, J. S. Chen, and Michael Hillman.

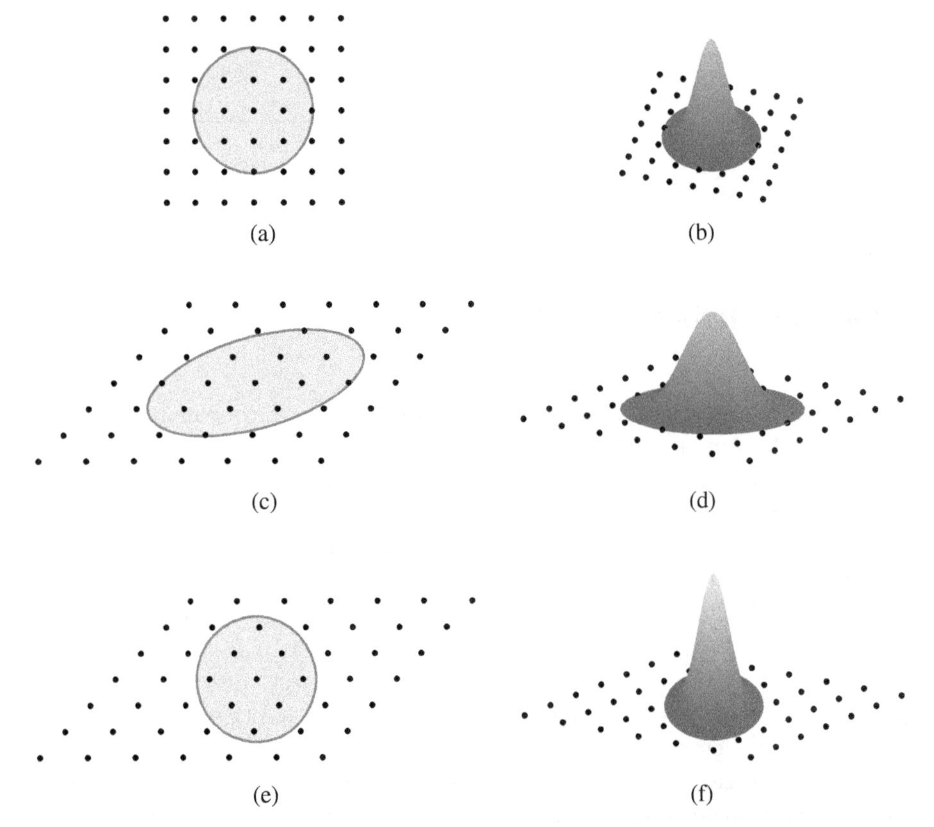

Figure 7.1 Lagrangian and semi-Lagrangian kernels and associated RK shape functions. The shape functions are built using a linear basis and a cubic B-spline kernel [2]: (a) Undeformed configuration, kernel support, (b) Undeformed configuration, RK shape function, (c) Deformed configuration, Lagrangian kernel support, (d) Deformed configuration, Lagrangian RK shape function, (e) Deformed configuration, semi-Lagrangian kernel support, (f) Deformed configuration, semi-Lagrangian RK shape function. *Source:* Pasetto et al. [2], figure 3, p. 8 / With permission of Elsevier.

7.1 Lagrangian Description of the Strong Form

Consider a body that initially occupies a region Ω^0 with boundary Γ^0 and undergoes a motion from its initial configuration to a deformed (current) configuration Ω with boundary Γ (see Section 2.3 for more background information on nonlinear problems). In the deformed configuration, the body is subjected to body force $\mathbf{b}$ in Ω, a surface traction $\bar{t}_i$ on the natural boundary Γ_t, and a prescribed displacement $\bar{u}_i$ on the essential boundary Γ_u, with $\Gamma_u \cup \Gamma_t = \Gamma$. We use a fixed Cartesian coordinate system and denote the particle positions in the initial configuration Ω^0 by $\mathbf{X}$, and those in the deformed configuration Ω at time t by a mapping function as follows:

$$\mathbf{x} = \boldsymbol{\varphi}(\mathbf{X}, t). \tag{7.1}$$

The mapping is one-to-one for each material particle under deformation, and the determinate of the Jacobian matrix associated with this mapping, $J = \det(\partial \mathbf{x}/\partial \mathbf{X})$, is neither zero nor infinite. The following equation of motion governs the body in the current configuration:

$$\rho \ddot{u}_i = \tau_{ij,j} + b_i, \quad \text{in } \Omega, \tag{7.2}$$

with boundary conditions

$$\tau_{ij} n_j = \bar{t}_i, \quad \text{on } \Gamma_t, \tag{7.3}$$

$$u_i = \bar{u}_i, \quad \text{on } \Gamma_u, \tag{7.4}$$

and initial conditions

$$u_i(\mathbf{X}, 0) = u_i^0(\mathbf{X}), \tag{7.5}$$

$$\dot{u}_i(\mathbf{X}, 0) = v_i^0(\mathbf{X}), \tag{7.6}$$

where ρ is the density, $u_i(\mathbf{X}, t) = \varphi_i(\mathbf{X}, t) - X_i = x_i(\mathbf{X}, t) - X_i$ is the material displacement, τ_{ij} is the Cauchy stress obtained from the constitutive law, n_i is the surface normal on the deformed boundary, u_i^0 is the initial displacement, and v_i^0 is the initial velocity. In (7.2) and (7.6), $\dot{()} \equiv D/Dt$ and $\ddot{()} \equiv D^2/Dt^2$ denote material time derivatives, which will be used in the developments in this chapter.

For a hyperelastic material, the second Piola–Kirchhoff stress S_{ij} is calculated from the strain energy density function W by

$$S_{ij} = \frac{\partial W}{\partial E_{ij}}, \tag{7.7}$$

where E_{ij} is the Green–Lagrange strain defined as

$$E_{ij} = \frac{1}{2}\left(F_{ki} F_{kj} - \delta_{ij}\right), \tag{7.8}$$

and F_{ij} is the deformation gradient defined as

$$F_{ij} = \frac{\partial x_i}{\partial X_j}. \tag{7.9}$$

It can be seen from the above that $J = \det(\mathbf{F})$.

The second Piola–Kirchhoff stress S_{ij} is related to the Cauchy stress τ_{ij} and the nominal stress P_{ij} (the transpose of the first Piola–Kirchhoff stress) by[1]

$$\tau_{ij} = \frac{1}{J} F_{im} S_{mn} F_{jn} = \frac{1}{J} F_{in} P_{nj}. \tag{7.10}$$

For path-dependent elastoplasticity, the Cauchy stress rate is obtained by the following equation when the Jaumann rate is employed for the constitutive relation:

$$\begin{aligned} \dot{\tau}_{ij} &= C^{c}_{ijkl} v_{(k,l)} + R_{ijkl} v_{[k,l]}, \\ v_{(k,l)} &= \frac{1}{2}\left(v_{k,l} + v_{l,k}\right), \\ v_{[k,l]} &= \frac{1}{2}\left(v_{k,l} - v_{l,k}\right). \end{aligned} \tag{7.11}$$

where $v_i \equiv \dot{u}_i$ is the velocity, $v_{(k,l)}$ and $v_{[k,l]}$ are the symmetric and antisymmetric parts of the velocity gradient $\partial v_i / \partial x_j$, respectively (this indicial bracket notation is adopted for all such symmetric and antisymmetric parts herein), and C^{c}_{ijkl} is the co-rotational material response tensor, which is material dependent, and R_{ijkl} is given by

$$R_{ijkl} = \frac{1}{2}\left(\tau_{il}\delta_{jk} + \tau_{jl}\delta_{ik} - \tau_{ik}\delta_{jl} - \tau_{jk}\delta_{il}\right). \tag{7.12}$$

We will assume that the test and trial function spaces for displacements are kinematically admissible for demonstration purposes. The variational equation (weak form) of the Lagrangian description of the equation of motion with reference to the deformed configuration is formulated as:

Find $u_i \in \mathrm{U}$, such that for all $\delta u_i \in \mathrm{U}_0$, the following equation is satisfied

$$\int_{\Omega} \delta u_i \rho \ddot{u}_i \, d\Omega + \int_{\Omega} \delta u_{(i,j)} \tau_{ij} d\Omega - \int_{\Omega} \delta u_i b_i d\Omega - \int_{\Gamma_t} \delta u_i \bar{t}_i \, d\Gamma d\Gamma = 0, \tag{7.13}$$

where U and U_0 are the spaces of admissible functions (see Chapter 2), and

$$u_i(\mathbf{X}, 0) = u_i^0(\mathbf{X}), \tag{7.14}$$

$$\dot{u}_i(\mathbf{X}, 0) = v_i^0(\mathbf{X}). \tag{7.15}$$

For the general case where the test and trial functional spaces for displacements are not kinematically admissible, Nitsche terms can be added to the variational equation, as discussed in Chapter 2. The equation given in (7.13) is with reference to the deformed configuration. This is termed the *Updated Lagrangian* description, where the Cauchy stress is employed as the stress measure, which is conveniently defined in the deformed configuration. This variational equation is commonly used for modeling path-dependent materials where the Cauchy stress is adopted in the constitutive laws [12].

The undeformed (initial) configuration can be conveniently used as the reference configuration for path-independent materials such as those with nonlinear elastic and hyperelastic laws. The variational equation of motion in (7.13) can be transformed to the undeformed configuration, which is called the *total Lagrangian* description, and will be discussed as follows.

In the Lagrangian formulation, mass conservation is automatically satisfied, and therefore it is more convenient to transform the first term in (7.13) to the initial configuration

$$\int_{\Omega} \delta u_i \rho \ddot{u}_i d\Omega = \int_{\Omega^0} \delta u_i \rho^0 \ddot{u}_i d\Omega, \tag{7.16}$$

where ρ^0 is the initial density. For hyperelastic materials, since the second Piola–Kirchhoff stress can be directly calculated following Eq. (7.7), it is more effective to express the second term in Eq. (7.13) using the nominal stress in the initial configuration by

$$\int_{\Omega} \delta u_{(i,j)} \tau_{ij} d\Omega = \int_{\Omega} \delta u_{i,j} \tau_{ij} d\Omega = \int_{\Omega^0} \delta F_{ij} P_{ji} d\Omega, \tag{7.17}$$

where $u_{i,j} = \partial u_i / \partial x_j$, and P_{ij} is the nominal stress.

With the aid of (7.16) and (7.17), the variational equation of motion with reference to the undeformed configuration can be expressed as

$$\int_{\Omega^0} \delta u_i \rho^0 \ddot{u}_i d\Omega + \int_{\Omega^0} \delta F_{ij} P_{ji} d\Omega = \int_{\Omega^0} \delta u_i b_i^0 d\Omega + \int_{\Gamma_t^0} \delta u_i \bar{t}_i^0 d\Gamma, \tag{7.18}$$

where $\bar{t}_i^0$ is the surface traction acting on the undeformed traction boundary Γ_t^0, mapped from the surface traction $\bar{t}_i$ acting on the deformed traction boundary using the following relationship:

$$\bar{t}_i^0 = \bar{t}_i J \left\| \mathbf{n}^0 \cdot \mathbf{F}^{-1} \right\|, \quad J = \det(\mathbf{F}), \tag{7.19}$$

where $\mathbf{n}^0$ is the surface normal in the undeformed configuration. In (7.18), b_i^0 is the body force acting in the undeformed domain Ω^0, mapped from the body force b_i acting in the deformed domain Ω, using the following relationship:

$$b_i^0 = J b_i. \tag{7.20}$$

Table 7.1 Total and updated Lagrangian formulations.

	Total Lagrangian	**Updated Lagrangian**
Equation of motion	$\int_{\Omega^0} \delta u_i \rho^0 \ddot{u}_i d\Omega + \int_{\Omega^0} \delta F_{ij} P_{ji} d\Omega = \int_{\Omega^0} \delta u_i b_i^0 d\Omega + \int_{\Gamma_t^0} \delta u_i \bar{t}_i^0 d\Gamma$	$\int_{\Omega} \delta u_i \rho \ddot{u}_i d\Omega + \int_{\Omega} \delta u_{(i,j)} \tau_{ij} d\Omega = \int_{\Omega} \delta u_i b_i d\Omega - \int_{\Gamma_t} \delta u_i \bar{t}_i d\Gamma$
Displacements	$u_i = x_i - X_i$: material displacement, $\ddot{u}_i$: material acceleration,	
Strains/ deformation	$F_{ij} = \frac{\partial x_i}{\partial X_j}$	$u_{(i,j)} = (\partial u_i / \partial x_j + \partial u_j / \partial x_i)/2$
Stresses	P_{ij}: Nominal stress	τ_{ij}: Cauchy stress
Other variables	ρ^0: initial density b_i^0: body force defined in the undeformed domain Ω^0 $\bar{t}_i^0$: surface traction mapped onto the undeformed traction boundary Γ_t^0	ρ: density at the current state b_i: body force defined in the deformed domain Ω $\bar{t}_i$: surface traction defined on the deformed traction boundary Γ_t
Domains and boundaries	Ω^0: undeformed domain Γ_t^0: undeformed traction boundary	Ω: deformed domain Γ_t: deformed traction boundary

The traction–stress relationships in the deformed and undeformed configurations are as follows:

$$n_j \tau_{ji} = \bar{t}_i \quad \text{or} \quad \mathbf{n} \cdot \boldsymbol{\tau} = \boldsymbol{\tau}^{\mathrm{T}} \cdot \mathbf{n} = \boldsymbol{\tau} \cdot \mathbf{n} = \bar{\mathbf{t}} \quad \text{on}\, \Gamma_t, \tag{7.21}$$

$$n_j^0 P_{ji} = \bar{t}_i^0 \quad \text{or} \quad \mathbf{n}^0 \cdot \mathbf{P} = \mathbf{P}^{\mathrm{T}} \cdot \mathbf{n}^0 = \bar{\mathbf{t}}^0 \quad \text{on}\, \Gamma_t^0. \tag{7.22}$$

The total Lagrangian and updated Lagrangian formulations shown in Table 7.1 are commonly used in computational solid mechanics. They are *mathematically equivalent*, and choosing one over the other is a matter of convenience depending on the material law employed.

7.2 Lagrangian Reproducing Kernel Approximation and Discretization

In Lagrangian RKPM, the Lagrangian RK shape functions, denoted as $\Psi_I^0(\mathbf{X})$, are constructed with similar discrete RK procedures as those described in Section 3.2.2, using the material coordinates $\mathbf{X}$ in the following form [1]:

$$\Psi_I^0(\mathbf{X}) = C(\mathbf{X}; \mathbf{X} - \mathbf{X}_I) \phi_a(\mathbf{X} - \mathbf{X}_I). \tag{7.23}$$

Here, $\phi_a(\mathbf{X} - \mathbf{X}_I)$ is the Lagrangian kernel function with distance measure defined in the undeformed configuration. For example, the Lagrangian cubic B-spline kernel function is given as

$$\phi_a(\mathbf{X} - \mathbf{X}_I) \equiv \phi_a(z) = \begin{cases} \frac{2}{3} - 4z^2 + 4z^3 & \text{for } 0 \le z < \frac{1}{2} \\ \frac{4}{3} - 4z + 4z^2 - \frac{4}{3}z^3 & \text{for } \frac{1}{2} \le z \le 1 \\ 0 & \text{for } z > 1 \end{cases}, \quad z = \|\mathbf{X} - \mathbf{X}_I\|/a. \tag{7.24}$$

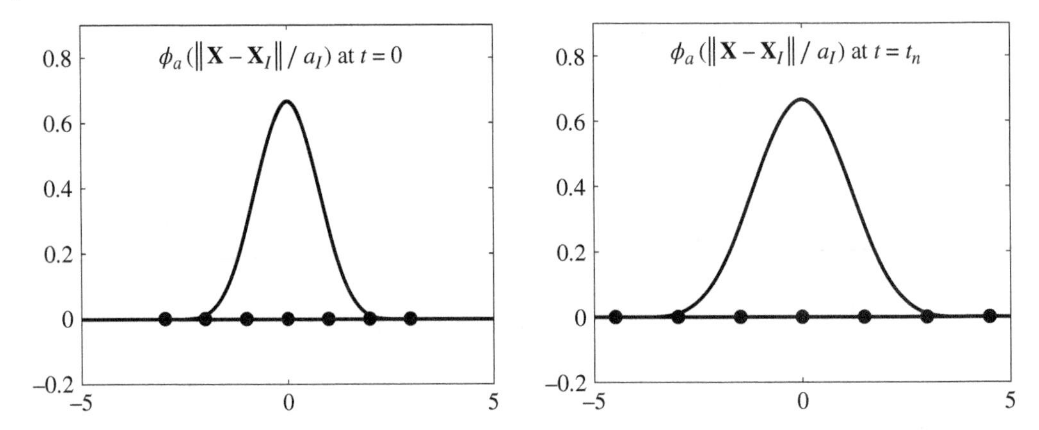

Figure 7.2 One-dimensional Lagrangian kernel function in the undeformed configuration ($t=0$) and its mapping to the current configuration ($t=t_n$).

As such, the kernel function $\phi_a(\mathbf{X}-\mathbf{X}_I)$ covers the same neighboring points under deformation, as shown in Figure 7.2 for one dimension and in Figure 7.1c and d for multidimensional geometries.

The correction function $C(\mathbf{X};\mathbf{X}-\mathbf{X}_I)$ in (7.23) is expressed as

$$C(\mathbf{X};\mathbf{X}-\mathbf{X}_I)=\mathbf{H}^{\mathrm{T}}(\mathbf{X}-\mathbf{X}_I)\mathbf{b}(\mathbf{X}), \tag{7.25}$$

$$\mathbf{H}^{\mathrm{T}}(\mathbf{X}-\mathbf{X}_I)=[1 \quad X_1-X_{1I} \quad X_2-X_{2I} \quad X_3-X_{3I} \quad \dots \quad (X_3-X_{3I})^n]. \tag{7.26}$$

Similar to the discussion in Section 3.2.2, the coefficient vector $\mathbf{b}(\mathbf{X})$ can be determined by imposing the following discrete reproducing conditions using the material coordinates:

$$\sum_{I=1}^{NP}\Psi_I^0(\mathbf{X})X_{1I}^iX_{2I}^jX_{3I}^k=X_1^iX_2^jX_3^k, \quad 0\le i+j+k\le n. \tag{7.27}$$

This reproducing condition then can be recast as

$$\sum_{I=1}^{NP}\Psi_I^0(\mathbf{X})(X_1-X_{1I})^i(X_2-X_{2I})^j(X_3-X_{3I})^k=\delta_{i0}\delta_{j0}\delta_{k0}, \quad 0\le i+j+k\le n, \tag{7.28}$$

or

$$\sum_{I=1}^{NP}\Psi_I^0(\mathbf{X})\mathbf{H}(\mathbf{X}-\mathbf{X}_I)=\mathbf{H}(\mathbf{0}). \tag{7.29}$$

Substituting (7.23) and (7.25) into (7.29) and solving for the coefficient vector $\mathbf{b}(\mathbf{X})$ yields the Lagrangian RK shape function:

$$\Psi_I^0(\mathbf{X})=\mathbf{H}^{\mathrm{T}}(\mathbf{0})\mathbf{M}^{-1}(\mathbf{X})\mathbf{H}(\mathbf{X}-\mathbf{X}_I)\phi_a(\mathbf{X}-\mathbf{X}_I), \tag{7.30}$$

where

$$\mathbf{M}(\mathbf{X})=\sum_{I=1}^{NP}\mathbf{H}(\mathbf{X}-\mathbf{X}_I)\mathbf{H}^{\mathrm{T}}(\mathbf{X}-\mathbf{X}_I)\phi_a(\mathbf{X}-\mathbf{X}_I). \tag{7.31}$$

As we have mentioned, support of the Lagrangian RK shape function $\Psi_I^0(\mathbf{X})$ is determined by the Lagrangian kernel function $\phi_a(\mathbf{X}-\mathbf{X}_I)$, which covers the same neighboring points under the deformation, as shown in Figure 7.2. This is a desirable property in large deformation analysis since, as long as the kernel support is sufficiently large to yield an invertible moment matrix **M** in the undeformed stage, the condition is satisfied at any deformation stage. The employment of the Lagrangian RK shape function thus avoids the so-called kernel instability in SPH-type methods in large deformation analysis; instability in SPH will be discussed further in Chapter 8.

The material displacement u_i in (7.18) is approximated by the Lagrangian RK shape functions (7.30) as

$$u_i^h(\mathbf{X}, t) = \sum_{I=1}^{NP} \Psi_I^0(\mathbf{X}) u_{iI}(t). \tag{7.32}$$

The corresponding approximations of velocity and acceleration are given by

$$\dot{u}_i^h(\mathbf{X}, t) = \sum_{I=1}^{NP} \Psi_I^0(\mathbf{X}) \dot{u}_{iI}(t), \tag{7.33}$$

$$\ddot{u}_i^h(\mathbf{X}, t) = \sum_{I=1}^{NP} \Psi_I^0(\mathbf{X}) \ddot{u}_{iI}(t). \tag{7.34}$$

Substituting (7.32)–(7.34) into (7.18) yields the following total Lagrangian discrete equation:

$$\mathcal{M}\ddot{\mathbf{d}} = \mathbf{f}^{\mathrm{ext}} - \mathbf{f}^{\mathrm{int}}, \tag{7.35}$$

where

$$\mathcal{M}_{IJ} = \int_{\Omega^0} \rho^0 \Psi_I^0(\mathbf{X}) \Psi_J^0(\mathbf{X}) \mathbf{I} d\Omega, \tag{7.36}$$

$$\mathbf{f}_I^{\mathrm{ext}} = \int_{\Omega^0} \Psi_I^0(\mathbf{X}) \mathbf{b}^0 d\Omega + \int_{\Gamma_t^0} \Psi_I^0(\mathbf{X}) \bar{\mathbf{t}}^0 d\Gamma, \tag{7.37}$$

$$\mathbf{f}_I^{\mathrm{int}} = \int_{\Omega^0} \left(\mathbf{B}_I^0\right)^{\mathrm{T}} \boldsymbol{\Sigma}^0 d\Omega. \tag{7.38}$$

Here, **I** denotes the identity matrix, **d** is the vector containing the nodal coefficients of displacements defined in (4.58) of Chapter 4.2, $\mathbf{B}_I^0$ is the gradient matrix associated with the deformation gradient **F** at node I, and $\boldsymbol{\Sigma}^0$ is the stress vector associated with the nominal stress **P**. For the two-dimensional plane strain case, for example, $\mathbf{B}_I^0$ and $\boldsymbol{\Sigma}^0$ are given as follows:

$$\mathbf{B}_I^0 = \begin{bmatrix} \dfrac{\partial \Psi_I^0}{\partial X_1} & 0 \\ 0 & \dfrac{\partial \Psi_I^0}{\partial X_2} \\ \dfrac{\partial \Psi_I^0}{\partial X_2} & 0 \\ 0 & \dfrac{\partial \Psi_I^0}{\partial X_1} \end{bmatrix}, \tag{7.39}$$

$$\boldsymbol{\Sigma}^0 = \begin{bmatrix} P_{11} \\ P_{22} \\ P_{12} \\ P_{21} \end{bmatrix}. \tag{7.40}$$

For path-dependent materials, the stress and strain calculation in (7.11) requires taking the spatial derivatives of $\Psi_I^0(\mathbf{X})$, which can be performed as follows [1]:

$$\frac{\partial \Psi_I^0(\mathbf{X})}{\partial x_i} = \frac{\partial \Psi_I^0(\mathbf{X})}{\partial X_j} F_{ji}^{-1}. \tag{7.41}$$

The deformation gradient $\mathbf{F}$ is first computed by using the material spatial derivatives of $\Psi_I^0(\mathbf{X})$

$$F_{ij}(\mathbf{X}, t) = \frac{\partial x_i^h}{\partial X_j} = \delta_{ij} + \sum_{I=1}^{NP} \frac{\partial \Psi_I^0(\mathbf{X})}{\partial X_j} u_{iI}(t), \tag{7.42}$$

and $\mathbf{F}^{-1}$ is obtained directly by the inversion of $\mathbf{F}$.

Both explicit and implicit time integration techniques can be employed to integrate the semi-discrete equation in (7.35) in the temporal domain. In the case of explicit time integration, such as the central difference method, lumped mass matrices are usually preferred. One of the popular mass lumping techniques, the row sum method, can be directly applied to the RK mass matrix:

$$\mathcal{M}_{IJ}^{\mathrm{L}} = \begin{cases} \sum_{J=1}^{NP} \mathcal{M}_{IJ} & \text{for } I = J \\ 0 & \text{for } I \neq J \end{cases}. \tag{7.43}$$

By considering the zeroth-order reproducing condition (the partition of unity) of the Lagrangian RK shape function, $\sum_{I=1}^{NP} \Psi_I^0(\mathbf{X}) = 1$, the diagonal terms of the lumped mass matrix can be simplified as

$$\mathcal{M}_{II}^{\mathrm{L}} = \sum_{J=1}^{NP} \mathcal{M}_{IJ} = \int_{\Omega^0} \rho^0 \Psi_I^0(\mathbf{X}) \left(\sum_{J=1}^{NP} \Psi_J^0(\mathbf{X}) \right) \mathbf{I} d\Omega = \int_{\Omega^0} \rho^0 \Psi_I^0(\mathbf{X}) \mathbf{I} d\Omega. \tag{7.44}$$

While the total Lagrangian formulation with the Lagrangian RK approximation is effective for large deformation analysis of solids without any kernel instability, this approach breaks down when the inverse of $\mathbf{F}$ does not exist. This may occur, for example, when the extreme deformation leads to a nonpositive definite $\mathbf{F}$, or when material separation takes place in the form of fractures or material damage. Thus, a semi-Lagrangian RK formulation has been introduced [8, 9] to address this issue in modeling solids and structures subjected to extreme deformations.

7.3 Semi-Lagrangian Reproducing Kernel Approximation and Discretization

The Lagrangian formulation breaks down when the mapping $\mathbf{x} = \boldsymbol{\varphi}(\mathbf{X}, t)$ or the inverse mapping $\mathbf{X} = \boldsymbol{\varphi}^{-1}(\mathbf{x}, t)$ is no longer regular (one-to-one). This happens in problems involving situations such as new free surface formation or free surface closure. A semi-Lagrangian RK approximation and discretization have been proposed to circumvent this difficulty [8, 9].

In the semi-Lagrangian RK formulation, the nodal point $\mathbf{x}_I$ in the current configuration follows the motion of the material point, that is, $\mathbf{x}_I = \boldsymbol{\varphi}(\mathbf{X}_I, t)$, and thus is "Lagrangian." However, the kernel function $\phi_a(\mathbf{x} - \mathbf{x}_I)$ is constructed with the distance measure $\|\mathbf{x} - \mathbf{x}_I\|$ defined in the current

Table 7.2 Lagrangian and Semi-Lagrangian RK approximation.

	Lagrangian	Semi-Lagrangian
RK approximation	$u_i^h(\mathbf{X}, t) = \sum_{I=1}^{NP} \Psi_I^0(\mathbf{X}) d_{iI}(t)$	$u_i^h(\mathbf{x}, t) = \sum_{I=1}^{NP} \Psi_I(\mathbf{x}) d_{iI}(t)$
Approximation function	$\Psi_I^0(\mathbf{X}) = \mathbf{H}^{\mathrm{T}}(\mathbf{0})\mathbf{M}^{-1}(\mathbf{X}) \times \mathbf{H}(\mathbf{X}-\mathbf{X}_I)\phi_a(\mathbf{X}-\mathbf{X}_I)$	$\Psi_I(\mathbf{x}) = \mathbf{H}^{\mathrm{T}}(\mathbf{0})\mathbf{M}^{-1}(\mathbf{x}) \times \mathbf{H}(\mathbf{x}-\mathbf{x}_I)\phi_a(\mathbf{x}-\mathbf{x}_I)$, $\mathbf{x}_I = \boldsymbol{\varphi}(\mathbf{X}_I, t)$
Kernel function	$\phi_a(z), \quad z = \Vert\mathbf{X}-\mathbf{X}_I\Vert/a_I$	$\phi_a(z), \quad z = \Vert\mathbf{x}-\boldsymbol{\varphi}(\mathbf{X}_I, t)\Vert/a_I$
Basis functions	$\mathbf{H}^{\mathrm{T}}(\mathbf{X}-\mathbf{X}_I) = [1 \quad X_1-X_{1I} \quad X_2-X_{2I} \quad \cdots \quad (X_3-X_{3I})^n]$	$\mathbf{H}^{\mathrm{T}}(\mathbf{x}-\mathbf{x}_I) = [1 \quad x_1-x_{1I} \quad x_2-x_{2I} \quad \cdots \quad (x_3-x_{3I})^n]$
Moment matrix	$\mathbf{M}(\mathbf{X}) = \sum_{I=1}^{NP} \mathbf{H}(\mathbf{X}-\mathbf{X}_I)\mathbf{H}^{\mathrm{T}}(\mathbf{X}-\mathbf{X}_I)\phi_a(\mathbf{X}-\mathbf{X}_I)$	$\mathbf{M}(\mathbf{x}) = \sum_{I=1}^{NP} \mathbf{H}(\mathbf{x}-\mathbf{x}_I)\mathbf{H}^{\mathrm{T}}(\mathbf{x}-\mathbf{x}_I)\phi_a(\mathbf{x}-\mathbf{x}_I)$

configuration, and the combined formulation is thus "semi-Lagrangian." For example, the cubic B-spline semi-Lagrangian kernel function is given as

$$\phi_a(\mathbf{x}-\mathbf{x}_I) \equiv \phi_a(z) = \begin{cases} \frac{2}{3} - 4z^2 + 4z^3 & \text{for } 0 \le z < \frac{1}{2} \\ \frac{4}{3} - 4z + 4z^2 - \frac{4}{3}z^3 & \text{for } \frac{1}{2} \le z \le 1, \\ 0 & \text{for } z > 1 \end{cases} \quad z = \Vert\mathbf{x}-\mathbf{x}_I\Vert/a. \tag{7.45}$$

As shown in Table 7.2, the main difference between the Lagrangian and the Semi-Lagrangian kernels is the distance measure defined in the undeformed and the deformed configurations, respectively. Under this definition, the material particles covered under the semi-Lagrangian kernel function support vary during material deformation, as shown in Figure 7.3 for the one-dimensional

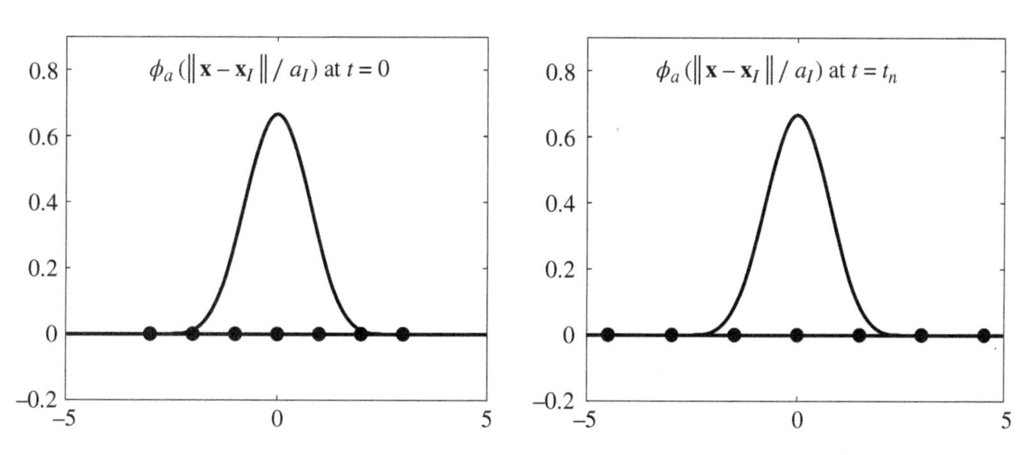

Figure 7.3 One-dimensional semi-Lagrangian kernel function in the undeformed configuration ($t = 0$) and the current configuration ($t = t_n$)

case and in Figure 7.1e and f for the multidimensional case, which is in contrast to the Lagrangian formulation (Figure 7.1c and d).

The semi-Lagrangian RK shape function $\Psi_I(\mathbf{x})$ is then formulated in the current configuration as [8, 9]

$$\Psi_I(\mathbf{x}) = C(\mathbf{x};\mathbf{x}-\mathbf{x}_I)\phi_a(\mathbf{x}-\mathbf{x}_I), \tag{7.46}$$

where $\mathbf{x}_I = \boldsymbol{\varphi}(\mathbf{X}_I, t)$. The correction function $C(\mathbf{x}; \mathbf{x}-\mathbf{x}_I)$ is expressed as

$$C(\mathbf{x};\mathbf{x}-\mathbf{x}_I) = \mathbf{H}^{\mathrm{T}}(\mathbf{x}-\mathbf{x}_I)\mathbf{b}(\mathbf{x}), \tag{7.47}$$

$$\mathbf{H}^{\mathrm{T}}(\mathbf{x}-\mathbf{x}_I) = [1 \quad x_1 - x_{1I} \quad x_2 - x_{2I} \quad \dots \quad (x_3 - x_{3I})^n]. \tag{7.48}$$

As in the discussion in Section 3.2.2, the coefficient vector $\mathbf{b}(\mathbf{x})$ can be determined by imposing the following discrete reproducing conditions in the current configuration:

$$\sum_{I=1}^{NP} \Psi_I(\mathbf{X}) x_{1I}^i x_{2I}^j x_{3I}^k = x_1^i x_2^j x_3^k, \quad 0 \le i+j+k \le n. \tag{7.49}$$

This reproducing condition in (7.49) can be recast as

$$\sum_{I=1}^{NP} \Psi_I(\mathbf{X})(x_1 - x_{1I})^i (x_2 - x_{2I})^j (x_3 - x_{3I})^k = \delta_{i0}\delta_{j0}\delta_{k0}, \quad 0 \le i+j+k \le n, \tag{7.50}$$

or

$$\sum_{I=1}^{NP} \Psi_I(\mathbf{x})\mathbf{H}(\mathbf{x}-\mathbf{x}_I) = \mathbf{H}(\mathbf{0}). \tag{7.51}$$

Solving for $\mathbf{b}(\mathbf{x})$ and substituting it into (7.46) yields the semi-Lagrangian RK shape function:

$$\Psi_I(\mathbf{x}) = \mathbf{H}^{\mathrm{T}}(\mathbf{0})\mathbf{M}^{-1}(\mathbf{x})\mathbf{H}(\mathbf{x}-\boldsymbol{\varphi}(\mathbf{X}_I,t))\phi_a(\mathbf{x}-\boldsymbol{\varphi}(\mathbf{X}_I,t)), \tag{7.52}$$

where

$$\mathbf{M}(\mathbf{x}) = \sum_{I=1}^{NP} \mathbf{H}(\mathbf{x}-\boldsymbol{\varphi}(\mathbf{X}_I,t))\mathbf{H}^{\mathrm{T}}(\mathbf{x}-\boldsymbol{\varphi}(\mathbf{X}_I,t))\phi_a(\mathbf{x}-\boldsymbol{\varphi}(\mathbf{X}_I,t)). \tag{7.53}$$

Or by using $\boldsymbol{\varphi}(\mathbf{X}_I, t) = \mathbf{x}_I$ we have simply

$$\Psi_I(\mathbf{x}) = \mathbf{H}^{\mathrm{T}}(\mathbf{0})\mathbf{M}^{-1}(\mathbf{x})\mathbf{H}(\mathbf{x}-\mathbf{x}_I)\phi_a(\mathbf{x}-\mathbf{x}_I), \tag{7.54}$$

where

$$\mathbf{M}(\mathbf{x}) = \sum_{I=1}^{NP} \mathbf{H}(\mathbf{x}-\mathbf{x}_I)\mathbf{H}^{\mathrm{T}}(\mathbf{x}-\mathbf{x}_I)\phi_a(\mathbf{x}-\mathbf{x}_I). \tag{7.55}$$

Let the velocity v_i be the primary variable in (7.18) and be approximated by the semi-Lagrangian RK shape functions:

$$v_i^h(\mathbf{x}, t) = \sum_{I=1}^{NP} \Psi_I(\mathbf{x}) v_{iI}(t). \tag{7.56}$$

The corresponding semi-Lagrangian approximation of acceleration is given by [8, 9]

$$\ddot{u}_i^h(\mathbf{x}, t) = \dot{v}_i^h(\mathbf{x}, t) = \sum_{I=1}^{NP} \left(\Psi_I(\mathbf{x}) \dot{v}_{iI}(t) + \Psi_I^*(\mathbf{x}) v_{iI}(t) \right). \tag{7.57}$$

Here, $\Psi_I^*(\mathbf{x})$ is the correction due to the time-dependent change of the semi-Lagrangian kernel $\dot{\phi}_a(\mathbf{x} - \mathbf{x}_I)$

$$\Psi_I^*(\mathbf{x}) = C(\mathbf{x}; \mathbf{x} - \boldsymbol{\varphi}(\mathbf{X}_I, t)) \dot{\phi}_a(\mathbf{x} - \boldsymbol{\varphi}(\mathbf{X}_I, t)), \tag{7.58}$$

and

$$\dot{\phi}_a(\mathbf{x} - \boldsymbol{\varphi}(\mathbf{X}_I, t)) = \dot{\phi}_a\left(\frac{\|\mathbf{x} - \boldsymbol{\varphi}(\mathbf{X}_I, t)\|}{a} \right) = \phi_a' \frac{\mathbf{q} \cdot (\mathbf{v} - \mathbf{v}_I)}{a}, \tag{7.59}$$

where

$$\mathbf{q} = (\mathbf{x} - \boldsymbol{\varphi}(\mathbf{X}_I, t)) / \|\mathbf{x} - \boldsymbol{\varphi}(\mathbf{X}_I, t)\|, \tag{7.60}$$

and $\|\cdot\|$ designates the length of a vector. Note that the correction function in (7.57) is used to ensure the reproducing condition of the time derivative of the semi-Lagrangian kernel $\dot{\phi}_a(\mathbf{x} - \boldsymbol{\varphi}(\mathbf{X}_I, t))$, and thus the time rate change of C is not considered.

Substituting (7.57) into the updated Lagrangian formulation in (7.13) yields the following semi-Lagrangian discrete equation:

$$\mathcal{M}\dot{\mathbf{v}} + \mathbf{N}\mathbf{v} = \mathbf{f}^{\text{ext}} - \mathbf{f}^{\text{int}}, \tag{7.61}$$

where

$$\mathcal{M}_{IJ} = \int_\Omega \rho \Psi_I(\mathbf{x}) \Psi_J(\mathbf{x}) \mathbf{I} d\Omega, \tag{7.62}$$

$$\mathbf{N}_{IJ} = \int_\Omega \rho \Psi_I(\mathbf{x}) \Psi_J^*(\mathbf{x}) \mathbf{I} d\Omega, \tag{7.63}$$

$$\mathbf{f}_I^{\text{ext}} = \int_\Omega \Psi_I(\mathbf{x}) \mathbf{b} d\Omega + \int_{\Gamma_t} \Psi_I(\mathbf{x}) \bar{\mathbf{t}} d\Gamma, \tag{7.64}$$

$$\mathbf{f}_I^{\text{int}} = \int_\Omega \mathbf{B}_I^{\text{T}} \mathbf{T} d\Omega. \tag{7.65}$$

Here, $\mathbf{I}$ denotes the identity matrix, $\mathbf{B}_I$ is the gradient matrix of $u_{(i,j)}^h = \left(\partial u_i^h / \partial x_j + \partial u_j^h / \partial x_i \right) / 2$ associated with node I, and $\mathbf{T}$ is the stress vector associated with the Cauchy stress $\boldsymbol{\tau}$. For a two-dimensional plane strain case, for example, $\mathbf{B}_I$ and $\mathbf{T}$ are given as

$$\mathbf{B}_I = \begin{bmatrix} \dfrac{\partial \Psi_I}{\partial x_1} & 0 \\ 0 & \dfrac{\partial \Psi_I}{\partial x_2} \\ \dfrac{\partial \Psi_I}{\partial x_2} & \dfrac{\partial \Psi_I}{\partial x_1} \end{bmatrix}, \tag{7.66}$$

$$\mathbf{T} = \begin{bmatrix} \tau_{11} \\ \tau_{22} \\ \tau_{12} \end{bmatrix}. \tag{7.67}$$

Note that the dimensions of $\mathbf{B}_I$ and $\mathbf{T}$ in the semi-Lagrangian discretization in (7.66) and (7.67) based on the updated Lagrangian formulation are smaller than $\mathbf{B}_I^0$ and $\mathbf{P}$ in the Lagrangian discretization in (7.39) and (7.40), which is based on the total Lagrangian formulation. This is due to the different stress and strain measures with different symmetry properties in the two formulations.

Similar to the Lagrangian RK discretization, a lumped mass matrix can be obtained by taking into account the partition of unity in the semi-Lagrangian RK shape functions

$$\sum_{I=1}^{NP} \Psi_I(\mathbf{x}) = 1 \tag{7.68}$$

to yield

$$\mathcal{M}_{IJ}^{\mathrm{L}} = \begin{cases} \sum\limits_{J=1}^{NP} \mathcal{M}_{IJ} & \text{for } I = J \\ 0 & \text{for } I \neq J \end{cases}, \tag{7.69}$$

where diagonal terms of the lumped mass matrix can be simplified as

$$\mathcal{M}_{II}^{\mathrm{L}} = \sum_{J=1}^{NP} \mathcal{M}_{IJ} = \int_{\Omega} \rho \Psi_I(\mathbf{x}) \left(\sum_{J=1}^{NP} \Psi_J(\mathbf{x}) \right) \mathbf{I} d\Omega = \int_{\Omega} \rho \Psi_I(\mathbf{x}) \mathbf{I} d\Omega. \tag{7.70}$$

Remark 7.1 *If a nodal integration scheme, such as direct nodal integration, stabilized conforming nodal integration, or stabilized nonconforming nodal integration as discussed in Chapter 6 is employed in Eqs. (7.61)–(7.65), $\Psi_I^*(\mathbf{x}_I) = 0$, and the remaining nodal integration terms of* **N** *have relatively negligible influence over (7.61). As such, the convective effect,* **Nv** *in (7.61), can be omitted in the semi-discrete equations of motion when nodal integration schemes are employed.*

Key Takeaways

For nonlinear analysis of solids, the two most popular implementations of RKPM use either the Lagrangian or semi-Lagrangian RK approximations. The key differences and notable features of each are as follows:

1) The Lagrangian RK approximation can be used with either path-dependent or path-independent materials. For path-dependent materials, the chain rule (7.41) must be employed to obtain the spatial derivatives commonly used in constitutive laws such as (7.11). The total or updated Lagrangian formulations (see Table 7.1) are both applicable to this formulation. As outlined in Section 7.1, they are mathematically equivalent, and the choice of one over the other is a matter of convenience.
2) In the Lagrangian RK formulation, the shape functions are constructed only once, and the formulation is very efficient. The increased cost of meshfree shape functions over FEM is not incurred during run-time. The only extra CPU time is due to the additional neighbors, which come into play in assembly-type operations and/or stiffness bandwidth considerations. On the other hand, semi-Lagrangian RK shape functions must be regularly reconstructed.

3) The condition under which the semi-Lagrangian RK approximation is necessary is when the mapping between the current and undeformed configurations breaks down, for instance, when damage and/or fracture occur.
4) It is possible to couple Lagrangian and semi-Lagrangian RK approximations to leverage the advantageous of both (efficiency of Lagrangian, capabilities of semi-Lagrangian), see [2] for details.
5) The Lagrangian RK approximation maintains the same kernel coverage and does not suffer from the kernel instability in traditional SPH, which is not purely Lagrangian.
6) The partition of unity is a necessary condition for row-sum mass lumping in both Lagrangian and semi-Lagrangian RKPM.

7.4 Stability of Lagrangian and Semi-Lagrangian Discretizations[2]

7.4.1 Stability Analysis for the Lagrangian RK Equation of Motion

The stability analyses for Lagrangian and semi-Lagrangian discretizations have been performed in [9]. For simplicity, consider the following one-dimensional Lagrangian discrete equation of motion constructed by introducing the Lagrangian RK approximation into the one-dimensional total Lagrangian formulation:

$$\mathcal{M}\ddot{\mathbf{d}} = \mathbf{f}^{\text{ext}} - \mathbf{f}^{\text{int}}, \tag{7.71}$$

where the internal forces $\mathbf{f}^{\text{int}}$ and external forces $\mathbf{f}^{\text{ext}}$ are

$$f_I^{\text{int}} = \int_{\Omega^0} \Psi_I^0(X)_{,X} P(X) d\Omega, \tag{7.72}$$

$$f_I^{\text{ext}} = \int_{\Omega^0} \Psi_I^0(X) b^0(X) d\Omega + \Psi_I^0(X)\bar{t}^0(X)\Big|_{\Gamma_t^0}. \tag{7.73}$$

Here, P, b^0, and $\bar{t}^0$ are the one-dimensional nominal stress, body force, and surface traction, respectively. Considering a lumped mass matrix with a uniform discretization in one dimension, we have using the partition of unity:

$$\mathcal{M}_{II}^{\text{L}} = \rho^0 \Delta X. \tag{7.74}$$

Introducing one-dimensional stabilized conforming nodal integration (SCNI) (see Section 6.6) into the integration of $\mathbf{f}^{\text{int}}$ and ignoring external force for stability analysis, the linearization of (7.71) in one-dimension can be rewritten in an explicit form as

$$\Delta \ddot{u}_I = -\frac{D^S F^2 + S}{\rho^0} \sum_{K=-m}^{m} \left[\bar{b}_I(X_{I+K}) \sum_{J=-m}^{m} \bar{b}_I(X_{I-J}) \Delta u_{I+J+K} \right]. \tag{7.75}$$

Here $\bar{b}_I$ is the smoothed gradient if SCNI is used and is the direct gradient $\Psi_{I,X}^0$ if direct nodal integration (DNI) is used. In (7.75), $D^S = 2\partial S/\partial(F^2 - 1)$; F is the one-dimensional deformation gradient, S is the one-dimensional second Piola–Kirchhoff stress, $m = \text{int}(R + 0.5)$ is an integer

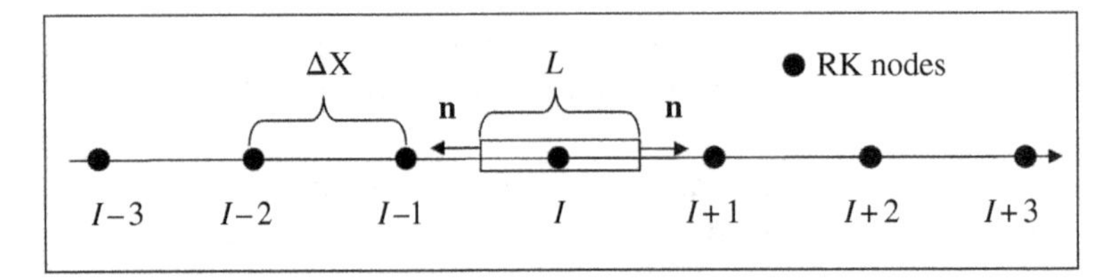

Figure 7.4 Representative domain for smoothed gradient operator in one dimension. *Source:* Guan et al. [9], figure 5, p. 674 / With permission of Elsevier.

determined by the normalized kernel support size R (the support size divided by the nodal distance). Assuming a uniform nodal distribution with nodal distance ΔX as shown in Figure 7.4, the following smoothed gradient for SCNI is obtained (see [13] for the nonlinear version of SCNI):

$$\overline{b}_I(X_J) = \frac{1}{\Delta X}\left[\Psi_I^0\left(X_J + \frac{\Delta X}{2}\right) - \Psi_I^0\left(X_J - \frac{\Delta X}{2}\right)\right]. \tag{7.76}$$

Now consider a plane waveform for the incremental displacement given as

$$\Delta u_I = \lambda e^{ikI\Delta X + i\omega t}, \tag{7.77}$$

where λ is the amplitude of the displacement increment, k is the wave number, and ω is the frequency. Stability requires that ω be real. Substituting (7.77) into (7.75) yields:

$$\omega^2 = \frac{D^S F^2 + S}{\rho_0}\sum_{K=1}^{2m+1}\left[\overline{b}_I(X_{I+K-m-1})\sum_{L=1}^{2m+1}\overline{b}_I(X_{I+L-m-1})\cos\left((K-L)k\Delta X\right)\right]. \tag{7.78}$$

The frequencies calculated based on (7.78) are shown in Figure 7.5 for various normalized support sizes R, with $c^2 = (D^S F^2 + S)/\rho_0$ the square of the wave speed. Here, linear basis and a cubic B-spline function are introduced for the RK shape functions. The results in Figure 7.5 show that the

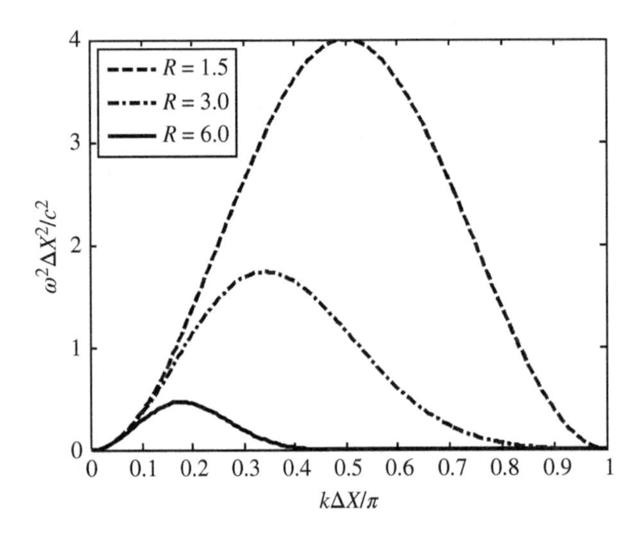

Figure 7.5 Frequency Characteristic of Lagrangian RK discretization with SCNI. *Source:* Guan et al. [9], figure 6, p. 675 / With permission of Elsevier.

Lagrangian RK approximation with SCNI is stable, as all frequencies are real. It is observed that the frequency is very close to zero near the cutoff point ($k\Delta X = \pi$) for the case with very large support size ($R = 6.0$). This implies that the stiffness matrix becomes more linearly dependent and ill-conditioned when a very large support size is used in the RK approximation; this should be avoided.

7.4.2 Stability Analysis for the Semi-Lagrangian RK Equation of Motion

The one-dimensional discrete equation of motion constructed by the semi-Lagrangian RK approximation in the one-dimensional updated Lagrangian formulation yields:

$$\mathcal{M}\dot{\mathbf{v}} + \mathbf{N}\mathbf{v} = \mathbf{f}^{\text{ext}} - \mathbf{f}^{\text{int}}, \tag{7.79}$$

where

$$\mathbf{N}_{IJ} = \int_\Omega \rho \Psi_I(x)\Psi_J^*(x)\mathbf{I}d\Omega, \tag{7.80}$$

$$f_I^{\text{int}} = \int_\Omega \Psi_{I,x}(x)\tau(x)d\Omega, \tag{7.81}$$

$$f_I^{\text{ext}} = \int_\Omega \Psi_I(x)b(x)d\Omega + \Psi_I(x)\bar{t}(x)\big|_{\Gamma_t}. \tag{7.82}$$

Here ρ is the density, $\Psi_J^*(x)$ is defined in Eq. (7.58), and $\tau(x)$ is the Cauchy stress. By considering the linearization of (7.79) with lumped mass $\mathcal{M}_{II}^{L} = \rho\Delta x$ for a uniform discretization in the current configuration, introducing SCNI for domain integration, assuming a small normalized support size ($R \le 1.5$), and letting $\Delta u_I = \lambda e^{ikI\Delta x + i\omega t}$, we have

$$\begin{aligned}
&\omega^2 - 4Q^2\frac{\left[C(\Delta x)\phi_a'(\Delta x)\right]^2}{R^2}(1 - \cos k\Delta x) \\
&- 2Q\omega\frac{C(\Delta x)\phi_a'(\Delta x)}{R}\left[\Psi_I(0)\sin k\Delta x + \Psi_I(\Delta x)\sin 2k\Delta x\right] \\
&- \frac{2(D^\tau + \tau(x_K))}{\rho\Delta x^2}\left[C\left(\frac{\Delta x}{2}\right)\phi_a\left(\frac{\Delta x}{2}\right)\right]^2(1 - \cos 2k\Delta x) = 0,
\end{aligned} \tag{7.83}$$

where D^τ is the material tangent modulus associated with Truesdell Cauchy stress rate, $C(\Delta x) \equiv C(x_I + \Delta x; \Delta x)$ is the correction function defined in (7.47) and

$$Q = \frac{\partial v}{\partial x}\bigg|_{x = x_I}, \quad \tau(x_K) = D^\tau \sum_{J=1}^{NP} \bar{b}_J(x_K)u_J, \tag{7.84}$$

where $\bar{b}_J(x_K)$ is the gradient, with $\bar{b}_J(x_K) = \frac{1}{\Delta x}\left[\Psi_J\left(x_K + \frac{\Delta x}{2}\right) - \Psi_J\left(x_K - \frac{\Delta x}{2}\right)\right]$ if SCNI is used and $\bar{b}_J(x_K) = \Psi_{J,x}(x_K)$ if DNI is employed. We now rewrite (7.83) as.

$$\omega^2 + \alpha\omega + \beta = 0, \tag{7.85}$$

where

$$\alpha = -2Q\frac{C(\Delta x)\phi_a'(\Delta x)}{R}\left[\Psi_I(0)\sin k\Delta x + \Psi_I(\Delta x)\sin 2k\Delta x\right], \tag{7.86}$$

$$\beta = -4Q^2\frac{[C(\Delta x)\phi_a'(\Delta x)]^2}{R^2}(1-\cos k\Delta x) - \frac{2(D^\tau + \tau(x_K))}{\rho\Delta x^2}\left[C\left(\frac{\Delta x}{2}\right)\phi_a\left(\frac{\Delta x}{2}\right)\right]^2(1-\cos 2k\Delta x). \tag{7.87}$$

If we define a frequency characteristic parameter γ as

$$\gamma = \alpha^2 - 4\beta, \tag{7.88}$$

it can be observed that if $\gamma \geq 0$, a real solution for ω is obtained and the discrete system is stable. Note that $\gamma \geq 0$ is satisfied as long as $D^\tau + \tau \geq 0$, where $D^\tau + \tau$ represents the tangent modulus including the effect of geometric nonlinearity. Therefore, under a Semi-Lagrangian discretization with SCNI, the sufficient condition for stability is when the tangent modulus is positive ($D^\tau + \tau \geq 0$), which represents a stable (hardening) material. Thus, the stability condition for the semi-Lagrangian discrete system integrated by SCNI is consistent with the stability condition of the material.

As a demonstration, consider a discretization with a uniform particle distribution with $\Delta x = 1.0$, a normalized support size of $R = 1.5$ and velocity gradient $Q = 1.0$. Figure 7.6a shows the comparison of frequency characteristic parameter γ of semi-Lagrangian RK with SCNI and DNI domain integration for the case where the material of the continuum is stable, that is, when $D^\tau + \tau \geq 0$. It can be observed from Figure 7.6a that in the case when $D^\tau + \tau \geq 0$ ($D^\tau/\tau \geq -1.0$), the frequency is real ($\gamma \geq 0$) for all wavelengths in SCNI. This represents a stable discrete system consistent with the continuum system when the semi-Lagrangian RK approximation with SCNI is used. Figure 7.6b shows the frequency characteristic parameter γ for the case where the continuum system exhibits material instability, such as materials undergoing softening or damaged states, that is, $D^\tau + \tau < 0$ ($D^\tau/\tau > -1.0$). It is seen in Figure 7.6b that $\gamma < 0$ (an imaginary angular frequency) occurs in a certain range of wavelengths for SCNI, and hence the discrete system is unstable, which is consistent with the physical material instability.

In contrast, when semi-Lagrangian RK with DNI is employed, the stability analysis in Figure 7.6a shows that for a *stable* continuum under tension ($D^\tau/\tau > -1.0$), an imaginary frequency ($\gamma < 0$)

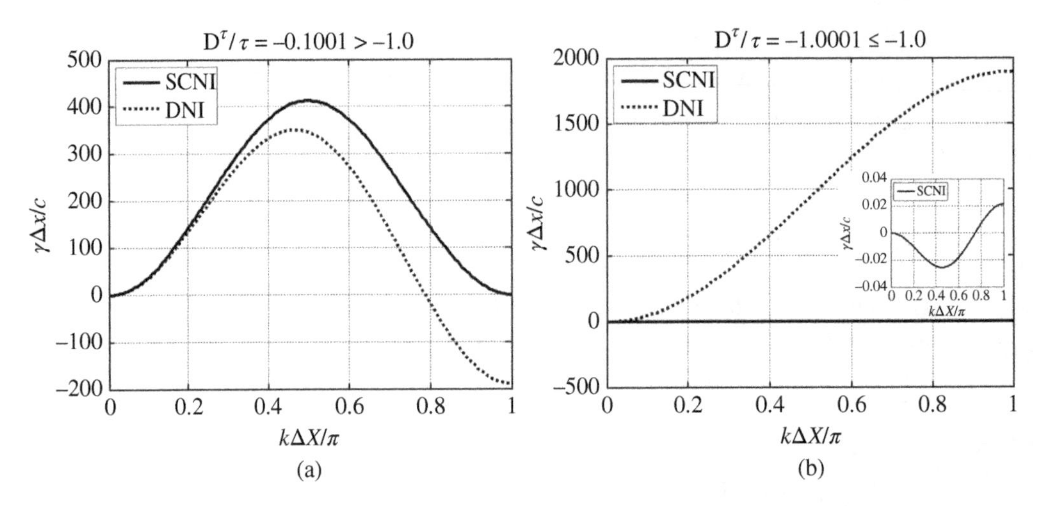

Figure 7.6 Frequency characteristic parameter for Semi-Lagrangian RK with SCNI and DNI for (a) a stable continuum and (b) an unstable (softening or damaged) continuum. *Source:* Guan et al. [9], figure 7, p. 676 / With permission of Elsevier.

occurs at certain wavelengths, showing an *unstable* discrete system. *This can be used to explain the tensile instability observed in SPH [14, 15].* Figure 7.6b further shows that when DNI is used, the-discrete system becomes *stable* ($\gamma > 0$) for softening or a damaged continuum ($D^{\tau}/\tau \leq -1.0$, *unstable*), showing nonphysical characteristics in the numerical solution.

Key Takeaways

In semi-Lagrangian analysis, SCNI plays a key role other than just providing accuracy and stability:

1) Direct nodal integration (DNI) is unstable for stable (hardening) materials and stable for unstable (softening) materials as shown in Figure 7.6; this is a direct contradiction to the mechanics at hand.
2) A common, and problematic manifestation of this issue is the inability of DNI (and some other related equivalent SPH formulations) to capture shear bands, damage localization, and other related processes.
3) SCNI, on the other hand, is stable for stable materials and unstable for unstable materials, which is physically correct, and precludes this problem. Therefore, strain-smoothing should always be adopted in semi-Lagrangian calculations (see Chapter 6 for more details).

7.4.3 Critical Time Step Estimation for the Lagrangian Formulation

In this section, we consider RK for the spatial discretization and central difference for the temporal discretization following [9]. The fully discrete Lagrangian RK equation of motion is

$$\mathbf{M}\left(\mathbf{d}^{n+1} - 2\mathbf{d}^{n} + \mathbf{d}^{n-1}\right) = (\Delta t)^2\left(\left(\mathbf{f}^{\text{ext}}\right)^{n+1} - \left(\mathbf{f}^{\text{int}}\right)^{n}\right), \tag{7.89}$$

where $\mathbf{f}^{\text{int}}$ and $\mathbf{f}^{\text{ext}}$ are defined in (7.72) and (7.73), respectively, for one-dimension using SCNI, and "n" denotes the time step. By considering normalized support size $R \leq 2.5$, the explicit form of the internal force $\mathbf{f}^{\text{int}}$ can be obtained as

$$\begin{aligned} f_I^{\text{int}} = \frac{c^2}{\Delta X^2}\Bigg\{ & \Psi^2\left(\frac{1}{2}\Delta X\right)[u_{I+2} - 2u_I + u_{I-2}] \\ & + \Psi\left(\frac{1}{2}\Delta X\right)\Psi\left(\frac{3}{2}\Delta X\right)[2u_{I+3} - 2u_{I+2} - 2u_{I+1} + 4u_I - 2u_{I-1} - 2u_{I-2} + 2u_{I-3}] \\ & + \Psi^2\left(\frac{3}{2}\Delta X\right)[u_{I+4} - 2u_{I+3} + u_{I+2} + u_{I-4} - 2u_{I-3} + u_{I-2} + 2u_{I+1} - 4u_I + 2u_{I-1}]\Bigg\}, \end{aligned} \tag{7.90}$$

where

$$\Psi(\Delta X) \equiv \Psi_I^0(\Delta X) = C(\Delta X)\phi_a(\Delta X), \quad C(\Delta X) \equiv C(X_I + \Delta X; \Delta X). \tag{7.91}$$

Consider the Fourier mode of the solution

$$u_I^n = (\lambda)^n e^{ik(I\Delta X)}, \tag{7.92}$$

where k is the wave number and λ is the amplification factor. By substituting (7.92) into (7.89), and ignoring the external force for stability analysis, the following equation is obtained:

$$\lambda - 2 + \frac{1}{\lambda} = -4\frac{c^2\Delta t^2}{\Delta X^2}\left\{\Psi^2\left(\frac{1}{2}\Delta X\right)\sin^2(k\Delta X) + 2\Psi^2\left(\frac{3}{2}\Delta X\right)\sin^2\left(\frac{k\Delta X}{2}\right)[1+\cos(3k\Delta X)] + 4\Psi\left(\frac{1}{2}\Delta X\right)\Psi\left(\frac{3}{2}\Delta X\right)\sin^2\left(\frac{k\Delta X}{2}\right)[\cos(2k\Delta X)+\cos(k\Delta X)]\right\}. \tag{7.93}$$

To have a stable solution of the numerical approximation (7.89), the roots of (7.93) need to be bounded by $|\lambda| \leq 1$. This yields the following von Neumann stability criterion

$$c^2\Delta t_{cr}^2 = \frac{\Delta X^2}{H}, \tag{7.94}$$

where

$$H = \Psi^2\left(\frac{1}{2}\Delta X\right)\sin^2(k\Delta X) + 2\Psi^2\left(\frac{3}{2}\Delta X\right)\sin^2\left(\frac{k\Delta X}{2}\right)[1+\cos(3k\Delta X)] + 4\Psi\left(\frac{1}{2}\Delta X\right)\Psi\left(\frac{3}{2}\Delta X\right)\sin^2\left(\frac{k\Delta X}{2}\right)[\cos(2k\Delta X)+\cos(k\Delta X)]. \tag{7.95}$$

The stability condition for (7.89), with domain integration by DNI, can be obtained using similar procedures. The equation for the amplification factor is obtained as follows for DNI:

$$\lambda - 2 + \frac{1}{\lambda} = -4c^2\Delta t^2\left[\Psi_{,X}^2(2\Delta X)\sin^2(2k\Delta X) + \Psi_{,X}^2(\Delta X)\sin^2(k\Delta X) + 4\Psi_{,X}(\Delta X)\Psi_{,X}(2\Delta X)\sin^2(k\Delta X)\cos(k\Delta X)\right], \tag{7.96}$$

and the corresponding H factor for the critical time step estimate of (7.94) is

$$H = \Delta X^2\left[\Psi_{,X}^2(2\Delta X)\sin^2(2k\Delta X) + \Psi_{,X}^2(\Delta X)\sin^2(k\Delta X) + 4\Psi_{,X}(\Delta X)\Psi_{,X}(2\Delta X)\sin^2(k\Delta X)\cos(k\Delta X)\right]. \tag{7.97}$$

The stability condition of the discrete equation integrated by DNI is associated with the shape function derivatives, as shown in (7.97). On the other hand, when the domain is integrated by SCNI, the stability condition is associated with the shape function itself, as shown in (7.95).

Similarly, for the discrete equation integrated by one-point Gauss quadrature, the H factor in (7.94) is

$$H = \Delta X^2\left\{\Psi_{,X}^2\left(\frac{1}{2}\Delta X\right)\sin^2\left(\frac{1}{2}k\Delta X\right) + \Psi_{,X}^2\left(\frac{3}{2}\Delta X\right)\sin^2\left(\frac{3}{2}k\Delta X\right) + 2\Psi_{,X}\left(\frac{3}{2}\Delta X\right)\Psi_{,X}\left(\frac{1}{2}\Delta X\right)\sin^2\left(\frac{1}{2}k\Delta X\right)[1+2\cos(k\Delta X)]\right\}. \tag{7.98}$$

The critical time steps associated with the use of different domain integration methods (DNI, SCNI, and one-point Gauss quadrature) are compared in Figure 7.7. In this study, a cubic B-spline kernel function and linear basis functions are used. The results show that the critical time steps for DNI and SCNI are generally larger than for one-point Gauss quadrature, especially when the kernel support size is small. Furthermore, SCNI offers better stability than DNI when kernel support size is selected as $R \geq 1.5$.

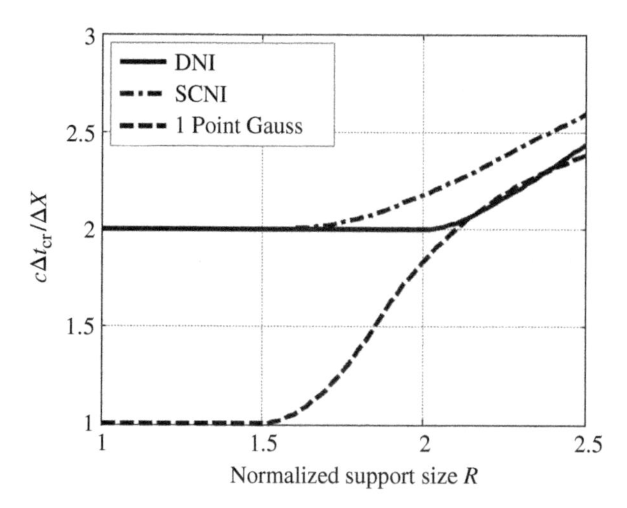

Figure 7.7 Critical time step estimation for different domain integration methods. *Source:* Guan et al. [9], figure 8, p. 677 / With permission of Elsevier.

7.4.4 Critical Time Step Estimation for the Semi-Lagrangian Formulation

The critical time step estimation is also performed following [9] for the fully discrete equation of motion of the Semi-Lagrangian formulation by introducing the reproducing kernel for the spatial discretization and central difference for the temporal discretization of (7.79). The fully discrete Semi-Lagrangian RK equation of motion is

$$\mathbf{M}\left(\mathbf{d}^{n+1} - 2\mathbf{d}^{n} + \mathbf{d}^{n-1}\right) = (\Delta t)^2\left(\left(\mathbf{f}^{\text{ext}}\right)^{n+1} - \left(\mathbf{f}^{\text{int}}\right)^{n} - \mathbf{N}\mathbf{v}^{n}\right), \tag{7.99}$$

where $\mathbf{f}^{\text{int}}$, $\mathbf{f}^{\text{ext}}$, and $\mathbf{N}$ are defined in (7.81), (7.82), and (7.80), respectively. The explicit form of (7.99) is obtained by introducing SCNI domain integration, and considering kernel support size $R \leq 2.5$ one obtains:

$$u_I^{n+1} = \left(2u_I^n - u_I^{n-1}\right) + \frac{\Delta t^2}{\Delta x^2}\frac{D^\tau}{\rho_I}A_1 + \rho_I \Delta x A_2 \tag{7.100}$$

where

$$\begin{aligned} A_1 = &\left\{\Psi^2\left(\frac{1}{2}\Delta x\right)\left[u_{I+2}^n - 2u_I^n + u_{I-2}^n\right]\right. \\ &+ \Psi^2\left(\frac{3}{2}\Delta x\right)\left[u_{I+4}^n - 2u_{I+3}^n + u_{I+2}^n + u_{I-4}^n - 2u_{I-3}^n + u_{I-2}^n + 2u_{I+1}^n - 4u_I^n + 2u_{I-1}^n\right] \\ &\left. + \Psi\left(\frac{1}{2}\Delta x\right)\Psi\left(\frac{3}{2}\Delta x\right)\left[2u_{I+3}^n - 2u_{I+2}^n - 2u_{I+1}^n + 4u_I^n - 2u_{I-1}^n - 2u_{I-2}^n + 2u_{I-3}^n\right]\right\}, \end{aligned} \tag{7.101}$$

$$\begin{aligned} A_2 = \Delta t\{&\Psi(2\Delta x)\left[\Psi^*(2\Delta x)\left(v_{I-4}^n - v_{I+4}^n\right) + \Psi^*(\Delta x)\left(v_{I-3}^n - v_{I+3}^n + v_{I+1}^n - v_{I-1}^n\right)\right] \\ &+ \Psi(\Delta x)\left[\Psi^*(2\Delta x)\left(v_{I-3}^n - v_{I+3}^n - v_{I+1}^n + v_{I-1}^n\right) + \Psi^*(\Delta x)\left(v_{I-2}^n - v_{I+2}^n\right)\right] \\ &+ \Psi(0)\left[\Psi^*(2\Delta x)\left(v_{I-2}^n - v_{I+2}^n\right) + \Psi^*(\Delta x)\left(v_{I-1}^n - v_{I+1}^n\right)\right]\}. \end{aligned} \tag{7.102}$$

In the above,

$$\Psi(\Delta x) \equiv \Psi_I(\Delta x) = C(\Delta x)\phi_a(\Delta x), \quad C(\Delta x) \equiv C(x_I + \Delta x; \Delta x), \tag{7.103}$$

$$\Psi^*(\Delta x) \equiv \Psi_I^*(x_{I+1}) = C(x_I + \Delta x; \Delta x)\phi_a'(\Delta x)\frac{(v_{I+1} - v_I)}{a_I}. \tag{7.104}$$

The additional term defined in (7.102) is due to the time derivative of the Semi-Lagrangian kernel function. Introducing the Fourier mode in (7.92) to (7.100), the equation for the amplification factor is obtained as

$$\lambda + 2S_1 + \frac{1}{\lambda}S_2 = 0, \tag{7.105}$$

where

$$\begin{aligned} S_1 &= -1 + 2\frac{\Delta t^2}{\Delta x^2}c^2 A_3 + \rho_I \Delta t \Delta x A_4, \\ S_2 &= 1 - 2\rho_I \Delta x \Delta t \Psi^*(\Delta x) A_4, \end{aligned} \tag{7.106}$$

and

$$\begin{aligned} A_3 = &\left\{\Psi^2\left(\frac{1}{2}\Delta x\right)\sin^2(k\Delta x) + 2\Psi^2\left(\frac{3}{2}\Delta x\right)\sin^2\left(\frac{k\Delta x}{2}\right)[1 + \cos(3k\Delta x)]\right. \\ &\left. + 4\Psi\left(\frac{1}{2}\Delta x\right)\Psi\left(\frac{3}{2}\Delta x\right)\sin^2\left(\frac{k\Delta x}{2}\right)[\cos(2k\Delta x) + \cos(k\Delta x)]\right\}, \end{aligned} \tag{7.107}$$

$$\begin{aligned} A_4 = &\, i\{\Psi^*(2\Delta x)[\Psi(2\Delta x)\sin(4k\Delta x) + \Psi(0)\sin(2k\Delta x) \\ &+ \Psi(\Delta x)[\sin(3k\Delta x) + \sin(k\Delta x)]] \\ &+ \Psi^*(\Delta x)[\Psi(2\Delta x)[\sin(3k\Delta x) - \sin(k\Delta x)] \\ &+ \Psi(\Delta x)\sin(2k\Delta x) + \Psi(0)\sin(k\Delta x)]\}. \end{aligned} \tag{7.108}$$

Note that the time derivative of the shape function, $\Psi^*(\Delta x)$, leads to a velocity difference between two particles as shown in (7.104), which affects the stability condition of the Semi-Lagrangian formulation.

For demonstration, consider cubic B-spline kernels with linear basis functions. The contour plots of the amplification factor λ, with kernel function normalized support sizes of $R = 1.5$ and $R = 2.0$, are shown in Figures 7.8 and 7.9, respectively. The area below contour line $\lambda = 1.0$ (shaded area) represents the stable region, and the contour line $\lambda = 1.0$ corresponds to the critical time step Δt_{cr}. It is observed that the critical time step decreases when the velocity difference between particles I and J increases. By comparing Figures 7.8 and 7.9, it can be observed that increasing the normalized support size improves the stability condition.

7.4.5 Numerical Tests of Critical Time Step Estimation

Let us now consider the following numerical tests to verify the stability analyses. First, consider a one-dimensional wave equation discretized with the Lagrangian RKPM formulation:

$$u_{tt} - c^2 u_{xx} = 0, x \in [0, L], \tag{7.109}$$

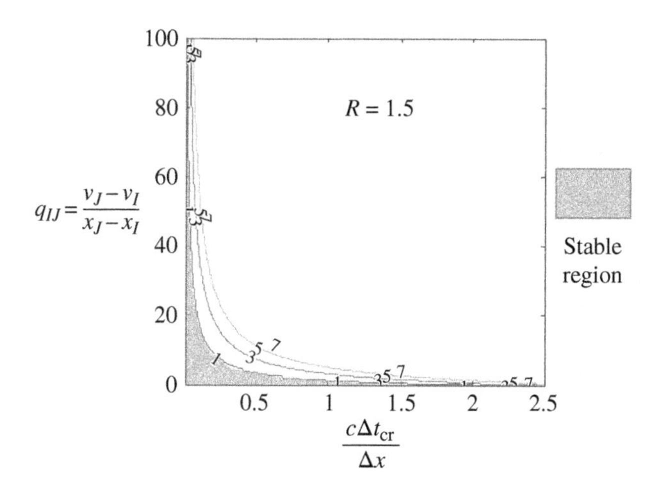

Figure 7.8 Contour plot for amplification factor λ with $R = 1.5$. *Source:* Guan et al. [9], figure 9, p. 678 / With permission of Elsevier.

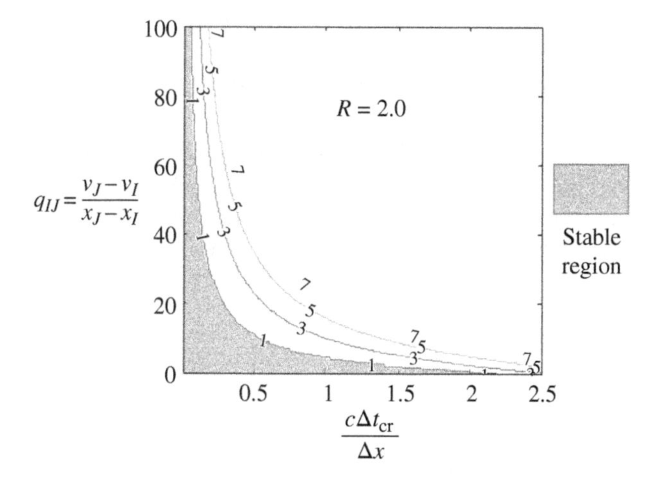

Figure 7.9 Contour plot for amplification factor λ with $R = 2.0$. *Source:* Guan et al. [9], figure 10, p. 678 / With permission of Elsevier.

with boundary conditions

$$u(0, t) = u(L, t) = 0, \tag{7.110}$$

and initial conditions

$$\begin{aligned} &u(x, 0) = \dot{u}(x, 0) = 0, \\ &u(0.5L, 0) = 0.01. \end{aligned} \tag{7.111}$$

A small perturbation in the middle of the domain is introduced into the initial condition to generate a wave. To be consistent with the stability analysis based on an infinite domain, we set L to be large, and numerically observe stability before the wave hits the boundaries. The cubic B-spline kernel and linear basis functions are used in the spatial discretization and central

Table 7.3 Critical time obtained from numerical tests.

Normalized support size R	Numerically tested critical time step for numerical result $\frac{c\Delta t_{cr}}{\Delta x}$		
	1 point Gauss	DNI	SCNI
1.001	1.0	2.0	2.0
1.5	1.0	2.0	2.0
2.0	1.837	2.0	2.174
2.5	2.38	2.42	2.59

Source: Guan et al. [9], Table 3, p. 678 / With permission of Elsevier.

difference is employed for temporal discretization. Table 7.3 shows the numerically characterized maximum allowable normalized time step for stable solutions using SCNI, DNI, and 1-point Gauss integration. The results show that the critical time step increases as the normalized support size increases. The analytical stability results in Eqs. (7.94)–(7.98) are also compared with the numerical stability results via numerical experiments. As shown in Figure 7.10, the analytical stability estimations in (7.94)–(7.98) agree quite well with the stability conditions obtained by numerical experiments.

Next, a numerical study is performed to evaluate the critical time step estimation for the semi-Lagrangian formulation. Since in the semi-Lagrangian formulation the stability is affected by the velocity gradient, the problem statement given in Figure 7.11 is introduced. In the numerical test, cubic B-spline kernel functions with linear basis are used for the spatial discretization, central

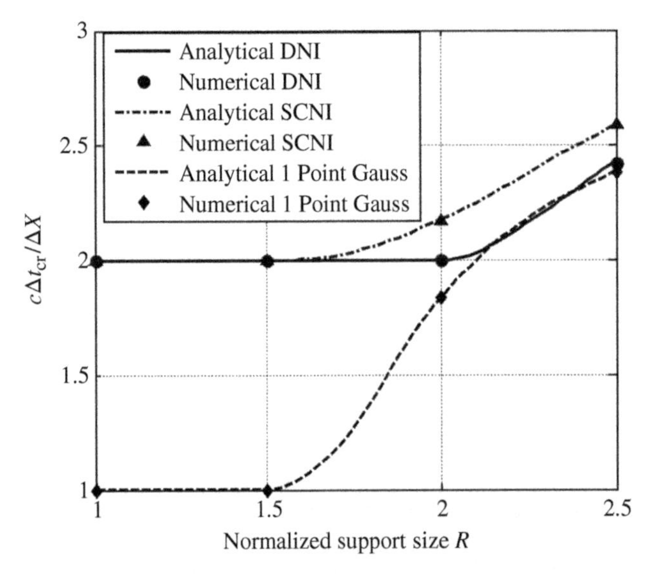

Figure 7.10 Comparison of analytical critical time step with the numerical test. *Source:* Guan et al. [9], figure 11, p. 678 / With permission of Elsevier.

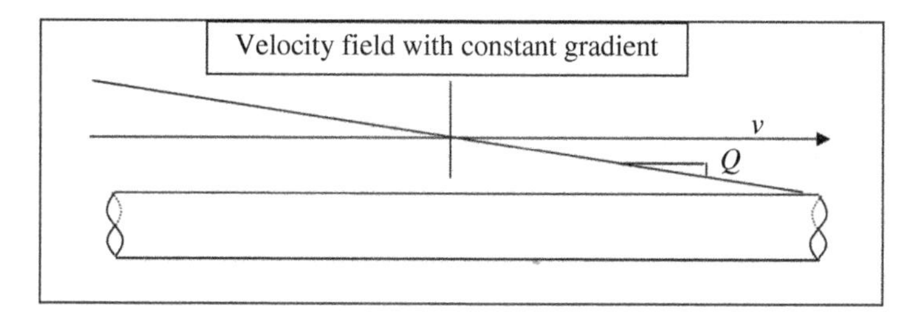

Figure 7.11 Problem statement for wave propagation with constant initial velocity gradient. *Source:* Guan et al. [9], figure 12, p. 679 / With permission of Elsevier.

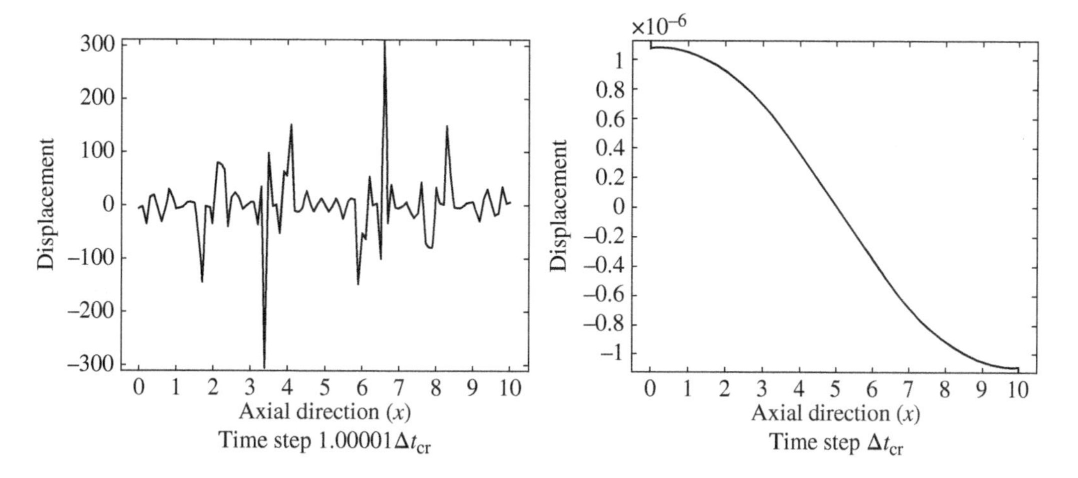

Figure 7.12 Numerical results at $t = 10$ for the semi-Lagrangian formulation using different time steps. *Source:* Guan et al. [9], figure 13, p. 679 / With permission of Elsevier.

difference is employed for the temporal discretization, and the kernel support size is fixed at $R = 1.5$ (not deforming with the material). The time step is updated based on the current nodal position and the local velocity gradient at each time step according to Eq. (7.105). The simulation result at $t = 10.0$ is shown in Figure 7.12. It is observed that using a time step slightly larger than the estimated Δt_{cr} ($1.0001\Delta t_{cr}$) leads to an unstable solution, while using the estimated Δt_{cr} yields a stable solution.

Key Takeaways

1) Nodally integrated RKPM greatly relaxes the stability condition of FEM and related methods, due to the increased influence of neighboring points. A time step of two times or more of that required in FEM can be employed in explicit dynamics when nodal integration or SCNI is employed, see Figure 7.10.
2) In semi-Lagrangian calculations, the temporal stability is dependent on the velocity gradient, and care should be taken to account for this condition in certain classes of problems where high gradients occur such as impact.

7.5 Neighbor Search Algorithms

It is important to note that in the semi-Lagrangian discretization, neighbor searches must be performed regularly. While several algorithms exist (see [16] for an overview), here we present an efficient bin-based strategy for nodally integrated equations.

1) **Construct bins.** The domain is first divided into "bins" as a preprocessing step, forming a Cartesian grid structure. Let the dimensions of the bins be $x_{\text{bin}} \times y_{\text{bin}} \times z_{\text{bin}}$; for RKPM with direct derivatives, the optimal dimensions of the bins for uniform support sizes in each direction is $a_x \times a_y \times a_z$ (for the general case including strain-smoothing, see Remark 7.2). The maximum and minimum of the nodal coordinates are used as the bounds for the grid, with the number of bins in each direction determined by the "ceiling" operation (rounding up to the nearest integer):

$$
\begin{aligned}
n_x^{\text{bin}} &= \text{ceiling}\left(\frac{x_{\max} - x_{\min}}{x_{\text{bin}}}\right), \quad x_{\max} = \max(x_I)\ \forall I,\ x_{\min} = \min(x_I)\ \forall I, \\
n_y^{\text{bin}} &= \text{ceiling}\left(\frac{y_{\max} - y_{\min}}{y_{\text{bin}}}\right), \quad y_{\max} = \max(y_I)\ \forall I,\ y_{\min} = \min(y_I)\ \forall I, \\
n_z^{\text{bin}} &= \text{ceiling}\left(\frac{z_{\max} - z_{\min}}{z_{\text{bin}}}\right), \quad z_{\max} = \max(z_I)\ \forall I,\ z_{\min} = \min(z_I)\ \forall I,
\end{aligned}
\tag{7.112}
$$

For search purposes, each bin is assigned a unique index using an (i, j, k) space as $M(i, j, k) = n_x^{\text{bin}} n_y^{\text{bin}}(k-1) + n_x^{\text{bin}}(j-1) + i$ as shown in Figure 7.13 for two dimensions (using (i, j) space, where $M(i, j) = n_x^{\text{bin}}(j-1) + i$).

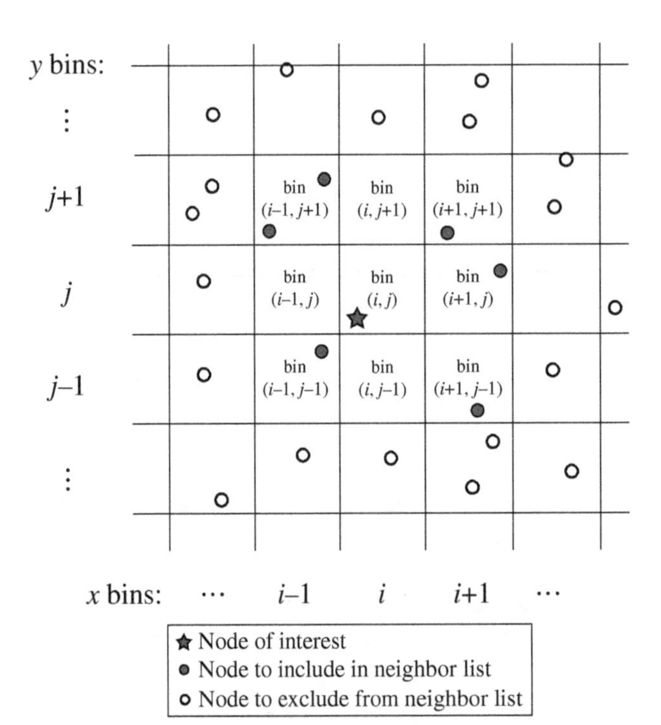

Figure 7.13 Illustration of bin search algorithm in two dimensions.

Once the bins are formed, neighbors for nodes during time integration can be identified quickly as follows:

2) **Bin the nodes.** Loop over nodes with index J:
   a) Identify the bin index (i, j, k) for each node J as follows using the "floor" operation (rounding down to the nearest integer):

$$i = \text{floor}\left(\frac{x_J - x_{\min}}{a_x}\right) + 1, \quad j = \text{floor}\left(\frac{y_J - y_{\min}}{a_y}\right) + 1, \quad k = \text{floor}\left(\frac{z_J - z_{\min}}{a_z}\right) + 1. \tag{7.113}$$

   b) Save the individual indices for this node as

$$\begin{aligned} \text{binx}(J) &= i, \\ \text{biny}(J) &= j, \\ \text{binz}(J) &= k. \end{aligned} \tag{7.114}$$

   c) Add this node to the list of nodes in this bin and update the number of nodes in the bin, e.g.

$$\begin{aligned} &\text{NodesInBin}(M) = \text{NodesInBin}(M) + 1, \\ &\text{BinList}(M, \text{NodesInBin}(M)) = J, \end{aligned} \tag{7.115}$$

   where $M = n_x^{\text{bin}} n_y^{\text{bin}}(k - 1) + n_x^{\text{bin}}(j - 1) + i$.

3) **Node search.** Loop over nodes with index J:
   a) Recall the box index $(i, j, k) = (\text{binx}(J), \text{biny}(J), \text{binz}(J))$ for this node
   b) Search this box and the adjacent boxes: this bin, one up/down, one left/right, and one back/front (see Figure 7.13 for a two-dimensional example):
      Loop: DO $\bar{i} = (i - 1, i, i + 1)$
      Loop: DO $\bar{j} = (j - 1, j, j + 1)$
      Loop: DO $\bar{k} = (k - 1, k, k + 1)$
      Identify the current bin:

$$M(\bar{i}, \bar{j}, \bar{k}) = n_x^{\text{bin}} n_y^{\text{bin}}(\bar{k} - 1) + n_x^{\text{bin}}(\bar{j} - 1) + \bar{i}$$

      Add the nodes in BinList(M, :) to the list of neighbor nodes for node J.

It can clearly be seen from the steps above that this algorithm is *linear* in NP, i.e., $O(NP)$ in operation count, whereas a naive search that loops over the nodes twice would scale as $O(NP^2)$ and would be prohibitively expensive for large systems.

Remark 7.2 *For general nonuniform discretizations where the supports are generally defined for each node, the dimensions of the bins can be constructed using the maximum support sizes for all nodes, or the number of adjacent bins in the search can be enlarged. This increases the computational complexity in the present algorithm, but more advanced searches can be employed for these situations such as KD-Tree algorithms* [17], *where* $O(NP \times \log(NP))$ *can still be achieved.*

Remark 7.3 *There are cases where the nodes inevitably fall outside the bin structure during time integration in semi-Lagrangian calculations and (7.113) fails to identify a meaningful bin. This can be taken care of by extending the first and last bins to infinity in their respective edge directions, i.e.,* $i > n_x^{\text{bin}} \rightarrow i = n_x^{\text{bin}}$ *and* $i < 1 \rightarrow i = 1$, *and so forth for the other directions. Step 1 can be repeated to* *save computational efficiency if the search becomes less efficient as more nodes move out of the grid structure.*

Remark 7.4 *For RKPM with the strain-smoothing methods discussed in Chapter 6, it should be noted that it is most effective to construct one list of neighbors for a node by taking into account the strain smoothing domains. Therefore, the dimensions of the bins should consider this, i.e., for uniform discretizations and conforming strain smoothing, the bin sizes should be at least* $1.5a_x \times 1.5a_y \times 1.5a_z$.

7.6 Smooth Contact Algorithm[3]

Contact problems are of significant importance in many engineering applications, ranging from structural impact, brake operations, metal forming processes, to bone joint mechanisms, and many others. In these problems, surface kinematics and discretization, and sliding contact under large deformations are among the major computational issues. Introductory and modern treatments of these can be found in the textbooks [18, 19] as well as more recent articles [20, 21]. Conventional finite element algorithms employ a piecewise linear approximation of kinematic variables and surface discretization based on node-to-segment [22, 23] or segment-to-segment contact [24, 25], where the second- and third-order derivative terms in the continuum formulation are ignored. The slope discontinuity leads to convergence problems in large sliding contact due to a jump of the force at the junction of two segments. Efforts have been devoted to the development of smooth surface discretization for use with continuum-based formulations under the finite element framework [18, 19, 25–35]. However, these approaches are often tedious, and the extension of these methods to three-dimensional contact is not trivial.

One of the most unique properties of meshfree approximations is the ease in constructing a smooth function. Specifically, the order of continuity is entirely independent of the order of monomial bases in MLS and RK (see Section 3.4). This unique property makes the meshfree approximation ideal for the discretization of the continuum-based contact formulation that requires at least C^2 continuity in the approximation of geometry and the displacement field [36]. This section presents a meshfree smooth formulation utilizing the higher-order continuity properties in meshfree approximations for large deformation simulations involving sliding contact, following [36].

7.6.1 Continuum-Based Contact Formulation

Let us consider the motion of two flexible bodies, denoted as $\Omega_0^{(1)}$ and $\Omega_0^{(2)}$, in the reference configuration shown in Figure 7.14. Superscripts (1) and (2) denote variables defined in body 1 and body 2, respectively, and we denote the initial configuration with the subscripts "0" in this section for notational clarity. It is assumed that no contact is present before any motion occurs. The subsequent configurations of the two bodies are denoted as $\Omega^{(1)}$ and $\Omega^{(2)}$, in which contact has been initiated. The initial boundaries of the two bodies are denoted as $\Gamma_0^{(1)}$ and $\Gamma_0^{(2)}$ and their current deformed boundaries as $\Gamma^{(1)}$ and $\Gamma^{(2)}$, respectively. The potential contact interfaces are expressed by $\Gamma_0^{C(1)}$ and $\Gamma_0^{C(2)}$, and their deformed configurations are denoted as $\Gamma^{C(1)}$ and $\Gamma^{C(2)}$, respectively. Material points in $\Omega_0^{(1)}$ are expressed as $\mathbf{X}^{(1)}$ and those in $\Omega_0^{(2)}$ as $\mathbf{X}^{(2)}$. The current locations of material points are denoted as $\mathbf{x}^{(1)} = \boldsymbol{\varphi}^{(1)}(\mathbf{X}^{(1)}, t)$ and $\mathbf{x}^{(2)} = \boldsymbol{\varphi}^{(2)}(\mathbf{X}^{(2)}, t)$, where "$t$" can be either pseudo-time in static problems or real time in dynamic problems.

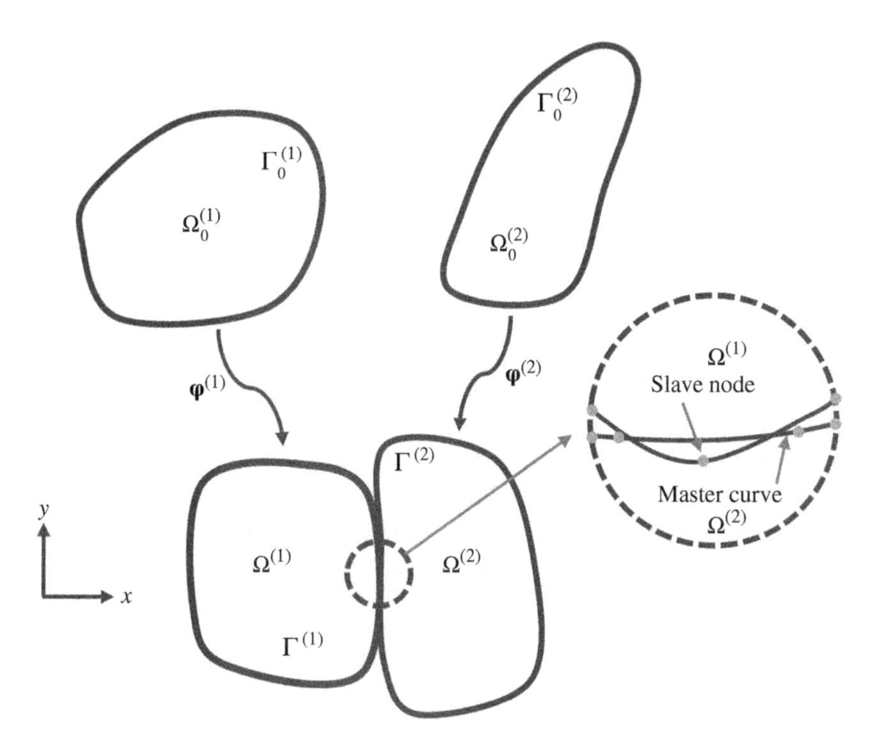

Figure 7.14 Illustration of contact kinematics. *Source:* Wang et al. [36], figure 1 p. 152 / Springer Nature.

Here we assign $\Gamma_0^{C(1)}$ as the slave surface and $\Gamma_0^{C(2)}$ as the master surface. In node-to-surface contact kinematics, a normal contact constraint (impenetrability) enforces the condition that points $\mathbf{x}^{(1)} = \boldsymbol{\varphi}^{(1)}(\mathbf{X}^{(1)}, t)$ on the slave surface $\Gamma^{C(1)}$ are not allowed to penetrate into the master surface $\Gamma^{C(2)}$. A gap g is defined in terms of the closest point projection of $\mathbf{x}^{(1)}$ onto surface $\Gamma^{C(2)}$, i.e.,

$$g\left(\mathbf{x}^{(1)}\right) = \alpha\left(\min_{\mathbf{x}^{(2)}} \|\mathbf{x}^{(1)} - \mathbf{x}^{(2)}\|\right), \quad \alpha = \begin{cases} -1 & \text{if no interpenetration} \\ 1 & \text{otherwise} \end{cases}. \tag{7.116}$$

The value of $\mathbf{x}^{(2)}$ that minimizes $\|\mathbf{x}^{(1)} - \mathbf{x}^{(2)}\|$ for a given $\mathbf{x}^{(1)}$, denoted as $\overline{\mathbf{x}}^{(2)}$, is the closest projection point of $\mathbf{x}^{(1)}$ onto $\Gamma^{C(2)}$. The impenetrability conditions for normal contact can be formulated in the following Karush–Kuhn–Tucker conditions:

$$\begin{aligned} &g(\mathbf{x}) \le 0, \\ &t_n(\mathbf{x}) \ge 0, \\ &t_n(\mathbf{x})g(\mathbf{x}) = 0, \end{aligned} \tag{7.117}$$

where the gap function $g(\mathbf{x})$ (also called the nonpenetration function) is always nonpositive, and normal contact traction measured on the slave point is $\mathbf{t}_n(\mathbf{x}) = t_n(\mathbf{x})\overline{\mathbf{n}}$, where $\overline{\mathbf{n}}$ is an outward normal to the master surface $\Gamma^{C(2)}$ at $\overline{\mathbf{x}}^{(2)}$, and t_n is the magnitude of normal contact stress to be solved for. Equation (7.117) represents: (i) if there is no contact (the gap is not zero), the normal contact force has to be zero and (ii) the contact force has a nonzero value only when there is contact (the gap is zero).

In the tangential direction of $\overline{\mathbf{x}}^{(2)}$ on the master contact surface, the Karush–Kuhn–Tucker condition is expressed as

$$\begin{aligned} &f(\mathbf{x}) = \|\mathbf{t}_\tau(\mathbf{x})\| - \mu_c t_n(\mathbf{x}) \le 0, \\ &\zeta(\mathbf{x}) \ge 0, \\ &f(\mathbf{x})\zeta(\mathbf{x}) = 0, \end{aligned} \tag{7.118}$$

with the regularized classical Coulomb friction law:

$$\mathbf{v}_\tau\left(\overline{\mathbf{x}}^{(2)}\right) + \zeta \frac{\mathbf{t}_\tau\left(\overline{\mathbf{x}}^{(2)}\right)}{\left\|\mathbf{t}_\tau\left(\overline{\mathbf{x}}^{(2)}\right)\right\|} = -\frac{1}{\varepsilon_\tau}\dot{t}_{\tau_\alpha}\left(\overline{\mathbf{x}}^{(2)}\right)\overline{\boldsymbol{\tau}}^\alpha\left(\overline{\mathbf{x}}^{(2)}\right), \tag{7.119}$$

where μ_c is the friction coefficient, ε_τ is the tangential penalty, ζ is the contact consistency parameter, $\mathbf{t}_\tau$ is the tangential traction with $\mathbf{t}_\tau\left(\overline{\mathbf{x}}^{(2)}\right) = t_{\tau_\alpha}\overline{\boldsymbol{\tau}}^\alpha\left(\overline{\mathbf{x}}^{(2)}\right)$, t_{τ_α} is the αth component of $\mathbf{t}_\tau$, $\overline{\boldsymbol{\tau}}^\alpha$ is the αth component of the tangential vector of the master surface, and $\mathbf{v}_\tau$ is the tangential relative velocity between the slave point and its corresponding master surface with

$$\mathbf{v}_\tau\left(\overline{\mathbf{x}}^{(1)}\right) = \left(\frac{\partial \mathbf{X}^{(2)}}{\partial \xi_\alpha} \cdot \frac{\partial \mathbf{X}^{(2)}}{\partial \xi_\beta}\right)\Bigg|_{\xi = \overline{\xi}} \dot{\overline{\xi}}_\beta \overline{\boldsymbol{\tau}}^\alpha, \tag{7.120}$$

where $\overline{\xi}_\beta$ denotes the βth component of the local parametric coordinate $\overline{\boldsymbol{\xi}}$ of the closest projection point on the master contact surface $\Gamma^{C(2)}$ that satisfies

$$\left[\mathbf{x}^{(1)} - \mathbf{x}^{(2)}\left(\overline{\boldsymbol{\xi}}\right)\right] \cdot \overline{\boldsymbol{\tau}}_\alpha = 0. \tag{7.121}$$

Note that $\overline{\boldsymbol{\tau}}_\alpha = \partial \mathbf{x}^{(2)}/\partial \xi_\alpha\big|_{\xi = \overline{\xi}}$, and $\overline{\boldsymbol{\tau}}^\alpha = \sum_\beta \overline{\boldsymbol{\tau}}_\beta / \left(\overline{\boldsymbol{\tau}}_\alpha \cdot \overline{\boldsymbol{\tau}}_\beta\right)$. Imposing the normal contact constraints by introducing a normal penalty ε_n to the potential energy function $\Pi(\mathbf{u})$, and considering the external work due to friction, the variation of the modified potential energy function for flexible-to-flexible body contact is

$$\delta\overline{\Pi}\left(\mathbf{u}^{(1)}, \mathbf{u}^{(2)}\right) = \delta\Pi\left(\mathbf{u}^{(1)}\right) + \delta\Pi\left(\mathbf{u}^{(2)}\right) + \int_{\Gamma^{C(1)}} \left[\varepsilon_n < g > \delta g - \sum_\alpha t_{\tau_\alpha}^{(1)} \overline{\boldsymbol{\tau}}^\alpha \cdot \left(\delta\mathbf{u}^{(1)} - \delta\mathbf{u}^{(2)}\right)\right] d\Gamma = 0, \tag{7.122}$$

where $\langle \cdot \rangle$ is the Macaulay bracket representing the positive part of its operand, $\Pi(\mathbf{u}^{(1)})$ and $\Pi(\mathbf{u}^{(2)})$ are the internal energy corresponding to body 1 and body 2, respectively, and $\mathbf{u}^{(1)}$ and $\mathbf{u}^{(2)}$ are the displacements of material points corresponding to body 1 and body 2, respectively.

With consideration of the parametric coordinate of the closest projection point $\overline{\boldsymbol{\xi}}$ and kinematic relationships, (7.122) can be reduced to

$$\delta\overline{\Pi}\left(\mathbf{u}^{(1)}, \mathbf{u}^{(2)}\right) = \delta\Pi\left(\mathbf{u}^{(1)}\right) + \delta\Pi\left(\mathbf{u}^{(2)}\right) + \delta\Pi_c\left(\mathbf{u}^{(1)}, \mathbf{u}^{(2)}\right) = 0, \tag{7.123}$$

$$\delta\Pi_c\left(\mathbf{u}^{(1)}, \mathbf{u}^{(2)}\right) = \int_{\Gamma^{C(1)}} \left[\varepsilon_n < g\left(\overline{\boldsymbol{\xi}}\right) > \delta g\left(\overline{\boldsymbol{\xi}}\right) - \sum_\alpha t_{\tau_\alpha}^{(1)} \delta\overline{\xi}_\alpha\right] d\Gamma. \tag{7.124}$$

The linearization of (7.123) for Newton iteration is

$$\Delta\left(\delta\Pi\left(\mathbf{u}^{(1)}\right)\right) + \Delta\left(\delta\Pi\left(\mathbf{u}^{(2)}\right)\right) + \Delta\left(\delta\Pi_c\left(\mathbf{u}^{(1)}, \mathbf{u}^{(2)}\right)\right) = -\left\{\delta\Pi\left(\mathbf{u}^{(1)}\right) + \delta\Pi\left(\mathbf{u}^{(2)}\right) + \delta\Pi_c\left(\mathbf{u}^{(1)}, \mathbf{u}^{(2)}\right)\right\}_m^{\nu}, \tag{7.125}$$

where m and ν denote load step and iteration counters, respectively. It can be shown that the evaluation of $\delta\Pi_c(\mathbf{u}^{(1)}, \mathbf{u}^{(2)})$ requires the evaluation of the first- and second-order derivatives of the contact surface function $\mathbf{x}(\boldsymbol{\xi})$ with respect to $\boldsymbol{\xi}$, such as $\mathbf{x}^{(2)}_{,\xi_\alpha}$ and $\mathbf{x}^{(2)}_{,\xi_\alpha\xi_\beta}$, and the evaluation of $\Delta\delta\Pi_c(\mathbf{u}^{(1)}, \mathbf{u}^{(2)})$ requires calculation of $\mathbf{x}^{(2)}_{,\xi_\alpha}$, $\mathbf{x}^{(2)}_{,\xi_\alpha\xi_\beta}$, and $\mathbf{x}^{(2)}_{,\xi_\alpha\xi_\beta\xi_\gamma}$. A C^2 contact surface representation and C^2 displacement approximation are desirable to ensure accurate computation of contact force vectors and contact stiffness matrices. This necessitates the shape functions used to approximate contact surfaces and its displacement field to be at least C^2 continuous, which can be achieved by introducing the MLS or RK approximation, see [36] for details.

7.6.2 Meshfree Smooth Curve Representation

To meet the continuity condition required by a consistent contact formulation, the meshfree kernel function can be selected to have at least C^2 continuity, such as a C^2 cubic B-spline. Consider a contact curve described by M discrete points as shown in Figure 7.15. A parametric coordinate system is defined as

$$\xi_I = \sum_{L=1}^{I-1} l_L \Big/ \sum_{L=1}^{M-1} l_L;\ l_L = \|\mathbf{X}_{L+1} - \mathbf{X}_L\|. \tag{7.126}$$

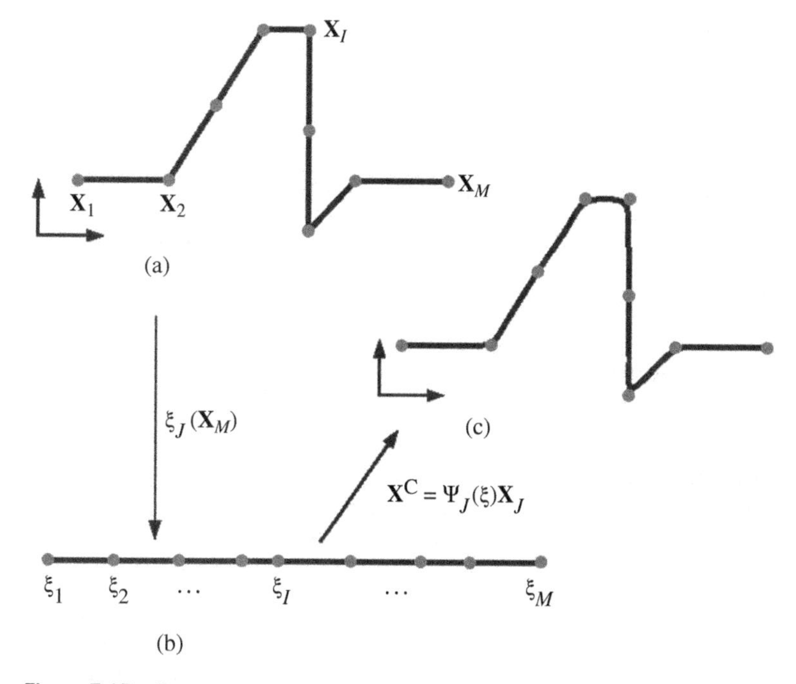

Figure 7.15 Contact curve representation by RK shape functions: (a) Master curve in spatial domain, (b) parametric discretization, and (c) approximation in spatial domain. *Source:* Wang et al. [36], figure 3 p. 156 / Springer Nature.

Local RK shape functions can be constructed using the same procedures discussed in Section 3.2.2 by satisfying nth order consistency in the parametric coordinate, i.e.,

$$\sum_{I=1}^{M} \Psi_I^{\mathrm{C}}(\xi)\xi_I^i = \xi^i, \quad 0 \leq i \leq n, \tag{7.127}$$

where M is the number of contact points. Subsequently, a set of local RK shape functions $\{\Psi_I^{\mathrm{C}}\}_{I=1}^{M}$ is constructed in the parametric domain

$$\Psi_I^{\mathrm{C}}(\xi) = \mathbf{H}^{\mathrm{T}}(0)\mathbf{M}^{-1}(\xi)\mathbf{H}(\xi - \xi_I)\phi_a(\xi - \xi_I), \tag{7.128}$$

$$\mathbf{H}(\xi - \xi_I) = \begin{bmatrix} 1 & \xi - \xi_I & \cdots & (\xi - \xi_I)^n \end{bmatrix}^{\mathrm{T}}, \tag{7.129}$$

$$\mathbf{M}(\xi) = \sum_{I=1}^{M} \mathbf{H}(\xi - \xi_I)\mathbf{H}^{\mathrm{T}}(\xi - \xi_I)\phi_a(\xi - \xi_I). \tag{7.130}$$

Employing this set of smooth local RK shape functions in the representation of the contact curve, we have

$$\mathbf{x}^{\mathrm{R}}(\xi) = \sum_{I=1}^{M} \Psi_I^{\mathrm{C}}(\xi)\mathbf{x}_I^{\mathrm{C}}, \tag{7.131}$$

where $\mathbf{x}_I^{\mathrm{C}}$ are the coordinates of the nodes defining the contact surface.

A similar approach can be applied to two-dimensional contact surfaces using two-dimensional parametric coordinates $\boldsymbol{\xi} = (\xi_1, \xi_2)$ and a set of RK shape functions $\{\Psi_I^{\mathrm{C}}(\boldsymbol{\xi})\}_{I=1}^{M}$ constructed for all contact points in the parametric coordinates of the contact surface, which satisfies the nth order reproducing conditions in the parametric coordinate:

$$\sum_{I=1}^{M} \Psi_I^{\mathrm{C}}(\boldsymbol{\xi})\xi_{1I}^i\xi_{2I}^j = \xi_1^i\xi_2^j, \quad 0 \leq i + j \leq n, \tag{7.132}$$

where $\{\xi_{1I}, \xi_{2I}\}_{I=1}^{M}$ can be obtained from CAD models.

The continuum form of the contact formulation in multiple dimensions requires the computation of the Jacobian components:

$$\frac{\partial \mathbf{x}^{\mathrm{R}}(\boldsymbol{\xi})}{\partial \xi_\alpha} = \sum_{I=1}^{M} \frac{\partial \Psi_I^{\mathrm{C}}(\boldsymbol{\xi})}{\partial \xi_\alpha}\mathbf{x}_I^{\mathrm{C}}. \tag{7.133}$$

Equation (7.131) is the meshfree smooth contact curve representation. The representation is exact for a line when a linear monomial basis is used in the shape function. The smoothness of the kernel function determines the smoothness of the representation. The geometric properties of the approximated contact curve are also influenced by the support size of the kernel functions. Curves represented by RK approximations with a cubic B-spline kernel function and various kernel support sizes are shown in Figure 7.16. As illustrated in this figure, an RK approximation similar to a piecewise linear approximation can be constructed when needed, by using a support size close to the distance of the adjacent particle. As shown in Figure 7.16, increasing the normalized support size provides smoothing at kinks. An approximation using the Gaussian kernel function with linear basis is shown in Figure 7.17, which provides quite smooth representations including derivatives. Figure 7.18 shows that a circular contact curve is well represented by RK shape functions constructed using a cubic B-spline kernel function with quadratic basis with a sufficiently large

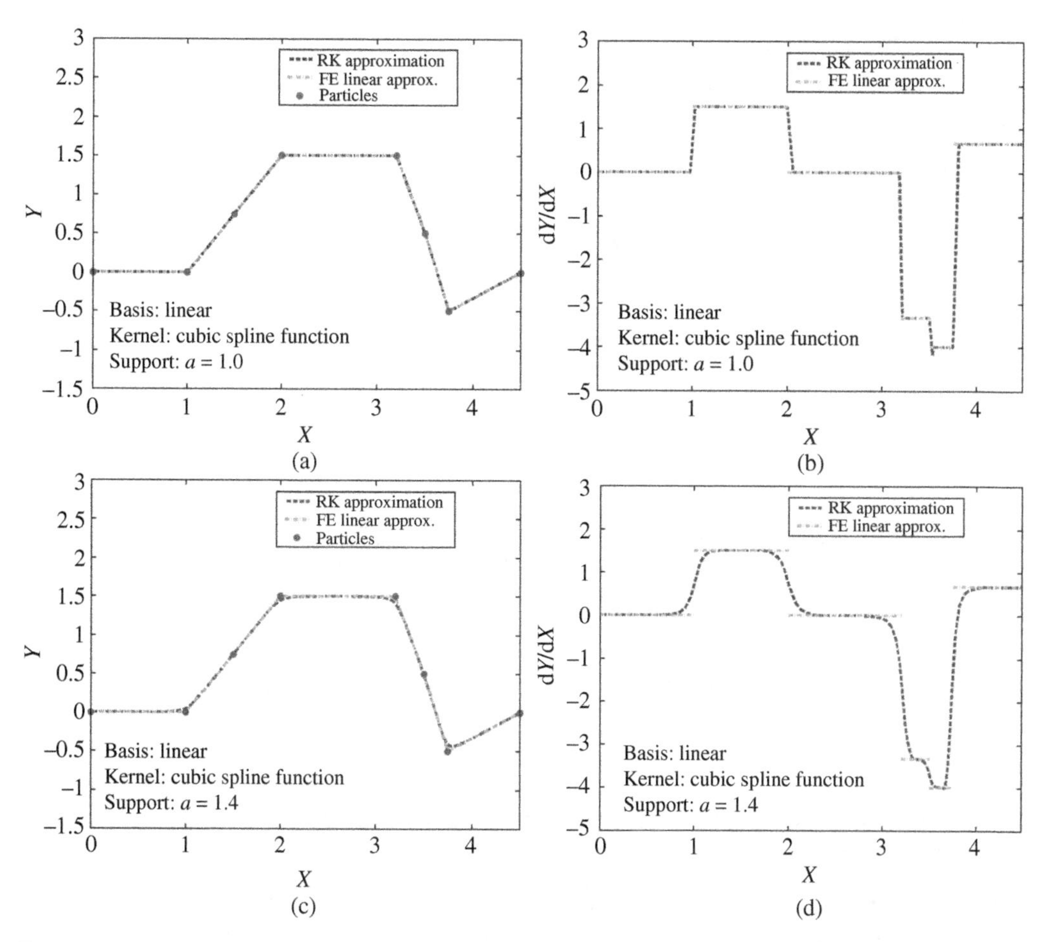

Figure 7.16 RK Approximation: Cubic B-spline with linear basis: (a) The approximated curves with $a = 1.0$, (b) the slope of the curves with $a = 1.0$, (c) The approximated curves with $a = 1.4$, (d) the slope of the curves with $a = 1.4$. *Source:* Wang et al. [36], figure 6 p. 157 / Springer Nature.

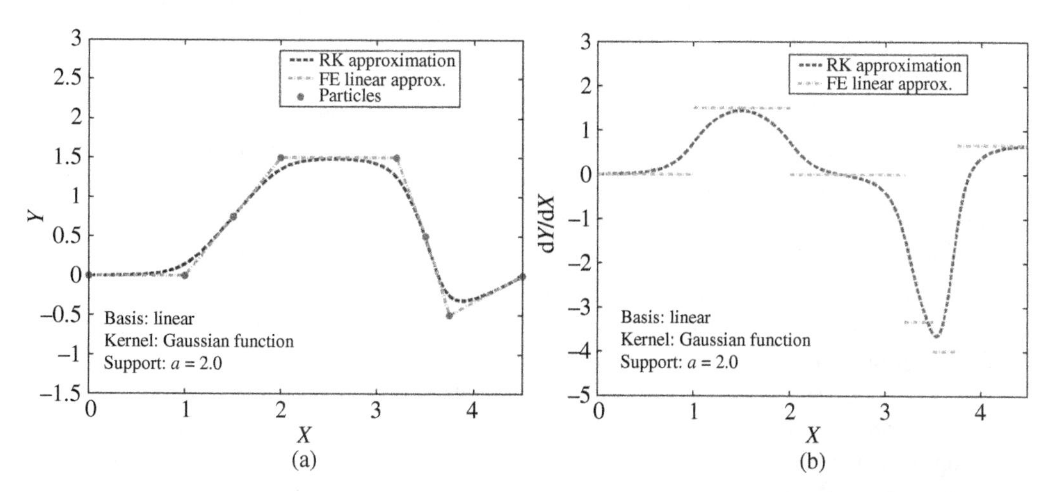

Figure 7.17 RK Approximation: Gaussian function with linear basis: (a) The approximated curves with $a = 2.0$, (b) the slope of the curves with $a = 2.0$. *Source:* Adapted from Chen et al. [37], figure 21 p. 38.

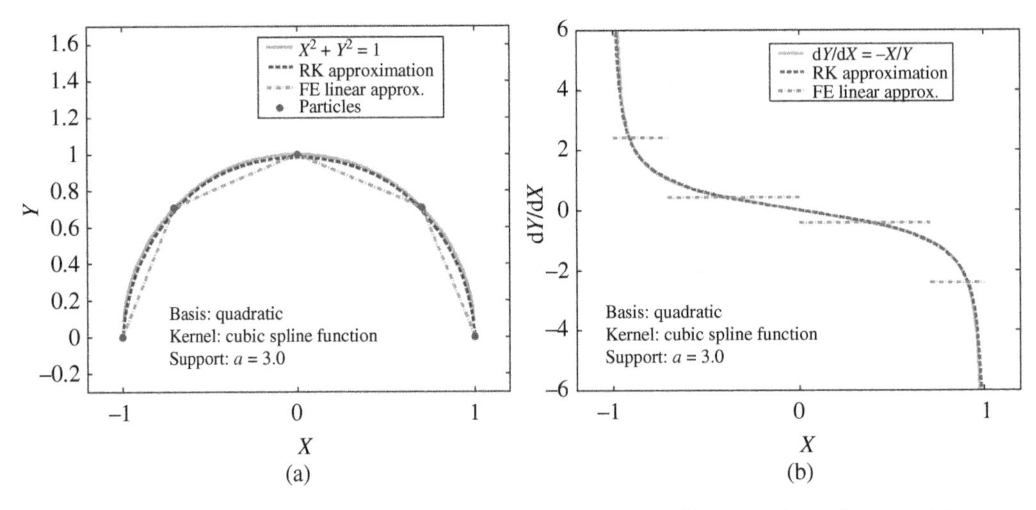

Figure 7.18 RK Approximation: Cubic B-spline with quadratic basis: (a) The approximated curves with $a = 3.0$, (b) the slope of the curves with $a = 3.0$. *Source:* Reproduced with permission from Wang et al. [36], figure 4 p. 157 / Springer Nature.

normalized support ($a/h \geq 2.0$). Note that a very coarse discretization is used in this curve representation.

7.6.3 Three-Dimensional Meshfree Smooth Contact Surface Representation and Contact Detection by a Nonparametric Approach

While parametric coordinates can be conveniently used for a smooth contact surface construction using RK shape functions in parametric coordinates, smooth contact surface representation and contact detection can be formulated directly in the physical coordinates, as will be discussed in this section. Consider two contacting bodies discretized by a set of points in three dimensions. One body is chosen as the slave body and the other is the master body. The master body is the target that a slave point cannot penetrate. The master surfaces are predefined by discrete points in the model. To identify the inward direction of the master surface, points in the interior of the master body are employed as reference points. To illustrate the method, consider a slave point $\mathbf{x}^{\text{sla}}$, a group of master points $\left\{\mathbf{x}_I^{\text{mas}}\right\}_{I=1}^{N_{\text{mas}}}$, and a group of reference points $\left\{\mathbf{x}_I^{\text{ref}}\right\}_{I=1}^{N_{\text{ref}}}$, located inside the master surface. A moving spherical (or circular in 2D) window is defined and centered on each slave point as shown in Figure 7.19. The window radius is selected large enough to contain a predefined $\overline{N}$ number of master points. The $\overline{N}$ master points located inside the moving window are called "the local smooth surface control points," which are used to define a local smooth contact surface using a reproducing kernel approximation. Generally, $\overline{N}$ should be chosen to be large enough so that during the iteration at each load step, the slave point will not slide outside of the master surface represented by $\overline{N}$ number of control points within the moving window.

Once a set of local smooth surface control points is collected by the moving window of each slave point, detection of potential contact is performed by employing a hierarchy-territory contact-searching algorithm [37]. Using this approach, a contact territory, T, is defined based on the current positions of the local smooth surface control points, $\mathbf{x}(x_1, x_2, x_3)$, as

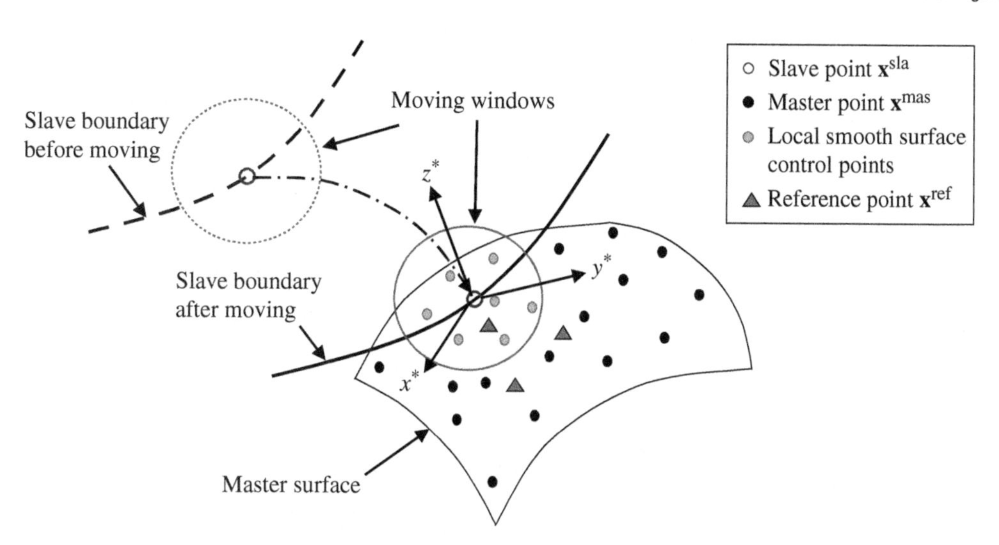

Figure 7.19 Illustration of a moving window in three-dimensional contact: the dashed circle stands for the original window position before moving; the solid circle denotes the new window position after moving. *Source:* Reproduced with permission from Wang et al. [36], figure 7 p. 158 / Springer Nature.

$$T = \left\{(x_1,\ x_2,\ x_3) \mid x_i^a - E \le x_i \le x_i^b + E,\ i = 1,\ 2,\ 3\right\}, \tag{7.134}$$

where

$$x_i^a = \min\left(x_{iI}, \text{ for } I = 1, 2, \ldots, \overline{N}\right) \text{ for } i = 1, 2, 3, \tag{7.135}$$

$$x_i^b = \max\left(x_{iI}, \text{ for } I = 1, 2, \ldots, \overline{N}\right) \text{ for } i = 1, 2, 3, \tag{7.136}$$

$$E = \max\left(\Delta u^J \mid \Delta u^J = \sqrt{(\Delta u_1^J)^2 + (\Delta u_2^J)^2 + (\Delta u_3^J)^2}\right) \text{ for } J = 1, 2, \ldots, n. \tag{7.137}$$

Here $\Delta u_i^J, i = 1, 2, 3$, is the incremental displacement of a slave point at the Jth load step. Figure 7.20 illustrates the definition of the contact territory in two dimensions. Only points contained in the territory T are considered potential contact points.

When $\mathbf{x}^{sla} \subset \mathrm{T}$, the local contact surface must be mathematically described so that the closest projection of the slave point onto the local contact surface can be determined, and consequently, the contact conditions evaluated. For surface construction, a set of local reproducing kernel shape functions are used based on the geometric information of the local smooth surface control points. To avoid a multivalued condition in the construction of the local contact surface represented in the

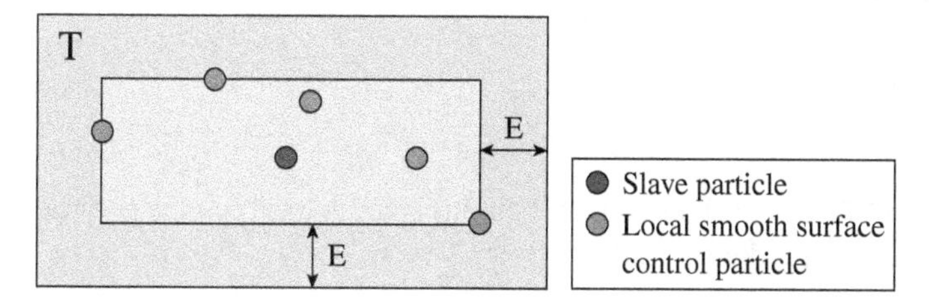

Figure 7.20 Illustration of a contact territory. *Source:* Adapted from Wang et al. [36], figure 8 p. 158 / Springer Nature.

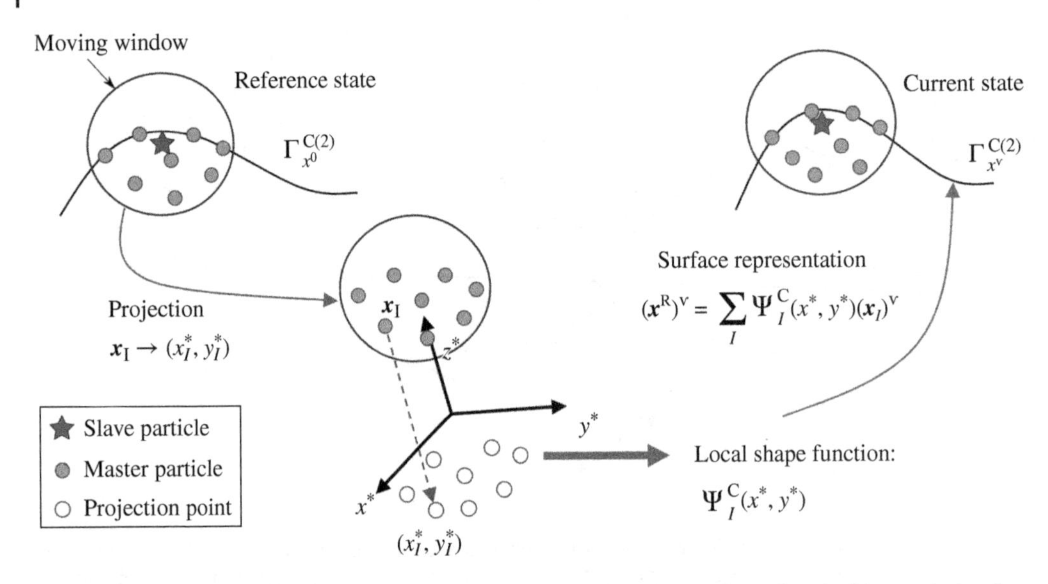

Figure 7.21 Illustration of local contact surface representation. *Source:* Reproduced with permission from Wang et al. [36], figure 9 p. 159 / Springer Nature.

form of $z = z(x, y)$, a projection plane, $a_1x + a_2y + a_3z = a_4$, is determined by the least-squares fit to the local smooth surface control points. To avoid nonuniqueness of a_i in the least-square minimization, the projection plane function is normalized with respect to a_4. Note that a_4 determines the distance of the projection plane to the origin of the coordinates, and it has no influence on the projection of local smooth control points. A local coordinate system (x^*, y^*, z^*) is then defined to construct a local contact surface in the form of $\mathbf{x} = \mathbf{x}(x^*, y^*)$, and the $x^* - y^*$ plane is parallel to $a_1x + a_2y + a_3z = 1$, which serves as a projection plane. Therefore, (x^*, y^*) is the projection of (x, y, z) onto the $x^* - y^*$ plane. Figure 7.21 illustrates the overall process of smooth contact surface representation (see [36] for more details).

With the set of projected nodal coordinates $\{x_I^*, y_I^*\}_{I=1}^{\overline{N}}$, a set of local RK shape functions $\{\Psi_I^C(x^*, y^*)\}_{I=1}^{\overline{N}}$ can be constructed. Consequently, the contact surface is reproduced, and the associated tangent vectors are obtained as

$$\mathbf{x}^R(x^*, y^*) = \sum_{I=1}^{\overline{N}} \Psi_I^C(x^*, y^*)\mathbf{x}_I, \tag{7.138}$$

$$\begin{aligned} \boldsymbol{\tau}_1 &= \frac{\partial \mathbf{x}^R}{\partial x^*} = \sum_{I=1}^{\overline{N}} \frac{\partial \Psi_I^C(x^*, y^*)}{\partial x^*}\mathbf{x}_I, \\ \boldsymbol{\tau}_2 &= \frac{\partial \mathbf{x}^R}{\partial y^*} = \sum_{I=1}^{\overline{N}} \frac{\partial \Psi_I^C(x^*, y^*)}{\partial y^*}\mathbf{x}_I. \end{aligned} \tag{7.139}$$

The closest projection point $\overline{\mathbf{x}} = \mathbf{x}^R(\overline{x}^*, \overline{y}^*)$ of slave point $\mathbf{x}^{sla}$ on this smooth local contact surface can be found by solving the following equations iteratively:

$$\begin{aligned} \left(\mathbf{x}^{sla} - \mathbf{x}^R(\overline{x}^*, \overline{y}^*)\right)\cdot\boldsymbol{\tau}_1(\overline{x}^*, \overline{y}^*) &= 0, \\ \left(\mathbf{x}^{sla} - \mathbf{x}^R(\overline{x}^*, \overline{y}^*)\right)\cdot\boldsymbol{\tau}_2(\overline{x}^*, \overline{y}^*) &= 0. \end{aligned} \tag{7.140}$$

The closest projection point $\overline{\mathbf{x}}$ of the slave point on the reproduced contact surface is obtained by

$$\overline{\mathbf{x}} = \sum_{I=1}^{\overline{N}} \Psi_I^{\mathrm{C}}(\overline{x}^*, \overline{y}^*)\mathbf{x}_I. \tag{7.141}$$

Once the closest projection point $\overline{\mathbf{x}}$ of the slave point on the reproduced contact surface is obtained, the master contact surface tangent and normal vectors at the closest projection point $\overline{\mathbf{x}}$ can be found as

$$\begin{aligned}
\overline{\boldsymbol{\tau}}_1 &= \sum_{I=1}^{N_s} \Psi_{I,x^*}^{\mathrm{C}}(\overline{x}^*, \overline{y}^*)\mathbf{x}_I, \\
\overline{\boldsymbol{\tau}}_2 &= \sum_{I=1}^{N_s} \Psi_{I,y^*}^{\mathrm{C}}(\overline{x}^*, \overline{y}^*)\mathbf{x}_I, \\
\overline{\mathbf{n}} &= \overline{\boldsymbol{\tau}}_1 \times \overline{\boldsymbol{\tau}}_2 / \|\overline{\boldsymbol{\tau}}_1 \times \overline{\boldsymbol{\tau}}_2\|.
\end{aligned} \tag{7.142}$$

The normal gap function g, the tangential friction law and regularization functions can then be computed. By introducing the closest projection point coordinate $\overline{\boldsymbol{\xi}} = \left[\overline{\xi}_1, \overline{\xi}_2\right]^{\mathrm{T}} = \left[\overline{x}^*, \overline{y}^*\right]^{\mathrm{T}}$ in the evaluation of the contact contributions $\delta\Pi_c(\mathbf{u}^{(1)}, \mathbf{u}^{(2)})$ and $\Delta(\delta\Pi_c(\mathbf{u}^{(1)}, \mathbf{u}^{(2)}))$ in the variational Eqs. (7.123)–(7.125), the incremental discrete equations for contact mechanics can then be obtained, see [36] for details.

In this continuum contact formulation, linearization of the penalized variational equation that accounts for impenetration and stick-slip conditions requires up to third-order derivatives of the contact surface function. A C^2 contact surface representation and displacement approximation are required to ensure a continuous contact force vector and obtain accurate contact stiffness matrices. This can be achieved using RK shape functions with at least C^2 continuity of the kernel functions (such as the cubic B-spline).

Now consider the following example: a 3D elastoplastic cylindrical billet is extruded through a rigid circular die (a quarter model) as shown in Figure 7.22. This problem is used to verify the performance of the meshfree contact formulation in dealing with three-dimensional contact with large sliding and large deformations. Two analyses are performed, one using a C^0 representation of the contact surface and the other with a smooth contact surface representation. The comparison of residual norms from both analyses is shown in Table 7.4, where the analysis with a piecewise (C^0) contact formulation diverges due to the oscillations in the residual norms, while the smooth contact analysis demonstrates good convergence.

7.7 Natural Kernel Contact Algorithm[4]

In extreme deformation problems with material separation, contact surfaces are unknown and are part of the solution. As a consequence, the conventional algorithms, in which all possible contacting surfaces are defined *a priori*, are ineffective in modeling such problems. A natural kernel contact (NKC) algorithm to approximate the contact condition without relying on the predefined surfaces at the preprocessing stage has been proposed [10] to address this difficult issue. The idea of NKC emanates from the inherent property of the semi-Lagrangian RK shape functions $\Psi_I(\mathbf{x})$ in (7.54), where the overlap between the kernel supports associated with particles from different bodies naturally serves as the impenetration condition, see Figure 7.23. By requiring the partition

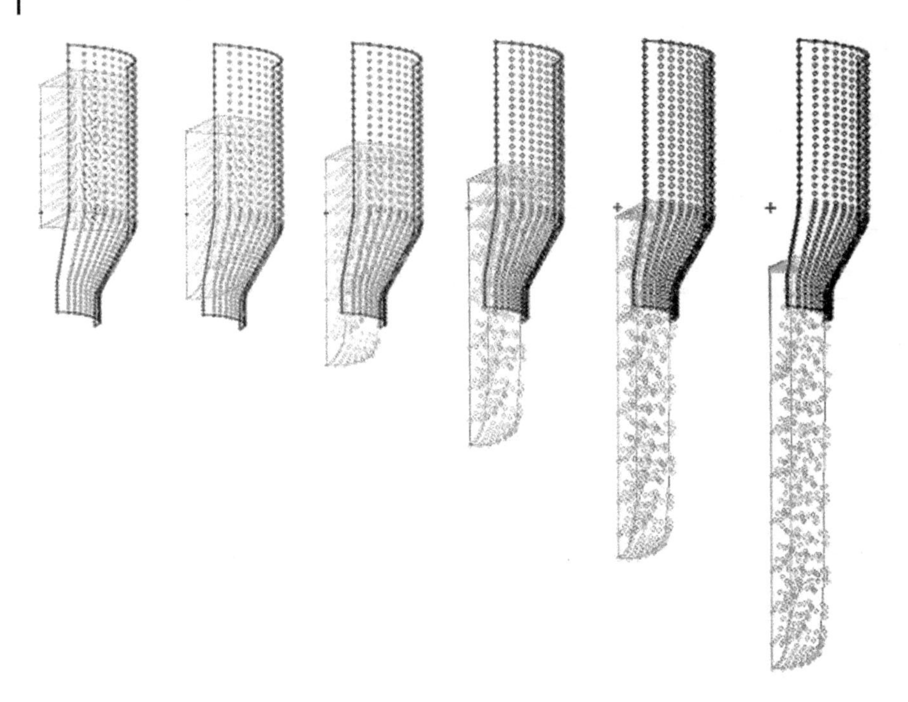

Figure 7.22 Progessive deformation of metal extrusion. *Source:* Reproduced with permission from Wang et al. [36], figure 22 p. 167 / Springer Nature.

Table 7.4 Comparison of residual norms at stage of 1.25%.

Iteration	Smooth contact	Piecewise contact
2	.19510E+01	.18000E+00
3	.31991E-01	.51980E-01
4	.90151E-02	.28985E+00
5	.78183E-04	.49532E-01
6		.39752E+00
7		.21982E-01
8		.48872E-01
9		.29390E+00
10		.49767E-01
11		.36867E+00

Source: Wang et al. [36], table 4, p. 167 / Springer Nature.

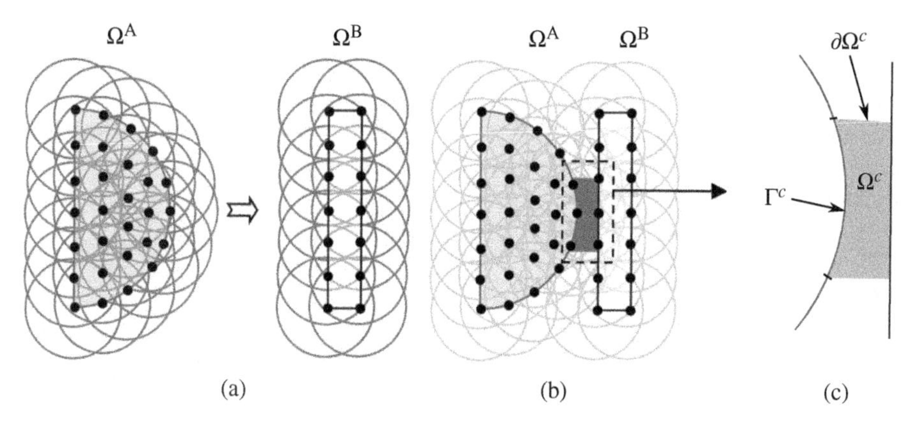

Figure 7.23 Natural kernel contact algorithm by kernel interaction between contacting bodies: (a) initial configuration, (b) deformed configuration, (c) kernel contact region. *Source:* Chen et al. [38], figure 24 p. 39 / John Wiley & Sons.

of unity condition at all particle locations of each of the contacting bodies, the overlap between the semi-Lagrangian RK shape functions induces internal forces between particles, ensuring the impenetrability between different bodies. A layer of a friction-like elastoplastic material occupying Ω^c, as shown in Figure 7.23c, has been introduced in the "contact processing zone" to mimic the friction law [10].

7.7.1 A Friction-like Plasticity Model

In the NKC, a friction-like material is introduced between contacting bodies, with domain Ω^c, as shown in Figure 7.23c, to mimic the frictional contact conditions. Based on the analogy between Coulomb's friction law and the elastoplasticity flow rule, the variational contact equation leads to the constitutive equation governing the stress–strain relationship of the friction-like material, such that Coulomb's friction is recovered [10]:

$$\boldsymbol{\tau}^c \cdot \mathbf{n} = t_N \mathbf{n} + \mathbf{t}_T, \tag{7.143}$$

where $\boldsymbol{\tau}^c$ is the Cauchy stress in Ω^c, $\mathbf{n}$ is the unit outward normal of the contact surface Γ^c, t_N is the normal component of the contact traction, and $\mathbf{t}_T$ is the tangential contact traction. Equation (7.143) indicates that the stresses in the friction-like material are in balance with t_N and $\mathbf{t}_T$ on Γ^c.

Therefore, an elastic perfectly plastic material, in which the stress $\boldsymbol{\tau}^c$ in Ω^c obeys (7.143), can be introduced into the contact processing zone to mimic Coulomb's friction law. To obtain $\boldsymbol{\tau}_c$, consider the following yield function and the associated Karush–Kuhn–Tucker conditions based on local coordinates where the 1-direction is aligned with the contact surface normal $\mathbf{n}$:

$$f(\boldsymbol{\chi}) = \|\boldsymbol{\chi}\| + \mu\hat{\tau}_{11} \leq 0, \tag{7.144}$$

$$\dot{\mathbf{e}} = \gamma \frac{\partial f}{\partial \boldsymbol{\chi}}, \tag{7.145}$$

$$\gamma \geq 0, \tag{7.146}$$

$$\gamma f = 0, \tag{7.147}$$

where $\boldsymbol{\chi} = [\hat{\tau}_{12} \quad \hat{\tau}_{13}]$, $\hat{\tau}_{11} \leq 0$ is the normal contact stress, $\dot{\mathbf{e}}$ is the tangential strain rate, and $\hat{\boldsymbol{\tau}} = \mathbf{L}\boldsymbol{\tau}\mathbf{L}^{\mathrm{T}}$ is the Cauchy stress tensor $\boldsymbol{\tau}$ rotated to the local coordinate system, the 2- and 3-directions are aligned with two mutually orthogonal unit vectors, $\mathbf{p}$ and $\mathbf{q}$, and $\mathbf{L} = [\mathbf{n}, \ \mathbf{p}, \ \mathbf{q}]^{\mathrm{T}}$. It is assumed that the normal contact stress $\hat{\tau}_{11}$ is known in (7.144). The yield stress $\mu|\hat{\tau}_{11}|$ mimics the frictional stress induced by the normal stress $\hat{\tau}_{11}$, and the slip condition is represented by the yield condition in the plasticity model:

$$
\begin{aligned}
&f < 0, \quad \text{stick condition,} && \text{(elastic)} \\
&f = 0, \ \|\boldsymbol{\chi}\| = -\mu\hat{\tau}_{11} \quad \text{slip condition.} && \text{(plastic)}
\end{aligned}
\tag{7.148}
$$

This approach can be implemented by a predictor-corrector algorithm, in which the stresses calculated based on the overlapping supports of the contacting bodies are obtained in the predictor step, and in the corrector step the tangential stresses are corrected according to Eqs. (7.144)–(7.147) with $\hat{\tau}_{11}$ fixed. To enhance the iteration convergence of the two-step approach, we introduce the radial return algorithm where the trial is nonslip (elastic trial) and the violation of the yield function (interpenetration) is corrected by the return mapping algorithm. Following radial return mapping, the corrected contact stresses $\hat{\boldsymbol{\tau}}^{\mathrm{c}}$ in the local coordinate system induced by the friction-like elastoplasticity model can be obtained as

$$
\hat{\boldsymbol{\tau}}^{\mathrm{c}} = \hat{\boldsymbol{\tau}}^{\mathrm{trial}} + \lambda \begin{bmatrix} 0 & \hat{\tau}_{12}^{\mathrm{trial}} & \hat{\tau}_{13}^{\mathrm{trial}} \\ \hat{\tau}_{12}^{\mathrm{trial}} & 0 & 0 \\ \hat{\tau}_{13}^{\mathrm{trial}} & 0 & 0 \end{bmatrix} \equiv \hat{\boldsymbol{\tau}}^{\mathrm{trial}} + \lambda\hat{\boldsymbol{\xi}},
\tag{7.149}
$$

where $\hat{\boldsymbol{\tau}}^{\mathrm{trial}}$ is the Cauchy stress in the local coordinate system, calculated by a standard stress calculation through particle interaction without considering the artificial friction-like elasto-plastic material, and $\lambda = 0$ if $f(\boldsymbol{\chi}^{\mathrm{trial}}) < 0$ and

$$
\lambda = \frac{\mu|\hat{\tau}_{11}^{\mathrm{trial}}| - \|\boldsymbol{\chi}^{\mathrm{trial}}\|}{\|\boldsymbol{\chi}^{\mathrm{trial}}\|} \quad \text{if} \quad f(\boldsymbol{\chi}^{\mathrm{trial}}) \geq 0.
\tag{7.150}
$$

Finally, the corrected contact stresses in the global coordinates can then be obtained by the inverse transformation:

$$
\boldsymbol{\tau}^{\mathrm{c}} = \mathbf{L}^{\mathrm{T}}\hat{\boldsymbol{\tau}}^{\mathrm{c}}\mathbf{L} \equiv \boldsymbol{\tau}^{\mathrm{trial}} + \lambda\boldsymbol{\varsigma},
\tag{7.151}
$$

where

$$
\boldsymbol{\varsigma} = \left(\mathbf{n}\otimes\boldsymbol{\tau}^{\mathrm{trial}}\cdot\mathbf{n} + \mathbf{n}\cdot\boldsymbol{\tau}^{\mathrm{trial}}\otimes\mathbf{n}\right) - 2t_{\mathrm{N}}\mathbf{n}\otimes\mathbf{n}.
\tag{7.152}
$$

Here, the orthogonality of $\mathbf{L}$ is applied to derive the above relationship. Equation (7.151) can then be directly used to calculate the contact forces described in the next section.

7.7.2 Semi-Lagrangian RK Discretization and Natural Kernel Contact Algorithms

This section describes the semi-Lagrangian-RK discretization and the contact force calculation in the proposed NKC algorithms. Consider continuum bodies Ω^{A} and Ω^{B} (Figure 7.23) discretized by a group of points $G^{\mathrm{A}} = \{\mathbf{x}_I \,|\, \mathbf{x}_I \in \Omega^{\mathrm{A}}\}$ and $G^{\mathrm{B}} = \{\mathbf{x}_I \,|\, \mathbf{x}_I \in \Omega^{\mathrm{B}}\}$, respectively, with each point at $\mathbf{x}_I$ associated with a nodal volume V_I and a kernel function $\phi_a(\mathbf{x} - \mathbf{x}_I)$ with a support of radius a independent of material deformation. When the two contacting bodies Ω^{A} and Ω^{B} approach each other and the semi-Lagrangian-RK shape functions form a partition of unity [9, 10], the interaction between the RK points from different bodies (Figure 7.23) induces stresses:

$$\boldsymbol{\tau}(\mathbf{x}) = \sum_{I \in N^{\mathrm{A}} \cup N^{\mathrm{B}}} \mathbf{D}(\mathbf{x})\overline{\mathbf{B}}_I(\mathbf{x})\mathbf{u}_I, \tag{7.153}$$

where $N^{\mathrm{A}} = \{I \mid \mathbf{x}_I \in G^{\mathrm{A}}\}$, $N^{\mathrm{B}} = \{I \mid \mathbf{x}_I \in G^{\mathrm{B}}\}$, $\mathbf{D}$ is the material response tensor of contacting bodies, and $\overline{\mathbf{B}}_I$ is the gradient matrix of the shape functions, ideally constructed by stabilized conforming or nonconforming nodal integration (see Section 6.6). The contact stresses between contacting bodies are obtained by (7.153) when $\mathbf{n} \cdot \boldsymbol{\tau} \cdot \mathbf{n} \leq 0$ in Ω^{c}. With the nodal integration schemes described in Chapter 6, the internal force acting on a point I resulting from the contact strain energy $\Pi_{\mathrm{c}} = \frac{1}{2}\int_{\Omega^{\mathrm{c}}} \boldsymbol{\varepsilon}(\mathbf{u}^{\mathrm{c}}) : \boldsymbol{\tau}^{\mathrm{c}} d\Omega$ with $\mathbf{u}^{\mathrm{c}}$ and $\boldsymbol{\tau}^{\mathrm{c}}$ the displacement and stress in the contact domain Ω^{c}, respectively, can then be obtained by

$$\mathbf{f}_I^{\mathrm{int}} = \sum_{J \in N_I^{\mathrm{c}}} \overline{\mathbf{B}}_I^{\mathrm{T}}(\mathbf{x}_J)\boldsymbol{\tau}^{\mathrm{c}}(\mathbf{x}_J)V_J, \tag{7.154}$$

where $N_I^{\mathrm{c}} = \left\{J \mid J \in N^{\mathrm{A}} \cup N^{\mathrm{B}},\ \ \phi_{a_I}(\mathbf{x}_J - \mathbf{x}_I) \neq 0,\ \ \mathbf{r}_{IJ} \cdot \boldsymbol{\tau}^{\mathrm{c}}(\mathbf{x}_J) \cdot \mathbf{r}_{IJ} < 0,\ \ \mathbf{r}_{IJ} = (\mathbf{x}_I - \mathbf{x}_J)/\left\|\mathbf{x}_I - \mathbf{x}_J\right\|\right\}$ is the set that contains the neighbor points under the support of point I, and the contact stress between those points and point I is in compression. The pair-wise interactions due to overlapping kernel functions naturally prevent interpenetration between different bodies. An artificial layer of material with the friction-like dissipating mechanism in the form of plasticity in the previous Section 7.7.1 is introduced. With the consideration of the frictional contact effect, the interactive forces associated with point I are corrected as

$$\mathbf{f}_I^{\mathrm{int}} = \sum_{J \in N_I^{\mathrm{c}}} \overline{\mathbf{B}}_I^{\mathrm{T}}(\mathbf{x}_J)\boldsymbol{\tau}^{\mathrm{c}}(\mathbf{x}_J)V_J = \sum_{J \in N_I^{\mathrm{c}}} (\mathbf{f}_{IJ} + \mathbf{g}_{IJ}), \tag{7.155}$$

where $\mathbf{f}_{IJ} = \overline{\mathbf{B}}_I^{\mathrm{T}}(\mathbf{x}_J)\boldsymbol{\tau}^{\mathrm{trial}}(\mathbf{x}_J)V_J$, and $\mathbf{g}_{IJ} = \overline{\mathbf{B}}_I^{\mathrm{T}}(\mathbf{x}_J)\lambda\boldsymbol{\varsigma}(\mathbf{x}_J)V_J$.

One remaining issue in implementing the NKC algorithm in the semi-Lagrangian formulation is determining the contact surface and surface normal from a purely point-based discretization. A level set-based method was introduced in [10] to obtain the contact surface and surface normal under the NKC contact framework. The level set function was chosen as the interpolant of the material ID (e.g. choosing -1 and $+1$ and identifying the contour of zero) using the semi-Lagrangian RK approximation. The interested reader is referred to [10] for more details.

Consider the dynamic impact of an elastic ring on a rigid surface modeled by the natural kernel contact algorithm with a frictionless condition [10]. The comparisons between the incident and reflection angles, the linear momentum, and the total energy histories are shown in Figure 7.24 and demonstrate the accuracy and effectiveness of the natural kernel contact algorithm.

The NKC algorithms under the semi-Lagrangian RKPM framework [10] have been employed for simulating a fragment-impact penetration experiment [39], where the progressive penetration process by the RKPM calculation as well as the evolutionary contact surface between the projectile and concrete pane are shown in Figure 7.25, and the comparisons of the velocity history between the experimental data [39] and the numerical predictions with various levels of discretization are shown in Figure 7.26. As observed, the proposed approaches achieve good agreement between the experimental and numerical results. Damage patterns on both impact and exit faces of the concrete panel are compared in Figure 7.27, where surfaces of the experimentally measured holes and craters are shown in solid and dashed lines, respectively. Further, the experiments' damage patterns extending beyond the hole surfaces match well with the numerical calculation. Comparisons for

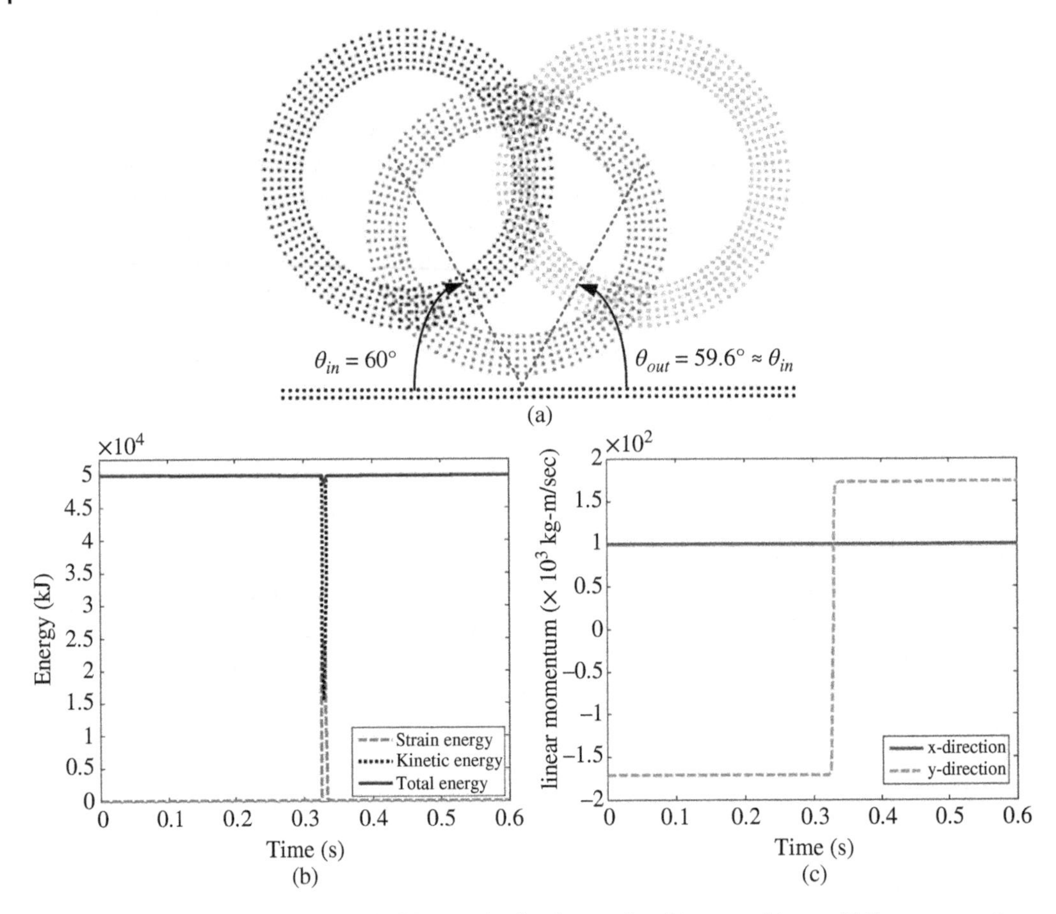

Figure 7.24 Elastic ring impact: (a) incident and reflection angles, (b) energy history, (c) linear momentum history. *Source:* Chi et al. [10], figure 15, p. 856 / With permission of John Wiley & Sons.

another set of experimental [41] numerical results using NKC for penetration of high-strength concrete are shown in Figure 7.28. The predicted hole and crater size and velocity reduction for several bullet grains and impact velocities match well.

Notes

1 For further discussions on the stress measures used in finite-deformation problems, see [12].
2 Section 7.4 includes content republished by permission from Elsevier: *Mechanics of Materials*, Semi-Lagrangian reproducing kernel formulation and application to modeling earth moving operations, Guan, Pai-Chen, et al., 2009.
3 Section 7.6 includes content republished by permission from Springer Nature: *Computational Mechanics*, A reproducing kernel smooth contact formulation for metal forming simulations, Hui-Ping Wang, et al., 2014.
4 Section 7.7 includes content republished by permission from John Wiley & Sons: *International Journal for Numerical Methods in Engineering*, A level set enhanced natural kernel contact algorithm for impact and penetration modeling, Chi, Sheng-Wei, et al., 2015.

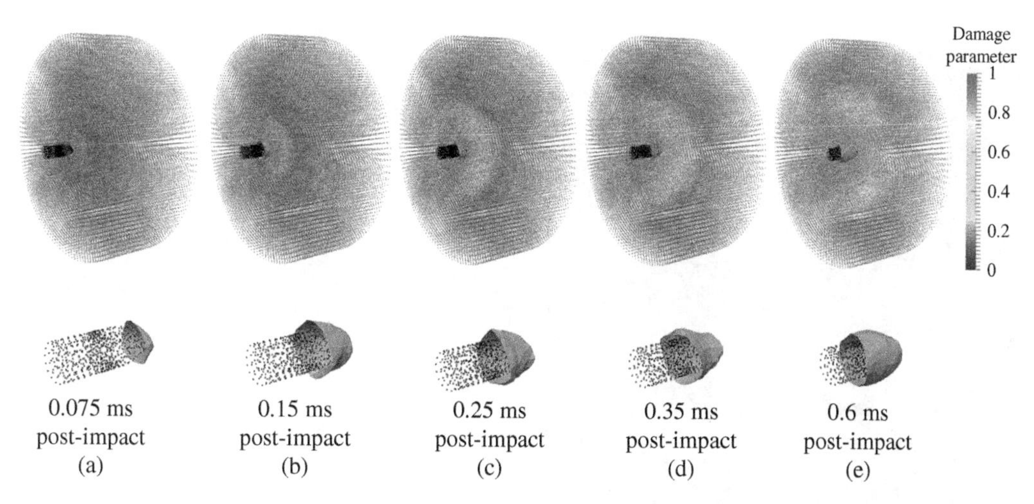

Figure 7.25 Progressive penetration processes and predicted contact surfaces by the RKPM semi-Lagrangian formulation with the NKC algorithms. (a) 0.075 ms, (b) 0.150 ms (c) 0.250 ms, (d) 0.350 ms, (e) 0.600 ms post-impact. *Source:* Chi et al. [10], figure 23, p. 861 / With permission of John Wiley & Sons.

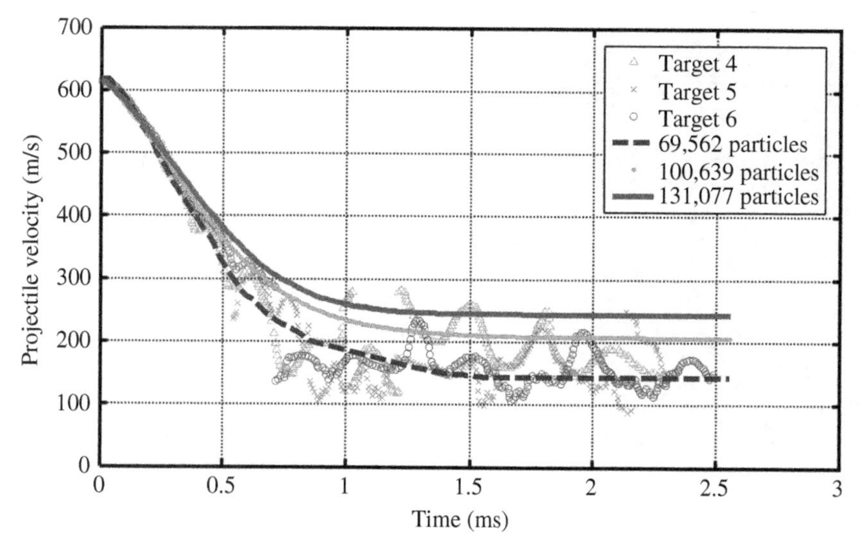

Figure 7.26 Comparison of the velocity history of the steel projectile between the numerical predictions and the experimental measurement [39]. *Source:* Chi et al. [10], figure 24, p. 862 / With permission of John Wiley & Sons.

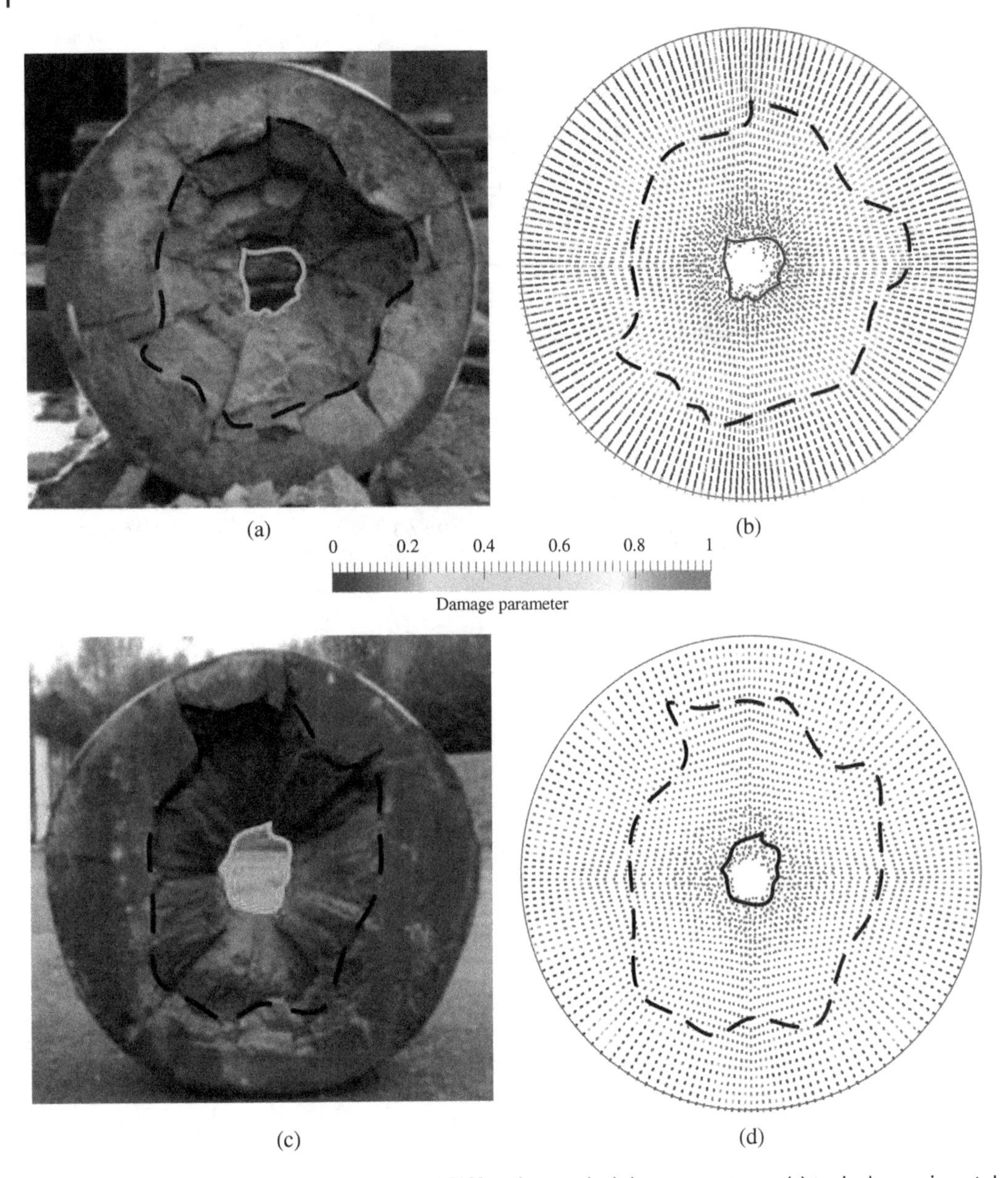

Figure 7.27 Comparison of the experimental [40] and numerical damage patterns: (a) typical experimental data (impact face), (b) RKPM numerical prediction (impact face), (c) typical experimental data (exit face), and (d) RKPM numerical prediction (exit face). *Source:* Chi et al. [10], figure 25, p. 862 / With permission of John Wiley & Sons.

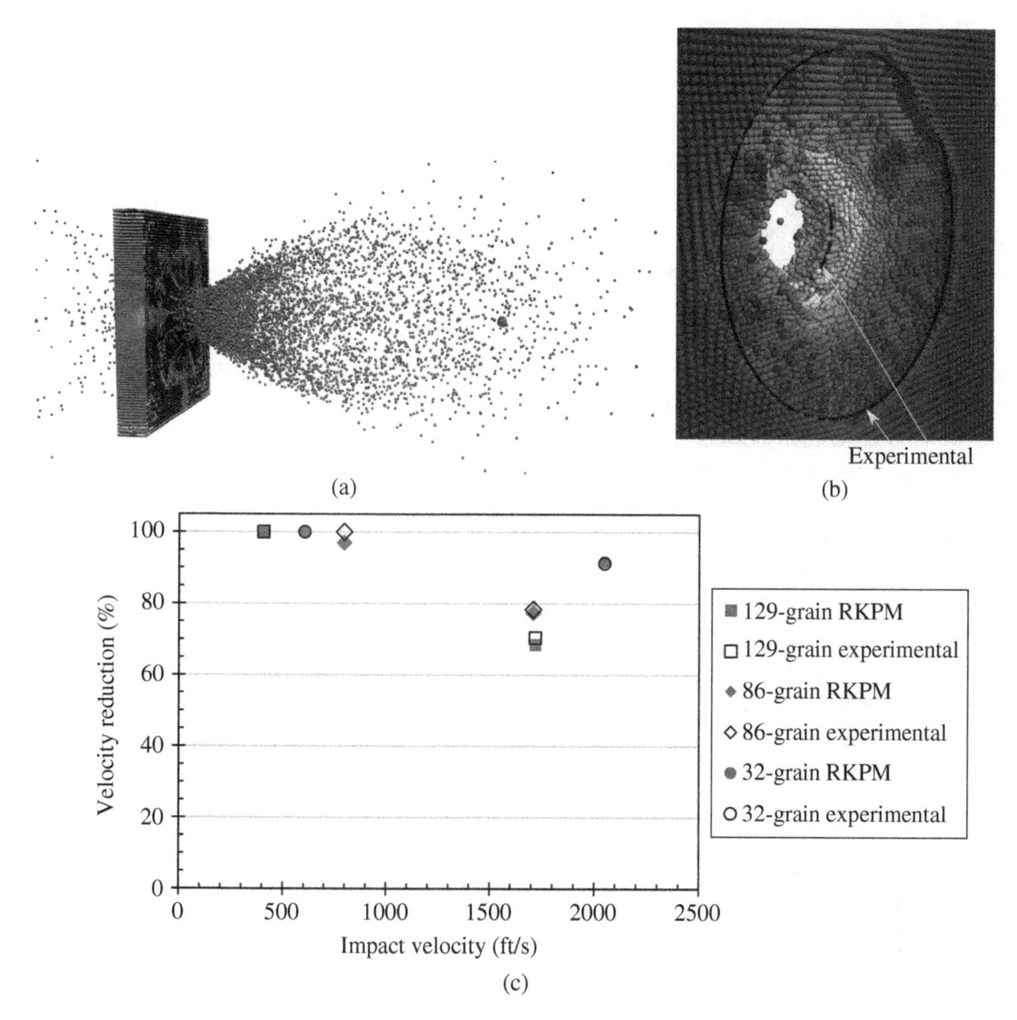

Figure 7.28 Comparison of the experimental [41] and numerical penetration tests for high-strength concrete: (a) typical numerical result, (b) RKPM numerical prediction compared with experimental measurements, (c) comparison of RKPM impact velocity with velocity reduction in experiments. *Source:* Reproduced with permission from Chen et al. [42], figure 25, p. 29 / With permission of ASCE.

References

1 Chen, J.-S., Pan, C., Wu, C.-T., and Liu, W.K. (1996). Reproducing kernel particle methods for large deformation analysis of non-linear structures. *Comput. Methods Appl. Mech. Eng.* 139 (1–4): 195–227.

2 Pasetto, M., Baek, J., Chen, J.S. et al. (2021). A Lagrangian/semi-Lagrangian coupling approach for accelerated meshfree modelling of extreme deformation problems. *Comput. Methods Appl. Mech. Eng.* 381: 113827.

3 Chen, J.-S. and Wu, C.-T. (1997). On computational issues in large deformation analysis of rubber bushings. *J. Struct. Mech.* 25 (3): 287–309.

4 Chen, J.-S., Wang, H.-P., Yoon, S., and You, Y. (2000). Some recent improvements in meshfree methods for incompressible finite elasticity boundary value problems with contact. *Comput. Mech.* 25: 137–156.

5 Chen, J.-S., Pan, C., Roque, C.M.O.L., and Wang, H.-P. (1998). A Lagrangian reproducing kernel particle method for metal forming analysis. *Comput. Mech.* 22 (3): 289–307.

6 Wu, C.-T., Chen, J.-S., Chi, L., and Huck, F. (2001). Lagrangian meshfree formulation for analysis of geotechnical materials. *J. Eng. Mech.* 127 (5): 440–449.

7 Chen, J.-S. and Wang, H.-P. (2000). Meshfree smooth surface contact algorithm for sheet metal forming. *Proceedings of the SAE*, Detroit (6–9 March 2000).

8 Guan, P.C., Chi, S.W., Chen, J.-S., and Slawson, T.R. (2011). Semi-Lagrangian reproducing kernel particle method for fragment-impact problems. *Int. J. Impact Eng.* 38 (12): 1033–1047.

9 Guan, P.C., Chen, J.-S., Wu, Y. et al. (2009). Semi-Lagrangian reproducing kernel formulation and application to modeling earth moving operations. *Mech. Mater.* 41 (6): 670–683.

10 Chi, S.-W., Lee, C.-H., Chen, J.-S., and Guan, P.-C. (2015). A level set enhanced natural kernel contact algorithm for impact and penetration modeling. *Int. J. Numer. Methods Eng.* 102 (3–4): 839–866.

11 Sherburn, J.A., Roth, M.J., Chen, J.-S., and Hillman, M. (2015). Meshfree modeling of concrete slab perforation using a reproducing kernel particle impact and penetration formulation. *Int. J. Impact Eng.* 86: 96–110.

12 Belytschko, T., Liu, W.K., Moran, B., and Elkhodary, K. (2013). *Nonlinear Finite Elements for Continua and Structures*. Chichester: John Wiley & Sons.

13 Chen, J.-S., Yoon, S., and Wu, C.-T. (2002). Non-linear version of stabilized conforming nodal integration for Galerkin mesh-free methods. *Int. J. Numer. Methods Eng.* 53 (12): 2587–2615.

14 Swegle, J.W., Hicks, D.L., and Attaway, S.W. (1995). Smoothed particle hydrodynamics stability analysis. *J. Comput. Phys.* 116 (1): 123–134.

15 Belytschko, T., Guo, Y., Liu, W.K., and Xiao, S.P. (2000). A unifieded stability analysis of meshless particle methods. *Int. J. Numer. Methods Eng.* 48 (9): 1359–1400.

16 Olliff, J., Alford, B., and Simkins, D.C. (2018). Efficient searching in meshfree methods. *Comput. Mech.* 62 (6): 1461–1483.

17 Bentley, J.L. (1975). Multidimensional binary search trees used for associative searching. *Commun. ACM.* 18 (9): 509–517.

18 Wriggers, P. (2006). *Computational Contact Mechanics*, 2e. Berlin Heidelberg: Springer.

19 Laursen, T.A. (2003). *Computational Contact and Impact Mechanics: Fundamentals of Modeling Interfacial Phenomena in Nonlinear Finite Element Analysis*. Springer Science & Business Media.

20 Puso, M.A. and Laursen, T.A. (2004). A mortar segment-to-segment frictional contact method for large deformations. *Comput. Methods Appl. Mech. Eng.* 193 (45–47): 4891–4913.

21 Hesch, C. and Betsch, P. (2011). Transient 3d contact problems-NTS method: mixed methods and conserving integration. *Comput. Mech.* 48 (4): 437–449.

22 Hughes, T.J.R., Taylor, R.L., Sackman, J.L. et al. (1976). A finite element method for a class of contact-impact problems. *Comput. Methods Appl. Mech. Eng.* 8 (3): 249–276.

23 Hallquist, J.O., Goudreau, G.L., and Benson, D.J. (1985). Sliding interfaces with contact-impact in large-scale Lagrangian computations. *Comput. Methods Appl. Mech. Eng.* 51 (1–3): 107–137.

24 Simo, J.C., Wriggers, P., and Taylor, R.L. (1985). A perturbed lagrangian formulation for the finite element solution of contact problems. *Comput. Methods Appl. Mech. Eng.* 50: 163–180.

25 Papadopoulos, P. and Taylor, R.L. (1992). A mixed formulation for the finite element solution of contact problems. *Comput. Methods Appl. Mech. Eng.* 94 (3): 373–389.

26 Belytschko, T. and Neal, M.O. (1991). Contact-impact by the pinball algorithm with penalty and Lagrangian methods. *Int. J. Numer. Methods Eng.* 31 (3): 547–572.

27 Rebelo, N., Nagtegaal, J.C., and Hibbitt, H.D. (1990). Finite element analysis of sheet forming processes. *Int. J. Numer. Methods Eng.* 30 (8): 1739–1758.

28 Wriggers, P., Krstulovic-Opara, L., and Korelc, J. (2001). Smooth C1-interpolations for two-dimensional frictional contact problems. *Int. J. Numer. Methods Eng.* 51 (12): 1469–1495.

29 Ping Wang, S. and Nakamachi, E. (1997). The inside-outside contact search algorithm for finite element analysis. *Int. J. Numer. Methods Eng.* 40: 3665–3685.

30 Bittencourt, E. and Creus, G.J. (1998). Finite element analysis of three-dimensional contact and impact in large deformation problems. *Comput. Struct.* 69 (2): 219–234.

31 Eterovic, A. and Bathe, K. (1991). An interface interpolation scheme for quadratic convergence in the finite element analysis of contact problems. In: *Nonlinear Computational Mechanics: State of the Art* (ed. P. Wriggers and W. Wagner), 703–715. Berlin: Springer-Verlag.

32 Wriggers, P. and Imhof, M. (1993). On the treatment of nonlinear unilateral contact problems. *Arch. Appl. Mech.* 63 (2): 116–129.

33 Heegaard, J.H. and Curnier, A. (1996). Geometric properties of 2D and 3D unilateral large slip contact operators. *Comput. Methods Appl. Mech. Eng.* 131 (3–4): 263–286.

34 Puso, M.A. and Laursen, T.A. (2002). A 3D contact smoothing method using Gregory patches. *Int. J. Numer. Methods Eng.* 54 (8): 1161–1194.

35 Krstulović-Opara, L., Wriggers, P., and Korelc, J. (2002). A C1-continuous formulation for 3D finite deformation frictional contact. *Comput. Mech.* 29 (1): 27–42.

36 Wang, H.-P., Wu, C.-T., and Chen, J.-S. (2014). A reproducing kernel smooth contact formulation for metal forming simulations. *Comput. Mech.* 54 (1): 151–169.

37 Zhong, Z.-H. (2002). *Finite Element Procedures for Contact-Impact Problems*. Springer: Berlin.

38 Chen, J.-S., Liu, W.K., Hillman, M. et al. (2017). Reproducing kernel approximation and discretization. In: *Encyclopedia of Computational Mechanics*, 2e (ed. E. Stein, R. de Borst and T.J.R. Hughes). Chichester: John Wiley & Sons.

39 Unosson, M. and Nilsson, L. (2006). Projectile penetration and perforation of high performance concrete: experimental results and macroscopic modelling. *Int. J. Impact Eng.* 32 (7): 1068–1085.

40 Unosson, M. (2000). Numerical simulations of penetration and perforation of high performance concrete with 75mm steel projectile, FOA Defense Research Establishment (FOA-R-00-01634-311-SE).

41 Chen, J.-S., Chi, S.W., Lee, C.-H. et al. (2011). *A multiscale meshfree approach for modeling fragment penetration into ultra high-strength concrete*. US Army Engineer Research and Development Center (ERDC/GSL TR-11-35).

42 Chen, J.-S., Hillman, M., and Chi, S.W. (2017). Meshfree methods: progress made after 20 years. *J. Eng. Mech.* 143 (4): 04017001.

8

Other Galerkin Meshfree Methods

Two types of meshfree discretizations of PDEs are most prevalent: those based on the weak form and those based on directly discretizing the strong form. Most meshfree methods have been formulated based on the weak form, as it assumes a very similar mathematical formulation and code implementation structure as the finite element methods. These methods have relaxed regularity requirements on the approximation, as the order of differentiation is reduced to half of that of the strong form. The resulting matrix equation is symmetric if the same shape functions are used to approximate test and trial functions. Furthermore, the natural boundary conditions are embedded in the formulation. The element-free Galerkin (EFG) and the reproducing kernel particle method (RKPM) are examples of meshfree methods constructed by the weak form.

This chapter presents other meshfree methods that utilize the weak form. They differ from EFG and RKPM in that they introduce a different set of test and trial functions in the Galerkin approximation. These methods include smoothed particle hydrodynamics (SPH), the partition of unity method and *h-p* clouds, and Sibson and non-Sibson interpolation used in the natural element method. We focus only on the construction and properties of the approximation functions, as the procedures for developing the discrete equations are identical to those in EFG and RKPM, which are described in Chapter 4.

8.1 Smoothed Particle Hydrodynamics

SPH, proposed in the late 1970s by Gingold and Monaghan [1] and Lucy [2], introduces the discretization of conservation laws by operations in terms of nodal data. Instead of using the strong form directly, they introduced a kernel estimate of the strong form (also called the smoothing operation) by a kernel function (often called the smoothing function in SPH), which plays the same role as the weight function in moving least-squares and the kernel in the reproducing kernel approximation. The kernel estimate of governing equations can be viewed as a weighted residual method and has a direct link to the Galerkin approximation. Construction of SPH equations typically involves performing integration by parts. The SPH literature attributes this operation to the desire to have the derivatives acting on the smoothing function rather than the physical variables such as stresses. This operation is exactly the one used in the Galerkin approximation to reduce the order of differentiation, but with certain assumptions on the boundary integral terms resulting from the divergence operations. The final discrete equation in SPH is obtained by collocation of the integral equation. We will show in the subsequent discussions that the SPH equations can be constructed from the Galerkin weak form with collocation.

Meshfree and Particle Methods: Fundamentals and Applications, First Edition.
Ted Belytschko, J. S. Chen, and Michael Hillman.

In this section, we first introduce the kernel estimate of a function. We then discuss the kernel estimate of conservation laws and the approximations involved that yield the SPH discrete equations for conservation laws.

8.1.1 Kernel Estimate

In the following, we will develop the discrete SPH equations for the PDEs that govern the large deformation, nonlinear behavior of solids and fluids. We will consider each equation in turn:

1) The mass conservation equation
2) The momentum conservation equation, which we will often simply call the momentum equation or the equation of motion
3) The energy conservation equation

These equations are given in Chapter 2, along with the associated assumptions. We will consider the Lagrangian form of SPH, which is the most common discretization. In this method, the SPH particles are attached to material points and remain coincident with those material points. This is directly analogous to the semi-Lagrangian RKPM approach introduced in Chapter 7, but for SPH, it is usually just called Lagrangian.

The earliest development in meshless methods can be attributed to Lucy [2] and Gingold and Monaghan [1] in their development of the SPH method for astrophysics modeling. The foundation of the SPH method is the kernel estimate [3, 4]. In the construction, the conservation laws are approximated by kernel estimates. We begin with the kernel estimate of a function $f(\mathbf{x})$:

$$f^{\mathrm{k}}(\mathbf{x}) = \int_{\Omega} f(\mathbf{s})w(\mathbf{x}-\mathbf{s}, h)d\mathbf{s}, \tag{8.1}$$

where $w(\mathbf{x}-\mathbf{s}, h)$ is called the smoothing function, h is the smoothing length, and $f^{\mathrm{k}}(\mathbf{x})$ is the kernel estimate of $f(\mathbf{x})$. In (8.1) and throughout this chapter, we have used the notation $d\mathbf{s}$ rather than $d\Omega$ to clearly specify the argument of integration, which will be important for later developments.

The smoothing function plays the same role as the weight function in the moving least-squares (MLS) approximation or the kernel function in the reproducing kernel (RK) approximation. It is a nonnegative function and is normalized, i.e.,

$$\begin{aligned} w(\mathbf{x}, h) &> 0 \quad \text{for} \quad \|\mathbf{x}\| < 2h, \\ w(\mathbf{x}, h) &= 0 \quad \text{for} \quad \|\mathbf{x}\| \geq 2h, \end{aligned} \tag{8.2}$$

with the normalization

$$\int_{\Omega} w(\mathbf{x}, h)d\mathbf{x} = 1. \tag{8.3}$$

Note that it is customary in SPH to define the smoothing function with a compact support of twice the smoothing length h as shown in (8.2).

It follows immediately that as $h \to 0$,

$$w(\mathbf{x}, h) \to \delta(\mathbf{x}), \tag{8.4}$$

and

$$f^{\mathrm{k}}(\mathbf{x}) \to f(\mathbf{x}). \tag{8.5}$$

Commonly used kernel functions include the Gaussian function and cubic B-spline.

Remark 8.1 *If the smoothing function is normalized and symmetric, the error of the kernel estimate is second order. This can be shown as follows. Assuming that f(s) is sufficiently smooth, its Taylor expansion is*

$$f(s) = f(x) + (s-x)f'(x) + \frac{(s-x)^2}{2}f''(x) + \cdots. \tag{8.6}$$

Substituting (8.6) into (8.1) we have

$$\begin{aligned} f^{\mathrm{k}}(x) &= \int_{\Omega} f(s)w(x-s,h)ds \\ &= f(x)\int_{\Omega} w(x-s,h)ds + f'(x)\int_{\Omega}(s-x)w(x-s,h)ds + \frac{1}{2}f''(x)\int_{\Omega}(s-x)^2 w(x-s,h)ds + \cdots \\ &= f(x) + \frac{\overline{m}_2}{2!}f''(x) + \cdots, \end{aligned} \tag{8.7}$$

where

$$\overline{m}_n = \int_{\Omega}(s-x)^n w(x-s,h)ds. \tag{8.8}$$

Here we have used the symmetric property of the smoothing function $w(x-s,h)$ and thus the first-order moment term vanishes: $\int_{\Omega}(s-x)w(x-s,h)ds = 0$. The error in the kernel estimate with a symmetric smoothing function is second order according to (8.7).

Remark 8.2 *In the SPH forms of conservation laws, the kernel estimate of the product of two functions is often considered as follows:*

$$\begin{aligned} (f(x)g(x))^{\mathrm{k}} &= \int_{\Omega} f(s)g(s)w(x-s,h)ds \\ &\approx \left(\int_{\Omega} f(s)w(x-s,h)ds\right)\left(\int_{\Omega} g(s)w(x-s,h)ds\right) = f^{\mathrm{k}}(x)g^{\mathrm{k}}(x). \end{aligned} \tag{8.9}$$

Considering the Taylor expansion of f(s) and g(s), we have

$$\begin{aligned} f(s)g(s) &= f(x)g(x) \\ &\quad + (s-x)[f'(x)g(x) + f(x)g'(x)] \\ &\quad + (s-x)^2\left[\frac{1}{2}f''(x)g(x) + \frac{1}{2}f(x)g''(x) + f'(x)g'(x)\right] \\ &\quad + \cdots. \end{aligned} \tag{8.10}$$

Using the normalized properties of the smoothing function $w(x-s, h)$, and if it is symmetric, the kernel estimate of the product of two functions is

$$(f(x)g(x))^{\mathrm{k}} = \int_{\Omega} f(s)g(s)w(x-s, h)ds$$
$$= f(x)g(x) + \overline{m}_2\left[\frac{1}{2}f''(x)g(x) + \frac{1}{2}f(x)g''(x) + f'(x)g'(x)\right] + \cdots. \tag{8.11}$$

The multiplication of two kernel-estimated functions using (8.7) is

$$f^{\mathrm{k}}(x)g^{\mathrm{k}}(x) = f(x)g(x) + \overline{m}_2\left[\frac{1}{2}f''(x)g(x) + \frac{1}{2}f(x)g''(x)\right] + \cdots. \tag{8.12}$$

Comparing (8.11) and *(8.12), the difference between $(f(x)g(x))^{\mathrm{k}}$ and $f(x)^{\mathrm{k}}g(x)^{\mathrm{k}}$ is second order, and the difference between $f(x)^{\mathrm{k}}g(x)^{\mathrm{k}}$ and $f(x)g(x)$ is also second order.*

In the SPH literature, physical terms such as "particle" and "mass" are used in the SPH equations for conservation laws, and we will follow their convention. Consider that a continuum is discretized into a set of N particles, with each particle I located at $\mathbf{x}_I$, and carrying a volume V_I, mass m_I, and density ρ_I, with $V_I = m_I/\rho_I$. The kernel estimate in (8.1) is discretized by collocation as follows:

$$f^{\mathrm{k}}(\mathbf{x}) = \int_{\Omega} f(\mathbf{s})w(\mathbf{x}-\mathbf{s}, h)d\mathbf{s} \approx \sum_{I=1}^{N} f_I w(\mathbf{x}-\mathbf{x}_I, h)V_I = \sum_{I=1}^{N} f_I w(\mathbf{x}-\mathbf{x}_I, h)\frac{m_I}{\rho_I}, \tag{8.13}$$

where f_I is the nodal value of f. As can be seen, under this construction, the contributions of each particle to the estimated function are weighted according to their distance to the point of evaluation and their density and mass. An interesting observation of the above equation is that by letting $f(\mathbf{x}) = \rho(\mathbf{x})$, we have

$$\rho^{\mathrm{k}}(\mathbf{x}) = \sum_{I=1}^{N} \rho_I w(\mathbf{x}-\mathbf{x}_I, h)\frac{m_I}{\rho_I} = \sum_{I=1}^{N} w(\mathbf{x}-\mathbf{x}_I, h)m_I, \tag{8.14}$$

Equation (8.14) is interpreted as obtaining a continuous density field by smoothing of a set of particle masses.

The kernel estimate in Eq. (8.13), however, does not reproduce a constant function, i.e., if $f(\mathbf{x}) = 1$, Eq. (8.13) yields

$$f^{\mathrm{k}}(\mathbf{x}) = \sum_{I=1}^{N} w(\mathbf{x}-\mathbf{x}_I, h)\frac{m_I}{\rho_I}. \tag{8.15}$$

Further, from (8.14) we require

$$\rho(\mathbf{x}_I) = \sum_{J=1}^{N} w(\mathbf{x}_I-\mathbf{x}_J, h)m_J. \tag{8.16}$$

Substituting (8.16) into (8.15) results in

$$f^{\mathrm{k}}(\mathbf{x}) = \sum_{I=1}^{N} \frac{m_I w(\mathbf{x}-\mathbf{x}_I, h)}{\sum_{J=1}^{N} m_J w(\mathbf{x}_I-\mathbf{x}_J, h)} \neq 1. \tag{8.17}$$

The above demonstration indicates that the SPH approximation cannot represent a constant field exactly, i.e., the approximation does not have the *partition of unity* property introduced in Chapter 3. The loss of partition of unity property in the approximation does not guarantee convergence, and convergence of SPH has only been proven under very restrictive conditions. This deficiency can be easily corrected by using MLS and RK shape functions as the smoothing functions.

In constructing the SPH equations for conservation laws, the kernel estimate of PDE involves the kernel estimate of spatial gradients. Let us start with the kernel estimate of the gradient of a function $f(\mathbf{x})$:

$$\left(\frac{\partial f(\mathbf{x})}{\partial x_i}\right)^{\mathrm{k}} = \int_{\Omega} \frac{\partial f(\mathbf{s})}{\partial s_i} w(\mathbf{x}-\mathbf{s}, h) d\mathbf{s}. \tag{8.18}$$

By integration by parts and the divergence theorem, and considering that w approaches zero fast enough so that the surface term vanishes, we have

$$\left(\frac{\partial f(\mathbf{x})}{\partial x_i}\right)^{\mathrm{k}} = -\int_{\Omega} f(\mathbf{s}) \frac{\partial w(\mathbf{x}-\mathbf{s}, h)}{\partial s_i} d\mathbf{s}. \tag{8.19}$$

The collocation form of Eq. (8.19) is expressed as

$$(\nabla f(\mathbf{x}))^{\mathrm{k}} = -\sum_{I=1}^{N} f_I \nabla w(\mathbf{x}-\mathbf{x}_I, h) \frac{m_I}{\rho_I}, \tag{8.20}$$

where

$$\nabla w(\mathbf{x}-\mathbf{x}_I, h) \equiv \left.\frac{\partial w(\mathbf{x}-\mathbf{s}, h)}{\partial \mathbf{s}}\right|_{\mathbf{s}=\mathbf{x}_I}. \tag{8.21}$$

Note that $w(\mathbf{x}-\mathbf{s}, h) \equiv w(|\mathbf{x}-\mathbf{s}|/h)$, and the gradient in (8.21) is computed by

$$\begin{aligned}\nabla w(\mathbf{x}-\mathbf{x}_I, h) &= \left.\frac{\partial w(|\mathbf{x}-\mathbf{s}|/h)}{\partial \mathbf{s}}\right|_{\mathbf{s}=\mathbf{x}_I} = \left.\frac{\partial w(|\mathbf{x}-\mathbf{s}|/h)}{\partial(|\mathbf{x}-\mathbf{s}|/h)} \frac{\partial(|\mathbf{x}-\mathbf{s}|/h)}{\partial \mathbf{s}}\right|_{\mathbf{s}=\mathbf{x}_I} \\ &= -w'(|\mathbf{x}-\mathbf{s}|/h) \left.\frac{\mathbf{x}-\mathbf{s}}{h|\mathbf{x}-\mathbf{s}|}\right|_{\mathbf{s}=\mathbf{x}_I} = -w'(\mathbf{x}-\mathbf{x}_I, h) \frac{\mathbf{x}-\mathbf{x}_I}{h|\mathbf{x}-\mathbf{x}_I|}.\end{aligned} \tag{8.22}$$

Here we have used the property $|\mathbf{x}-\mathbf{s}| = [(\mathbf{x}-\mathbf{s})\cdot(\mathbf{x}-\mathbf{s})]^{1/2}$ and thus $\partial|\mathbf{x}-\mathbf{s}|/\partial\mathbf{s} = -[(\mathbf{x}-\mathbf{s})\cdot(\mathbf{x}-\mathbf{s})]^{-1/2}(\mathbf{x}-\mathbf{s}) = -(\mathbf{x}-\mathbf{s})/|\mathbf{x}-\mathbf{s}|$. If one further defines $\nabla_{\mathbf{x}} w(\mathbf{x}-\mathbf{x}_I, h) = \frac{\partial w(\mathbf{x}-\mathbf{x}_I, h)}{\partial \mathbf{x}}$, and with $\nabla w(\mathbf{x}-\mathbf{x}_I, h)$ defined in (8.21), it can be shown that $\nabla_{\mathbf{x}} w(\mathbf{x}-\mathbf{x}_I, h) = -\nabla w(\mathbf{x}-\mathbf{x}_I, h)$, and hence (8.20) can be expressed as

$$(\nabla f(\mathbf{x}))^{\mathrm{k}} = \sum_{I=1}^{N} f_I \nabla_{\mathbf{x}} w(\mathbf{x}-\mathbf{x}_I, h) \frac{m_I}{\rho_I}. \tag{8.23}$$

As can be seen in (8.23), the kernel estimate of the function gradient $\nabla f(\mathbf{x})$ is obtained by the nodal values of $f(\mathbf{x})$, f_I, and the values of the smoothing function gradient $\nabla_{\mathbf{x}} w(\mathbf{x}-\mathbf{x}_I, h)$ centered at $\mathbf{x}_I$. It is interesting to observe that taking the gradient of (8.13) with respect to $\mathbf{x}$ yields (8.23). If one views $w(\mathbf{x}-\mathbf{x}_I, h)m_I/\rho_I$ as a shape function in an approximation, it is natural to obtain the approximation of the gradient of a function by the gradient of the shape function, i.e., $\nabla_{\mathbf{x}} w(\mathbf{x}-\mathbf{x}_I, h)m_I/\rho_I$.

Note that there is a sign difference between (8.20) and (8.23). The SPH literature [5, 6] commonly uses the kernel estimate gradient form in (8.20) to express the conservation laws, as will be discussed in the following sections.

8.1.2 SPH Conservation Equations

8.1.2.1 Mass Conservation (Continuity Equation)

We demonstrate below how to construct the SPH equation for mass conservation. The governing equation is (see Chapter 2):

$$\dot{\rho} + \rho v_{i,i} = 0. \tag{8.24}$$

By introducing a kernel estimate of the above equation, we have

$$\int_\Omega w(\mathbf{x}-\mathbf{s}, h)\left(\dot{\rho}(\mathbf{s}) + \rho(\mathbf{s})\frac{\partial v_i(\mathbf{s})}{\partial s_i}\right)d\mathbf{s} = 0. \tag{8.25}$$

Remark 8.3 *It can be shown that (8.25) is a Galerkin approximation of (8.24). The Galerkin approximation can be obtained by a weighted residual of (8.24) with a test function* $\delta u = \sum_{I=1}^{N} \phi_I(\mathbf{x})\delta u_I$, *where* $\phi_I(\mathbf{x})$ *is the shape function:*

$$\int_\Omega \left(\sum_{I=1}^{N} \phi_I(\mathbf{x})\delta u_I\right)\left(\dot{\rho}(\mathbf{x}) + \rho(\mathbf{x})\frac{\partial v_i(\mathbf{x})}{\partial x_i}\right)d\mathbf{x} = 0. \tag{8.26}$$

For (8.26) to hold for an arbitrary test function $\delta u = \sum_{I=1}^{N} \phi_I(\mathbf{x})\delta u_I$, *that is, for arbitrary* δu_I, *we have*

$$\int_\Omega \phi_I(\mathbf{x})\left(\dot{\rho}(\mathbf{x}) + \rho(\mathbf{x})\frac{\partial v_i(\mathbf{x})}{\partial x_i}\right)d\mathbf{x} = 0. \tag{8.27}$$

Let the shape function $\phi_I(\mathbf{x})$ *be the smoothing function centered at* $\mathbf{x}_I$, and let $\phi_I(\mathbf{x}) = w(\mathbf{x}_I - \mathbf{x}, h)$; *changing the integration coordinate* $\mathbf{x} \to \mathbf{s}$, we have

$$\int_\Omega w(\mathbf{x}_I-\mathbf{s}, h)\left(\dot{\rho}(\mathbf{s}) + \rho(\mathbf{s})\frac{\partial v_i(\mathbf{s})}{\partial s_i}\right)d\mathbf{s} = 0. \tag{8.28}$$

Evaluating $\mathbf{x} = \mathbf{x}_I$ *in (8.25) yields (8.28).*

Now, the kernel estimate of mass conservation in (8.25) can be expressed as

$$\int_\Omega \dot{\rho}(\mathbf{s})w(\mathbf{x}-\mathbf{s}, h)d\mathbf{s} = -\int_\Omega \rho(\mathbf{s})\frac{\partial v_i(\mathbf{s})}{\partial s_i}w(\mathbf{x}-\mathbf{s}, h)d\mathbf{s}. \tag{8.29}$$

An approximation that is commonly used in SPH equations, introduced previously in (8.9), is the following:

$$\int_\Omega f(\mathbf{s})g(\mathbf{s})w(\mathbf{x}-\mathbf{s}, h)d\mathbf{s} \approx \left(\int_\Omega f(\mathbf{s})w(\mathbf{x}-\mathbf{s}, h)d\mathbf{s}\right)\left(\int_\Omega g(\mathbf{s})w(\mathbf{x}-\mathbf{s}, h)d\mathbf{s}\right). \tag{8.30}$$

The above equation possesses second-order accuracy as shown in Remark 8.2. This approximation is commonly used when $f(\mathbf{s})$ or $g(\mathbf{s})$ involves derivatives. The same approximation is applied to the right-hand side of (8.29) to yield:

$$\int_{\Omega}\rho(\mathbf{s})\frac{\partial v_i(\mathbf{s})}{\partial s_i}w(\mathbf{x}-\mathbf{s},h)d\mathbf{s}\approx\left(\int_{\Omega}\rho(\mathbf{s})w(\mathbf{x}-\mathbf{s},h)d\mathbf{s}\right)\left(\int_{\Omega}\frac{\partial v_i(\mathbf{s})}{\partial s_i}w(\mathbf{x}-\mathbf{s},h)d\mathbf{s}\right). \tag{8.31}$$

Recalling that $\int_{\Omega}\rho(\mathbf{s})w(\mathbf{x}-\mathbf{s},h)d\mathbf{s}$ is the kernel estimate of $\rho(\mathbf{x})$, (8.31) becomes

$$\int_{\Omega}\rho(\mathbf{s})\frac{\partial v_i(\mathbf{s})}{\partial s_i}w(\mathbf{x}-\mathbf{s},h)d\mathbf{s}\approx\rho(\mathbf{x})\int_{\Omega}\frac{\partial v_i(\mathbf{s})}{\partial s_i}w(\mathbf{x}-\mathbf{s},h)d\mathbf{s}. \tag{8.32}$$

Substituting (8.32) into (8.29), introducing the kernel estimate on the left-hand side of (8.29), considering the kernel estimate of a gradient in (8.20), and performing collocation following (8.20), we have

$$\dot{\rho}(\mathbf{x})\approx\rho(\mathbf{x})\sum_{J=1}^{N}v_{iJ}w_{J,i}(\mathbf{x})\frac{m_J}{\rho_J}, \tag{8.33}$$

where $w_{J,i}(\mathbf{x})\equiv\partial w(\mathbf{x}-\mathbf{s},h)/\partial s_i|_{\mathbf{s}=\mathbf{x}_J}$ is a commonly used notation in SPH. Note that according to (8.22), $\partial w(\mathbf{x}-\mathbf{s},h)/\partial s_i|_{\mathbf{s}=\mathbf{x}_J}=-w'(\mathbf{x}-\mathbf{x}_J,h)(x_i-x_{iJ})/(h|\mathbf{x}-\mathbf{x}_J|)$. Evaluating (8.33) at $\mathbf{x}_I$ reads

$$\dot{\rho}_I=\rho_I\sum_{J=1}^{N}v_{iJ}w_{J,i}(\mathbf{x}_I)\frac{m_J}{\rho_J}\equiv\rho_I\sum_{J=1}^{N}v_{iJ}w_{IJ,i}\frac{m_J}{\rho_J}, \tag{8.34}$$

where $w_{IJ,i}\equiv w_{J,i}(\mathbf{x}_I)=\partial w(\mathbf{x}_I-\mathbf{s},\ h)/\partial s_i|_{\mathbf{s}=\mathbf{x}_J}=-w'(\mathbf{x}_I-\mathbf{x}_J,\ h)(x_{iI}-x_{iJ})/(h|\mathbf{x}_I-\mathbf{x}_J|)$. To maintain symmetry in the SPH equation, one antisymmetric property of $w_{I,i}(\mathbf{x})$ is further invoked

$$\int_{\Omega}w_{I,i}(\mathbf{x})d\mathbf{x}\approx 0. \tag{8.35}$$

In collocation form, it can be written as

$$\sum_{J=1}^{N}w_{JI,i}\frac{m_J}{\rho_J}=0\rightarrow\sum_{J=1}^{N}w_{IJ,i}\frac{m_J}{\rho_J}=0. \tag{8.36}$$

Multiplying (8.36) by $-\rho_I v_{iI}$ and adding it to (8.34), one obtains

$$\dot{\rho}_I=\rho_I\sum_{J=1}^{N}(v_{iJ}-v_{iI})w_{IJ,i}\frac{m_J}{\rho_J}. \tag{8.37}$$

The above is the symmetrized form of the SPH mass conservation equation. Another commonly used SPH mass conservation equation is the direct kernel estimate shown in (8.14), where no PDE is needed.

8.1.2.2 Equation of Motion

The SPH equation of motion will be constructed following similar procedures as in the previous section. Consider here a momentum equation (equation of motion) without a body force for simplicity of illustration:

$$\rho\dot{v}_i=\sigma_{ij,j}. \tag{8.38}$$

Applying the kernel estimate to (8.38), we have

$$\int_{\Omega} \dot{v}_i(\mathbf{s}) w(\mathbf{x}-\mathbf{s}, h) d\mathbf{s} = \int_{\Omega} \frac{1}{\rho(\mathbf{s})} \frac{\partial \sigma_{ij}(\mathbf{s})}{\partial s_j} w(\mathbf{x}-\mathbf{s}, h) d\mathbf{s}. \tag{8.39}$$

The kernel estimate of the left-hand side of (8.39) is

$$\int_{\Omega} \dot{v}_i(\mathbf{s}) w(\mathbf{x}-\mathbf{s}, h) d\mathbf{s} \approx \dot{v}_i(\mathbf{x}). \tag{8.40}$$

The right-hand side of (8.39) is further manipulated to yield

$$\begin{aligned}
&\int_{\Omega} \frac{1}{\rho(\mathbf{s})} \frac{\partial \sigma_{ij}(\mathbf{s})}{\partial s_j} w(\mathbf{x}-\mathbf{s}, h) d\mathbf{s} \\
&= \int_{\Omega} \frac{\partial}{\partial s_j}\left(\frac{\sigma_{ij}(\mathbf{s})}{\rho(\mathbf{s})}\right) w(\mathbf{x}-\mathbf{s}, h) d\mathbf{s} - \int_{\Omega} \sigma_{ij}(\mathbf{s}) \frac{\partial}{\partial s_j}\left(\frac{1}{\rho(\mathbf{s})}\right) w(\mathbf{x}-\mathbf{s}, h) d\mathbf{s} \\
&= \int_{\Omega} \frac{\partial}{\partial s_j}\left(\frac{\sigma_{ij}(\mathbf{s})}{\rho(\mathbf{s})}\right) w(\mathbf{x}-\mathbf{s}, h) d\mathbf{s} + \int_{\Omega} \frac{\sigma_{ij}(\mathbf{s})}{\rho^2(\mathbf{s})} \frac{\partial \rho(\mathbf{s})}{\partial s_j} w(\mathbf{x}-\mathbf{s}, h) d\mathbf{s}.
\end{aligned} \tag{8.41}$$

By the kernel estimate of the gradient in (8.19), the first term in the last line of (8.41) is approximated as

$$\int_{\Omega} \frac{\partial}{\partial s_j}\left(\frac{\sigma_{ij}(\mathbf{s})}{\rho(\mathbf{s})}\right) w(\mathbf{x}-\mathbf{s}, h) d\mathbf{s} \approx -\int_{\Omega} \frac{\sigma_{ij}(\mathbf{s})}{\rho(\mathbf{s})} \frac{\partial w(\mathbf{x}-\mathbf{s}, h)}{\partial s_j} d\mathbf{s}, \tag{8.42}$$

and the second term on the last line of (8.41) becomes

$$\begin{aligned}
&\int_{\Omega} \frac{\sigma_{ij}(\mathbf{s})}{\rho^2(\mathbf{s})} \frac{d\rho(\mathbf{s})}{ds_j} w(\mathbf{x}-\mathbf{s}, h) d\mathbf{s} \\
&\approx \left(\int_{\Omega} \frac{\sigma_{ij}(\mathbf{s})}{\rho^2(\mathbf{s})} w(\mathbf{x}-\mathbf{s}, h) d\mathbf{s}\right)\left(\int_{\Omega} \frac{d\rho(\mathbf{s})}{ds_j} w(\mathbf{x}-\mathbf{s}, h) d\mathbf{s}\right) \\
&\approx -\frac{\sigma_{ij}(\mathbf{x})}{\rho^2(\mathbf{x})} \int_{\Omega} \rho(\mathbf{s}) \frac{\partial w(\mathbf{x}-\mathbf{s}, h)}{\partial s_j} d\mathbf{s}.
\end{aligned} \tag{8.43}$$

Substituting (8.41)–(8.43) into (8.39), and with a kernel estimate on the left-hand side term, we have

$$\dot{v}_i(\mathbf{x}) = -\int_{\Omega} \frac{\sigma_{ij}(\mathbf{s})}{\rho(\mathbf{s})} \frac{\partial w(\mathbf{x}-\mathbf{s}, h)}{\partial s_j} d\mathbf{s} - \frac{\sigma_{ij}(\mathbf{x})}{\rho^2(\mathbf{x})} \int_{\Omega} \rho(\mathbf{s}) \frac{\partial w(\mathbf{x}-\mathbf{s}, h)}{\partial s_j} d\mathbf{s}. \tag{8.44}$$

Evaluating (8.44) at nodal points and performing collocation on the right-hand side following (8.20), we obtain the following SPH equation:

$$\dot{v}_{iI} = -\sum_{J=1}^{N} \frac{\sigma_{ij}^J}{\rho_J} w_{IJ,j} \frac{m_J}{\rho_J} - \frac{\sigma_{ij}^I}{\rho_I^2}\left(\sum_{J=1}^{N} \rho_J w_{IJ,j} \frac{m_J}{\rho_J}\right) = -\sum_{J=1}^{N}\left(\frac{\sigma_{ij}^I}{\rho_I^2} + \frac{\sigma_{ij}^J}{\rho_J^2}\right) w_{IJ,j} m_J, \tag{8.45}$$

where $\sigma_{ij}^I \equiv \sigma_{ij}(\mathbf{x}_I)$. It can be seen that the right-hand side of (8.45) possesses a symmetric structure.

8.1.2.3 Energy Conservation Equation

Consider here the energy conservation equation under isothermal conditions:

$$\rho \dot{E} = \sigma_{ij} v_{i,j}, \tag{8.46}$$

where E is the specific internal energy. The kernel estimate of (8.46) gives

$$\begin{aligned}\int_{\Omega} \dot{E}(\mathbf{s}) w(\mathbf{x}-\mathbf{s}, h) d\mathbf{s} &= \int_{\Omega} \frac{\sigma_{ij}(\mathbf{s})}{\rho(\mathbf{s})} \frac{\partial v_i(\mathbf{s})}{\partial s_j} w(\mathbf{x}-\mathbf{s}, h) d\mathbf{s} \\ &= \int_{\Omega} \frac{\sigma_{ij}(\mathbf{s})}{\rho^2(\mathbf{s})} \left(\frac{\partial(\rho(\mathbf{s}) v_i(\mathbf{s}))}{\partial s_j} - v_i \frac{\partial \rho(\mathbf{s})}{\partial s_j} \right) w(\mathbf{x}-\mathbf{s}, h) d\mathbf{s}.\end{aligned} \tag{8.47}$$

Introducing the kernel estimate of a function on the left-hand side, and the kernel estimate of the product of two functions on the right-hand side of (8.47), we have

$$\dot{E}(\mathbf{x}) \approx \frac{\sigma_{ij}(\mathbf{x})}{\rho^2(\mathbf{x})} \int_{\Omega} \frac{\partial(\rho(\mathbf{s}) v_i(\mathbf{s}))}{\partial s_j} w(\mathbf{x}-\mathbf{s}, h) d\mathbf{s} - \frac{\sigma_{ij}(\mathbf{x}) v_i(\mathbf{x})}{\rho^2(\mathbf{x})} \int_{\Omega} \frac{\partial \rho(\mathbf{s})}{\partial s_j} w(\mathbf{x}-\mathbf{s}, h) d\mathbf{s}. \tag{8.48}$$

Further considering the kernel estimate of the gradient in (8.19), (8.48) yields

$$\dot{E}(\mathbf{x}) \approx -\frac{\sigma_{ij}(\mathbf{x})}{\rho^2(\mathbf{x})} \int_{\Omega} \rho(\mathbf{s}) v_i(\mathbf{s}) \frac{\partial w(\mathbf{x}\text{-}\mathbf{s}, h)}{\partial s_j} d\mathbf{s} + \frac{\sigma_{ij}(\mathbf{x}) v_i(\mathbf{x})}{\rho^2(\mathbf{x})} \int_{\Omega} \rho(\mathbf{s}) \frac{\partial w(\mathbf{x}\text{-}\mathbf{s}, h)}{\partial s_j} d\mathbf{s}. \tag{8.49}$$

Evaluating (8.49) at nodal points and performing collocation on the right-hand side, we obtain the following SPH equation for energy conservation:

$$\dot{E}_I = -\frac{\sigma_{ij}^I}{\rho_I^2} \sum_{J=1}^{N} (v_{iJ} - v_{iI}) w_{IJ,j} m_J. \tag{8.50}$$

The SPH equations for the cases discussed above can be obtained by making the approximations listed in **Box 8.1**.

Box 8.1 Assumptions in SPH Equations

1) Kernel estimate of a function:

$$f(\mathbf{x}) \approx \int_{\Omega} f(\mathbf{s}) w(\mathbf{x}-\mathbf{s}, h) d\mathbf{s}. \tag{8.51}$$

2) Kernel estimate of a PDE $\mathcal{L}(\mathbf{u}(\mathbf{x}, t)) = 0$ (weighted residual):

$$\int_{\Omega} w(\mathbf{x}-\mathbf{s}, h) \mathcal{L}(\mathbf{u}(\mathbf{x}, t)) d\mathbf{s} = 0 \tag{8.52}$$

(Continued)

Box 8.1 (Continued)

3) Kernel estimate of two functions:

$$\int_{\Omega} f(\mathbf{s})g(\mathbf{s})w(\mathbf{x}-\mathbf{s},h)d\mathbf{s} \approx \left(\int_{\Omega} f(\mathbf{s})w(\mathbf{x}-\mathbf{s},h)d\mathbf{s}\right)\left(\int_{\Omega} g(\mathbf{s})w(\mathbf{x}-\mathbf{s},h)d\mathbf{s}\right). \tag{8.53}$$

4) Kernel estimate of function gradients:

$$\int_{\Omega} \nabla f(\mathbf{s})w(\mathbf{x}-\mathbf{s},h)d\mathbf{s} = -\int_{\Omega} f(\mathbf{s})(\nabla w(\mathbf{x}-\mathbf{s},h))d\mathbf{s}. \tag{8.54}$$

5) Collocation of domain integration:

$$\int_{\Omega} f(\mathbf{s})w(\mathbf{x}-\mathbf{s},h)d\mathbf{s} \approx \sum_{I=1}^{N} f_I w(\mathbf{x}-\mathbf{x}_I,h)\frac{m_I}{\rho_I}. \tag{8.55}$$

8.1.3 Stability of SPH

Although the SPH method is often classified as a Lagrangian formulation in the SPH community, the method is not strictly Lagrangian. The term "Lagrangian" refers to the fact that the smoothing function centroid is attached to the same material particle throughout the deformation. However, the support of the smoothing function does not necessarily cover the same group of material particles in the SPH construction. If the smoothing length does not deform with the materials, the SPH discrete equation is not a pure Lagrangian formulation (see Chapter 7). Similar discussions have been made in connection to RKPM [7] where a kernel function may be classified in the following ways:

1) A spatial kernel (or Eulerian kernel) if the distance measure in the kernel function evaluation is based on the spatial coordinate of discrete points fixed in space, i.e., $w(\mathbf{x}-\mathbf{x}_I,h) \equiv w(d)$, $d = \|\mathbf{x}-\mathbf{x}_I\|/h$;
2) A material kernel (or Lagrangian kernel) if the distance measure in the kernel function evaluation is based on the material coordinate of discrete points attached to material points, i.e., $w(\mathbf{X}-\mathbf{X}_I,h) \equiv w(d)$, $d = \|\mathbf{X}-\mathbf{X}_I\|/h$; or
3) A semi-Lagrangian kernel defined in Guan et al. [8, 9] if the distance measure in the kernel function evaluation is based on the spatial coordinate of discrete points attached to material points, i.e., $w(\mathbf{x}-\mathbf{x}(\mathbf{X}_I,t),h) \equiv w(d)$, $d = \|\mathbf{x}-\mathbf{x}(\mathbf{X}_I,t)\|/h$.

The semi-Lagrangian kernel in RKPM plays a similar role as the smoothing function in SPH. More discussions on Lagrangian kernels and semi-Lagrangian kernels and their stability are given in Chapter 7.

It can be easily understood that for proper evaluation of interactions between paired discrete particles, any evaluation point has to be covered by at least two smoothing functions. If the SPH smoothing function is evaluated using spatial coordinates with a fixed smoothing length, some evaluation points (SPH nodal points) may not be covered by at least two smoothing functions under excessive tensile deformation. This yields the so-called tensile instability. Belytschko et al. [10]

showed that the tensile instability in SPH can be removed by using a Lagrangian smoothing function $w(d)$, $d = \|\mathbf{X} - \mathbf{X}_I\|/h$, in which the smoothing function support covers the same group of material particles throughout the deformation and thus prevents tensile instability. A conservative smoothing approach was proposed by Swegel et al. [5] to resolve tensile instability in SPH. The early development of RKPM [7] for large deformation problems was based on a Lagrangian kernel, and no tensile instability was ever observed. An artificial stress was introduced by Monaghan [11] to remove the tensile instability in SPH.

Swegle et al. [5, 12] studied the tensile instability in SPH in the tensile region. A von Neumann stability analysis on the SPH algorithm was performed, and the result yielded a simple stability condition at the shortest wavelength in terms of stress state and the second derivative of the kernel function as follows:

$$\begin{aligned} w''\sigma &> 0, \quad \text{(unstable)} \\ w''\sigma &\leq 0. \quad \text{(stable)} \end{aligned} \tag{8.56}$$

This von Neumann stability condition is consistent with the physical justification from the SPH conservation of momentum. In one dimension, the equation of motion implies

$$\ddot{u} \propto \Delta\sigma, \tag{8.57}$$

where $\Delta\sigma$ is the increment of stress. The one-dimensional SPH equation of motion at the minimum wavelength can be observed in (8.45), with consideration of the sign difference in w' at the left and right neighboring points [5, 12]:

$$\ddot{u} \propto \Delta(-\sigma w'). \tag{8.58}$$

A comparison of (8.57) and (8.58) shows that the SPH effective stress is proportional to $-\sigma w'$:

$$\sigma^{\text{eff}} \propto -\sigma w'. \tag{8.59}$$

The instability is a result of an effective stress with a negative modulus (imaginary sound speed), being produced by the interaction between the constitutive relation and the kernel function, and is not caused by the numerical time integration algorithm [5, 12]. Following the analogy of [5, 12], Figure 8.1 shows that if the stress is tensile ($\sigma > 0$), and the slope of $-w'$ is positive ($w'' < 0$), the effective tensile stress increases as particles separate and decreases as particles approach, which is stable. When the slope of $-w'$ is negative ($w'' > 0$), the effective tensile stress decreases as particles separate and increases as particles approach, which is unstable. Conversely, when the stress is compressive ($\sigma < 0$), and when the slope of w' is positive ($w'' > 0$), the effective compressive stress decreases as particles separate, and increases as particles approach, which is stable. When the slope of w' is negative ($w'' < 0$), the effective compressive stress increases as particles separate, and decreases as particles approach, which is unstable. These stability observations from a physics point of view are consistent with the von Neumann stability condition shown in (8.56).

In the case of a cubic B-spline smoothing function with smoothing length h equal to the nodal distance (that is, the compact support is twice the nodal distance according to (8.2)), the nearest-neighbor particle is unstable in tension and stable in compression as shown in Figure 8.2. This instability leads to particles clumping together in stable configurations, yielding artifacts such as spurious fracture and fragmentation [5, 12]. One way to remedy the tensile instability is to increase the smoothing length h. According to Swegle et al. [5], in the case of an arbitrary number of neighbors, the stability condition can be obtained by assuming the same stress state at the neighbors to yield

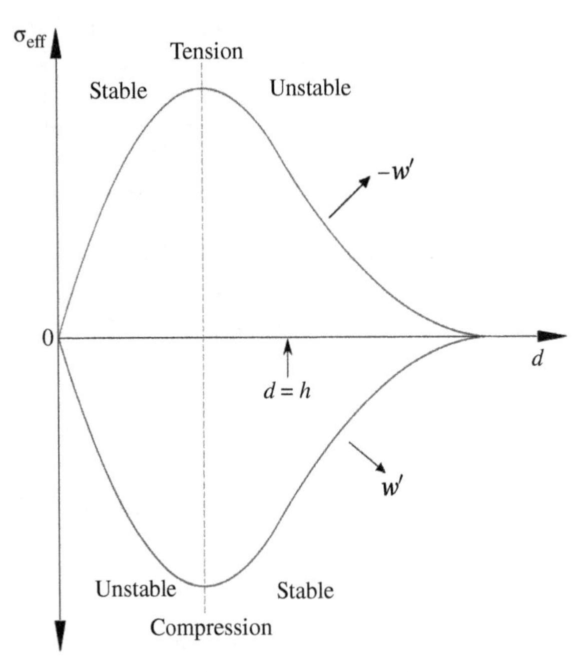

Figure 8.1 SPH effective stress and stability condition for a cubic B-spline smoothing function with compact support of twice the nodal distance. *Source:* Swegle et al. [12], figure 4.2, p. 129/ with permission from Elsevier.

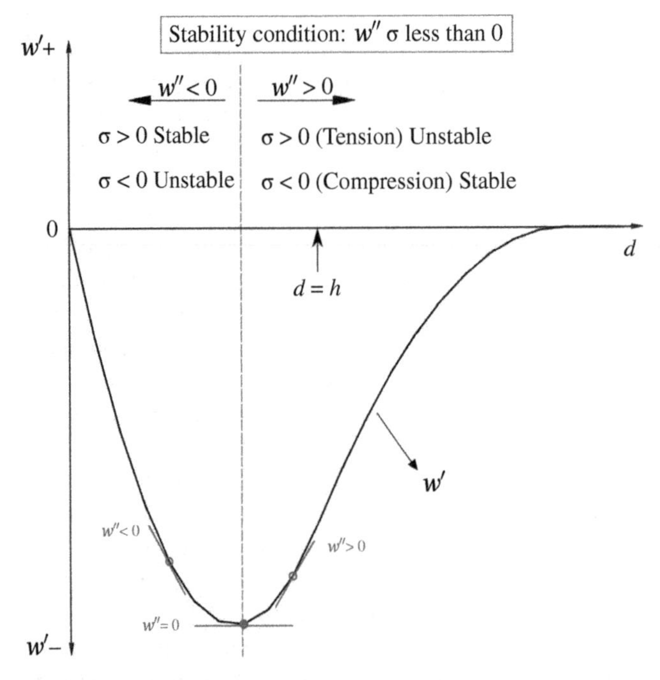

Figure 8.2 Stability regimes for a cubic B-spline smoothing function. *Source:* Swegle et al. [12], figure 3.1, p. 127/ with permission from Elsevier.

$$\begin{aligned} &\left(\sum_{\text{odd spacings}} w''\right)\sigma > 0, \quad \text{(unstable)} \\ &\left(\sum_{\text{odd spacings}} w''\right)\sigma \leq 0. \quad \text{(stable)} \end{aligned} \tag{8.60}$$

The stability condition in (8.60) involves the sum of the values of w'' at all particles under the weight function support and are with odd separations from the particle at the center of the support, as shown in Figure 8.3. It is thus possible to choose a proper smoothing length for stability, for example, $h = 1.5\Delta x$ where the contributions from the odd particles have zero slope on w'.

Another type of instability in the SPH method is rank instability due to the use of collocation in the domain integration. Dyka and Ingel [13] introduced a stress point approach and suggested that the stress points resolved tensile instability. It appears that by employing more stress points, the chance for any evaluation point to be covered by less than two smoothing functions is somewhat

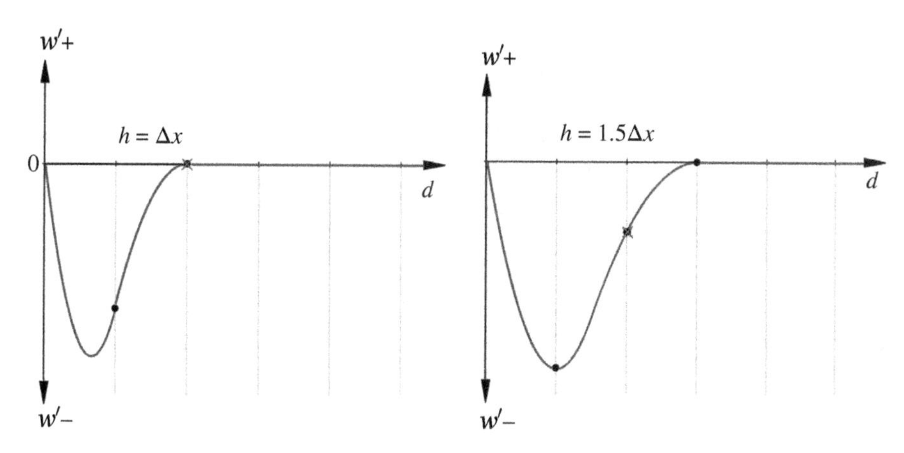

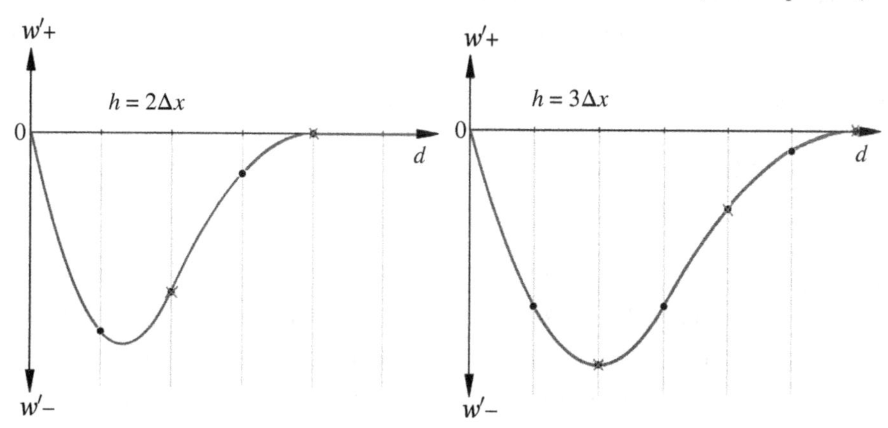

Figure 8.3 Neighbors under smoothing function with support $2h$, where particles without cross symbols do not contribute to the stability condition. *Source:* Swegle et al. [12], figure 5.1, p. 130/ with permission from Elsevier.

reduced (but not completely avoided), thus giving improved stability under tension. However, in the latter study by Belytschko et al. [10], it was pointed out that the stress point approach also improved the rank instability, which was not mentioned in the original work by Dyka and Ingel [13]. In fact, the paper by Randles and Libersky [6] also attributed to the elimination of tensile instability to the use of the stress point approach. Other methods for suppressing the spurious singular modes due to rank instability include least-squares stabilization by Beissel and Belytschko [14] and Bonet and Kulasegaram [15] and a stabilized conforming nodal integration (SCNI) by Chen et al. [16], among several newer variants (see Chapter 6).

8.2 Partition of Unity Finite Element Method and *h-p* Clouds

In mathematics, the "partition of unity" refers to the ability of a set of approximation functions to represent a constant field, that is, the sum of approximation functions must yield unity (one) at any location in the domain where the approximation functions are defined. The most important property of the partition of unity is that it allows the approximation of global functions by local functions. In developing finite element and meshfree methods, the partition of unity finite element method (PUFEM) serves as a fundamental building block for constructing test and trial spaces. It can also be understood as a generalization of the classical *h*, *p*, and *h-p* versions of the finite element method (see the work by Melenk and Babuska [17]). A good review of partition of unity and *h-p* clouds can be found in Belytschko et al. [18].

The partition of unity is a paradigm in which *NP* overlapping patches cover a domain, or subdomains Ω_I, $\Omega_I \cap \Omega \neq \emptyset$, each of which is associated with a function $\Psi_I^0(\mathbf{x})$ which is nonzero only in Ω_I and has the property that

$$\sum_{I=1}^{NP} \Psi_I^0(\mathbf{x}) = 1 \qquad \left(\Omega \subset \cup_{I=1}^{NP} \Omega_I\right). \tag{8.61}$$

Here, $\left\{\Psi_I^0\right\}_{I=1}^{NP}$ is called a partition of unity subordinate to $\{\Omega_I\}_{I=1}^{NP}$. A partition of unity $\left\{\Psi_I^0\right\}_{I=1}^{NP}$ can be used to patch together local functions to represent global functions. The smoothness of the local functions $\Psi_I^0(\mathbf{x})$ determines the smoothness of the approximated functions. Typically, additional regularity conditions are imposed on the norm of the first-order derivatives (the H^1 semi-norm) of the partition of unity if one uses $\Psi_I^0(\mathbf{x})$ to construct test and trial functions for solving second-order PDEs. An example of a partition of unity function is the Shepard function:

$$\Psi_I^0(\mathbf{x}) = \frac{\phi_a(\mathbf{x} - \mathbf{x}_I)}{\sum_{J=1}^{NP} \phi_a(\mathbf{x} - \mathbf{x}_J)}. \tag{8.62}$$

Another possible choice of partition of unity is the collection of the standard family of finite element shape functions. Melenk and Babuska [17] pointed out that PUFEM has approximation properties very similar to the usual *h* and *p* version if the local approximation functions $\left\{\Psi_I^0\right\}_{I=1}^{NP}$ are chosen as polynomials. By increasing the number of patches $\{\Omega_I\}_{I=1}^{NP}$ with a fixed degree *p* in $\left\{\Psi_I^0\right\}_{I=1}^{NP}$, the method behaves like the *h* version. If the patches are kept fixed while increasing the degree *p* in $\left\{\Psi_I^0\right\}_{I=1}^{NP}$, the method behaves like the *p* version. In this sense, the PUFEM is a generalization of the *h* and *p* versions.

Unlike the usual construction of finite element shape functions, partition of unity functions are global functions with compact supports, and the supports can overlap. This removes the major restriction in the construction of finite element shape functions. In finite elements, "parent domains" are used to construct the shape functions and compatibility of shape functions is required on the element boundaries. The order of continuity along the element boundaries is dependent on the order of continuity one chooses in approximating the functions. Unlike the construction of conventional finite element shape functions, PUFEM avoids mapping to the parent domain and therefore eliminates compatibility requirement; that is, PUFEM combines a given set of local approximation spaces together to form a conforming global space. This property significantly simplifies h refinement in any dimension and is a major breakthrough over the conventional finite element method.

The partition of unity can be used as a framework for the construction of approximation functions with a desired order of completeness, or with enrichment of special bases representing characteristics of the PDEs. An example of a partition of unity method for the Helmholtz equation in one dimension is to introduce the following approximation [17]:

$$u^h(x) = \sum_{I=1}^{NP} \Psi_I^0(x)\left(a_{0I} + a_{1I}x + \cdots + a_{kI}x^k + b_{1I}\sinh mx + b_{2I}\cosh mx\right), \tag{8.63}$$

where $\{\Psi_I^0(x)\}_{I=1}^{NP}$ is the partition of unity with compact support, the polynomials are used to impose consistency, and sinhmx and coshmx are enhancement functions with m associated with the parameter in the Helmholtz equation. The use of enrichment functions in constructing the approximation in Eq. (8.63) is called extrinsic enrichment. While the approximation in (8.63) achieves better accuracy by introducing extrinsic bases, it increases degrees of freedom (a_{0I}, a_{1I}, ..., a_{kI}, b_{1I}, b_{2I}) when solving PDEs. One way to avoid this is to introduce the enrichment functions in the approximation which are called "intrinsic bases," which are the bases functions employed in Chapter 3. In the MLS and RK approximations, the enrichment functions are embedded in the local construction of the partition of unity function Ψ_I^0 with the corresponding coefficients obtained from the local equations, and extra degrees of freedom are avoided. For example, in the RK approximation in (3.77)–(3.79) in Chapter 3, the coefficients $b_{ijk}(\mathbf{x})$ of the monomial bases $(x_1 - x_{1I})^i$ $(x_2 - x_{2I})^j(x_3 - x_{3I})^k$ are solved locally from the reproducing conditions in (3.80), and they are not degrees of freedom in the solution of PDEs. These monomial bases $(x_1 - x_{1I})^i(x_2 - x_{2I})^j(x_3 - x_{3I})^k$ in the RK approximation are called intrinsic bases; other functions like those in (8.63) can also be included.

It can be seen that the partition of unity approximation, in general, does not possess Kronecker delta properties. That is, the coefficients associated with the approximation functions are not nodal values. Babuska and Melenk [19] introduced the following approximation to recover the Kronecker delta properties:

$$\begin{aligned} u^h(\mathbf{x}) &= \sum_{J=1}^{NP} \Psi_J^0(\mathbf{x})\left(\sum_I b_I L_{JI}(\mathbf{x})\right) \\ &= \sum_I \sum_{J:\mathbf{x}_I \in \Omega_J} \Psi_J^0(\mathbf{x}) L_{JI}(\mathbf{x}) b_I, \end{aligned} \tag{8.64}$$

where $L_{JI}(\mathbf{x})$ are Lagrange interpolants, therefore $L_{JI}(\mathbf{x}_K) = \delta_{IK}$, for any J, and b_I is the approximation coefficient associated with node I. Now define the following function

$$\Psi_I(\mathbf{x}) = \sum_{J:\mathbf{x}_I \in \Omega_J} \Psi_J^0(\mathbf{x}) L_{JI}(\mathbf{x}). \tag{8.65}$$

Using the Kronecker delta property $L_{JI}(\mathbf{x}_K) = \delta_{IK}$, we have

$$\begin{aligned}\Psi_I(\mathbf{x}_K) &= \sum_{J:\mathbf{x}_I \in \Omega_J} \Psi_J^0(\mathbf{x}_K) L_{JI}(\mathbf{x}_K) \\ &= \sum_{J:\mathbf{x}_I \in \Omega_J} \Psi_J^0(\mathbf{x}_K)\delta_{IK} = \delta_{IK}.\end{aligned} \tag{8.66}$$

Although straightforward, the above construction is difficult for arbitrary geometries, as the Lagrange interpolants do not exist for a general geometries in multiple dimensions. Methods to achieve interpolation properties discussed in Chapter 5 can be applied to PUFEM.

Duarte and Oden [20, 21] introduced the following *h-p* clouds approximation as a generalization of PUFEM:

$$u^h(\mathbf{x}) = \sum_{J=1}^{NP} \Psi_J^k(\mathbf{x}) \left(u_J + \sum_i b_{iJ} q_i(\mathbf{x}) \right), \tag{8.67}$$

where $\{\Psi_I^k(\mathbf{x})\}_{I=1}^{NP}$ is the partition of unity with kth order completeness, and $q_i(x)$ is the extrinsic basis which can be a monomial of order greater than k, or a special enhancement function. The partition of unity $\{\Psi_I^k(\mathbf{x})\}_{I=1}^{NP}$ can be constructed by using, for example, the MLS or RK approximation with complete kth-order monomials. The coefficients associated with the basis functions used in the approximation to form $\{\Psi_I^k(\mathbf{x})\}_{I=1}^{NP}$ are solved locally by minimizing the least-squares errors in MLS or by satisfying the reproducing conditions in the RK approximation, and therefore these basis functions are called the intrinsic bases. Thus, the *h-p* cloud approximation in (8.67) can be viewed as an approximation with both intrinsic and extrinsic enrichments. The extrinsic bases are allowed to vary from node to node, whereas the intrinsic bases in MLS and RK cannot vary in space without introducing a discontinuity in the approximation.

8.3 Natural Element Method

Another class of meshfree methods in which the approximation functions are constructed based on a set of scattered points is the natural element method (NEM). In this approach, the partition of unity is constructed based on Sibson interpolation. Sibson interpolation was formulated based on natural neighbor coordinates originally introduced by Sibson [22, 23] and has been widely used for multivariate data fitting and reconstruction. Natural neighbor interpolation (or Sibson interpolation) was first introduced for solving PDEs by Braun and Sambridge [24] and for solving mechanics problems by Sukumar, Moran, and Belytschko [25]. A comprehensive review of NEM can be found in [26].

8.3.1 First-Order Voronoi Diagram and Delaunay Triangulation

The natural neighbor interpolation is constructed by using the geometry of the first- and second-order Voronoi diagrams. To start, let the domain Ω be discretized by a set S of NP scattered nodes $S = \{\mathbf{x}_1, \mathbf{x}_2, \ldots, \mathbf{x}_{NP}\}$ as in Chapter 3. The Voronoi diagram $\mathcal{V}(S)$ of the set S is a subdivision of the domain into regions $\mathcal{V}_I$, such that any point in $\mathcal{V}_I$ is closer to node I than to any other $\mathbf{x}_J \in S, J \neq I$,

$$\mathcal{V}_I = \{\mathbf{x} | d(\mathbf{x}, \mathbf{x}_I) \leq d(\mathbf{x}, \mathbf{x}_J), \forall J \neq I\}, \ \forall I. \tag{8.68}$$

where d is the distance function, usually the Euclidean distance. The subdivision $\mathcal{V}_I$, $\mathbf{x}_I \in S$, forms the first-order Voronoi diagram as shown in Figure 8.4.

The Delaunay tessellation (DT) is the dual of the Voronoi diagram constructed by connecting nodes that have a common Voronoi facet as shown in Figure 8.4. Important properties of the DT are

1) The circumcircle of DT contains no other nodes in S.
2) Given any nodes in the set S, the Voronoi diagram is unique, whereas the DT is not.
3) Let l_{IJ} be the common edge facet of $\mathcal{V}_I$ and $\mathcal{V}_J$, where $\mathbf{x}_J$ is one of the natural neighbors of $\mathbf{x}_I$, i.e., $\mathcal{V}_I \cap \mathcal{V}_J \neq \emptyset$. Then $l_{IJ} \perp \overline{\mathbf{x}_I\mathbf{x}_J}$ and l_{IJ} intersects with $\overline{\mathbf{x}_I\mathbf{x}_J}$ at the midpoint of $\overline{\mathbf{x}_I\mathbf{x}_J}$.

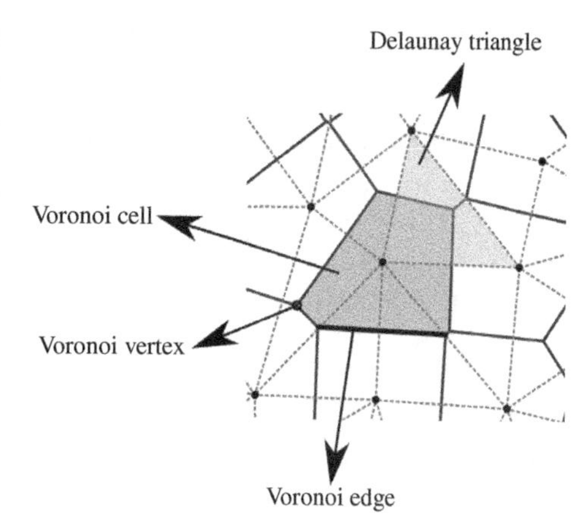

Figure 8.4 First-order Voronoi diagram and Delaunay triangulation.

8.3.2 Second-Order Voronoi Cell and Sibson Interpolation

To define the Sibson interpolation at $\mathbf{x}$, we further define a second-order Voronoi cell $\mathcal{V}_{IJ}$ $(I \neq J)$:

$$\mathcal{V}_{IJ} = \{\mathbf{x} | d(\mathbf{x}, \mathbf{x}_I) \leq d(\mathbf{x}, \mathbf{x}_J) \leq d(\mathbf{x}, \mathbf{x}_K), \quad \forall K \neq I, J\}. \tag{8.69}$$

Here $\mathcal{V}_{IJ}$ is the locus of all points that have $\mathbf{x}_I$ as the nearest neighbor, and $\mathbf{x}_J$ as the second nearest neighbor.

The natural neighbor (Sibson) shape function is constructed by first inserting "point of evaluation" P at $\mathbf{x}$ into the Voronoi diagram of a set of nodes S. Then, the shape function of node I evaluated at the location P is defined as the ratio of the area of the second-order Voronoi cell ($A_I(\mathbf{x})$= area of $\mathcal{V}_{PI}$), i.e., the overlapping region shown in Figure 8.5, to the total area of the first-order Voronoi cell ($A(\mathbf{x})$ = area of $\mathcal{V}_P$):

$$\phi_I(\mathbf{x}) = \frac{A_I(\mathbf{x})}{A(\mathbf{x})}, \tag{8.70}$$

where $A(\mathbf{x}) = \sum_{J \in G_P} A_J(\mathbf{x})$, and $G_P = \{I \mid \mathcal{V}_I \cap \mathcal{V}_P \neq \emptyset, \ \forall I \in S\}$ is the set of node numbers in which the corresponding Voronoi cells $\mathcal{V}_I$ have interaction with the Voronoi cell $\mathcal{V}_P$ associated with the evaluation point P at $\mathbf{x}$. This set of Sibson shape functions has the partition of unity property:

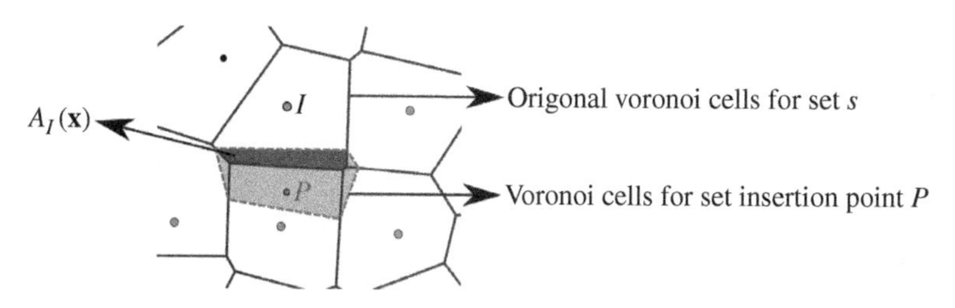

Figure 8.5 Natural neighbors of an inserted point P and the second-order Voronoi diagram associated with the evaluation point P, with definition of the associated area.

$$\sum_{I \in G_P} \phi_I(\mathbf{x}) = 1. \tag{8.71}$$

The derivatives of the Sibson shape function are

$$\phi_{I,j}(\mathbf{x}) = \frac{A_{I,j}(\mathbf{x}) - \phi_I(\mathbf{x}) A_{,j}(\mathbf{x})}{A(\mathbf{x})}. \tag{8.72}$$

The important properties of the Sibson shape function are summarized as follows:

1) Positivity: $0 \leq \phi_I(\mathbf{x}) \leq 1$.
2) Interpolation: $\phi_I(\mathbf{x}_J) = \delta_{IJ}$.
3) Partition of unity: $\sum_{I=1}^{NP} \phi_I(\mathbf{x}) = 1$.
4) First-order consistency: $\sum_{I=1}^{NP} \phi_I(\mathbf{x}) \mathbf{x}_I = \mathbf{x}$.
5) Compact support: the support of ϕ_I is the union of Delaunay circumcircles about node I, as shown in Figure 8.6.
6) Continuity: $\phi_I(\mathbf{x}) \in C^1$ for $\mathbf{x} \neq \mathbf{x}_K$, and $\phi_I(\mathbf{x}_K) \in C^0, \quad \forall K$.

8.3.3 Laplace Interpolant (Non-Sibson Interpolation)

Computing $A_I(\mathbf{x})$, $A(\mathbf{x})$, and their derivatives in the Sibson interpolants are computationally costly. In three dimensions, computing volumes is even more demanding. The Laplace shape functions have been introduced to reduce the order of geometry involved in the shape function construction by one. That is, it requires line calculation in the construction of two-dimensional

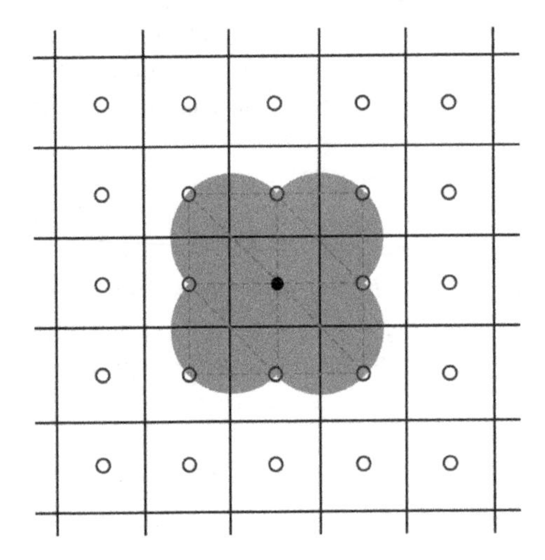

Figure 8.6 Compact support of Sibson shape function.

shape functions and areas in the construction of three-dimensional shape functions. We demonstrate the main ideas by using two-dimensional shape functions. With reference to Figure 8.7, the Laplace shape function is defined as [26]

$$\phi_I(\mathbf{x}) = \frac{\alpha_I(\mathbf{x})}{\alpha(\mathbf{x})}, \quad \alpha_I(\mathbf{x}) = \frac{S_I(\mathbf{x})}{h_I(\mathbf{x})}, \quad \alpha(\mathbf{x}) = \sum_J \alpha_J(\mathbf{x}), \tag{8.73}$$

where $S_I(\mathbf{x})$ is the length (or area in 3D) of the Voronoi edge (facet) associated with the point of evaluation P at $\mathbf{x}$ and node I and $h_I(\mathbf{x})$ is the associated perpendicular distance to the edge. The derivatives of the Laplace shape function are

$$\phi_{I,j}(\mathbf{x}) = \frac{\alpha_{I,j}(\mathbf{x}) - \phi_I(\mathbf{x})\alpha_{,j}(\mathbf{x})}{\alpha(\mathbf{x})}, \tag{8.74}$$

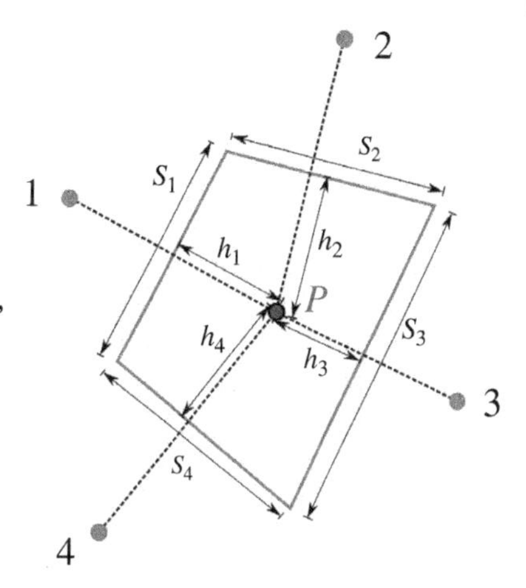

Figure 8.7 Geometry parameters in the Laplace interpolant.

The properties of the Laplace interpolation function are the same as that of the Sibson interpolation function, except that the Sibson interpolation function is C^1 everywhere except at nodes and boundaries of the support, whereas only C^0 continuity is achieved for the Laplace interpolation function.

Solving PDEs under the Galerkin weak form approach using natural neighbor interpolants as test and trial functions is called the natural element method (or NEM), first introduced by Braun and Sambridge [24]. Sukumar et al. [25] extended the application of NEM to solid mechanics and showed that C^1 continuity can be achieved through use of Bézier splines [27]. Since these Sibson and non-Sibson type NEM shape functions are not strictly linear over nonconvex boundaries, the approach does not make the test functions vanish over the whole essential boundary region. A modification of the initial NEM version was introduced by Cueto et al. [28] based on α-shapes to ensure the linear completeness of the interpolant over convex and nonconvex boundaries. Based on the non-Sibsonian interpolant proposed by Belikov et al. [29], which was linearly complete on nonconvex boundaries, Sukumar et al. [30] presented a new implementation of NEM using the non-Sibsonian interpolant.

The SCNI quadrature scheme presented in Chapter 6 [16, 31] is particularly well suited for its application to the NEM, since Voronoi cells employed in the natural neighbor shape functions are readily available for SCNI strain smoothing. Yoo et al. [32] introduced SCNI for domain integration in NEM, called nodal-NEM, and illustrated improved performance and significant advantages over NEM integrated by Gauss quadrature. This nodal-NEM also showed substantial promise for problems with material incompressibility, large deformations, and for the computation of higher order gradients [32].

References

1 Gingold, R.A. and Monaghan, J.J. (1977). Smoothed particle hydrodynamics: theory and application to non-spherical stars. *Mon. Not. R. Astron. Soc.* 181 (3): 375–389.

2 Lucy, L.B. (1977). A numerical approach to the testing of the fission hypothesis. *Astron. J.* 82: 1013–1024.

3 Monaghan, J.J. (1982). Why particle methods work. *SIAM J. Sci. Stat. Comput.* 3 (4): 422–433.
4 Monaghan, J.J. (1988). An introduction to SPH. *Comput. Phys. Commun.* 48 (1): 89–96.
5 Swegle, J.W., Attaway, S.W., Heinstein, M.W. et al. An analysis of smoothed particle hydrodynamics. Technical Report SAND93-2513, Sandia National Laboratories, 1994.
6 Randles, P.W. and Libersky, L.D. (1996). Smoothed particle hydrodynamics: some recent improvements and applications. *Comput. Methods Appl. Mech. Eng.* 139 (1–4): 375–408.
7 Chen, J.-S., Pan, C., Wu, C.-T., and Liu, W.K. (1996). Reproducing kernel particle methods for large deformation analysis of non-linear structures. *Comput. Methods Appl. Mech. Eng.* 139 (1–4): 195–227.
8 Guan, P.C., Chen, J.-S., Wu, Y. et al. (2009). Semi-Lagrangian reproducing kernel formulation and application to modeling earth moving operations. *Mech. Mater.* 41 (6): 670–683.
9 Guan, P.C., Chi, S.W., Chen, J.-S. et al. (2011). Semi-Lagrangian reproducing kernel particle method for fragment-impact problems. *Int. J. Impact Eng.* 38 (12): 1033–1047.
10 Belytschko, T., Guo, Y., Liu, W.K., and Xiao, S.P. (2000). A unifieded stability analysis of meshless particle methods. *Int. J. Numer. Methods Eng.* 48 (9): 1359–1400.
11 Monaghan, J.J. (2000). SPH without a tensile instability. *J. Comput. Phys.* 159 (2): 290–311.
12 Swegle, J.W., Hicks, D.L., and Attaway, S.W. (1995). Smoothed particle hydrodynamics stability analysis. *J. Comput. Phys.* 116 (1): 123–134.
13 Dyka, C.T. and Ingel, R.P. (1995). An approach for tension instability in smoothed particle hydrodynamics (SPH). *Comput. Struct.* 57 (4): 573–580.
14 Beissel, S.R. and Belytschko, T. (1996). Nodal integration of the element-free Galerkin method. *Comput. Methods Appl. Mech. Eng.* 139: 49–74.
15 Bonet, J. and Kulasegaram, S. (2000). Correction and stabilization of smooth particle hydrodynamics methods with applications in metal forming simulations. *Int. J. Numer. Methods Eng.* 47: 1189–1214.
16 Chen, J.-S., Wu, C.-T., and Yoon, S. (2001). A stabilized conforming nodal integration for Galerkin mesh-free methods. *Int. J. Numer. Methods Eng.* 50 (2): 435–466.
17 Melenk, J.M. and Babuška, I. (1996). The partition of unity finite element method: basic theory and applications. *Comput. Methods Appl. Mech. Eng.* 139 (1–4): 289–314.
18 Belytschko, T., Krongauz, Y., Organ, D. et al. (1996). Meshless methods: An overview and recent developments. *Comput. Methods Appl. Mech. Eng.* 139: 3–47.
19 Babuška, I. and Melenk, J.M. (1997). The partition of unity method. *Int. J. Numer. Methods Eng.* 40 (4): 727–758.
20 Duarte, C.A.M. and Oden, J.T. (1996). *H-p* clouds — an *h-p* meshless method. *Numer. Methods Partial Differ. Equ.* 12 (6): 673–705.
21 Duarte, C.A.M. and Oden, J.T. (1996). An *h-p* adaptive method using clouds. *Comput. Methods Appl. Mech. Eng.* 139 (1–4): 237–262.
22 Sibson, R. (1980). A vector identity for the Dirichlet tessellation. *Math. Proc. Camb. Philos. Soc.* 87 (1): 151.
23 Sibson, R. (1981). A brief description of natural neighbour interpolation. In: *Interpreting Multivariate Data* (ed. V. Barnett), 21–36. Chichester: John Wiley.
24 Braun, J. and Sambridge, M. (1995). A numerical method for solving partial differential equations on highly irregular evolving grids. *Nature* 376 (6542): 655–660.
25 Sukumar, N., Moran, B., and Belytschko, T. (1998). The natural element method in solid mechanics. *Int. J. Numer. Methods Eng.* 43 (5): 839–887.
26 Cueto, E., Sukumar, N., Calvo, B. et al. (2003). Overview and recent advances in natural neighbour Galerkin methods. *Arch. Comput. Methods Eng.* 10 (4): 307–384.
27 Sukumar, N. and Moran, B. (1999). C1 natural neighbor interpolant for partial differential equations. *Numer. Methods Partial Differ. Equations An Int. J.* 15 (4): 417–447.

28 Cueto, E., Doblaré, M., and Gracia, L. (2000). Imposing essential boundary conditions in the natural element method by means of density-scaled α-shapes. *Int. J. Numer. Methods Eng.* 49 (4): 519–546.

29 Belikov, V.V., Ivanov, V.D., Kontorovich, V.K. et al. (1997). The non-Sibsonian interpolation: a new method of interpolation of the values of a function on an arbitrary set of points. *Comput. Math. Math. Phys.* 37 (1): 9–15.

30 Sukumar, N., Moran, B., Yu Semenov, A., and Belikov, V.V. (2001). Natural neighbour Galerkin methods. *Int. J. Numer. Methods Eng.* 50 (1): 1–27.

31 Chen, J.-S., Yoon, S., and Wu, C.-T. (2002). Non-linear version of stabilized conforming nodal integration for Galerkin mesh-free methods. *Int. J. Numer. Methods Eng.* 53 (12): 2587–2615.

32 Yoo, J.W., Moran, B., and Chen, J.-S. (2004). Stabilized conforming nodal integration in the natural-element method. *Int. J. Numer. Methods Eng.* 60 (5): 861–890.

9

Strong Form Collocation Meshfree Methods

While there are many attractive features of Galerkin-based meshfree methods, an alternative approach is to utilize the smoothness of the approximations to discretize partial differential equations (PDEs) directly. These are called strong form collocation methods. The governing equation and boundary conditions (the strong form) are evaluated (collocated) at a collection of points in the domain and on the boundaries. In contrast to some misconceptions, the number of collocation points does not have to be equal to the number of nodes, and in fact, it is often desirable to have more collocation points to achieve better accuracy and convergence.

One advantage of this approach is that it obviates the need for quadrature: no weak (integral) form is involved, and the complexity of numerical integration in Galerkin meshfree methods is precluded. Essential boundary conditions are also directly enforced (they are collocated), and there is also no reason to develop special variational principles or procedures to enforce these conditions.

The generalized finite difference method developed by Jenson [1] in 1972 could be considered the earliest strong-form meshfree approach that employs scattered points, and even the first meshfree method as it predates smoothed particle hydrodynamics. Perrone, Kao, Liszka, and Orkisz then refined it in the late 70s and early 1980s [2–4] making it much more practical (see Section 3.5.4). In any case, these methods can be considered a type of (or closely related to) an implicit gradient approach, as discussed in Section 3.5, so we will not explicitly cover them. They are also generally not based on formal functional approximations, so the enforcement of boundary conditions complicates the presentation.

Although methods for interpolating scattered data had existed for decades (see [5]), using these functional approximations for strong form solutions of PDEs did not emerge until Kansa's seminal work in 1990 [6, 7]. He employed global radial basis functions (RBFs), and the method has since been referred to as Kansa's method, or more commonly, the radial basis collocation method (RBCM). Subsequently, the compactly supported meshfree approximations discussed in Chapter 3 were employed. Oñate developed the finite point method in 1996 using the weighted-least squares approximation [8], and Aluru employed point collocation with the reproducing kernel approximation [9], which was later termed the reproducing kernel collocation method (RKCM) and extensively refined by Hu, Chen, and co-workers [10–12].

While there have been many developments in the collocation literature, in this chapter, we have selected topics that are most relevant to a reader interested in meshfree formulations (although they have applicability to other methods). Section 9.1 first describes the collocation formulation, and commonly used approximation functions are then discussed in Section 9.2. Hu et al. [13] derived weights in the least-squares solution of overdetermined systems that yield optimal convergence and this is discussed in Section 9.3. A gradient reproducing kernel collocation method developed by Chi

Meshfree and Particle Methods: Fundamentals and Applications, First Edition.
Ted Belytschko, J. S. Chen, and Michael Hillman.

and co-workers [11, 14] is then presented in Section 9.4, which alleviates the computational complexity in RKCM by employing implicit gradients (see Section 3.5.3). This method also allows for optimal convergence with nodal collocation and simplifies computer implementation. Section 9.5 then describes a subdomain method for problems with heterogeneities (weak discontinuities) and cracks (strong discontinuities). Section 9.6 concludes the chapter with a performance comparison of node-based Galerkin and collocation methods.

9.1 The Meshfree Collocation Method

For demonstration, consider the application of strong form collocation to the boundary value problem for elasticity (see Chapter 2):

$$\begin{aligned} \nabla\cdot\boldsymbol{\sigma}(\mathbf{u}(\mathbf{x})) &= -\mathbf{b}(\mathbf{x}) && \forall\mathbf{x}\in\Omega,\\ \mathbf{n}\cdot\boldsymbol{\sigma}(\mathbf{u}(\mathbf{x})) &= \bar{\mathbf{t}}(\mathbf{x}) && \forall\mathbf{x}\in\Gamma_t,\\ \mathbf{u}^h(\mathbf{x}) &= \bar{\mathbf{u}}(\mathbf{x}) && \forall\mathbf{x}\in\Gamma_u. \end{aligned} \tag{9.1}$$

In the above, we have explicitly written out the elasticity Eqs. (2.69)–(2.71) in terms of their dependence on **u** and **x**, and moved the given data to the right-hand side.

Now consider enforcing the residuals of the above to be zero at Nc collocation points $\{\mathbf{x}_{\hat{J}}\}_{\hat{J}=1}^{Nc}\in\bar{\Omega}$ where $\bar{\Omega}\equiv\Omega\cup\Gamma$ with $\Gamma=\Gamma_u\cup\Gamma_t$, i.e., the collocation points are distributed on each domain and boundary in (9.1):

$$\begin{aligned} \nabla\cdot\boldsymbol{\sigma}(\mathbf{u}(\mathbf{x}_{\hat{J}})) &= -\mathbf{b}(\mathbf{x}_{\hat{J}}) && \forall\mathbf{x}_{\hat{J}}\in\Omega,\\ \mathbf{n}\cdot\boldsymbol{\sigma}(\mathbf{u}(\mathbf{x}_{\hat{J}})) &= \bar{\mathbf{t}}(\mathbf{x}_{\hat{J}}) && \forall\mathbf{x}_{\hat{J}}\in\Gamma_t,\\ \mathbf{u}(\mathbf{x}_{\hat{J}}) &= \bar{\mathbf{u}}(\mathbf{x}_{\hat{J}}) && \forall\mathbf{x}_{\hat{J}}\in\Gamma_u. \end{aligned} \tag{9.2}$$

To complete the process and form an algebraic system we can solve, let the approximation $\mathbf{u}^h$ of **u** be expressed as a linear combination of basis functions:

$$\mathbf{u}^h(\mathbf{x}) = \sum_{I=1}^{Ns} g_I(\mathbf{x})\mathbf{u}_I, \tag{9.3}$$

where Ns is the number of nodes (called source points in the collocation literature), g_I is the meshfree shape function associated with $\mathbf{x}_I$, and $\mathbf{u}_I$ is the corresponding coefficient. In this chapter, we will adopt the standard terminology of collocation methods and often refer to the set of nodes $\{\mathbf{x}_I\}_{I=1}^{Ns}$ as *source points*.

Introducing the approximation (9.3) into (9.2) yields a discrete system of equations of size $Nc\times Ns$:

$$\begin{aligned} \nabla\cdot\boldsymbol{\sigma}(\mathbf{u}^h(\mathbf{x}_{\hat{J}})) &= -\mathbf{b}(\mathbf{x}_{\hat{J}}) && \forall\mathbf{x}_{\hat{J}}\in\Omega,\\ \mathbf{n}\cdot\boldsymbol{\sigma}(\mathbf{u}^h(\mathbf{x}_{\hat{J}})) &= \bar{\mathbf{t}}(\mathbf{x}_{\hat{J}}) && \forall\mathbf{x}_{\hat{J}}\in\Gamma_t,\\ \mathbf{u}^h(\mathbf{x}_{\hat{J}}) &= \bar{\mathbf{u}}(\mathbf{x}_{\hat{J}}) && \forall\mathbf{x}_{\hat{J}}\in\Gamma_u. \end{aligned} \tag{9.4}$$

An illustration of the collocation discretization process is shown in Figure 9.1: meshfree nodes (source points) can be distributed as desired, and the collocation points are then chosen. The two most common approaches in selecting these points are:

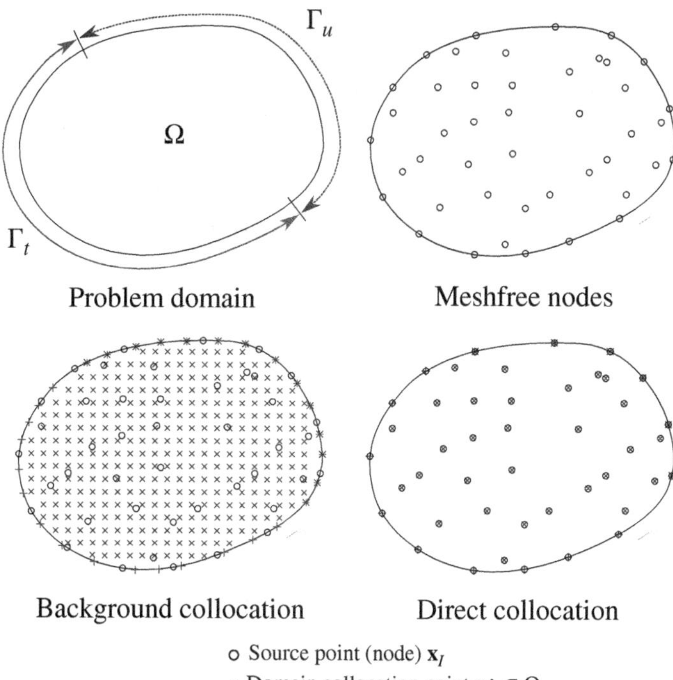

Figure 9.1 Illustration of meshfree domain discretization and collocation point distribution: two schemes are shown: background and direct.

1) **Background collocation.** The source and collocation points are independent. Here, the source points need not be in any particular location, e.g. on the essential or natural boundary. Typically, $Nc > Ns$.
2) **Direct collocation.** The source points also serve as collocation points, yielding $Nc = Ns$. This is termed the direct collocation method [13].

When $Nc > Ns$, as is often the case in background collocation, (9.4) leads to an overdetermined system, and its solution can be obtained by using a least-squares method. The solution using $Nc > Ns$ usually offers better accuracy and the solution is less sensitive to the nodal distribution; however, to achieve optimal accuracy in general, the least-squares system for $Nc > Ns$ needs to be properly weighted. This is referred to as the weighted collocation method, which will be discussed in Section 9.3. On the other hand, the gradient collocation method presented in Section 9.4 allows for optimal convergence with $Nc = Ns$.

Remark 9.1 *It can be noted that the collocation in Eq. (9.2) is equivalent to the weighted residual (see Chapter 2) of (2.69)–(2.71), i.e., seeking* $\mathbf{u} \in \mathrm{H}^2$ *such that* $\forall \mathbf{v},\ \mathbf{v}_t,\ \mathbf{v}_u \in \mathrm{L}_2$:

$$
\begin{aligned}
&\int_{\Omega} \mathbf{v}(\mathbf{x}) \cdot (\nabla \cdot \boldsymbol{\sigma}(\mathbf{u}(\mathbf{x})) + \mathbf{b}(\mathbf{x})) d\Omega, \\
&+ \int_{\Gamma_t} \mathbf{v}_t(\mathbf{x}) \cdot (\bar{\mathbf{t}}(\mathbf{x}) - \mathbf{n} \cdot \boldsymbol{\sigma}(\mathbf{u}(\mathbf{x}))) d\Gamma + \int_{\Gamma_u} \mathbf{v}_u(\mathbf{x}) \cdot (\mathbf{u}(\mathbf{x}) - \bar{\mathbf{u}}(\mathbf{x})) d\Gamma = 0.
\end{aligned}
\tag{9.5}
$$

The weighted residual (9.5) leads to (9.2) when $\mathbf{v} = \mathbf{v}_t = \mathbf{v}_u = \sum_{\hat{J}=1}^{Nc} \delta(\mathbf{x} - \mathbf{x}_{\hat{J}})\mathbf{v}_{\hat{J}}$ *where* $\delta(\cdot)$ *is the Dirac Delta function in d-dimensional space and* $\mathbf{v}_{\hat{J}}$ *is the associated arbitrary coefficient. Note that in this approach the admissible approximation* $\mathbf{u}^h$ *is required to be in* H^2*, which is difficult for conventional finite elements to achieve. However, for the general meshfree approximations discussed in Chapter 3, the regularity requirement can easily be satisfied.*

Following the ideas laid out in Chapter 3, the summation in (9.3) can be condensed when compactly supported basis functions are used:

$$\mathbf{u}^h(\mathbf{x}) = \sum_{I=1}^{Ns} g_I(\mathbf{x})\mathbf{u}_I = \sum_{I \in S_\mathbf{x}} g_I(\mathbf{x})\mathbf{u}_I, \tag{9.6}$$

where $S_\mathbf{x} = \{I \mid \mathbf{x} \in \operatorname{supp} g_I(\mathbf{x})\}$. Then (9.2) can rewritten as

$$\begin{aligned} &\sum_{I \in S_{\mathbf{x}_{\hat{J}}}} \mathcal{L} g_I(\mathbf{x}_{\hat{J}})\mathbf{u}_I = -\mathbf{b}(\mathbf{x}_{\hat{J}}) \quad \forall \mathbf{x}_{\hat{J}} \in \Omega, \\ &\sum_{I \in S_{\mathbf{x}_{\hat{J}}}} \mathcal{B}_t g_I(\mathbf{x}_{\hat{J}})\mathbf{u}_I = \bar{\mathbf{t}}(\mathbf{x}_{\hat{J}}) \quad \forall \mathbf{x}_{\hat{J}} \in \Gamma_t, \\ &\sum_{I \in S_{\mathbf{x}_{\hat{J}}}} \mathcal{B}_u g_I(\mathbf{x}_{\hat{J}})\mathbf{u}_I = \bar{\mathbf{u}}(\mathbf{x}_{\hat{J}}) \quad \forall \mathbf{x}_{\hat{J}} \in \Gamma_u, \end{aligned} \tag{9.7}$$

where, for two dimensions in elasticity

$$\begin{aligned} \mathcal{L} &= \begin{bmatrix} (\lambda + 2\mu)\dfrac{\partial^2}{\partial x^2} + \mu\dfrac{\partial^2}{\partial y^2} & (\lambda + \mu)\dfrac{\partial^2}{\partial x \partial y} \\ (\lambda + \mu)\dfrac{\partial^2}{\partial x \partial y} & \mu\dfrac{\partial^2}{\partial x^2} + (\lambda + 2\mu)\dfrac{\partial^2}{\partial y^2} \end{bmatrix}, \\ \mathcal{B}_t &= \begin{bmatrix} (\lambda + 2\mu)n_x\dfrac{\partial}{\partial x} + \mu n_y\dfrac{\partial}{\partial y} & \mu n_y\dfrac{\partial}{\partial x} + \lambda n_x\dfrac{\partial}{\partial y} \\ \lambda n_y\dfrac{\partial}{\partial x} + \mu n_x\dfrac{\partial}{\partial y} & \mu n_x\dfrac{\partial}{\partial x} + (\lambda + 2\mu)n_y\dfrac{\partial}{\partial y} \end{bmatrix}, \\ \mathcal{B}_u &= \begin{bmatrix} 1 & 0 \\ 0 & 1 \end{bmatrix}. \end{aligned} \tag{9.8}$$

A matrix form for (9.7) can be obtained as

$$\mathbf{K}\mathbf{d} = \mathbf{f}, \tag{9.9}$$

where

$$\mathbf{K} = \begin{bmatrix} \mathbf{K}^{\mathcal{L}} \\ \mathbf{K}^{\mathcal{B}_t} \\ \mathbf{K}^{\mathcal{B}_u} \end{bmatrix}, \quad \mathbf{f} = \begin{bmatrix} \mathbf{f}^{\mathcal{L}} \\ \mathbf{f}^{\mathcal{B}_t} \\ \mathbf{f}^{\mathcal{B}_u} \end{bmatrix}, \tag{9.10}$$

with the matrices and vectors constructed as $\mathbf{K}^{\mathcal{L}} = [\mathcal{L} g_J(\mathbf{x}_{\hat{I}})]$, $\mathbf{K}^{\mathcal{B}_t} = [\mathcal{B}_t g_J(\mathbf{x}_{\hat{I}})]$, $\mathbf{K}^{\mathcal{B}_u} = [\mathcal{B}_u g_J(\mathbf{x}_{\hat{I}})]$, $\mathbf{f}^{\mathcal{L}} = \{-\mathbf{b}(\mathbf{x}_{\hat{I}})\}$, $\mathbf{f}^{\mathcal{B}_t} = \{\bar{\mathbf{t}}(\mathbf{x}_{\hat{I}})\}$, $\mathbf{f}^{\mathcal{B}_u} = \{\bar{\mathbf{u}}(\mathbf{x}_{\hat{I}})\}$, and $\mathbf{d} = \{\mathbf{u}_I\}$. It should be apparent from (9.6) that when compactly supported shape functions are employed, the left-hand side matrix in (9.9) is sparse; it is dense when the approximations are global.

As previously discussed, when $Nc > Ns$, the overdetermined matrix system in (9.9) can be solved by weighted least squares:

$$\mathbf{K}^{\mathrm{T}}\mathbf{W}\mathbf{K}\mathbf{d} = \mathbf{K}^{\mathrm{T}}\mathbf{W}\mathbf{f}, \tag{9.11}$$

where $\mathbf{W}$ is a matrix of weights, which should be chosen according to the discussions in Section 9.3 in order to obtain optimal convergence.

Key Takeaways

1) The strong form can directly be solved using the smoothness of meshfree approximations.
2) The evaluation points (called the collocation points) can be chosen to be independent of the nodal positions (called source points). $Nc > Ns$ results in an over-determined system which can be solved by using weighted least squares, with optimal convergence rates obtained by using weights discussed in Section 9.3.
3) The collocation points can also be chosen as the source points ($Nc = Ns$), which yields a determined system and relative simplicity, but generally at the cost of reduced accuracy and convergence. Nevertheless, the implicit gradient approach in Section 9.4 still allows for optimal convergence with $Nc = Ns$.

9.2 Approximations and Convergence for Strong Form Collocation

Next, we will discuss several meshfree techniques to construct the smooth approximation $\mathbf{u}^h$. Two popular approaches are first given: the global RBFs, and the compactly supported moving least squares (MLS) and reproducing kernel (RK) approximations introduced in Chapter 3. We will then introduce a technique for localizing the RBFs using RK shape functions.

9.2.1 Radial Basis Functions

Although any C^2 approximation can be used in the collocation methods, the RBFs are popular for the solution of PDEs, tracing back to the work of Kansa [6, 7]. The theoretical foundation of the RBF method for solving PDEs has been established by Franke and Schaback [15], and error estimates for the solution of smooth problems have been derived by Wendland [16]. RBFs have been used in many applications, such as surface fitting, turbulence analysis, neural networks, meteorology, and so forth. Hardy [17, 18] investigated multiquadric RBFs for scattered data interpolation problems, and good performance using multiquadric and thin-plate spline RBFs has been observed [15]. Since then, the advances in applying RBFs to various problems have progressed consistently.

A few commonly used RBFs are given below:

Multiquadrics (MQs):

$$g_I(\mathbf{x}) = \left(r_I^2 + \delta^2\right)^{m-\frac{3}{2}}, m = 1, 2, \ldots \tag{9.12}$$

Gaussian:

$$g_I(\mathbf{x}) = \exp\left(-\frac{r_I^2}{\delta^2}\right), \tag{9.13}$$

Thin plate splines:

$$g_I(\mathbf{x}) = \begin{cases} r_I^{2m} \ln r_I, & (2-\text{dimensional}) \\ r_I^{2m-1}, & (3-\text{dimensional}) \end{cases} \tag{9.14}$$

Logarithmic:

$$g_I(\mathbf{x}) = r_I^m \ln r_I, \tag{9.15}$$

In the above, $r_I = \|\mathbf{x} - \mathbf{x}_I\|$ with $\|\cdot\|$ the Euclidean distance norm, δ is a constant, m is an integer, and $\mathbf{x}_I$ is the source point. For a function $u(\mathbf{x})$, the approximation denoted by $u^h(\mathbf{x})$ is expressed as in (9.3), with $g_I(\mathbf{x})$ selected as an RBF function. The application to the strong form of partial differential equations is natural as the they are infinitely differentiable ($g_I(\mathbf{x}) \in C^\infty$). When used in the collocation Eq. (9.7), the method is called the radial basis collocation method (RBCM).

The constant δ in (9.12) and (9.13) is called the *shape parameter*. The MQ RBF in (9.12) is the most popular function used in the solution of PDEs; the function is called reciprocal if $m = 1$, linear if $m = 2$, cubic if $m = 3$, and so on. Error bounds for MQs have been established by Madych [19], local errors of scattered data interpolation by RBFs in suitable variational formulations have been investigated [20], and the convergence of RBFs in Sobolev spaces has been demonstrated [21]. All these studies present exponential convergence of RBFs. It has also been demonstrated that the convergence rate is accelerated for monotonically increasing values of δ [22].

Madych [19] showed that there exists an exponential convergence rate given by

$$\|u - u^h\|_{H^s} \leq c_\nu \eta^{\delta/H} \|u\|_{H^r}, \tag{9.16}$$

where $\|\cdot\|_{H^s}$ is the Sobolov norm of degree s (see the Glossary of Notation), c_ν is a generic constant where the subscript ν denotes that it is dependent on the Poisson's ratio ν, $0 < \eta < 1$ is a real number, H is the radial distance defined as $H = \sup_{\mathbf{x} \in \Omega} \left(\inf_{\mathbf{x}_I \in S} \|\mathbf{x}_I - \mathbf{x}\| \right)$, and $\|\cdot\|_{H^r}$ is induced from the regularity requirements of the approximated function u and the RBFs, see [19, 23]. The accuracy and rate of convergence of MQ-RBF approximations are determined by the shape parameter δ and the number of basis functions Ns (the number of source points). That is, the rate *increases* as the refinement increases, as seen in (9.16), which is referred to as exponential convergence. This is in contrast to nth order monomial reproducing compactly supported approximations such as MLS and RK, which offer algebraic convergence rates (constant for a given n); see the following subsection. Nevertheless, the shape functions in (9.12)–(9.15) are global in nature, making the resulting system of equations prone to linear dependence and ill conditioning.

9.2.2 Moving Least Squares and Reproducing Kernel Approximations

The MLS and RK approximations described in Chapter 3 can easily be adopted in collocation of the strong form. When using the RK approximation, it is termed the reproducing kernel collocation method (or RKCM) [9, 10, 12]. In addition to first derivatives, higher-order derivatives are mandatory when using strong form collocation. They can be obtained by direct differentiation of the MLS or RK shape functions or approximated by other approaches like those discussed in Section 3.5. The use of implicit gradients is discussed later in this chapter in Section 9.4.

Although the employment of C^2 continuous kernel or weight functions would satisfy the regularity requirements of strong form collocation, higher-order continuous functions offer better numerical stability, especially when the point density is high. Therefore, a quintic B-spline kernel is often adopted in RKCM:

$$\phi_a(z) = \begin{cases} \frac{11}{20} - \frac{9}{2}z^2 + \frac{81}{4}z^4 - \frac{81}{4}z^5 & \text{for } 0 \le z < \frac{1}{3} \\ \frac{17}{40} + \frac{15}{8}z - \frac{63}{4}z^2 + \frac{135}{4}z^3 - \frac{243}{8}z^4 + \frac{81}{8}z^5 & \text{for } \frac{1}{3} \le z < \frac{2}{3} \\ \frac{81}{40} - \frac{81}{8}z + \frac{81}{4}z^2 - \frac{81}{4}z^4 + \frac{81}{8}z^4 - \frac{81}{40}z^5 & \text{for } \frac{2}{3} \le z \le 1 \\ 0 & \text{for } z > 1 \end{cases} \tag{9.17}$$

For a smooth function $u(\mathbf{x})$, the approximation, denoted by $u^h(\mathbf{x})$, can be expressed as the linear combination of MLS/RK shape functions as in (9.6). Solving the PDE by collocation (9.4) with the MLS or RK approximation has been termed RKCM [10], and there exists an algebraic convergence rate as shown by Hu et al. [10, 12]:

$$\left\|u - u^h\right\|_E \le c\chi a^{n-1}|u|_{\mathrm{H}^{n+1}}, \tag{9.18}$$

where $|\cdot|_{\mathrm{H}^s}$ denotes the H^s semi-norm, c is a generic constant, χ is called the overlapping parameter which is equal to the maximum number of kernels covering a point, a is the maximum support measure, n is the order of complete monomials in the shape functions, and

$$\|v\|_E \equiv \left(\|v\|_{1,\Omega}^2 + \|\mathcal{L}v\|_{0,\Omega}^2 + \|\mathcal{B}_t v\|_{0,\Gamma_t}^2 + \|\mathcal{B}_u v\|_{0,\Gamma_u}^2\right)^{\frac{1}{2}}, \tag{9.19}$$

where the subscripts denote the Sobolev norm on a certain domain, $\mathcal{L}$, $\mathcal{B}_t$, and $\mathcal{B}_u$ denote the differential operators associated with the domain, Neumann boundary, and Dirichlet boundary, respectively, for general second order PDEs (e.g. see the tensorial operators in (9.8)). Contrasting (9.18) with the error estimates for Galerkin meshfree methods discussed in Section 3.4, it can be seen that the solution will not converge when $n = 0$ or $n = 1$. An order n of *at least two is mandatory for convergence* in strong-form collocation of second-order PDEs [10, 12].

9.2.3 Reproducing Kernel Enhanced Local Radial Basis

The commonly used RBF approximation functions in the strong form collocation method offers exponential convergence, however the method suffers from large condition numbers due to its "nonlocal" approximation. The MLS and RK functions, on the other hand, provide polynomial reproducibility in a "local" approximation, and the corresponding discrete systems are relatively well-conditioned. Nonetheless, RKCM produces only algebraic convergence as seen in (9.18). An approach has been proposed by Chen et al. [24] to combine the advantages of RBF and RK functions to yield a local approximation that is better conditioned than that of the RBF, while at the same time offering a higher rate of convergence than that of RK:

$$u^h(\mathbf{x}) = \sum_{I \in S_\mathbf{x}} \left[\Psi_I(\mathbf{x})\left(d_I + \sum_{J=1}^{M} \tilde{g}_I^J(\mathbf{x})\tilde{d}_I^J\right)\right], \tag{9.20}$$

where $\Psi_I(\mathbf{x})$ is the RK function with compact support, $\tilde{g}_I(\mathbf{x})$ is an RBF (e.g. (9.12)–(9.15)), and d_I and $\tilde{d}_I^J$ are coefficients to be sought. Applying the approximation in (9.20) to strong form collocation is called the localized radial basis collocation method (L-RBCM).

It can be noted that the function in (9.20) utilizes the compactly supported partition of unity to "patch" the global RBFs together (see Chapter 8). The error analysis shows that if the error due to the RK approximation is sufficiently small, the method maintains the exponential convergence of RBFs, while significantly improving the condition of the discrete system, and yields a banded matrix as discussed in [24]:

1) Using the partition of unity properties of the RK localizing function, there exists the following error bound:

$$\left\|u - u^h\right\|_{H^s} \leq \chi c \eta^{\delta/H} \|u\|_{H^r}, \tag{9.21}$$

where c is a constant, δ is the RBF shape parameter, and χ is the overlapping parameter as in (9.18). Other parameters are defined the same as in (9.16).
2) The enhanced stability in the L-RBCM can be demonstrated by a perturbation analysis of the strong form collocation equations expressed as the linear system (9.9), and the stability can be measured by the condition number of the stiffness matrix **K**. The following estimation of the condition number of L-RBCM has been obtained:

$$\text{Cond}(\mathbf{K}) \approx O\left(a^{-3d/2}\right), \tag{9.22}$$

where d is the spatial dimension. In two-dimensional elasticity, the following condition numbers can be obtained for collocation using RBFs, RK, and RK localized RBFs in (9.20):

$$\begin{aligned} &\text{RBCM}: \quad \text{Cond}(\mathbf{K}) \approx O\left(h^{-8}\right), \\ &\text{RKPM}: \quad \text{Cond}(\mathbf{K}) \approx O\left(h^{-2}\right), \\ &\text{L-RBCM}: \quad \text{Cond}(\mathbf{K}) \approx O\left(h^{-3}\right) \end{aligned} \tag{9.23}$$

The L-RBCM approach thus offers a significant improvement in stability over RBCM. Although the discrete system of L-RBCM is slightly less well-conditioned than that of RKCM, it offers exponential convergence rates similar to those obtained using RBCM.

Key Takeaways

1) MLS or RK used in strong form collocation results in algebraic convergence, but unlike the Galerkin method, it requires the basis order $n \geq 2$ for convergence. The system is sparse and well-conditioned since the functions are compactly supported.
2) RBFs in strong form collocation offers exponential convergence but can yield ill-conditioned systems of equations.
3) RBFs can be localized by the RK approximation, which results in exponential convergence and a significant improvement in conditioning of the equations over the global RBFs, combining the advantage of both approaches.

9.3 Weighted Collocation Methods and Optimal Weights

When $Nc > Ns$, the collocation Eq. (9.4) recast in a matrix form as (9.9) leads to an overdetermined system, and a least-squares method can be applied for seeking the solution, equivalent to minimizing a weighted residual. The residual is defined as $e(\mathbf{d}) = 1/2(\mathbf{Kd} - \mathbf{f})^{\mathrm{T}}\mathbf{W}(\mathbf{Kd} - \mathbf{f})$, where $\mathbf{W}$ is a symmetric weighting matrix. Minimizing $e(\mathbf{d})$ yields

$$\mathbf{K}^{\mathrm{T}}\mathbf{W}\mathbf{K}\mathbf{d} = \mathbf{K}^{\mathrm{T}}\mathbf{W}\mathbf{f}. \tag{9.24}$$

It has been shown that solving strong form collocation equations by a least-squares method is equivalent to minimizing a least-squares functional with quadrature [13]. For elasticity, the problem is to find $\mathbf{u}^h$ such that

$$E\left(\mathbf{u}^h\right) = \inf_{\mathbf{v}\in \mathrm{U}} E(\mathbf{v}), \tag{9.25}$$

where U is the admissible finite-dimensional space spanned by meshfree shape functions, and

$$E(\mathbf{v}) = \frac{1}{2}\hat{\int_{\Omega}}(\mathcal{L}\mathbf{v} + \mathbf{b})^2 d\Omega + \frac{1}{2}\hat{\int_{\Gamma_t}}(\mathcal{B}_t\mathbf{v} - \overline{\mathbf{t}})^2 d\Gamma + \frac{1}{2}\hat{\int_{\Gamma_u}}(\mathcal{B}_u\mathbf{v} - \overline{\mathbf{u}})^2 d\Gamma. \tag{9.26}$$

Recall that $\hat{\int}$ denotes integration with quadrature. It has been shown that the errors from the domain and boundary integrals in (9.26) are unbalanced [13]. Therefore, a weighted least-squares functional should be introduced:

$$E(\mathbf{v}) = \frac{1}{2}\hat{\int_{\Omega}}(\mathcal{L}\mathbf{v} + \mathbf{b})^2 d\Omega + \frac{\alpha_t}{2}\hat{\int_{\Gamma_t}}(\mathcal{B}_t\mathbf{v} - \overline{\mathbf{t}})^2 d\Gamma + \frac{\alpha_u}{2}\hat{\int_{\Gamma_u}}(\mathcal{B}_u\mathbf{v} - \overline{\mathbf{u}})^2 d\Gamma. \tag{9.27}$$

Here the weights α_t and α_u are determined by considering error balancing of the weighted least-squares functional associated with the domain and boundary equations [13]:

$$\sqrt{\alpha_t} \approx O(1), \quad \sqrt{\alpha_u} \approx O(\kappa Ns), \tag{9.28}$$

where $\kappa = \max(\lambda, \mu)$, or more generally, the maximum coefficient involved in the differential operator and boundary operator for the problem at hand. It is also noted that when dealing with nearly incompressible problems, $\kappa = \mu$ has been suggested [25] as λ grows unbounded in the incompressible limit.

For practical implementation, minimizing (9.27) is equivalent to solving the following weighted collocation equations by the least-squares method:

$$\begin{aligned}
&\sum_{I\in S_{\mathbf{x}_{\hat{J}}}} \mathcal{L}g_I\left(\mathbf{x}_{\hat{J}}\right)\mathbf{u}_I = -\mathbf{b}\left(\mathbf{x}_{\hat{J}}\right) && \forall \mathbf{x}_{\hat{J}} \in \Omega,\\
&\sqrt{\alpha_t}\sum_{I\in S_{\mathbf{x}_{\hat{J}}}} \mathcal{B}_t g_I\left(\mathbf{x}_{\hat{J}}\right)\mathbf{u}_I = \sqrt{\alpha_t}\overline{\mathbf{t}}\left(\mathbf{x}_{\hat{J}}\right) && \forall \mathbf{x}_{\hat{J}} \in \Gamma_t,\\
&\sqrt{\alpha_u}\sum_{I\in S_{\mathbf{x}_{\hat{J}}}} \mathcal{B}_u g_I\left(\mathbf{x}_{\hat{J}}\right)\mathbf{u}_I = \sqrt{\alpha_u}\overline{\mathbf{u}}\left(\mathbf{x}_{\hat{J}}\right) && \forall \mathbf{x}_{\hat{J}} \in \Gamma_u.
\end{aligned} \tag{9.29}$$

Then, the above can be recast as $\mathbf{Kd} = \mathbf{f}$ as in (9.9), with the solution obtained by $\mathbf{K}^{\mathrm{T}}\mathbf{Kd} = \mathbf{K}^{\mathrm{T}}\mathbf{f}$.

To illustrate the importance of using a properly weighted system (9.29) over the least-squares solution of (9.7), consider an infinitely long (plane-strain) tube subject to uniform internal pressure. The tube is made of an elastic material with Young's modulus $E = 3 \times 10^7$ Pa, and Poisson's ratio $\nu = 0.25$. The inner and outer radii of the tube are 4 and 10 m, respectively, and the inner surface of the tube is subjected to a pressure $p = 100$ N/m. Due to symmetry, only a quarter of the model, as shown in Figure 9.2a, is discretized by RBCM with MQ RBFs, with proper symmetric boundary conditions specified. The corresponding boundary value problem can be expressed as

$$\nabla \cdot \boldsymbol{\sigma} = \mathbf{0} \text{ in } \Omega, \tag{9.30}$$

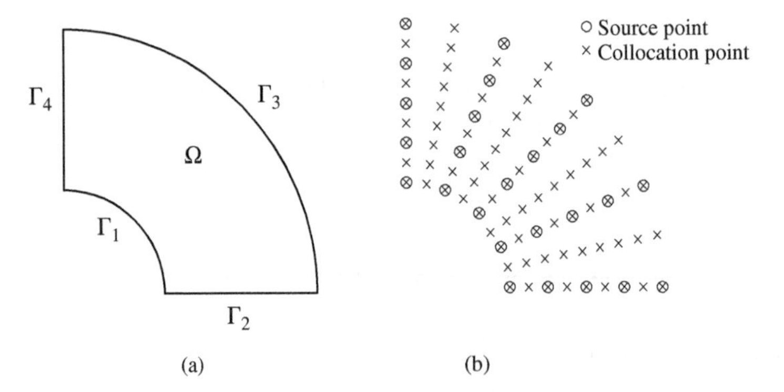

Figure 9.2 Tube problem: (a) Quarter model, and (b) distribution of source points and collocation points. *Source:* Reproduced with permission from ASCE Chen et al. [26], figure 17, p. 22.

with boundary conditions:

$$\begin{aligned} h_i &= -Pn_i && \text{on } \Gamma_1, \\ h_1 &= 0,\ u_2 = 0 && \text{on } \Gamma_2, \\ h_i &= 0 && \text{on } \Gamma_3, \\ h_2 &= 0,\ u_1 = 0 && \text{on } \Gamma_4, \end{aligned} \tag{9.31}$$

where $h_i = \sigma_{ij} n_j$.

In this problem, both source points and collocation points are uniformly distributed as shown in Figure 9.2b. Three different discretizations, 7×7, 9×9, and 11×11 source points, are used, and the shape parameters δ for the three discretizations are 10.0, 7.5, and 6.0, respectively. The number of corresponding collocation points is $(2N_1 - 1)(2N_2 - 1)$, where N_1 is the number of source points along the radial direction and N_2 is the number of source points along the angular direction.

RBCM and weighted RBCM (W-RBCM) are used in the numerical test. Weights for Dirichlet collocation equations $\sqrt{\alpha_u} = 10^9$ and Neumann collocation equations $\sqrt{\alpha_t} = 1$ are selected based on (9.28). The convergence of the two methods in the L_2 norm and H^1 semi-norm are compared in Figure 9.3. As is shown in the numerical results, the collocation method with proper weights for

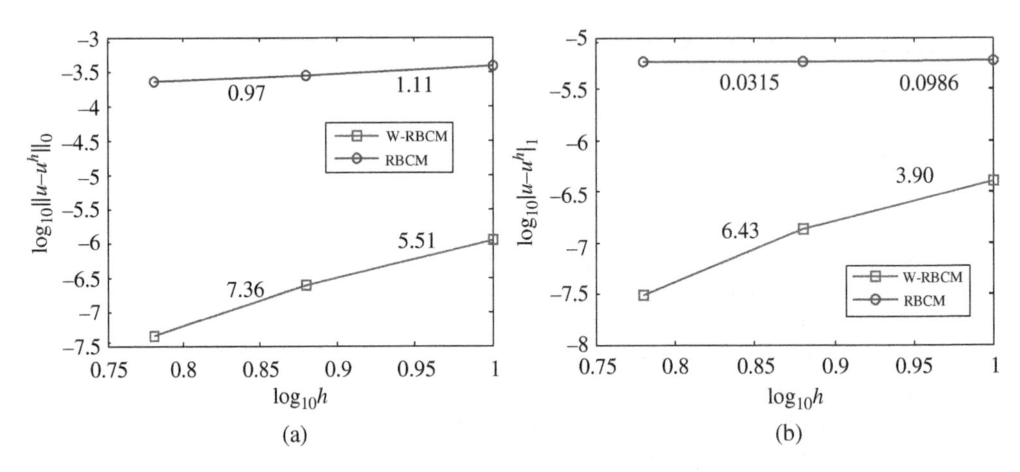

Figure 9.3 Convergence for the tube problem in (a) the L_2 error norm, and (b) the H^1 semi-norm. *Source:* Reproduced with permission from ASCE Chen et al. [26], figure 18, p. 23.

Dirichlet and Neumann boundaries offers a far superior solution. In standard RBCM, the error on the essential boundary dominates the solution error and convergence is not achieved.

9.4 Gradient Reproducing Kernel Collocation Method

While MLS/RK approximation functions can be arbitrarily smooth, taking derivatives of these functions is computationally expensive. In particular, the high complexity in RKCM is caused by matrix-vector operations and computing the derivatives of the inverse moment matrix in the multi-dimensional MLS/RK shape functions [10]. Further, for optimal convergence in RKCM, using a number of collocation points much larger than the number of source points is needed, and this adds additional computational effort [10, 12]. To enhance the computational efficiency of RKCM, an implicit gradient approximation (see Section 3.5.3) has been introduced for solving second-order PDEs with strong form collocation, which has been termed the gradient reproducing kernel collocation method (GRKCM) [11, 14].

Consider the two-dimensional version of the elastic model problem in (9.1). The approximations of the derivatives $\mathbf{u}_{,x}$ and $\mathbf{u}_{,y}$ are constructed by employing first-order implicit gradients $\left\{\Psi_I^{(\alpha)}(\mathbf{x})\right\}_{|\alpha|=1}$ where $\Psi_I^{(\alpha)}(\mathbf{x})$ is given in Section 3.5.3:

$$
\begin{aligned}
\mathbf{u}_{,x} \approx \mathbf{u}_x^h &= \sum_{I \in S_{\mathbf{x}}} \Psi_I^{(10)}(\mathbf{x})\mathbf{u}_I, \\
\mathbf{u}_{,y} \approx \mathbf{u}_y^h &= \sum_{I \in S_{\mathbf{x}}} \Psi_I^{(01)}(\mathbf{x})\mathbf{u}_I.
\end{aligned}
\tag{9.32}
$$

The approximation of second order derivatives of $\mathbf{u}$ is obtained by taking direct derivatives of $\mathbf{u}_x^h$ and $\mathbf{u}_y^h$, e.g.

$$
\begin{aligned}
\mathbf{u}_{,xx} \approx \mathbf{u}_{x,x}^h &= \sum_{I \in S_{\mathbf{x}}} \Psi_{I,x}^{(10)}(\mathbf{x})\mathbf{u}_I, \\
\mathbf{u}_{,yy} \approx \mathbf{u}_{y,y}^h &= \sum_{I \in S_{\mathbf{x}}} \Psi_{I,y}^{(01)}(\mathbf{x})\mathbf{u}_I.
\end{aligned}
\tag{9.33}
$$

Introducing (9.32) and (9.3) in the discretization in the strong form (9.1) leads to:

$$
\begin{aligned}
&\mathcal{L}^1\mathbf{u}_x^h + \mathcal{L}^2\mathbf{u}_y^h = -\mathbf{b} && \text{in } \Omega, \\
&\mathcal{B}_t^1\mathbf{u}_x^h + \mathcal{B}_t^2\mathbf{u}_y^h = \bar{\mathbf{t}} && \text{on } \Gamma_t, \\
&\mathcal{B}_u\mathbf{u}^h = \bar{\mathbf{u}} && \text{on } \Gamma_u,
\end{aligned}
\tag{9.34}
$$

where

$$
\begin{aligned}
&\mathcal{L}^1 = \begin{bmatrix} (\lambda + 2\mu)\dfrac{\partial}{\partial x} & \mu\dfrac{\partial}{\partial y} \\ \lambda\dfrac{\partial}{\partial y} & \mu\dfrac{\partial}{\partial x} \end{bmatrix}, \quad \mathcal{L}^2 = \begin{bmatrix} \mu\dfrac{\partial}{\partial y} & \lambda\dfrac{\partial}{\partial x} \\ \mu\dfrac{\partial}{\partial x} & (\lambda + 2\mu)\dfrac{\partial}{\partial y} \end{bmatrix}, \\
&\mathcal{B}_t^1 = \begin{bmatrix} (\lambda + 2\mu)n_x & \mu n_y \\ \lambda n_y & \mu n_x \end{bmatrix}, \quad \mathcal{B}_t^2 = \begin{bmatrix} \mu n_y & \lambda n_x \\ \mu n_x & (\lambda + 2\mu)n_y \end{bmatrix}.
\end{aligned}
\tag{9.35}
$$

When $Nc > Ns$, the overdetermined system can be obtained by a least-squares method with proper weights to achieve optimal solution accuracy. Introducing (9.32) into (9.34) with the weights from Section 9.3, we have

$$\begin{aligned}
&\sum_{I \in S_{\mathbf{x}_{\hat{J}}}} \left[\mathcal{L}^1 \Psi_I^{(10)}(\mathbf{x}_{\hat{J}}) + \mathcal{L}^2 \Psi_I^{(01)}(\mathbf{x}_{\hat{J}})\right] \mathbf{u}_I = -\mathbf{b}(\mathbf{x}_{\hat{J}}) && \forall \mathbf{x}_{\hat{J}} \in \Omega, \\
&\sqrt{\alpha_t} \sum_{I \in S_{\mathbf{x}_{\hat{J}}}} \left[\mathcal{B}_t^1 \Psi_I^{(10)}(\mathbf{x}_{\hat{J}}) + \mathcal{B}_t^2 \Psi_I^{(01)}(\mathbf{x}_{\hat{J}})\right] \mathbf{u}_I = \sqrt{\alpha_t}\, \bar{\mathbf{t}}(\mathbf{x}_{\hat{J}}) && \forall \mathbf{x}_{\hat{J}} \in \Gamma_t, \\
&\sqrt{\alpha_u} \sum_{I \in S_{\mathbf{x}_{\hat{J}}}} \mathcal{B}_u \Psi_I(\mathbf{x}_{\hat{J}}) \mathbf{u}_I = \sqrt{\alpha_u}\, \bar{\mathbf{u}}(\mathbf{x}_{\hat{J}}) && \forall \mathbf{x}_{\hat{J}} \in \Gamma_u.
\end{aligned} \tag{9.36}$$

For balance of the errors between the domain and boundary equations, the following weights should be selected:

$$\sqrt{\alpha_t} \approx O(1), \quad \sqrt{\alpha_u} \approx O\left(\kappa a^{p-n-1}\right), \tag{9.37}$$

where $\kappa = \max(\lambda, \mu)$, a is the kernel support measure, n is the MLS/RK order in $\mathbf{u}^h$, and p is the order of the RK approximation used to construct the implicit gradients $\mathbf{u}_x^h$ and $\mathbf{u}_y^h$. The convergence properties of GRKCM are [13]:

$$\begin{aligned}
&\left\|u - u^h\right\|_{H^1} \approx O\left(a^{p-1}\right), \\
&\left\|u_{,x} - u_{,x}^h\right\|_{H^1} + \left\|u_{,y} - u_{,y}^h\right\|_{H^1} \approx O\left(a^{p-1}\right),
\end{aligned} \tag{9.38}$$

$$\begin{aligned}
&\left\|u - u^h\right\|_{L_2} \approx O\left(a^{p}\right), \\
&\left\|u_{,x} - u_{,x}^h\right\|_{L_2} + \left\|u_{,y} - u_{,y}^h\right\|_{L_2} \approx O\left(a^{p}\right).
\end{aligned} \tag{9.39}$$

Remark 9.2 *The error estimates in (9.38)* and *(9.39) indicate that the convergence of GRKCM is only dependent on the polynomial degree p in the implicit approximation of* $\mathbf{u}_{,x}$ *and* $\mathbf{u}_{,y}$*, and is independent of the polynomial degree n in the approximation of* $\mathbf{u}$*. Further,* $p \geq 2$ *is mandatory for convergence, as in the direct gradient approach.*

Remark 9.3 *GRKCM allows the use of* $Nc = Ns$ *for sufficient accuracy* [11], *which allows optimal solutions even with direct (node-based, see Figure 9.1) collocation. This is in contrast to the use of explicitly differentiated of the shape functions, which in general, requires* $Nc > Ns$ *for sufficient accuracy, see* [13] *for a discussion.*

Mahdavi et al. [14] introduced implicit gradients to directly approximate the derivatives in second-order PDEs to reduce the computational burden even further. That is,

$$\begin{aligned}
\mathbf{u}_{,xx} \approx \mathbf{u}_{,xx}^h &= \sum_{I \in S_{\mathbf{x}}} \Psi_I^{(20)}(\mathbf{x}) \mathbf{u}_I, \\
\mathbf{u}_{,xy} \approx \mathbf{u}_{,xy}^h &= \sum_{I \in S_{\mathbf{x}}} \Psi_I^{(11)}(\mathbf{x}) \mathbf{u}_I, \\
\mathbf{u}_{,yy} \approx \mathbf{u}_{,yy}^h &= \sum_{I \in S_{\mathbf{x}}} \Psi_I^{(02)}(\mathbf{x}) \mathbf{u}_I.
\end{aligned} \tag{9.40}$$

The above employed in (9.4) leads to:

$$
\begin{aligned}
&\overline{\mathcal{L}}^1\mathbf{u}_{xx}^h + \overline{\mathcal{L}}^2\mathbf{u}_{yy}^h + \overline{\mathcal{L}}^3\mathbf{u}_{xy}^h = -\mathbf{b} && \text{in } \Omega, \\
&\mathcal{B}_t^1\mathbf{u}_x^h + \mathcal{B}_t^2\mathbf{u}_y^h = \bar{\mathbf{t}} && \text{on } \Gamma_t, \\
&\mathcal{B}_u\mathbf{u}^h = \bar{\mathbf{u}} && \text{on } \Gamma_u,
\end{aligned} \tag{9.41}
$$

where

$$
\overline{\mathcal{L}}^1 = \begin{bmatrix} (\lambda + 2\mu) & 0 \\ 0 & \mu \end{bmatrix}, \ \overline{\mathcal{L}}^2 = \begin{bmatrix} \mu & 0 \\ 0 & (\lambda + 2\mu) \end{bmatrix}, \ \overline{\mathcal{L}}^3 = \begin{bmatrix} 0 & \lambda + \mu \\ \lambda + \mu & 0 \end{bmatrix}. \tag{9.42}
$$

While these implicit gradient methods maintain the same convergence rates as their explicit counterparts, depending on the problem, the error can be higher in some cases. Nevertheless, the reduction in computational cost offsets this error when both are taken into consideration, i.e., it takes less CPU time for GRKCM to achieve the same level of error by RKCM. By the same token, for a given CPU time, GRKCM provides better accuracy. More details can be found in [14].

9.5 Subdomain Collocation for Heterogeneity and Discontinuities

Due to the overlapping supports of meshfree shape functions (particularly RBFs) and the high smoothness required in the strong form collocation methods, special treatments are required for problems with weak or strong discontinuities (e.g. material heterogeneity or cracks). The subdomain collocation method [27, 28] has been introduced for this purpose.

Take the elastic domain shown in Figure 9.4 as an example. The domain has two subdivisions, Ω^+ and Ω^-, with boundaries Γ^+ and Γ^-, respectively, with $\overline{\Omega}^{\pm} = \Omega^{\pm} \cup \Gamma^{\pm}$, and a shared interface Γ_{int}. Heterogeneity can result from discontinuities in material properties, body forces, and in some other special cases. For illustration, here we consider that each subdomain has its own material properties $\lambda^{\pm}$ and $\mu^{\pm}$, and this introduces a weak discontinuity along the interfaces with $\Gamma_{\text{int}} = \Gamma_{\text{int}}^{\text{W}}$ (see Chapter 2).

First, the approximation of $\mathbf{u}$ in each subdomain and on their boundaries is performed using separate sets of basis functions:

$$
\mathbf{u}^h(\mathbf{x}) = \begin{cases} \mathbf{u}^{h+}(\mathbf{x}) = g_1^+(\mathbf{x})\mathbf{u}_1^+ + \cdots + g_{N_S^+}^+(\mathbf{x})\mathbf{u}_{N_S^+}^+, & \mathbf{x} \in \overline{\Omega}^+, \\ \mathbf{u}^{h-}(\mathbf{x}) = g_1^-(\mathbf{x})\mathbf{u}_1^- + \cdots + g_{N_S^-}^-(\mathbf{x})\mathbf{u}_{N_S^-}^-, & \mathbf{x} \in \overline{\Omega}^-. \end{cases} \tag{9.43}
$$

The collocation of the governing equations for each subdomain is then carried out independently as follows:

$$
\begin{cases} \mathcal{L}^+\mathbf{u}^{h+} = -\mathbf{b}^+ & \text{in } \Omega^+, \\ \mathcal{B}_t^+\mathbf{u}^{h+} = \bar{\mathbf{t}}^+ & \text{on } \Gamma^+ \cap \Gamma_t, \\ \mathcal{B}_u^+\mathbf{u}^{h+} = \bar{\mathbf{u}}^+ & \text{on } \Gamma^+ \cap \Gamma_u, \end{cases} \tag{9.44}
$$

$$
\begin{cases} \mathcal{L}^-\mathbf{u}^{h-} = -\mathbf{b}^- & \text{in } \Omega^-, \\ \mathcal{B}_t^-\mathbf{u}^{h-} = \bar{\mathbf{t}}^- & \text{on } \Gamma^- \cap \Gamma_t, \\ \mathcal{B}_u^-\mathbf{u}^{h-} = \bar{\mathbf{u}}^- & \text{on } \Gamma^- \cap \Gamma_t, \end{cases} \tag{9.45}
$$

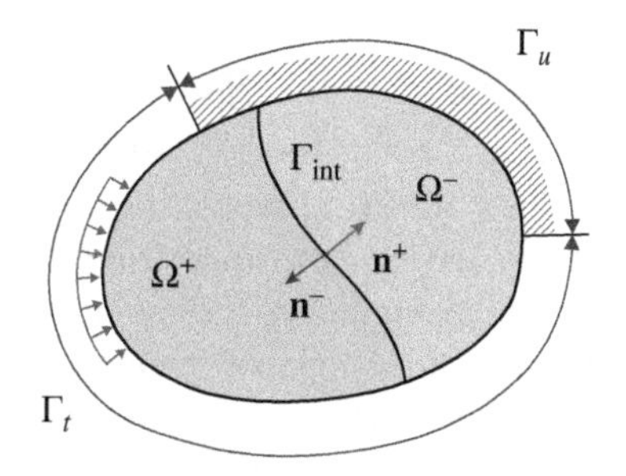

Figure 9.4 Two subdomains of a problem with material heterogeneity.

where the operators $\mathcal{L}^{\pm}$, $\mathcal{B}_t^{\pm}$, and $\mathcal{B}_u^{\pm}$ are defined by (9.8) using $\lambda^{\pm}$ and $\mu^{\pm}$, and the data on the right-hand side is defined with respect to each subdomain $\overline{\Omega}^{\pm}$.

The Neumann-type traction equilibrium condition for weak discontinuities in elasticity can be expressed as $\mathcal{B}_t^+ \mathbf{u}^{h+} + \mathcal{B}_t^- \mathbf{u}^{h-} = \mathbf{0}$, on $\Gamma_{\text{int}}^{\text{W}}$ (see Chapter 2), which is collocated as follows:

$$\mathcal{B}_t^+ \mathbf{u}^{h+} + \mathcal{B}_t^- \mathbf{u}^{h-} = \mathbf{0} \text{ on } \Gamma_{\text{int}}^{\text{W}}. \tag{9.46}$$

Dirichlet-type conditions can be further introduced and collocated on the interface to allow for optimal convergence [27]:

$$\mathcal{B}_u^+ \mathbf{u}^{h+} - \mathcal{B}_u^- \mathbf{u}^{h-} = \mathbf{0} \text{ on } \Gamma_{\text{int}}^{\text{W}}. \tag{9.47}$$

That is, in [27], it was shown that *both Neumann and Dirichlet boundary conditions should be imposed on the interface to achieve optimum convergence*. Strong discontinuities can also be considered, see [28] for more details.

As before, if $Nc > Ns$ a weighted least-squared method can be applied, and the weighted discretized collocation equations read:

$$\begin{aligned} &\begin{bmatrix} \mathbf{K}^+ \\ \mathbf{K}^- \\ \mathbf{A} \end{bmatrix} \mathbf{d} = \begin{bmatrix} \mathbf{f}^+ \\ \mathbf{f}^- \\ \mathbf{0} \end{bmatrix}, \\ &\rightarrow \mathbf{K}\,\mathbf{d} = \mathbf{f}, \\ &\rightarrow \mathbf{K}^{\mathrm{T}}\mathbf{W}\mathbf{K}\,\mathbf{d} = \mathbf{K}^{\mathrm{T}}\mathbf{W}\mathbf{f}, \end{aligned} \tag{9.48}$$

with submatrices defined as

$$\mathbf{K}^{\pm} = \begin{bmatrix} \mathbf{K}^{\mathcal{L}\pm} \\ \sqrt{\alpha_t^{\pm}}\,\mathbf{K}^{\mathcal{B}_t\pm} \\ \sqrt{\alpha_u^{\pm}}\,\mathbf{K}^{\mathcal{B}_u\pm} \end{bmatrix}, \ \mathbf{f}^{\pm} = \begin{bmatrix} \mathbf{f}^{\mathcal{L}\pm} \\ \sqrt{\alpha_t^{\pm}}\,\mathbf{f}^{\mathcal{B}_t\pm} \\ \sqrt{\alpha_u^{\pm}}\,\mathbf{f}^{\mathcal{B}_u\pm} \end{bmatrix}, \ \mathbf{A} = \begin{bmatrix} \sqrt{\overline{\alpha}_t}\mathbf{A}_t \\ \sqrt{\overline{\alpha}_u}\mathbf{A}_u \end{bmatrix}, \tag{9.49}$$

where $\mathbf{K}^{\mathcal{L}\pm}$, $\mathbf{K}^{\mathcal{B}_t\pm}$, and $\mathbf{K}^{\mathcal{B}_u\pm}$ are the matrices associated with the differential operators following (9.10); $\mathbf{A}_t$ and $\mathbf{A}_u$ are associated with the Neumann and Dirichlet-type interface conditions in (9.46) and (9.47), respectively. For balanced errors for all the different terms associated with domains, boundaries, and the interface, the following weights have been derived [13, 27]:

$$\begin{aligned} &\sqrt{\alpha_u^+} = \sqrt{\alpha_u^-} = \sqrt{\overline{\alpha}_u} = \mathrm{O}\left(\overline{c}_{\mathrm{MAX}} \cdot \overline{Ns}\right), \\ &\sqrt{\alpha_t^+} = \mathrm{O}\left(\hat{c}^+\right), \\ &\sqrt{\alpha_t^-} = \mathrm{O}\left(\hat{c}^-\right), \\ &\sqrt{\overline{\alpha}_t} = \mathrm{O}\left(1\right), \end{aligned} \tag{9.50}$$

where $\overline{c}_{\mathrm{MAX}} = \max\left(c_{\mathrm{MAX}}^+, c_{\mathrm{MAX}}^-\right)$; $c_{\mathrm{MAX}}^{\pm} = \max\left(\lambda^{\pm}, \mu^{\pm}\right)$, $\overline{Ns} = \max\left(Ns^+, Ns^-\right)$, $\hat{c}^{\pm} = \overline{c}_{\mathrm{MAX}}/c_{\mathrm{MAX}}^{\pm}$, and $Ns^{\pm}$ is the number of source points in $\overline{\Omega}^{\pm}$.

The L-RBCM approach (9.20) combined with the subdomain collocation method has been successfully applied to problems with heterogeneities (weak discontinuities) [27]. For problems with strong discontinuities such as cracks, a domain decomposition of near-tip and far-field subdomains, with the employment of separate approximations in each subdomain with proper interface conditions has been introduced [28]. This method allows (i) the natural representation of the discontinuities across crack surfaces and (ii) the enrichment with the crack-tip solution in a local subdomain. With the proper decomposition and interface conditions, exponential convergence rates can be achieved while keeping the discrete system well-conditioned.

9.6 Comparison of Nodally-Integrated Galerkin Meshfree Methods and Nodally Collocated Strong Form Meshfree Methods

Galerkin meshfree methods rely chiefly on nodal integration of the weak form for a truly meshfree technique. In the case of strong form collocation meshfree methods, direct collocation at the nodes can be employed. This section compares the relative performance of the node-based Galerkin meshfree methods discussed in Chapters 4–7 and the node-based direct collocation method in terms of accuracy, efficiency, and stability. Considering both accuracy and efficiency, the overall effectiveness in terms of CPU time versus error is also assessed.

The minimum order of approximation required for convergence in both the L_2 and H^1 norms is employed for each method: for Galerkin methods the minimum is linear (see Section 3.4), while for strong form collocation the order is quadratic (see Section 9.2.2). Due to the reduced performance of using collocation at nodes (versus using more collocation points), and the superconvergence observed in the gradient-smoothing Galerkin methods, both are similar in terms of rates of convergence and yield a fair comparison. The nodal integration methods in Chapter 6 are employed for the Galerkin method, while direct collocation at the nodes is employed for collocation of the strong form as described in Section 9.1.

9.6.1 Performance of Galerkin and Collocation Methods

Consider the Poisson equation (see Chapter 2) with diffusivity $k = 1$ in a domain $\overline{\Omega} = [-1, 1] \times [-1, 1]$, with $\Gamma_u = \Gamma$, source term $s = \sin(\pi x)\sin(\pi y)$ and $\overline{u} = 0$. The exact solution of this problem is

$$u = -\frac{1}{2\pi^2}\sin(\pi x)\sin(\pi y). \tag{9.51}$$

The problem is solved using a linear RK approximation in the Galerkin method and a quadratic RK approximation in the strong form collocation method, with cubic B-spline kernels employed with normalized dilations of 1.75 and 2.75, respectively.

The solutions by meshfree Galerkin methods are particularly sensitive to the uniformity of the discretization. Thus, to truly evaluate the performance of the methods discussed, nonuniform discretizations must be considered. The nonuniform node distributions shown in Figure 9.5 are thus used in this study.

Figure 9.6 shows the error plotted against the nodal spacing h for each of the methods. It can be seen that Galerkin methods yield optimal convergence (rates of two in the L_2 norm and one in the H^1 semi-norm) so long as variationally consistent (VC) integration is employed (see Chapter 6): these

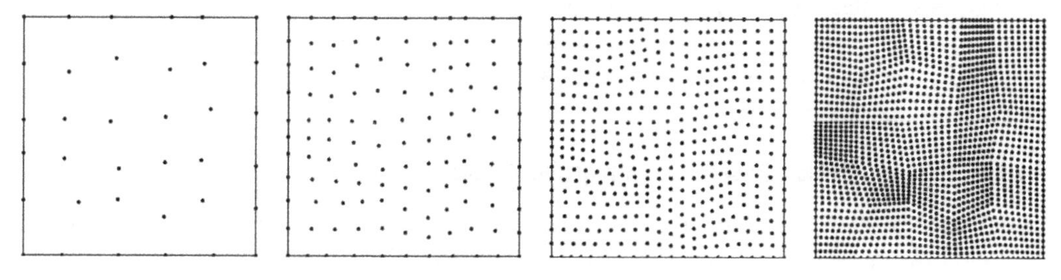

Figure 9.5 Refinements for convergence test in non-uniform discretizations. *Source:* Hillman and Chen [29], figure 5, p. 157 / Springer Nature.

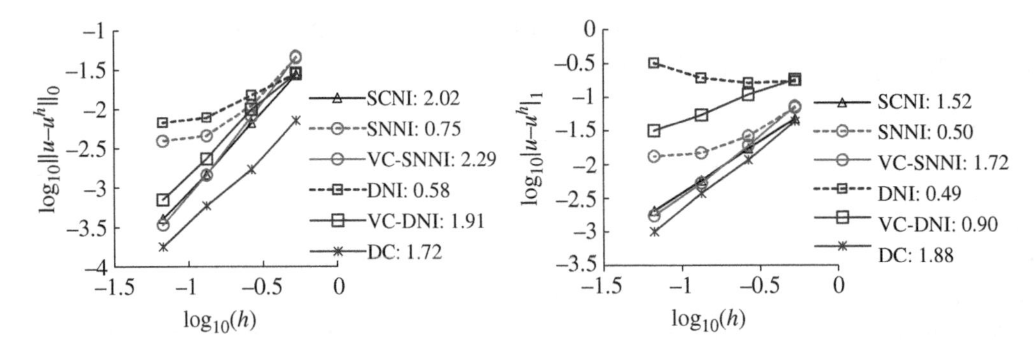

Figure 9.6 Convergence of nodal Galerkin and collocation methods. *Source:* Hillman and Chen [29], figure 6, p. 158 / Springer Nature.

methods are stabilized conforming nodal integration (SCNI), and VC-corrected stabilized non-conforming nodal integration (SNNI), denoted VC-SNNI. It can be noted that the gradient smoothing methods of SCNI and SNNI achieve superconvergent rates in derivatives.

The direct collocation (DC) method in the strong form method exhibits rates of two in the L_2 norm, lower than optimal rate of three for the quadratic basis employed [10]. This can be attributed to the lower accuracy of using very few collocation points, which can be explained by the equivalent least-squares residual of the strong form collocation method [13]. This reduced convergence in nodal collocation is referred to as the odd-even phenomenon in the literature, where the rate of n is observed for the L_2 and H^1 error norms for even orders of n, and $n-1$ for odd orders [9, 11, 30, 31]. Therefore, the present study makes linear basis with RKPM and quadratic basis with RKCM a fair comparison: the expected convergence rates in these node-based methods are the same. It can be noted that a recursive gradient formulation [30] has been introduced that exhibits superconvergence in DC, with rates of n and $n+1$ for even and odd orders of approximations, respectively.

Overall, the variationally consistent Galerkin methods that employ strain smoothing (SCNI and VC-SNNI) and the DC method yield comparable rates of error and convergence. The DC method yields slightly lower error in both norms.

Now comparing the Galerkin methods and the strong form collocation methods in terms of effectiveness (CPU time versus error), it can be seen in Figure 9.7 that the variationally consistent Galerkin methods and the strong form with DC have similar effectiveness in the L_2 norm. However, due to the lower accuracy in derivatives in VC-DNI, only SCNI, VC-SNNI and DC have comparable results considering both L_2 and H^1 norms.

The similarity in the performance of these three methods is striking considering the vastly different approaches, including the (expensive) higher-order derivatives, higher-order bases and thus also larger dilations employed for strong form collocation. This cost, however, is somewhat alleviated by the savings in constructing the matrix equations with one neighbor loop, rather than the two required for assembly in the Galerkin weak form-based methods.

9.6.2 Stability of Node-Based Galerkin and Collocation Methods

Spatial stability is an important consideration in node-based Galerkin methods, as outlined in Chapter 6. To examine the stability of the present methods, which include direct nodal collocation, we examine the solution derivatives for the last refinement in the previous example. For reference, the derivatives of the exact solution in Eq. (9.51) are shown in Figure 9.8.

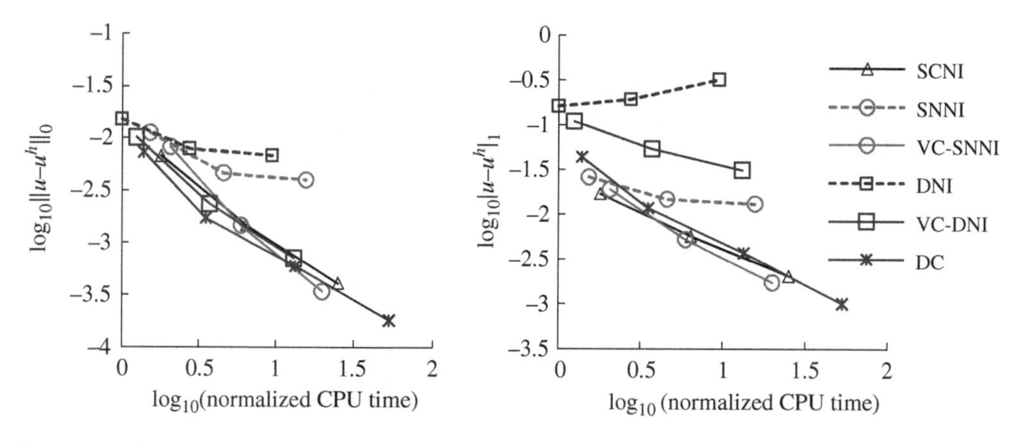

Figure 9.7 Relative performance of nodal Galerkin and collocation methods. *Source:* Hillman and Chen [29], figure 7, p. 158 / Springer Nature.

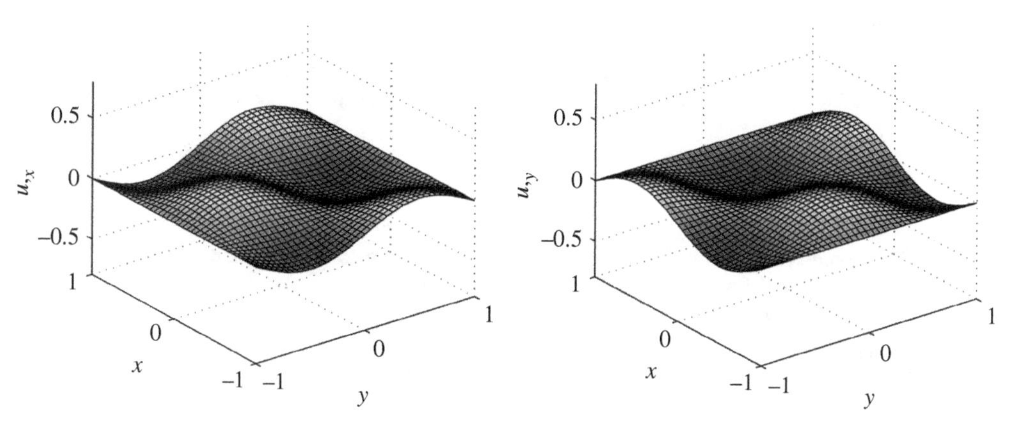

Figure 9.8 Derivatives of exact solution in test problem. *Source:* Hillman and Chen [29], figure 8, p. 158 / Springer Nature.

The numerical solutions are shown in Figure 9.9 for all the methods considered herein. Several conclusions can be drawn from the figures. First, it can be seen that the VC correction remarkably stabilizes the solution of the nodally-integrated Galerkin methods, which has been observed in other contexts as well [32]. However, it is apparent that unstable modes still exist in direct nodal integration, which in turn can lead to poorer rates of convergence and poorer levels of error versus the other nodally-integrated Galerkin methods, which is observed in Figure 9.6. It can also be seen that the VC correction is sufficient to stabilize SNNI, which could explain its apparently good performance in terms of convergence rates and error, yielding solutions similar to SCNI. Finally, the only stable nodally-integrated Galerkin methods tested are VC-SNNI and SCNI, while the DC Strong Form Collocation method also gives a stable solution. It would appear that collocation methods do not suffer from instability when evaluating key quantities at nodal locations and do not require special treatment. It can be noted that the second-order derivatives at nodal locations are nonzero, in contrast to the first-order ones, which are problematic in Galerkin methods (see Section 6.9).

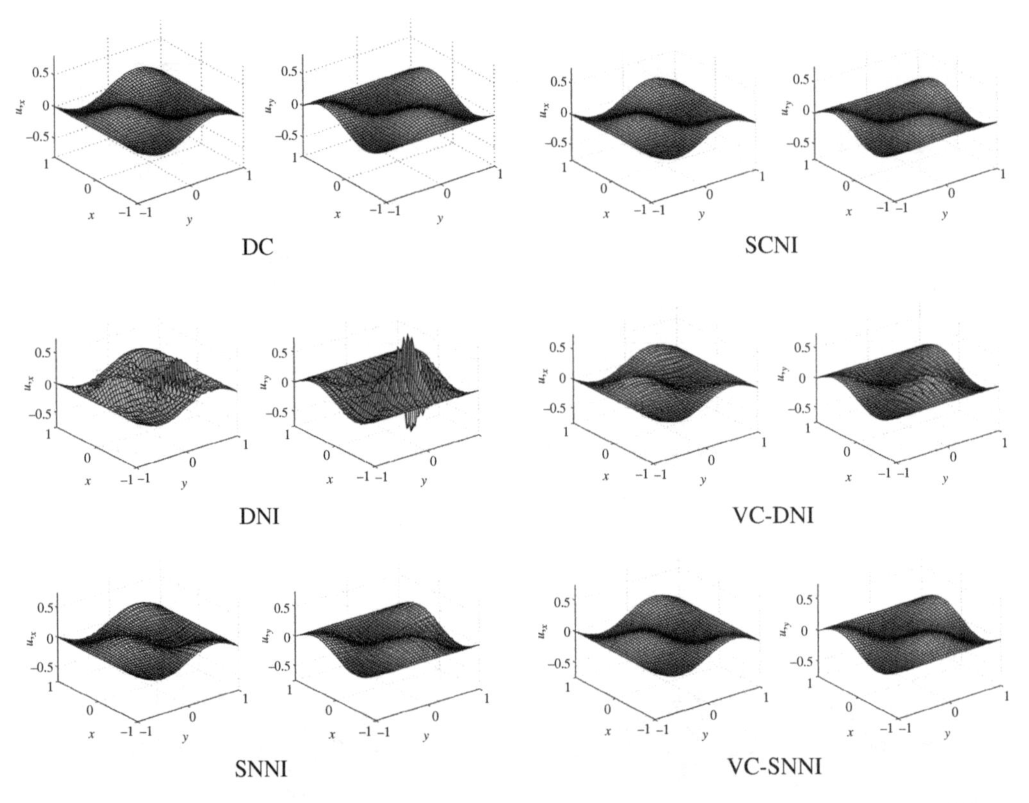

Figure 9.9 Solution derivatives obtained by nodal-based Galerkin and collocation methods. *Source:* Adapted from Hillman and Chen [29], figure 9, pp. 158-160 / Springer Nature.

Overall, considering both accuracy, convergence, and efficiency, SCNI, VC-SNNI, and DC are comparable methods. Of these, SCNI in the Galerkin method, and DC in the strong form method, both offer relative ease of implementation. With all other aspects equal, from an implementation standpoint, DC (strong form collocation with nodes) could arguably be considered preferential because of the simplicity of the formulation. Nevertheless, the employment of collocation for nonlinear problems is complicated by the use of the strong form, see the Ph.D. dissertation of Basava [33].

The Galerkin-based methods offer other advantages over strong form collocation methods due to the legacy of codes, and their algorithms and subroutines (in particular, for linearization and inelasticity in general), which are based on the Galerkin formulation and structure. Stability for node-based collocation in time has been discussed in Chi et al. [34], and some open questions also remain, as reported recently by Wang and Hillman [35].

References

1 Jensen, P.S. (1972). Finite difference techniques for variable grids. *Comput. Struct.* 2: 17–29.

2 Perrones, N. and Kao, R. (1974). A general-finite difference method for arbitrary meshes. *Comput. Struct.* 5.

3 Liszka, T.J. and Orkisz, J. (1980). The finite difference method at arbitrary irregular grids and its application in applied mechanics. *Comput. Struct.* 11: 83–95.

4 Liszka, T.J. (1984). An interpolation method for an irregular net of nodes. *Int. J. Numer. Methods Eng.* 20: 1599–1612.

5 Franke, R. (1982). Scattered data interpolation: tests of some methods. *Math. Comput.* 38: 181–181.

6 Kansa, E.J. (1990). Multiquadrics – a scattered data approximation scheme with applications to computational fluid-dynamics – II solutions to parabolic, hyperbolic and elliptic partial differential equations. *Comput. Math. Appl.* 19: 147–161.

7 Kansa, E.J. (1990). Multiquadrics – a scattered data approximation scheme with applications to computational fluid-dynamics – I surface approximations and partial derivative estimates. *Comput. Math. Appl.* 19: 127–145.

8 Oñate, E., Idelsohn, S.R., Zienkiewicz, O.C., and Taylor, R.L. (1996). A finite point method in computational mechanics. Applications to convective transport and fluid flow. *Int. J. Numer. Methods Eng.* 39: 3839–3866.

9 Aluru, N.R. (2000). A point collocation method based on reproducing kernel approximations. *Int. J. Numer. Methods Eng.* 47: 1083–1121.

10 Hu, H.-Y., Lai, C.-K., and Chen, J.-S. (2009). A study on convergence and complexity of reproducing kernel collocation method. *Interact. Multiscale Mech.* 2: 295–319.

11 Chi, S.-W., Chen, J.-S., Hu, H.-Y., and Yang, J.P. (2013). A gradient reproducing kernel collocation method for boundary value problems. *Int. J. Numer. Methods Eng.* 93: 1381–1402.

12 Hu, H.-Y., Chen, J.-S., and Hu, W. (2011). Error analysis of collocation method based on reproducing kernel approximation. *Numer. Methods Partial Differ. Equ.* 27: 554–580.

13 Hu, H.-Y., Chen, J.-S., and Hu, W. (2007). Weighted radial basis collocation method for boundary value problems. *Int. J. Numer. Methods Eng.* 69: 2736–2757.

14 Mahdavi, A., Chi, S.-W., and Zhu, H. (2019). A gradient reproducing kernel collocation method for high order differential equations. *Comput. Mech.* 64: 1421–1454.

15 Franke, C. and Schaback, R. (1998). Solving partial differential equations by collocation using radial basis functions. *Appl. Math. Comput.* 93: 73–82.

16 Wendland, H. (1999). Meshless Galerkin methods using radial basis functions. *Math. Comput. Am. Math. Soc.* 68: 1521–1531.

17 Hardy, R.L. (1971). Multiquadric equations of topography and other irregular surfaces. *J. Geophys. Res.* 76: 1905.

18 Hardy, R.L. (1990). Theory and applications of the multiquadric-biharmonic method. *Comput. Math. Appl.* 19: 163–208.

19 Madych, W.R. (1992). Miscellaneous error bounds for multiquadric and related interpolators. *Comput. Math. Appl.* 24: 121–138.

20 Wu, Z. and Schaback, R. (1993). Local error estimates for radial basis function interpolation of scattered data. *IMA J. Numer. Anal.* 13: 13–27.

21 Yoon, J. (2001). Spectral approximation orders of radial basis function interpolation on the Sobolev space. *SIAM J. Math. Anal.* 33: 946–958.

22 Buhmann, M.D. and Micchelli, C.A. (1992). Multiquadric interpolation improved. *Comput. Math. Appl.* 24: 21–25.

23 Madych, W.R. and Nelson, S.A. (1990). Multivariate interpolation and conditionally positive definite functions II. *Math. Comput.* 54: 211–230.

24 Chen, J.-S., Hu, W., and Hu, H.-Y. (2008). Reproducing kernel enhanced local radial basis collocation method. *Int. J. Numer. Methods Eng.* 75: 600–627.

25 Chi, S.-W., Chen, J.-S., and Hu, H.-Y. (2014). A weighted collocation on the strong form with mixed radial basis approximations for incompressible linear elasticity. *Comput. Mech.* 53: 309–324.

26 Chen, J.-S., Hillman, M., and Chi, S.W. (2017). Meshfree methods: progress made after 20 years. *J. Eng. Mech.* 143(4): 04017001.

27 Wang, L., Chen, J.-S., Hu, H.-Y. (2010). Subdomain radial basis collocation method for heterogeneous media. *Int. J. Numer. Methods Eng.* 80: 163–190.

28 Wang, L., Chen, J.-S., and Hu, H.-Y. (2010). Subdomain radial basis collocation method for fracture mechanics. *Int. J. Numer. Methods Eng.* 83: 851–876.

29 Hillman, M. and Chen, J.S. (2018). Performance comparison of nodally integrated Galerkin meshfree methods and nodally collocated strong form meshfree methods. In: *Advances in Computational Plasticity* (ed. E. Onate), 145–164. Springer.

30 Wang, D., Wang, J., and Wu, J. (2018). Superconvergent gradient smoothing meshfree collocation method. *Comput. Methods Appl. Mech. Eng.* 340: 728–766.

31 Qi, D., Wang, D., Deng, L. et al. (2019). Reproducing kernel mesh-free collocation analysis of structural vibrations. *Eng. Comput.* 36: 734–764.

32 Hillman, M., Chen, J.-S., and Chi, S.-W. (2014). Stabilized and variationally consistent nodal integration for meshfree modeling of impact problems. *Comp. Part. Mech.* 1: 245–256.

33 Basava, R.R. (2015). Meshfree Image-Based Reduced Order Modeling of Multiple Muscle Components with Connective Tissue and Fat. Ph.D. Dissertation, University of California, San Diego.

34 Chi, S.-W., Chen, J.-S., Luo, H. et al. (2013). Dispersion and stability properties of radial basis collocation method for elastodynamics. *Numer. Methods Partial Differ. Equ.* 29: 818–842.

35 Wang, J. and Hillman, M.C. (2022). Temporal stability of collocation, Petrov–Galerkin, and other non-symmetric methods in elastodynamics and an energy conserving time integration. *Comput. Methods Appl. Mech. Eng.* 393: 114738.

10

RKPM2D: A Two-Dimensional Implementation of RKPM

This chapter details an open-source software called RKPM2D for solving partial differential equations (PDEs) under the reproducing kernel particle method (RKPM) meshfree computational framework. Two-dimensional linear elastostatics (static elasticity) is chosen as the model problem, but the extension of the RKPM open-source software for solving other types of PDEs is straightforward; an example is given in the Appendix. The RKPM2D code is implemented under the MATLAB environment [1] with fully integrated preprocessing, solver, and postprocessing. The software package aims to support reproducible research and serve as an efficient test platform for further development of meshfree methods, but we hope that the reader will also find it very useful for understanding the practical implementation of these methods.

The RKPM2D software consists of a set of data structures and subroutines to provide a complete meshfree solver, including discretizing two-dimensional domains of arbitrary geometry, creating nodal representative domains by Voronoi diagram partitioning, generating reproducing kernel shape functions, and providing visualization and post-processing.

First, an overview covering the theoretical aspects of RKPM2D is given, including the reproducing kernel approximation, weak form, domain integration, and fully discrete equations. Various aspects of the computer implementation of RKPM2D are then discussed in detail. Example problems for RKPM2D are provided.

Nitsche's method (see Chapters 2 and 4) is adopted for the imposition of essential boundary conditions. For domain integration, the variationally consistent and stabilized nodal integration methods, modified stabilized conforming nodal integration (MSCNI), and naturally stabilized nodal integration (NSNI), are implemented into RKPM2D, along with conventional Gauss quadrature (see Chapter 6). The latest version of the source code can be downloaded from the Internet by visiting https://tinyurl.com/RKPM2D. The code described in this chapter is with reference to Version 3.

10.1 Reproducing Kernel Particle Method: Approximation and Weak Form

10.1.1 Reproducing Kernel Approximation

In RKPM, the numerical approximation is constructed based upon a set of scattered points called nodes. A set of NP nodes $\{\mathbf{x}_1, \mathbf{x}_2, \ldots \mathbf{x}_{NP}\}$ discretizes a domain of interest $\overline{\Omega} = \Omega \cup \Gamma$, where $\mathbf{x}_I$ is the position vector of node I. The nth-order RK approximation $\mathbf{u}^h$ of a vector function $\mathbf{u}$ (such as the displacement in elasticity) in two dimensions is expressed as (see Section 3.2.2)

Meshfree and Particle Methods: Fundamentals and Applications, First Edition.
Ted Belytschko, J. S. Chen, and Michael Hillman.

$$u_i^h(\mathbf{x}) = \sum_{I \in S_\mathbf{x}} \Psi_I(\mathbf{x}) u_{iI},$$
or
$$\mathbf{u}^h(\mathbf{x}) = \sum_{I \in S_\mathbf{x}} \Psi_I(\mathbf{x}) \mathbf{u}_I, \tag{10.1}$$

where $\mathbf{x}$ is the spatial coordinate, u_{iI} is the ith component of the associated nodal coefficient $\mathbf{u}_I$ to be determined, $S_\mathbf{x}$ is the set of nodal indices for which $\Psi_I(\mathbf{x})$ has a nonzero contribution to $\mathbf{u}^h$ at $\mathbf{x}$ (see Chapter 3), and $\Psi_I(\mathbf{x})$ is the reproducing kernel (RK) shape function associated with node I:

$$\Psi_I(\mathbf{x}) = \mathbf{H}^\mathrm{T}(\mathbf{0})\mathbf{M}^{-1}(\mathbf{x})\mathbf{H}(\mathbf{x}-\mathbf{x}_I)\phi_{a_I}(\mathbf{x}-\mathbf{x}_I), \tag{10.2}$$

$$\mathbf{M}(\mathbf{x}) = \sum_{I=1}^{NP} \mathbf{H}(\mathbf{x}-\mathbf{x}_I)\mathbf{H}^\mathrm{T}(\mathbf{x}-\mathbf{x}_I)\phi_{a_I}(\mathbf{x}-\mathbf{x}_I), \tag{10.3}$$

$$\mathbf{H}^\mathrm{T}(\mathbf{x}-\mathbf{x}_I) = [1 \;\; x_1 - x_{1I} \;\; x_2 - x_{2I} \;\; \ldots \;\; (x_2 - x_{2I})^n], \tag{10.4}$$

where $\phi_{a_I}(\mathbf{x}-\mathbf{x}_I)$ is the kernel function centered at $\mathbf{x}_I$ with compact support size a_I defined as

$$a_I = \tilde{a} h_I. \tag{10.5}$$

In the above equation, $\tilde{a}$ is the normalized support size, and h_I is the nodal spacing associated with nodal point $\mathbf{x}_I$ defined in RKPM2D as

$$h_I = \max\{\|\mathbf{x}_J - \mathbf{x}_I\|\}, \;\; \forall \mathbf{x}_J \in B_I, \tag{10.6}$$

in which the set B_I contains the four nodes that are closest to point $\mathbf{x}_I$. The kernel function controls the smoothness of the approximation, i.e., if $\{\phi_{a_I}(\mathbf{x})\}_{I=1}^{NP} \in C^k$, then $\{\Psi_I(\mathbf{x})\}_{I=1}^{NP} \in C^k$ (see Chapter 3).

10.1.2 Galerkin Formulation

The RKPM2D code solves static linear elastic problems; the equilibrium equation for the displacement in Ω is (see Chapter 2)

$$\sigma_{ij,j} + b_i = 0 \;\; \text{in } \Omega, \tag{10.7}$$

where σ_{ij} is the stress and b_i is the body force.

The code considers the subdivision of the boundaries of Ω according to components where the ith displacement component u_i is prescribed on Γ_u^i, and the ith traction component $t_i = \sigma_{ij}n_j$ is prescribed on Γ_t^i as follows:

$$u_i = \bar{u}_i \;\; \text{on} \;\; \Gamma_u^i, \tag{10.8}$$

$$\sigma_{ij}n_j = \bar{t}_i \;\; \text{on} \; \Gamma_t^i, \tag{10.9}$$

where $\bar{u}_i$ and $\bar{t}_i$ are prescribed values of displacement and traction, respectively. The boundaries are complementary: $\Gamma_u^i \cup \Gamma_t^i = \Gamma^i$ and $\Gamma_u^i \cap \Gamma_t^i = \emptyset$.

Nitsche's method is employed to enforce the essential boundary conditions in (10.8). The discrete equations are developed following Section 4.2.2:

$$\sum_{J \in S} (\mathbf{K}_{IJ} + \mathbf{K}_{IJ}^\mathrm{N})\mathbf{u}_J = \mathbf{f}_I + \mathbf{f}_I^\mathrm{N} \quad \forall I \in S, \tag{10.10}$$

where S is the set of all NP nodal indices,

$$\begin{aligned} \mathbf{K}_{IJ}^{\mathrm{N}} &= -\mathbf{H}_{IJ} - \mathbf{H}_{IJ}^{\mathrm{T}} + \mathbf{A}_{IJ}, \\ \mathbf{f}_{I}^{\mathrm{N}} &= -\mathbf{h}_{I} + \mathbf{g}_{I}, \end{aligned} \tag{10.11}$$

and the Nitsche matrices in Chapter 4 are modified to include a switch matrix $\mathbf{S}$ to specify essential boundary condition components as follows:

$$\mathbf{K}_{IJ} = \int_{\Omega} \mathbf{B}_I^{\mathrm{T}} \mathbf{D} \mathbf{B}_J d\Omega, \tag{10.12}$$

$$\mathbf{f}_I = \int_{\Omega} \Psi_I \mathbf{b} d\Omega + \int_{\Gamma_t} \Psi_I \bar{\mathbf{t}} d\Gamma, \tag{10.13}$$

$$\mathbf{H}_{IJ} = \int_{\Gamma_u} \Psi_I \mathbf{S} \boldsymbol{\eta}^{\mathrm{T}} \mathbf{D} \mathbf{B}_J d\Gamma, \tag{10.14}$$

$$\mathbf{A}_{IJ} = \beta \int_{\Gamma_u} \Psi_I \Psi_J \mathbf{S} d\Gamma, \tag{10.15}$$

$$\mathbf{h}_I = \int_{\Gamma_u} \mathbf{B}_I^{\mathrm{T}} \mathbf{D} \boldsymbol{\eta} \mathbf{S} \bar{\mathbf{u}} d\Gamma, \tag{10.16}$$

$$\mathbf{g}_I = \beta \int_{\Gamma_u} \Psi_I \mathbf{S} \bar{\mathbf{u}} d\Gamma, \tag{10.17}$$

where the components in the above are

$$\mathbf{B}_I(\mathbf{x}) = \begin{bmatrix} \Psi_{I,1}(\mathbf{x}) & 0 \\ 0 & \Psi_{I,2}(\mathbf{x}) \\ \Psi_{I,2}(\mathbf{x}) & \Psi_{I,1}(\mathbf{x}) \end{bmatrix}, \tag{10.18}$$

$$\boldsymbol{\eta} = \begin{bmatrix} n_1 & 0 \\ 0 & n_2 \\ n_2 & n_1 \end{bmatrix}, \quad \mathbf{S} = \begin{bmatrix} s_1 & 0 \\ 0 & s_2 \end{bmatrix}, \quad \mathbf{b} = \begin{bmatrix} b_1 \\ b_2 \end{bmatrix}, \quad \bar{\mathbf{t}} = \begin{bmatrix} \bar{t}_1 \\ \bar{t}_2 \end{bmatrix}, \quad \bar{\mathbf{u}} = \begin{bmatrix} \bar{u}_1 \\ \bar{u}_2 \end{bmatrix}, \tag{10.19}$$

where n_i is the ith component of the surface unit outward normal on the essential boundary, and

$$s_i = \begin{cases} 1 & \text{if } u_i = \bar{u}_i \text{ on } \Gamma_u^i \\ 0 & \text{else} \end{cases}, \tag{10.20}$$

serves as a switch for imposing each component of the boundary displacement, see [2]. The parameter β in (10.15) and (10.17) is normalized as described in Section 4.2.2: $\beta = \beta_{\mathrm{nor}} E h_{\max}^{-1}$ where β_{nor} is an independent constant, E is Young's modulus, and $h_{\max} = \max(h_I)$, $\forall\, \mathbf{x}_I$ gives a measure of h for arbitrary discretizations, with h_I computed from (10.6). The value β_{nor} is the input penalty parameter `Model.Beta_Nor` for the code (see Section 10.4.1).

Note that the consistent weak form can be employed in RKPM2D by omitting (10.15) and (10.17) and implementing strong-type essential boundary condition if desired (see Chapter 5).

10.2 Domain Integration

Domain integration plays an important role in accuracy, stability, and convergence of meshfree methods. Unlike the finite element method (FEM) which utilizes the element topology for integration, quadrature domains for meshfree methods can be chosen as cells that are either independent from the nodal locations, or directly associated with nodal representative domains. The former scheme is commonly adopted for Gauss quadrature and the latter is used for nodal integration schemes; both have been implemented in RKPM2D and are discussed in this section.

10.2.1 Gauss Integration

When Gauss quadrature is adopted, quadrature points are generated based upon background cells [3, 4], as shown in Figure 10.1, where only the quadrature points inside the physical domain are considered for domain integration. Gauss points for contour integrals are generated along the natural and essential boundaries, also shown in Figure 10.1.

The domain and boundary integration in (10.12)–(10.17) is carried out numerically as follows:

$$\int_{\Omega} f_{\Omega}(\mathbf{x}) d\Omega \approx \sum_{L=1}^{NG} f_{\Omega}(\mathbf{x}_L) W_L, \tag{10.21}$$

$$\int_{\Gamma} f_{\Gamma}(\mathbf{x}) d\Gamma \approx \sum_{L=1}^{NGb} f_{\Gamma}(\hat{\mathbf{x}}_L) \hat{W}_L, \tag{10.22}$$

where f_{Ω} and f_{Γ} denote integrands in the domain and boundary integrals in (10.12)–(10.17); $\mathbf{x}_L$, W_L, and NG are the domain Gauss points, weights, and the number of domain Gauss points, respectively, and $\hat{\mathbf{x}}_L$, $\hat{W}_L$, and NGb are boundary Gauss points, weights, and the number of boundary Gauss points, respectively. The same integration rules are used for all types of boundary integration in RKPM2D (both natural and essential). To use Gauss integration in RKPM2D, set `Quadrature.Integration = 'GAUSS'` (see Section 10.4.1).

Since RK shape functions are rational functions and their supports overlap, the misalignment between Gauss integration cells and shape function supports leads to large quadrature errors unless high-order integration schemes are adopted (see Section 6.3). Nevertheless, Gauss integration is made available as a reference quadrature scheme.

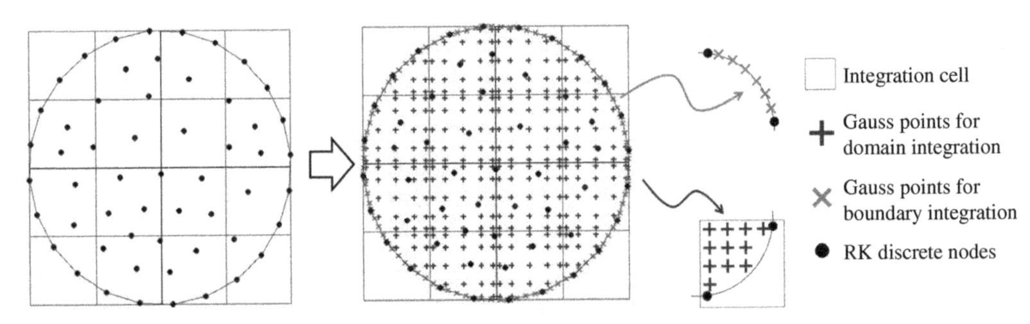

Figure 10.1 Meshfree nodes and background Gauss quadrature points for an arbitrary two-dimensional domain Ω. Source: Huang et al. [20], figure 5, p. 398 / Springer Nature.

10.2.2 Variationally Consistent Nodal Integration

The simplest quadrature rule in meshfree methods is direct nodal integration (DNI), where all quantities are evaluated and integrated at nodal locations. Using this scheme, the domain and boundary terms in (10.12)–(10.17) are computed as follows:

$$\int_{\Omega} f_{\Omega} d\Omega \approx \sum_{L=1}^{NP} f_{\Omega}(\mathbf{x}_L) A_L, \tag{10.23}$$

$$\int_{\Gamma} f_{\Gamma}(\mathbf{x}) d\Gamma \approx \sum_{L=1}^{NPb} f_{\Gamma}(\hat{\mathbf{x}}_L) l_L, \tag{10.24}$$

where $\mathbf{x}_L$, A_L and NP are the RK node locations, nodal representative domain areas, and the number of RK nodes, respectively, and $\hat{\mathbf{x}}_L$, l_L, and NPb are nodal boundary integration points, length of the nodal representative domain segments intersecting with the boundary, and the number of boundary integration points, respectively. Set `Quadrature.Integration = 'DNI'` to use DNI as the base nodal integration rule in RKPM2D (see Section 10.4.1).

Direct nodal integration is notorious for spurious zero-energy modes and nonconvergent numerical solutions. To ensure linear variational consistency, i.e., the ability of numerical methods to pass the linear patch test and subsequently converge, Chen et al. [5] showed that the quadrature rules need to meet the following first-order integration constraint for the shape function gradient (see Section 6.5 for more details):

$$\hat{\int_{\Omega}} \Psi_{I,i} d\Omega = \hat{\int_{\Gamma}} \Psi_I n_i d\Gamma, \tag{10.25}$$

In the above, the "∧" over the integral symbols denotes numerical integration. For nodal integration of gradient terms over the domain, e.g., the left-hand side of (10.25), Chen et al. [5] introduced the following smoothed strain approximation at nodal points $\mathbf{x}_L$ (see Section 6.6):

$$\begin{gathered}\tilde{\mathcal{E}}(\mathbf{x}_L) = \sum_{I=1}^{NP} \tilde{\mathbf{B}}_I(\mathbf{x}_L)\mathbf{u}_I, \\ \tilde{\mathbf{B}}_I(\mathbf{x}_L) = \begin{bmatrix} \tilde{b}_{1I}^L & 0 \\ 0 & \tilde{b}_{2I}^L \\ \tilde{b}_{2I}^L & \tilde{b}_{1I}^L \end{bmatrix}, \qquad \tilde{b}_{iI}^L = \frac{1}{A_L} \hat{\int_{\Gamma_L}} \Psi_I(\mathbf{x}) n_i(\mathbf{x}) d\Gamma,\end{gathered} \tag{10.26}$$

where A_L denotes the area of the nodal representative domain Ω_L associated with node L, and n_i denotes the ith component of the outward unit normal vector to the smoothing domain boundary Γ_L as shown in Figure 10.2. In the above, $\tilde{\mathcal{E}}(\mathbf{x}_L)$ and $\tilde{\mathbf{B}}_I(\mathbf{x}_L)$ are the smoothed strain vector and smoothed strain gradient matrix, respectively.

When evaluating (10.25) with nodal integration using the smoothed gradient $\tilde{b}_{iI}^L$ in (10.26), the first-order integration constraint in (10.25) is exactly satisfied as long as the same boundary integral quadrature rules are used for both contour integrals in Eqs. (10.24) and (10.26).

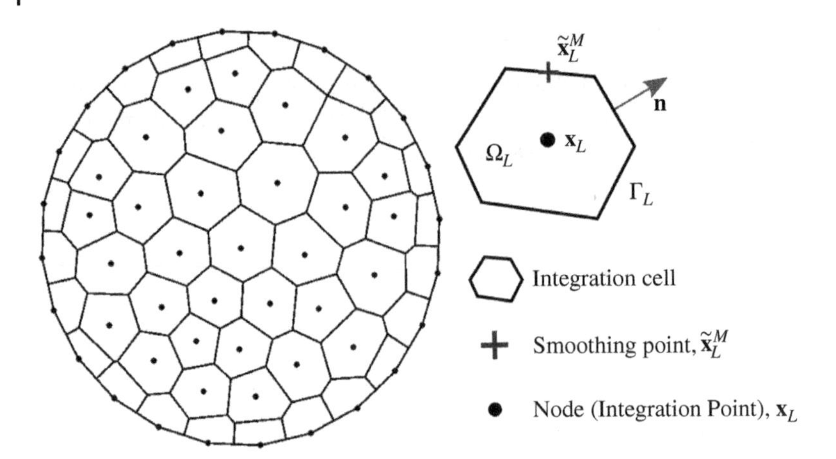

Figure 10.2 Voronoi cell diagram in two-dimensional domain Ω. Source: Huang et al. [20], figure 6, p. 399 / Springer Nature.

As discussed in [6], in order to maintain gradient consistency of a smoothed linearly consistent shape function, a simple one-point Gauss integration rule can be used for the contour integral in (10.26):

$$\tilde{b}_{iI}^{L} = \frac{1}{A_L}\int_{\Gamma_L} \Psi_I(\mathbf{x})n_i(\mathbf{x})d\Gamma \approx \frac{1}{A_L}\sum_{M\in S_L}\Psi_I(\tilde{\mathbf{x}}_L^M)n_i(\tilde{\mathbf{x}}_L^M)l_M, \tag{10.27}$$

where $S_L = \{M \mid \tilde{\mathbf{x}}_L^M \in \Gamma_L\}$ contains all center points of each boundary segment associated with node $\mathbf{x}_L$ (see Figure 10.2), and the integration weight l_M is the length of the Mth segment of the smoothing cell boundary Γ_L. By employing smoothed shape function gradients in the strain approximation, the stiffness matrix and force vectors in (10.12), (10.14), and (10.16) are re-formulated as follows:

$$\mathbf{K}_{IJ} = \int_{\Omega} \mathbf{B}_I^{\mathrm{T}}(\mathbf{x})\mathbf{D}\mathbf{B}_J(\mathbf{x})d\Omega \approx \sum_{L=1}^{NP} \tilde{\mathbf{B}}_I^{\mathrm{T}}(\mathbf{x}_L)\mathbf{D}\tilde{\mathbf{B}}_J(\mathbf{x}_L)A_L, \tag{10.28}$$

$$\mathbf{H}_{IJ} = \int_{\Gamma_u} \Psi_I(\mathbf{x})\mathbf{S}\boldsymbol{\eta}^{\mathrm{T}}(\mathbf{x})\mathbf{D}\mathbf{B}_J(\mathbf{x})d\Gamma \approx \sum_{L=1}^{NPu} \Psi_I(\mathbf{x}_L)\mathbf{S}\boldsymbol{\eta}^{\mathrm{T}}(\mathbf{x}_L)\mathbf{D}\tilde{\mathbf{B}}_J(\mathbf{x})l_L, \tag{10.29}$$

$$\mathbf{h}_I = \int_{\Gamma_u} \mathbf{B}_I^{\mathrm{T}}(\mathbf{x})\mathbf{D}\boldsymbol{\eta}(\mathbf{x})\mathbf{S}\bar{\mathbf{u}}d\Gamma \approx \sum_{L=1}^{NPu} \tilde{\mathbf{B}}_I^{\mathrm{T}}(\mathbf{x}_L)\mathbf{D}\boldsymbol{\eta}(\mathbf{x}_L)\mathbf{S}\bar{\mathbf{u}}l_L, \tag{10.30}$$

where NPu is the number of essential boundary integration points and $\tilde{\mathbf{B}}_I(\mathbf{x}_L)$ is defined in (10.26). To use SCNI as the base integration rule in RKPM2D, set `Quadrature.Integration = 'SCNI'` (see Section 10.4.1).

10.2.3 Stabilized Nodal Integration Schemes

As discussed in Section 6.9, spurious oscillatory modes can be triggered in nodal integration methods, even in SCNI under certain conditions. Therefore, additional stabilization techniques are needed to eliminate these low-energy modes, which will be described in this subsection.

10.2.3.1 Modified Stabilized Nodal Integration

The first stabilization technique employed here for eliminating spurious low-energy modes is called modified stabilized nodal integration proposed by Chen, Puso, and co-workers [7, 8], where a least-squares type stabilization term is introduced into the stiffness matrix (see Section 6.10.2) with:

$$\overline{\mathbf{K}}_{IJ} = \sum_{L=1}^{NP}\left(\underbrace{\tilde{\mathbf{B}}_I^{\mathrm{T}}(\mathbf{x}_L)\mathbf{D}\tilde{\mathbf{B}}_J(\mathbf{x}_L)A_L}_{\text{SCNI}} + \underbrace{c_S\sum_{K=1}^{NS}\left(\tilde{\mathbf{B}}_I^{\mathrm{T}}(\mathbf{x}_L) - \tilde{\mathbf{B}}_I^{\mathrm{T}}\left(\hat{\mathbf{x}}_L^K\right)\right)\mathbf{D}\left(\tilde{\mathbf{B}}_J(\mathbf{x}_L) - \tilde{\mathbf{B}}_J\left(\hat{\mathbf{x}}_L^K\right)\right)A_L^K}_{\text{stabilization}}\right), \tag{10.31}$$

where $\overline{\mathbf{K}}$ denotes the stabilized nodally integrated stiffness matrix, NS denotes the number of sub-cells associated with each nodal integration cell, $\hat{\mathbf{x}}_L^K$ denotes the centroid of the Kth subcell of node L, as shown in Figure 10.3, $\tilde{\mathbf{B}}_I\left(\hat{\mathbf{x}}_L^K\right)$ is the smoothed gradient evaluated at $\hat{\mathbf{x}}_L^K$ as in (10.27):

$$\tilde{\mathbf{B}}_I\left(\hat{\mathbf{x}}_L^K\right) = \begin{bmatrix} \tilde{b}_{1I}^{KL} & 0 \\ 0 & \tilde{b}_{2I}^{KL} \\ \tilde{b}_{2I}^{KL} & \tilde{b}_{1I}^{KL} \end{bmatrix}, \qquad \tilde{b}_{iI}^{KL} = \frac{1}{A_L^K}\hat{\int_{\Gamma_L^K}}\Psi_I(\mathbf{x})n_i(\mathbf{x})d\Gamma, \tag{10.32}$$

and A_L^K denotes the area of the Kth sub-cell of the Lth node with boundary Γ_L^K. Here, $0 \le c_S \le 1$ is a stabilization parameter, which is chosen to be $c_S = 1$ for elasticity based on the study of Puso et al. [8]. If *direct* gradients $\mathbf{B}_I(\mathbf{x}_L)$ and $\mathbf{B}_I\left(\hat{\mathbf{x}}_L^K\right)$ are instead used for nodal integration and the stabilization terms in (10.31), the stiffness matrix is formulated as

$$\overline{\mathbf{K}}_{IJ} = \sum_{L=1}^{NP}\left(\underbrace{\mathbf{B}_I^{\mathrm{T}}(\mathbf{x}_L)\mathbf{D}\mathbf{B}_J(\mathbf{x}_L)A_L}_{\text{DNI}} + \underbrace{c_S\sum_{K=1}^{NS}\left(\mathbf{B}_I^{\mathrm{T}}(\mathbf{x}_L) - \mathbf{B}_I^{\mathrm{T}}\left(\hat{\mathbf{x}}_L^K\right)\right)\mathbf{D}\left(\mathbf{B}_J(\mathbf{x}_L) - \mathbf{B}_J\left(\hat{\mathbf{x}}_L^K\right)\right)A_L^K}_{\text{stabilization}}\right), \tag{10.33}$$

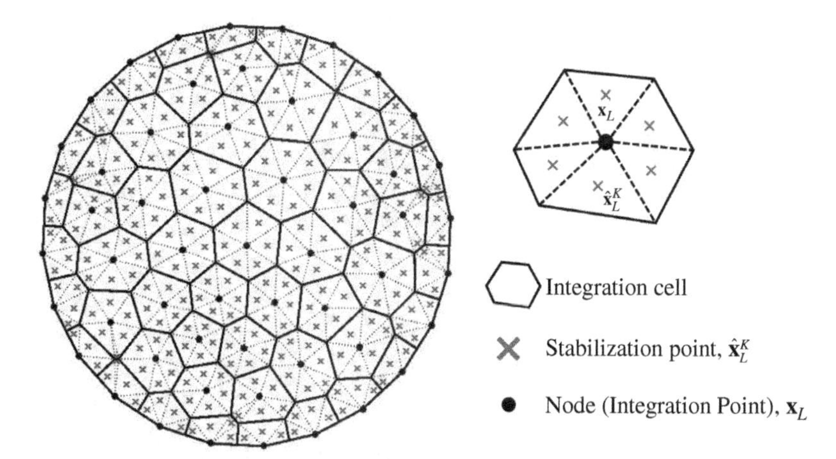

Figure 10.3 Illustration of nodal integration cells of the modified stabilized nodal integration. Source: Huang et al. [20], figure 7, p. 399 / Springer Nature.

which can be considered a modified stabilization for the DNI method. The least-squares type stabilization terms in (10.31) and (10.33) enhance the coercivity (stability) of the discrete formulation and suppresses spurious low-energy modes, but meanwhile, a number of additional shape function evaluations are required. As noted in Section 6.10.2, the SCNI-based formulation (10.31) retains the ability to pass the pass test.

In RKPM2D, the modified type stabilization is activated when `Quadrature.Stabilization = 'M'` (see Section 10.4.1). If SCNI is employed, then (10.31) is used. When DNI is chosen, then the solution is stabilized by (10.33).

10.2.3.2 Naturally Stabilized Nodal Integration

The other stabilized integration technique employed in RKPM2D is *naturally stabilized nodal integration* (NSNI) proposed by Hillman and Chen [9], where an implicit gradient expansion of the strain field is introduced (see Section 6.10.3):

$$\boldsymbol{\varepsilon}(\mathbf{u}^h(\mathbf{x})) \approx \boldsymbol{\varepsilon}_L(\mathbf{u}^h) + (x - x_L)\boldsymbol{\varepsilon}_L(\mathbf{u}_x^h) + (y - y_L)\boldsymbol{\varepsilon}_L\left(\mathbf{u}_y^h\right), \tag{10.34}$$

where $\boldsymbol{\varepsilon}_L(\mathbf{u}^h) \equiv \boldsymbol{\varepsilon}(\mathbf{u}^h(\mathbf{x}_L))$ and

$$\begin{aligned} \mathbf{u}_x^h(\mathbf{x}_L) &= \sum_{I=1}^{NP} \Psi_I^{(10)}(\mathbf{x}_L)\mathbf{u}_I, \\ \mathbf{u}_y^h(\mathbf{x}_L) &= \sum_{I=1}^{NP} \Psi_I^{(01)}(\mathbf{x}_L)\mathbf{u}_I, \end{aligned} \tag{10.35}$$

are the implicit derivatives of the displacement, with $\Psi_I^{(\alpha)}$ the implicit gradient shape function (See Section 3.5.3):

$$\Psi_I^{(\alpha)}(\mathbf{x}) = \mathbf{H}^{(\alpha)^{\mathrm{T}}}(\mathbf{0})\mathbf{M}^{-1}(\mathbf{x})\mathbf{H}(\mathbf{x} - \mathbf{x}_I)\phi_{a_I}(\mathbf{x} - \mathbf{x}_I), \tag{10.36}$$

where the vector $\mathbf{H}^{(\alpha)}$ takes on the following values for linear basis:

$$\begin{aligned} \mathbf{H}^{(10)} &= [0 \quad -1 \quad 0]^{\mathrm{T}}, \\ \mathbf{H}^{(01)} &= [0 \quad 0 \quad -1]^{\mathrm{T}}. \end{aligned} \tag{10.37}$$

Introducing the gradient expansion terms (10.34) into the variational equations, the stiffness matrix for SCNI with natural stabilization is obtained as

$$\overline{\mathbf{K}}_{IJ} = \sum_{L=1}^{NP} \left(\underbrace{\tilde{\mathbf{B}}_I^{\mathrm{T}}(\mathbf{x}_L)\mathbf{D}\tilde{\mathbf{B}}_J(\mathbf{x}_L)A_L}_{\text{SCNI}} + \underbrace{\left\{ \left\{\mathbf{B}_I^{(10)}(\mathbf{x}_L)\right\}^{\mathrm{T}} \mathbf{D}\mathbf{B}_J^{(10)}(\mathbf{x}_L)M_{Lx} + \left\{\mathbf{B}_I^{(01)}(\mathbf{x}_L)\right\}^{\mathrm{T}} \mathbf{D}\mathbf{B}_J^{(01)}(\mathbf{x}_L)M_{Ly} \right\}}_{\text{stabilization}} \right), \tag{10.38}$$

where $\mathbf{B}_I^{(10)}(\mathbf{x}_L)$ and $\mathbf{B}_I^{(01)}(\mathbf{x}_L)$ are defined as follows:

$$\mathbf{B}_I^{(10)}(\mathbf{x}_L) = \begin{bmatrix} \Psi_{I,1}^{(10)}(\mathbf{x}_L) & 0 \\ 0 & \Psi_{I,2}^{(10)}(\mathbf{x}_L) \\ \Psi_{I,2}^{(10)}(\mathbf{x}_L) & \Psi_{I,1}^{(10)}(\mathbf{x}_L) \end{bmatrix}, \quad \mathbf{B}_I^{(01)}(\mathbf{x}_L) = \begin{bmatrix} \Psi_{I,1}^{(01)}(\mathbf{x}_L) & 0 \\ 0 & \Psi_{I,2}^{(01)}(\mathbf{x}_L) \\ \Psi_{I,2}^{(01)}(\mathbf{x}_L) & \Psi_{I,1}^{(01)}(\mathbf{x}_L) \end{bmatrix}, \tag{10.39}$$

and

$$M_{Lx_i} = \int_{\Omega_L} (x_i - x_{iL})^2 d\Omega. \tag{10.40}$$

From Eqs. (10.38)–(10.40), it can be seen that no subdivision of integration cells is required for stabilization, and no additional sampling points are needed. This combined with the fact that (10.36) and (10.2) are nearly identical, NSNI has a very little cost associated with it. Note that the computation of stabilization terms in (10.39) could be further accelerated and unified with SCNI in (10.38) by the employment of strain smoothing, as described in Section 6.10.4, but this is not done in the RKPM2D code.

Similar to the implementation of the modified-type stabilization, this technique can also be employed with DNI by replacing the smoothed gradients in (10.38) with direct gradients:

$$\overline{\mathbf{K}}_{IJ} = \sum_{L=1}^{NP} \left(\underbrace{\mathbf{B}_I^{\mathrm{T}}(\mathbf{x}_L)\mathbf{D}\mathbf{B}_J(\mathbf{x}_L)A_L}_{\text{DNI}} + \underbrace{\left\{ \left\{\mathbf{B}_I^{(10)}(\mathbf{x}_L)\right\}^{\mathrm{T}} \mathbf{D}\mathbf{B}_I^{(10)}(\mathbf{x}_L)M_{Lx} + \left\{\mathbf{B}_I^{(01)}(\mathbf{x}_L)\right\}^{\mathrm{T}} \mathbf{D}\mathbf{B}_I^{(01)}(\mathbf{x}_L)M_{Ly} \right\}}_{\text{stabilization}} \right), \tag{10.41}$$

which is the original form presented in [9], but nevertheless requires a correction to pass the patch test.

The natural-type stabilization is implemented for SCNI using (10.38) in RKPM2D, while DNI is stabilized using (10.41). These formulations are denoted with a prefix "N," e.g., NSCNI. To activate the natural stabilization in RKPM2D, set `Quadrature.Stabilization = 'N'` (see Section 10.4.1).

10.3 Computer Implementation

In this section, the basic data structures and main functions of RKPM2D will be explained, which entail domain discretization, RK shape functions and gradient computation, domain integration, assembly of stiffness matrices and force vectors, and visualization of the numerical results. Also, the key differences between an RKPM and an FEM program are highlighted.

A general flowchart of the code is given in Figure 10.4. Unlike in FEM where the element order dictates the mesh/node generation, the order of basis (and smoothness) is independent of the domain discretization in RKPM. Nevertheless, the general program functionalities (such as matrix assembly and solver) are quite similar.

As shown in Figure 10.4, the numerical procedures for the problem described in Eqs. (10.10)–(10.17) consist of input file generation, domain discretization, quadrature rule definition, shape function construction, matrix assembly, solver, and postprocessing. The corresponding subroutine names in RKPM2D are also shown in the figure.

10.3.1 Domain Discretization

Using the basic model parameters input by the user (`RK`, `Quadrature`, `Model`), the function `Pre_GenerateDiscretization(Model)` is first called for domain discretization. The function will return the structure `Discretization` which consists of the following fields:

- `xI_Boundary`: coordinates of the boundary RK nodes
- `xI_Interior`: coordinates of the interior RK nodes
- `nP`: the total number of RK nodes, *NP*.

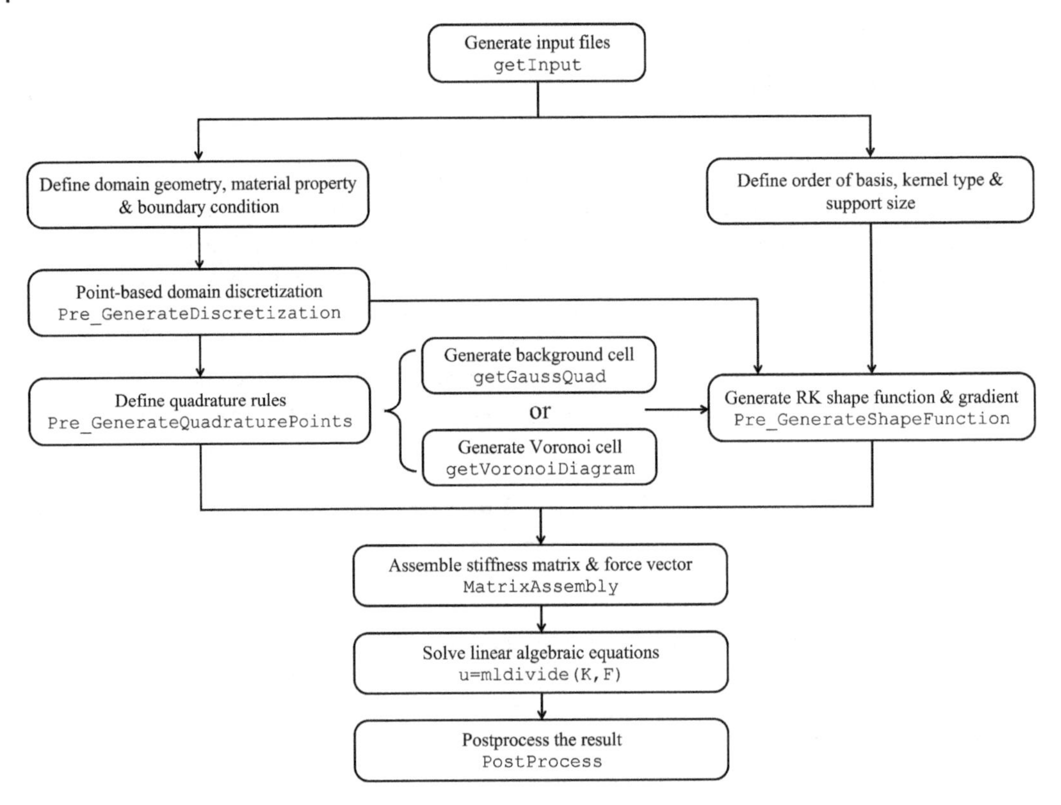

Figure 10.4 Flowchart of the RKPM2D code. Source: Huang et al. [20], figure 8, p. 401 / Springer Nature.

As shown in the command lines in Listing 10.1, the MATLAB built-in function `geometryFromEdges` and `generateMesh` are employed to decompose the domain into conforming subdomains (that can be considered an FEM mesh), and the resulting vertices are directly employed as meshfree nodes with the mesh discarded.

Listing 10.1 Command lines of domain discretization.

```
% create PDE model object for MATLAB default mesh generator
% for more information, please refer to
%https://www.mathworks.com/help/pde/ug/pde.pdemodel.
geometryfromedges.html
model = createpde;

% read in the domain vertices from Model of getInput
x1_vertices = Model.xVertices(:,1);
x2_vertices = Model.xVertices(:,2);

% Create polygon based on edge of the vertices
```

Listing 10.1 (Continued)

```
R1 = [3,length(x1_vertices),x1_vertices',x2_vertices']';

% gm is the geometry based on vertices,
% sf is a set formula created to define if there is any subtraction
% between geometry, eg sf = R1-C1 where C1 may be a circle
gm = [R1]; sf = 'R1';

% create geometry
ns = char('R1'); ns = ns';

% Decompose constructive geometry into minimal regions
g = decsg(gm,sf,ns);

% create geometry for Model object
geoModel = geometryFromEdges(model,g);

% generate FE mesh for Model by MATLAB
FEmesh = generateMesh(model);

% Use FE mesh nodes as RK nodes.
RKnodes = FEmesh.Nodes';
```

Alternatively, users of RKPM2D can also define the meshfree nodes manually or by importing an FEM mesh from CAD/FEM software instead of using the MATLAB built-in meshing function. Such a subroutine, `sub_ReadNeutralInputFiles`, is given in Listing 10.2 which reads in nodal coordinates from a Patran neutral file. The example in the folder "05_FEA_Model" illustrates this functionality.

Listing 10.2 Command lines for reading nodal coordinates from a Patran neutral file.

```
function [xI] = sub_ReadNeutralInputFiles(NeutralFileName)

% Read-in the CAD/FEA neutral file in Patran neutral format
% e.g., NeutralFileName = 'FE_Neutral.dat';
filename = fullfile(NeutralFileName);  % open the full neutral file
T = readtable(filename); % read in table format in MATLAB
C = table2cell(T);        % convert table format to cell format

% Convert cell format to double precision to obtain coordinates
nLine = length(C); % number of lines in neutral file
```

(Continued)

Listing 10.2 (Continued)

```
C_double = cellfun(@str2num,C,'UniformOutput',false);

% Read discretization information
LineOfC = C_double{2};
NNODE = LineOfC(5); % number of nodes
NELEM = LineOfC(6); % number of elements
xI = zeros(NNODE,2); % nodal coordinates initialization
disp(['From ',filename,' file: #node is ',num2str(NNODE),',
#element is ',num2str(NELEM)])

% Read the file
iL = 2; idx_node = 1;

while (iL <= nLine) % read line by line from the input files

      LineMatrix = C_double{iL};

      if ~isempty(LineMatrix) % no empty line is read

      IDCARD = LineMatrix(1);

      %% IDCARD = 01 is the coordinates list; Read coordinates
      if (IDCARD == 1) && length(LineMatrix) == 9

      iL = iL + 1;  % read next line
      xI(idx_node,1:2) = C_double{iL}(1:2);
      idx_node = idx_node + 1; % next node
      iL = iL + 1;  % next line

      end

      end % end if ~isempty

      iL = iL + 1; % next line

end % end of reading each line

end % end of the function
```

10.3.2 Quadrature Point Generation

The quadrature points are generated in the function `Pre_GenerateQuadraturePoints` which supports the following two domain integration schemes:

- Nodal integration where the Voronoi cells are defined.
- Gauss integration where the background integration cells are defined.

Gauss–Legendre rules for coordinates and weights are employed for Gauss integration in each cell and along the boundary segments. The quadrature points and weights are based on Voronoi cells for nodal integration. The construction of the Voronoi diagram is achieved by employing the function `getVoronoiDiagram`, which is modified from the open-source code `VoronoiLimit` [10]. By calling `getVoronoiDiagram`, the structure `VoronoiDiagram` will be returned, which contains the following fields:

- `VerticeCoordinates`: all vertices' coordinates of Voronoi cells.
- `VoronoiCell`: indexes of vertices within each Voronoi cell.

The `VoronoiDiagram` structure defines the quadrature rules required for nodal integration as described in Eqs. (10.23) and (10.24). By looping over the Voronoi cells, the following two classes are added by `Pre_GenerateQuadraturePoints` to the structure `Quadrature` which defines the quadrature rules to compute the RK shape functions for domain and boundary integration:

- `Domain`: class that defines the variables required for domain integration.
  - `nQuad`: the total number of quadrature points.
  - `xQuad`: the coordinates of quadrature points.
  - `Weight`: quadrature weights for domain integral.
- `BC`: class that defines the variables required for boundary integration.
  - `nQuad_onBoundary`: the number of quadrature points on the boundary.
  - `xQuad_onBoundary`: coordinates of quadrature points on the boundary.
  - `Weight_onBoundary`: quadrature weights for contour integrals.
  - `Normal_onBoundary`: outward unit normal vectors at quadrature points along the boundary.

10.3.3 RK Shape Function Generation

In FEM programing, double loops are required to evaluate the stiffness matrix and force vector: one loop over all the elements and another loop over all the quadrature points within each element. In RKPM, only one loop is required if a nodal integration method is used: a loop over all the nodes. As shown in Figure 10.5, at each quadrature point, the RKPM shape functions and their derivatives $\Psi_I(\mathbf{x}_L)$, $\Psi_{I,1}(\mathbf{x}_L)$, and $\Psi_{I,2}(\mathbf{x}_L)$ (and/or smooth and implicit gradients given in (10.26) and (10.36)) are evaluated in the physical domain. In contrast, FEM shape functions and derivatives are constructed in the parametric domain and require an isoparametric mapping process which may introduce large errors into the solution when the elements are severely distorted.

Another difference in computing RKPM and FEM shape functions is the determination of nodal neighbors. FEM relies on using an element mesh to define the shape functions; thus, the neighbors for a given evaluation point are simply the nodes of the element. In RKPM, the shape functions are instead defined directly at nodes without the element connectivity. Consequently, a neighbor search is necessary to determine the neighbors of a given evaluation point. The neighbor search algorithm can be CPU intensive if care is not taken, especially for 3D simulations, so efficient spatial search algorithms such as KD-Tree [11], DESS [12], among others [13–16] have been developed.

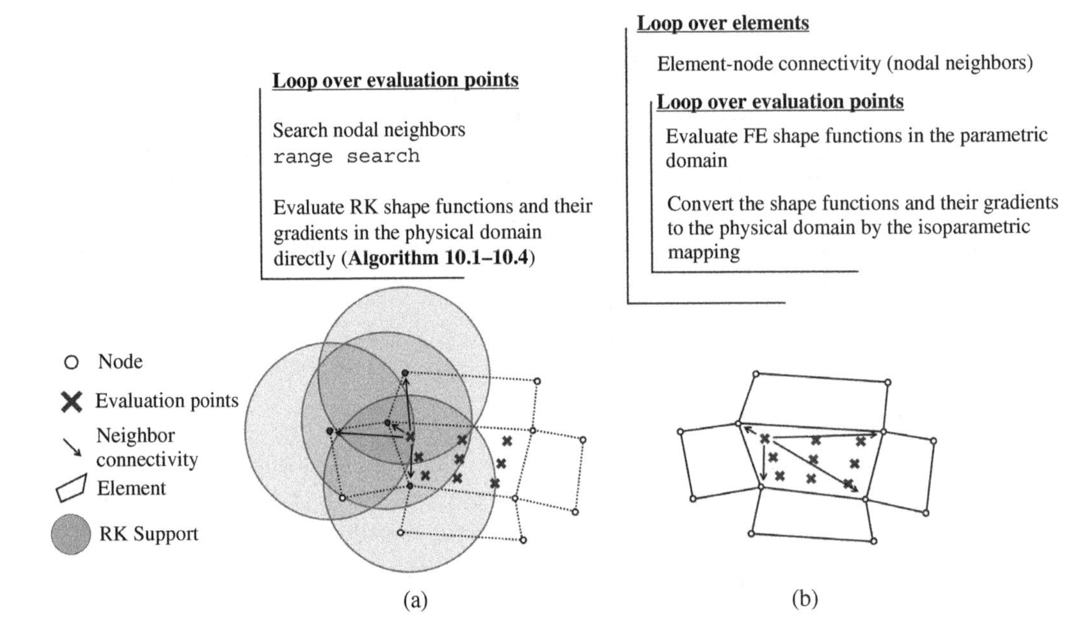

Figure 10.5 Flowchart and overview of the domain integration algorithms in (a) RKPM and (b) FEM. Source: Huang et al. [20], figure 10, p. 408 / Springer Nature.

RKPM2D uses the MATLAB built-in function `rangesearch` with the efficient KD-Tree algorithm [11]. The command line for constructing the neighbor list is:

```
[NeighborList] = rangesearch(xI,x,SupportSize,'Distance','euclidean');
```

Here, the inputs of `rangesearch` are as follows

- `xI`: coordinates of RK nodes $\mathbf{x}_I$.
- `x`: coordinates of evaluation points $\mathbf{x}$.
- `SupportSize`: the support size a_I.

Evaluation of RK shape functions with respect to RK nodes at $\mathbf{x}_I$ evaluated at point $\mathbf{x}$ is performed in `getRKShapeFunction`, for which the program structure is illustrated in **Algorithms 10.1–10.4**. Note that, in the evaluation of the derivatives of circular kernel functions in **Algorithm 10.3**, the derivative of the distance $z_{I,i} = \frac{(x_i - x_{iI})}{(z_I a_I^2)}$ becomes singular when the evaluation point $\mathbf{x}$ approaches node $\mathbf{x}_I$ (i.e., $\|\mathbf{x} - \mathbf{x}_I\| \to 0$). This can be avoided by setting $z_I = 0, \nabla z_I = \mathbf{0}$ if $\|\mathbf{x} - \mathbf{x}_I\| < eps$, where *eps* is the default positive machine precision number in MATLAB, which works well for symmetric and smooth kernels. Alternatively, we can set $z_{I,1} = \frac{(x_1 - x_{1I})}{(z_I a_I^2 + \text{eps})}$, to keep the denominator of $z_{I,1}$ positive (as with $z_{I,2}$). The latter approach is implemented in RKPM2D.

The evaluation of direct derivatives $\Psi_{I,1}(\mathbf{x}_L)$ and $\Psi_{I,2}(\mathbf{x}_L)$ for DNI and GI at quadrature points $\mathbf{x}_L$ is straightforward, and the direct derivatives are computed by **Algorithms 10.1–10.4**. In SCNI, $\Psi_{I,1}(\mathbf{x}_L)$ and $\Psi_{I,2}(\mathbf{x}_L)$ are replaced with smoothed derivatives $\tilde{b}_{1I}^L$ and $\tilde{b}_{2I}^L$, respectively (see (10.26)), for which the smoothing procedure is given in **Algorithm 10.5**. **VoronoiCell**{*I*}(*K*) denotes a cell structure for Voronoi cells to define the K*th* index of cell vertices for the I*th* Voronoi cell, and **Vertices** is a vector that defines the Cartesian coordinates of Voronoi cell vertices.

Algorithm 10.1 RK shape function $\Psi_I(\mathbf{x})$ evaluation

1. **function** $[\Psi_I, \Psi_{I,1}, \Psi_{I,2}] = getRKShapeFunction(\mathbf{RK}, \{\mathbf{x}_I\}_{I=1}^{NP}, \mathbf{x})$
2. $\{\%\}$ find $\mathbf{x} \in \text{supp}(\{\mathbf{x}_I\}_{I=1}^{NP})$, order & physical nodal kernel support size
3. $n \leftarrow$ **RK.Order**; $\{a_I\}_{I=1}^{NP} \leftarrow$ **RK.SupportSize**;
4. Neighbor List: $S_\mathbf{x} \leftarrow rangesearch(\{\mathbf{x}_I\}_{I=1}^{NP}, \mathbf{x}, \{a_I\}_{I=1}^{NP})$
5. $\mathbf{M} = \mathbf{0}$; $\mathbf{M}_{,1} = \mathbf{0}$; $\mathbf{M}_{,2} = \mathbf{0}$; $\{\%\}$ Initialize the moment matrix
6. **for** $I \in S_\mathbf{x}$ **do**
7. $[\mathbf{H}, \mathbf{H}_{,1}, \mathbf{H}_{,2}] = Basis(n, \mathbf{x}_I, \mathbf{x})$ $\{\%\}$ Eq. (10.4)
8. $[\phi_a, \phi_{a,1}, \phi_{a,2}] = Kernel(a_I, \mathbf{x}_I, \mathbf{x})$
9. $[\mathbf{M}, \mathbf{M}_{,1}, \mathbf{M}_{,2}]$ $\{\%\}$ Eq. (10.3), sum moment matrix $= Moment(\mathbf{M}, \mathbf{M}_{,1}, \mathbf{M}_{,2}, \mathbf{H}, \mathbf{H}_{,1}, \mathbf{H}_{,2}, \phi_a, \phi_{a,1}, \phi_{a,2})$
10. **end for**
11. $\{\Psi_I\}_{I=1}^{NP} = \mathbf{H}_\mathbf{0}^\mathrm{T}\mathbf{M}^{-1}\mathbf{H}\phi_a$ $\{\%\}$ Sparse vectorized Eq. (10.2), $\mathbf{H}_\mathbf{0} \equiv \mathbf{H}(\mathbf{0})$
12. $\mathbf{M}_{,1}^{-1} = -\mathbf{M}^{-1}\mathbf{M}_{,1}\mathbf{M}^{-1}$; $\mathbf{M}_{,2}^{-1} = -\mathbf{M}^{-1}\mathbf{M}_{,2}\mathbf{M}^{-1}$;
13. $\{\Psi_{I,1}\}_{I=1}^{NP} = \mathbf{H}_\mathbf{0}^\mathrm{T}\mathbf{M}_{,1}^{-1}\mathbf{H}\phi_a + \mathbf{H}_\mathbf{0}^\mathrm{T}\mathbf{M}^{-1}\mathbf{H}_{,1}\phi_a + \mathbf{H}_\mathbf{0}^\mathrm{T}\mathbf{M}^{-1}\mathbf{H}\phi_{a,1}$
14. $\{\Psi_{I,2}\}_{I=1}^{NP} = \mathbf{H}_\mathbf{0}^\mathrm{T}\mathbf{M}_{,2}^{-1}\mathbf{H}\phi_a + \mathbf{H}_\mathbf{0}^\mathrm{T}\mathbf{M}^{-1}\mathbf{H}_{,2}\phi_a + \mathbf{H}_\mathbf{0}^\mathrm{T}\mathbf{M}^{-1}\mathbf{H}\phi_{a,2}$
15. **end function**

Algorithm 10.2 Basis vector computation

1. **function** $[\mathbf{H}, \mathbf{H}_{,1}, \mathbf{H}_{,2}] = Basis(n, \mathbf{x}_I, \mathbf{x})$
2. $x_1 = \mathbf{x}(1)$; $x_2 = \mathbf{x}(2)$
3. $x_{1I} = \mathbf{x}_I(1)$; $x_{2I} = \mathbf{x}_I(2)$
4. $count = 1$
5. **for** $k = 0:n$ **do**
6. **for** $j = 0:k$ **do**
7. $i = k - j$
8. $\mathbf{H}(count, I) = (x_1 - x_{1I})^i(x_2 - x_{2I})^j$
9. $\mathbf{H}_{,1}(count, I) = i(x_1 - x_{1I})^{i-1}(x_2 - x_{2I})^j$
10. $\mathbf{H}_{,2}(count, I) = j(x_1 - x_{1I})^i(x_2 - x_{2I})^{j-1}$
11. $count = count + 1$
12. **end for**
13. **end for**
14. **end function**

The smoothing in **Algorithm 10.5** is computed by the function `getSmoothedDerivative` as given in the Listing 10.3. For given inputs of a smoothing cell vertices' coordinates $\{\mathbf{x}_v\}_{v=1}^{LV}$ (denoted as "xV" in the Listing 10.3) and nodal coordinates $\mathbf{x}_I$ (denoted as "`xI`" in Listing 10.3), the function

Algorithm 10.3 Kernel function evaluation (Cubic B-spline with circular support)

1. **function** $[\phi_a, \phi_{a,1}, \phi_{a,2}] = Kernel(a_I, \mathbf{x}_I, \mathbf{x})$
2. {%} different types of kernels are implemented in a similar manner
3. $z_I = \frac{\|\mathbf{x}-\mathbf{x}_I\|}{a_I}$; $z_{I,1} = \frac{(x_1 - x_{1I})}{(z_I a_I^2)}$; $z_{I,2} = \frac{(x_2 - x_{2I})}{(z_I a_I^2)}$
4. **if** $0 \leq z_I < \frac{1}{2}$ **then**
5. $\phi_a = \frac{2}{3} - 4z_I^2 + 4z_I^3$; $\phi_{a,z} = -8z_I + 12z_I^2$
6. **elseif** $½ \leq z_I < 1$ **then**
7. $\phi_a = \frac{4}{3} - 4z_I + 4z_I^2 - \frac{4z_I^3}{3}$; $\phi_{a,z} = -4 + 8z_I - 4z_I^2$
8. **else**
9. $\phi_a = 0$; $\phi_{a,z} = 0$
10 **end if**
11. $\phi_{a,1} = \phi_{a,z} z_{,1}$
12. $\phi_{a,2} = \phi_{a,z} z_{,2}$
13. **end function**

Algorithm 10.4 Moment matrix computation

1. **function** $[\mathbf{M}, \mathbf{M}_{,1}, \mathbf{M}_{,2}] = Moment(\mathbf{M}, \mathbf{M}_{,1}, \mathbf{M}_{,2}, \mathbf{H}, \mathbf{H}_{,1}, \mathbf{H}_{,2}, \phi_a, \phi_{a,1}, \phi_{a,2})$
2. $\mathbf{M} = \mathbf{M} + \mathbf{H}\mathbf{H}^{\mathrm{T}}\phi_a(\boldsymbol{x})$
3. $\mathbf{M}_{,1} = \mathbf{M}_{,1} + \mathbf{H}_{,1}\mathbf{H}^{\mathrm{T}}\phi_a + \mathbf{H}\mathbf{H}_{,1}^{\mathrm{T}}\phi_a + \mathbf{H}\mathbf{H}^{\mathrm{T}}\phi_{a,1}$
4. $\mathbf{M}_{,2} = \mathbf{M}_{,2} + \mathbf{H}_{,2}\mathbf{H}^{\mathrm{T}}\phi_a + \mathbf{H}\mathbf{H}_{,2}^{\mathrm{T}}\phi_a + \mathbf{H}\mathbf{H}^{\mathrm{T}}\phi_{a,2}$
5. **end function**

Algorithm 10.5 SCNI smoothed derivatives

1. **for** $L = 1 : NP$ **do** {%} Loop over quadrature points (nodal points)
2. $LV \leftarrow$ **VoronoiCell**$\{L\}(:)$ {%} Number of cell edge for Voronoi cell L
3. $\{\mathbf{x}_v\}_{v=1}^{LV} \leftarrow$ **Vertices**(**VoronoiCell**$\{L\}(:)$) {%} Voronoi vertices coordinates
4. $A_L \leftarrow \{\mathbf{x}_v\}_{v=1}^{LV}$ {%} Calculate domain area of Voronoi cell L
5. $\mathbf{x}_I \leftarrow$ Discretization {%} RK nodal points
6. $\tilde{b}_{iI}^L = \mathbf{0}$; {%} initialization for shape function evaluated at smoothing point
7. **for** $M = 1 : LV$ **do** {%} Loop over cell edges of Voronoi cell L
8. find Voronoi cell edge quadrature points $\tilde{\mathbf{x}}_L^M$ from $\{\mathbf{x}_v\}_{v=M}^{M+1}$
9. find Voronoi cell edge normal n_{iM} from $\{\mathbf{x}_v\}_{v=M}^{M+1}$
10. find Voronoi cell edge length l_M from $\{\mathbf{x}_v\}_{v=M}^{M+1}$

Algorithm 10.5 (Continued)

11. $[\Psi_I(\tilde{\mathbf{x}}_L^M)] = getRKShapeFunction(\mathbf{RK}, \{\mathbf{x}_I\}_{I=1}^{NP}, \tilde{\mathbf{x}}_L^M)$
12. $\tilde{b}_{iI}^L = \tilde{b}_{iI}^L + \Psi_I(\tilde{\mathbf{x}}_L^M) n_{iM} l_M$
13. **end for**
14. $\tilde{b}_{iI}^L = \frac{1}{A_L}\tilde{b}_{iI}^L$ {%} Smoothed derivative, Eq. (10.26)
15. **end for**

Listing 10.3 Command lines for smoothing procedure in each cell.

```
function [SHPDX1_smoothed,SHPDX2_smoothed,Area_Cell] =
getSmoothedDerivative(RK,xI,xV)

% Input : xI, RK nodes; xV, vertices of the cell;
% Output: Smoothed Derivative SHPDX1/2 and cell area
% smoothing point sequence for SCNI cell

nP = length(xI); % number of node
nV = length(xV); % number of vertices

% obtain the area of each voronoi cell
Area_Cell = polyarea(xV(:,1),xV(:,2))+eps;

% initiate shape function at the smoothed points
SHP_Smoothed_local = sparse(2,nP);

for k = 0:nV-1 % loop over each edge for each cell

    % find the two ends of the edge
    if k == 0
        Vertex1 = xV(end,:);
    else
        Vertex1 = xV(k,:);
    end
    Vertex2 = xV(k+1,:);

    % Cell edge length Lk_Cell and normal Nk_Cell
    Lk_Cell = norm(Vertex2-Vertex1,2);
    xv21 = Vertex2 - Vertex1;
    xv21_Normal = xv21*[cos(pi/2) -sin(pi/2); sin(pi/2) cos(pi/2)];
```

(Continued)

Listing 10.3 (Continued)

```
        % calculate the unit normal, eps is machine precision
        Nk_Cell = xv21_Normal/(norm(xv21_Normal,2)+eps);
        if norm(Lk_Cell,2) < eps || norm(Nk_Cell,2) < eps
            Lk_Cell = eps; Nk_Cell = [eps eps];
        End

        % Smoothing Points of each segment
        xtilde_CellEdge = (Vertex1 + Vertex2)/2;

        % RK shape function evaluation at smoothing point
        [SHP_xtilde] = ...
        getRKShapeFunction(RK,xI,xtilde_CellEdge(1:2),[1,0,0]);
        SHP_Smoothed_local = SHP_Smoothed_local + ...
        (Nk_Cell'*SHP_xtilde)*Lk_Cell;

end % end of each cell boundary

% evaluate the smoothed derivative
SHPDX1_smoothed = (1/Area_Cell)*SHP_Smoothed_local(1,:);
SHPDX2_smoothed = (1/Area_Cell)*SHP_Smoothed_local(2,:);

end
```

`getSmoothedDerivative` gives the corresponding smoothed derivatives $\tilde{b}_{1I}^{L}$ and $\tilde{b}_{2I}^{L}$ and the smoothing domain area A_L.

The computed shape functions and their derivatives are stored in the structure `Quadrature` as

- `SHP`: matrix of size `nQuad×nP` that stores the shape functions $\Psi_I(\mathbf{x}_L)$ of all nodes.
- `nQuad`: number of quadrature points for domain integration, the value $=NG$ for Gauss integration and $=NP$ for nodal integration
- `SHPDX1`: matrix of size `nQuad×nP` that stores the shape function derivatives $\Psi_{I,1}(\mathbf{x}_L)$ or $\tilde{b}_{1I}^{L}$ of all nodes.
- `SHPDX2`: matrix of size `nQuad×nP` that stores the shape function derivatives $\Psi_{I,2}(\mathbf{x}_L)$ or $\tilde{b}_{2I}^{L}$ of all nodes.

10.3.4 Stabilization Methods

The stabilization terms associated with modified- and natural-type stabilization are computed in `Pre_GenerateShapeFunction` where RK shape functions and direct/smoothed gradients are also evaluated. For the modified-type stabilization method discussed in Section 10.2.3, the nodal representative domain is divided into subcells by using the MATLAB built-in function `delaunay-Triangulation`, which divides the polygon Voronoi domain into several triangular subcells.

Listing 10.4 Command lines for subdivision of gradient smoothing cells for the modified-type stabilization.

```
% M-type Stabilization by subdivision of integration cell

xVerices_MSCNI = unique([xV; xQuad(idx_nQuad,:)],'rows');

% use delauny triangle as subcell
DT_MSCNI = delaunayTriangulation(xVerices_MSCNI(:,1),...
                                 xVerices_MSCNI(:,2));
[N_subcell,~] = size(xNs_MSCNI);

for s = 1 : N_subcell % loop over the sub cell

% vertices coordinates of the sub cell
xV_SubCell = DT_MSCNI.Points(DT_MSCNI.ConnectivityList(s,:),:);

% evaluate area of the sub cell
AREA_Is{idx_nQuad}(s,1) = polyarea(xV_SubCell(:,1),xV_SubCell
(:,2));

end
```

The procedures for generating the subcells and associated evaluation points for the additional stabilization terms are shown in Listing 10.4. By looping over each subcell, the shape function gradients with associated integration weights (i.e., the area of each subcell) are computed. At each subcell, `getSmoothedDerivative` in Listing 10.3 is employed to evaluate the smoothed gradient to construct the modified-type stabilization terms in (10.31) for SCNI. An alternative way is to evaluate direct shape function gradient as in Eq. (10.33) by **Algorithms 10.1–10.4** at the centroid of each subcell for the modified-type stabilization (see Section 10.2.3), which is adopted in RKPM2D for DNI-based methods.

Once all stabilization terms are computed, they are stored under the structure `Quadrature.MtypeStabilization` with the following fields:

- `nS`: number of sub-cells NS for the Ith Voronoi cell.
- `SHPDX1_Is`: cell structure contains a matrix of size `nSxnP` that stores the shape function derivatives $\Psi_{I,1}(\hat{\mathbf{x}}_L^K)$ or $\tilde{b}_{1I}^{KL}$ in the Ith Voronoi cell
- `SHPDX2_Is`: cell structure contains a matrix of size `nSxnP` that stores the shape function derivatives $\Psi_{I,2}(\hat{\mathbf{x}}_L^K)$ or $\tilde{b}_{2I}^{KL}$ in the Ith Voronoi cell
- `AREA_Is`: cell structure that stores area of the Mth subcell associated with the Ith Voronoi cell

Unlike modified-type stabilization, the subdivision of the Voronoi diagrams is not required in the natural-type stabilization method discussed in Section 10.2.3. The second-order shape function derivatives are computed by taking direct derivatives of the first-order implicit gradients, which can be easily achieved by replacing $\mathbf{H_0}(=\mathbf{H}(\mathbf{0}))$ in **Algorithm 10.1** with $\mathbf{H}^{(10)}$ or $\mathbf{H}^{(01)}$ shown in (10.37). The corresponding output $\Psi_{I,1}(\mathbf{x})$ and $\Psi_{I,2}(\mathbf{x})$ becomes $\Psi_{I,1}^{(10)}$ and $\Psi_{I,2}^{(10)}$ when $\mathbf{H}^{(10)}$ is used, and $\Psi_{I,1}^{(01)}$ and $\Psi_{I,2}^{(01)}$ when $\mathbf{H}^{(01)}$ is used. The second moments of inertia M_{1L}, M_{2L} in each nodal integration domain are computed straightforwardly by the formula in [17] from the Voronoi vertices' coordinates as shown in Listing 10.5.

Listing 10.5 Command lines for evaluating $\Psi_{I,1}^{(10)}, \Psi_{I,2}^{(10)}, \Psi_{I,1}^{(01)} \approx \Psi_{I,2}^{(10)}, \Psi_{I,2}^{(01)}, M_{1L}, M_{2L}$ in natural-type stabilization for each node.

```
% NSNI stabilization for calculating moment inertia

for idx_nQuad = 1:nQuad % loop over quadrature points

% xQuad(idx_nQuad,:) is quadrature points (nodal points)
% Evaluation of second order gradient of shape function

IG = 1; % where 'IG' is the switch to turn on/off implicit gradient
[~,~,~,SHPDX1X1(idx_nQuad,:),SHPDX2X2(idx_nQuad,:),SHPDX1X2
(idx_nQuad,:)] = getRKShapeFunction(RK,xI,xQuad(idx_nQuad,:),IG);

% normalized the vertices coordinates by mean coordinates
Vertices1 = xV - ones(nV,1)*mean(xV);
Vertices2 = circshift(xV - ones(nV,1)*mean(xV),[-1 0]);

% using formula to calculate inertia for cell by vertices
M1N(idx_nQuad) = (1/12)*abs(sum((Vertices1(:,2).^2+Vertices1(:,2).
*Vertices2(:,2)+Vertices2(:,2).^2).*(Vertices1(:,1).*Vertices2
(:,2) - Vertices2(:,1).*Vertices1(:,2))));
M2N(idx_nQuad) = (1/12)*abs(sum((Vertices1(:,1).^2+Vertices1(:,1).
*Vertices2(:,1)+Vertices2(:,1).^2).*(Vertices1(:,1).*Vertices2
(:,2) - Vertices2(:,1).*Vertices1(:,2))));

% Parallel axis theorem for shift it back to global coordinates
dx1 = sqrt((xQuad(idx_nQuad,1)-mean(xV(:,1))).^2);
dx2 = sqrt((xQuad(idx_nQuad,2)-mean(xV(:,2))).^2);
M1N(idx_nQuad) = M1N(idx_nQuad) + Area_VoronoiCell(idx_nQuad)
*dx1^2;
M2N(idx_nQuad) = M2N(idx_nQuad) + Area_VoronoiCell(idx_nQuad)
*dx2^2;

end
```

Once all stabilization terms are computed, they are stored under the structure `Quadrature.NtypeStabilization` with the following fields:

- `SHPDX1X1`: matrix of size `nPxnP` that stores the shape function second derivatives $\Psi_{I,1}^{(10)}$ of all nodes.
- `SHPDX1X2`: matrix of size `nPxnP` that stores the shape function second derivatives $\Psi_{I,2}^{(10)}$ of all nodes.
- `SHPDX2X1`: matrix of size `nPxnP` that stores the shape function second derivatives $\Psi_{I,1}^{(01)}$ of all nodes.
- `SHPDX2X2`: matrix of size `nPxnP` that stores the shape function second derivatives $\Psi_{I,2}^{(01)}$ of all nodes.
- `M`: second moments of inertia M_{1L}, M_{2L} in each nodal integration domain.

10.3.5 Matrix Evaluation and Assembly

Once RK shape functions are computed, the function `MatrixAssmebly` is called to evaluate and assemble the stiffness matrix and force vector. Utilizing the MATLAB built-in sparse matrix data structure [18], the sparse storage scheme is employed for matrix evaluation and assembly to reduce the memory requirements. However, other storage schemes that avoid operations on elements of zero can also be adopted. The program structure of `MatrixAssmebly` is given in **Algorithm 10.6.**

Algorithm 10.6 Stiffness matrix and force vector assembly

1. **function** $[\mathbf{K}_{IJ}, \mathbf{f}_I]$ = *MatrixAssembly*(**Quadrature, Model**)
2. {%} initialization of stiffness matrix and force vector
3. $\mathbf{K}_{IJ}, \mathbf{H}_{IJ}, \mathbf{A}_{IJ} = \mathbf{0}$; $\mathbf{f}_I^b, \mathbf{f}_I^t, \mathbf{h}_I, \mathbf{g}_{IJ} = \mathbf{0}$;
4. $\mathbf{D}, \beta \leftarrow$ **Model**; {%} elastic tensor and penalty parameter
5. {%} assemble the stiffness matrix and body force vector
6. *nQuad* = **Quadrature.Domain.nQuad** {%} # of quadrature points
7. **for** $L = 1:nQuad$ **do**
8. $\mathbf{x}_L, A_L, \mathbf{B}_I, \Psi_I \leftarrow$ **Quadrature.Domain**
9. $\mathbf{K}_{IJ} = \mathbf{K}_{IJ} + \mathbf{B}_I^{\mathrm{T}}\mathbf{D}\mathbf{B}_J A_L$ {%} tangent stiffness by Eq. (10.21)/(10.23)/(10.28)
10. $\mathbf{K}_{IJ} = \mathbf{K}_{IJ} + \mathbf{K}_{IJ}^{\text{stab}}$
11. {%} $\mathbf{K}_{IJ}^{\text{stab}}$: additional stabilization in Eq. (10.31)/(10.33), (10.38)/(10.41)
12. $\mathbf{b} \leftarrow$ *getBodyForce*($\mathbf{x}_L$) {%} obtain body force $\mathbf{b}$
13. $\mathbf{f}_I^b = \mathbf{f}_I^b + \Psi_I \mathbf{b} A_L$ {%} body force by Eq. (10.21)/(10.23)
14. **end for**
15. {%} boundary integration for natural and essential boundary conditions
16. $nQuad_{BC}$ = **Quadrature.BC.nQuadonBoundary**
17. **for** $L = 1:nQuad_{BC}$ **do**

(Continued)

Algorithm 10.6 (Continued)

18. $\mathbf{x}_L,\ A_L,\ \boldsymbol{\eta},\ \mathbf{B}_I,\ \Psi_I \leftarrow$ **Quadrature.BC**
19. **if** $\mathbf{x}_L \in$ *Essential Boundary Conditions* **then**
20. $\bar{\mathbf{u}} \leftarrow getGEBC(\mathbf{x}_L)$ {%} obtain prescribed displacement $\bar{\mathbf{u}}$
21. $\mathbf{S} \leftarrow getSEBC(\mathbf{x}_L)$ {%} obtain switch matrix $\mathbf{S}$
22. {%} Eqs. (10.15) and (10.14) by (10.22)/(10.24), (10.29)/(10.30)
23. $\mathbf{A}_{IJ} = \mathbf{A}_{IJ} + \beta \boldsymbol{\Psi}_I \mathbf{S} \boldsymbol{\Psi}_J l_L$
24. $\mathbf{H}_{IJ} = \mathbf{H}_{IJ} + \mathbf{B}_I^T \mathbf{D} \boldsymbol{\eta} \mathbf{S} \boldsymbol{\Psi}_J l_L$
25. $\mathbf{g}_I = \mathbf{g}_I + \beta \boldsymbol{\Psi}_I \mathbf{S} \bar{\mathbf{u}} l_L$
26. $\mathbf{h}_I = \mathbf{h}_I + \mathbf{B}_I^{\mathrm{T}} \mathbf{D} \boldsymbol{\eta} \mathbf{S} \bar{\mathbf{u}} l_L$
27. **elseif** $\mathbf{x}_L \in$ *Natural Boundary Conditions* **then**
28. $\bar{\mathbf{t}} \leftarrow getTraction(\boldsymbol{x}_N)$ {%} obtain surface traction $\bar{\mathbf{t}}$
29. $\mathbf{f}_I^t = \mathbf{f}_I^t + \Psi_I \mathbf{t} l_L$
30. **end if**
31. **end for**
32. $\mathbf{K}_{IJ} = \mathbf{K}_{IJ} + \mathbf{A}_{IJ} - \left(\mathbf{H}_{IJ} + \mathbf{H}_{IJ}{}^{\mathrm{T}}\right);\ \mathbf{f}_I = \mathbf{f}_I^b + \mathbf{f}_I^t + \mathbf{g}_I - \mathbf{h}_{IJ};$
33. **end function**

Listing 10.6 gives the MATLAB command lines for the assembly of the stiffness matrix $\mathbf{K}_{IJ}$ and body force vector $\mathbf{f}_I^b$ in **Algorithm 10.6**.

Listing 10.6 Command lines of stiffness matrix $\mathbf{K}_{IJ}$ and body force vector $\mathbf{f}_I^b$ assembly for elasticity problems.

```
DOFu = Model.DOFu; % load degrees of freedom

% B matrix and Psi matrix initiation
B = sparse(3,nP*DOFu) ;
PSI = sparse(2,nP*DOFu) ;

% Load shape function from Quadrature
SHP = Quadrature.SHP;
SHPDX1 = Quadrature.SHPDX1;
SHPDX2 = Quadrature.SHPDX2;

for idx_quad = 1:nQuad % loop over quadrature points

 B(:,:) = 0; PSI(:,:) = 0;
```

(Continued)

Listing 10.6 (Continued)

```
% Load B and Psi matrix with shape function and their derivative
B(1,1:2:end) = SHPDX1(idx_nQuad,:);
B(2,2:2:end) = SHPDX2(idx_nQuad,:);
B(3,1:2:end) = SHPDX2(idx_nQuad,:);
B(3,2:2:end) = SHPDX1(idx_nQuad,:);
PSI(1,1:2:end) = SHP(idx_nQuad,:);
PSI(2,2:2:end) = SHP(idx_nQuad,:);

% Stiffness matrix assembly
KIJ_c = KIJ_c + (B')*(C)*B*(Weight(idx_nQuad,1));

% Obtain the body force at quadrature point's coordinates
b = f_b(xQuad(idx_nQuad,1),xQuad(idx_nQuad,2));

% Body force vector assembly
FI_b = FI_b + (PSI')*b*(Weight(idx_nQuad,1));

end
```

The command lines for the assembly of the traction force $\mathbf{f}_I^t$ as well as the Nitsche's term, $\mathbf{A}_{IJ}$, $\mathbf{H}_{IJ}$, $\mathbf{g}_I$, $\mathbf{h}_I$, in **Algorithm 10.6** are given in Listing 10.7.

Listing 10.7 Command lines of assembly of the traction force and the Nitsche's term for elasticity problems.

```
nP = Quadrature.nP; % number of nodal points
DOFu = Model.DOFu; % degrees of freedom of u

% Load Model, Young's modulus
E = Model.E;

% Load function handle for evaluating S,g,t
f_S = Model.ExactSolution.S;
f_g = Model.ExactSolution.g;
f_t = Model.ExactSolution.t;

% Beta Parameter for Nitsche's Method
beta = Model.Beta_Nor*E/sqrt((Model.DomainArea/nP)); % Penalty
Number
```

(Continued)

Listing 10.7 (Continued)

```
% Load information required for boundary integration
nQuad_BC = Quadrature.BC.nQuad_onBoundary;
xQuad_BC = Quadrature.BC.xQuad_onBoundary;
Weight_BC = Quadrature.BC.Weight_onBoundary;
Normal_BC = Quadrature.BC.Normal_onBoundary;
SHP_onBC = Quadrature.BC.SHP_BC;
SHPDX1_onBC = Quadrature.BC.SHPDX1_BC;
SHPDX2_onBC = Quadrature.BC.SHPDX2_BC;

for idx_nQuad = 1:nQuad_BC % for loop of quadrature points at
boundary

 % normal at quadrature points
 n1 = Normal_BC(idx_nQuad,1);
 n2 = Normal_BC(idx_nQuad,2);

 % PSI matrix at quadrature points
 PSI(:,:) = 0;
 PSI(1,1:2:end) = SHP_onBC(idx_nQuad,:);
 PSI(2,2:2:end) = SHP_onBC(idx_nQuad,:);

 % switch to different boundary conditions
 switch Quadrature.BC.EBCtype{idx_nQuad}

  case {'NBC'}

  % surface traction
  t = f_t(xQuad_BC(idx_nQuad,1),xQuad_BC(idx_nQuad,2),n1,n2);
  FI_t = FI_t + PSI'*t*Weight_BC(idx_nQuad);

  case {'EBC'}

  % Surface normal Eta
  ETA = [n1 0 n2; 0 n2 n1]';

  % Load B Matrix
  B(:,:) = 0;
  B(1,1:2:end) = SHPDX1_onBC(idx_nQuad,:);
  B(2,2:2:end) = SHPDX2_onBC(idx_nQuad,:);
  B(3,1:2:end) = SHPDX2_onBC(idx_nQuad,:);
  B(3,2:2:end) = SHPDX1_onBC(idx_nQuad,:);

  % Essential boundary condition g
```

(Continued)

Listing 10.7 (Continued)

```
    g = f_g(xQuad_BC(idx_nQuad,1),xQuad_BC(idx_nQuad,2));

    % Switch s
    S = f_S(xQuad_BC(idx_nQuad,1),xQuad_BC(idx_nQuad,2));

    % Nitche's terms
    KIJ_g = KIJ_g + (PSI'*S'*ETA*C*B)*Weight_BC(idx_nQuad);
    FI_g = FI_g + (B'*C*ETA*S*g)*Weight_BC(idx_nQuad);
    KIJ_beta = KIJ_beta + beta*(PSI'*S*PSI)*Weight_BC(idx_nQuad);
    FI_beta = FI_beta + beta*(PSI'*S*g)*Weight_BC(idx_nQuad);

  end % end swtich different boundary conditions

end % end for loop of quadrature points at boundary
```

As seen from **Algorithm 10.6**, Listings 10.6 and 10.7, the function `MatrixAssembly` only involves shape functions and quadrature rules under the data structure `Quadrature`, whereas the material parameters and evaluation of traction $\bar{\mathbf{t}}$, body force $\mathbf{b}$, essential boundary conditions $\bar{\mathbf{u}}$, and switch $\mathbf{S}$ are obtained under the data structure `Model`. After the assembly, the resultant system of linear equations $\sum_J \mathbf{K}_{IJ}\mathbf{u}_J = \mathbf{f}_I$ can be solved by either direct methods such as LU factorization, Gauss elimination or non-stationary iterative methods such as PCG, GMRES, and BICGSTAB [18]. Many of these methods are available in public domain or open-source software and can be readily integrated or linked into RKPM2D. Here, the MATLAB function `mldivide` is adopted, which takes advantage of matrix symmetries and automatically assigns an appropriate matrix solver.

Remark 10.1 *Marginal changes in inputs from* `Quadrature` *and* `Model` *and slight modifications in Listings 10.6 and 10.7 can be performed to convert the code to solve a different type of equation. For demonstration, an example of modifying RKPM2D to solve a diffusion equation is given in the Appendix.*

10.3.6 Description of subroutines in RKPM2D

The subroutines for all of the different functionalities of RKPM are included in the folder "MAIN_PROGRAM," as shown in Figure 10.6. For reference purposes, the function of each subroutine is described as follows.

`getBackgroundIntegrationCell.m`: This subroutine constructs the rectangular background Gauss integration cells if Gauss integration is adopted.

`getBoundaryConditions.m`: This subroutine outputs the functions of the prescribed boundary conditions $\bar{\mathbf{t}} = \boldsymbol{\eta}^{\mathrm{T}}\mathbf{D}\mathcal{E}^{\mathrm{exact}}$, $\bar{\mathbf{u}} = \mathbf{u}^{\mathrm{exact}}$, $\mathbf{S}$ and the body force $\mathbf{b} = -\nabla \bullet \boldsymbol{\sigma}^{\mathrm{exact}}$ according to given expressions of displacement $\mathbf{u}^{\mathrm{exact}}$ in terms of x_1 and x_2 in a symbolic format.

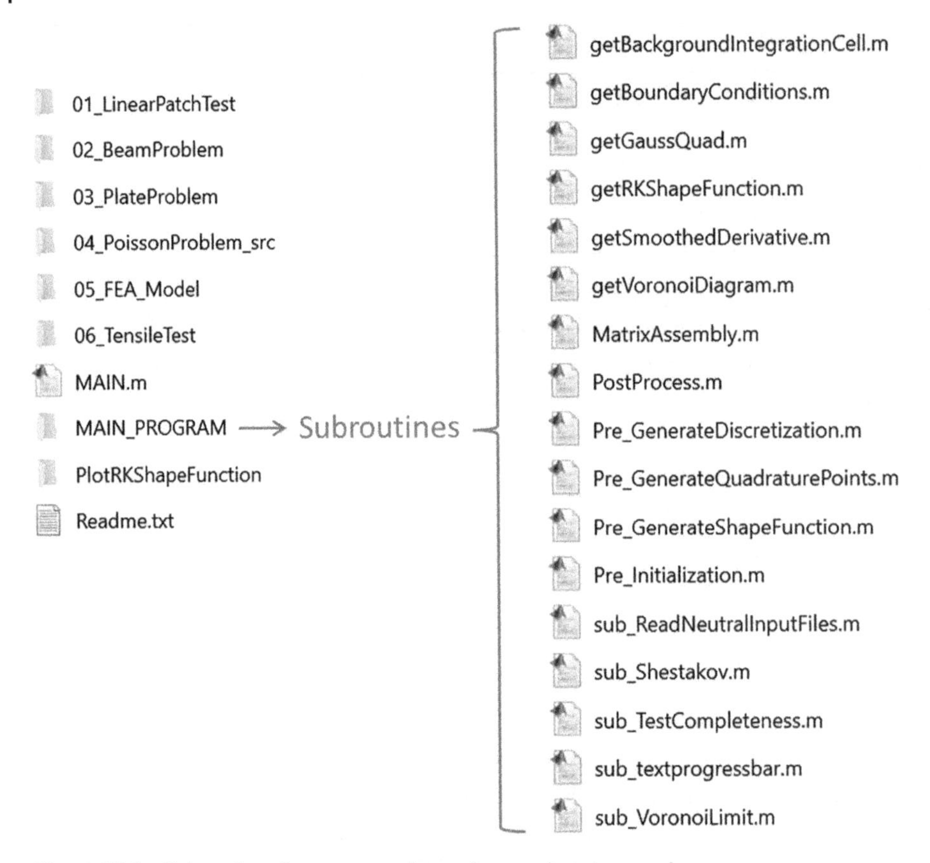

Figure 10.6 Subroutines for preprocessing, solver, and postprocessing.

`getGaussQuad.m`: This subroutine outputs Gaussian quadrature points and weights for any 1D domain. For example, `[Xgp,Wgp] = getGaussQuad(N,a,b)` gives with point location `Xgp` and weight `Wgp` of `N` Gauss points in a line segment with two ends `a` and `b`, where `a` and `b` denote the 1D coordinates of the two end points.

`getRKShapeFunction.m`: This subroutine computes the 2D RK shape function and its direct gradients at a given evaluation point.

`getSmoothedDerivetive.m`: This subroutine computes smoothed gradients for a given representative domain.

`getVoronoidDiagram.m`: This subroutine calls `sub_VoronoiLimit.m` to generate Voronoi cells and then re-arranges the Voronoi cell IDs and coordinates.

`MatrixAssembly.m`: This subroutine assembles the stiffness matrix and force vector.

`PostProcess.m`: This subroutine visualizes the calculated displacement, strain and stress fields.

`Pre_GenerateDiscretization.m`: This subroutine generates point-based domain discretization based on domain vertices' coordinates and the chosen discretization method.

`Pre_GenerateQuadraturePoint.m`: This subroutine computes the quadrature points for domain integration and boundary integration for nodal integration or Gauss integration.

`Pre_GenerateShapeFunction.m`: This subroutine generates the RK shape functions and gradients at quadrature points.

`Pre_Initialization.m`: This subroutine initializes the matrices and vectors for the simulation. It also displays the input information in the MATLAB command window.

`sub_Shestakov.m`: This subroutine generates highly distorted meshes using Shestakov's algorithm [19] for a rectangular domain. This function is called in `Pre_GenerateDiscretization.m` when discretization method 'D' is used.

`sub_TestCompleteness.m`: This subroutine tests the reproducing condition of RK shape functions and gradients.

`sub_textprogressbar.m`: This subroutine generates the simulation progress bar in MATLAB command window.

`sub_VoronoiLimit.m`: This subroutine calls a library that computes vertices' coordinates of Voronoi cells and indices of vertices within each Voronoi cell.

10.4 Getting Started

Running RKPM2D requires the following environment:

- MATLAB version (R2018a) or higher
- MATLAB Partial Differential Equation Toolbox™
- MATLAB Mapping Toolbox™
- Statistics and Machine Learning Toolbox™ or Text Analytics Toolbox™

For the general purpose of solving two-dimensional elasticity problems, only the input file `getInput.m` needs to be modified. As shown in Figure 10.4, this subroutine defines the problem at hand (geometry, boundary conditions, and material properties), and the methods used in the RKPM discretization. Sample input files for various linear elasticity problems are provided in main directory as shown in Figure 10.7. Note that "04_PoissonProblem_src" does not contain an input file; it is the code converted to solve a Poisson problem, as described in the Appendix. The rest of the subroutines are located in the folder "MAIN_PROGRAM." The description for each subroutine was given in Section 10.3.6. To run the simulation, one needs to execute the main script `MAIN.m` after setting up the model parameters in `getInput.m`. Finally, we note that the folder "PlotRKShapeFunction" contains a separate set of scripts to plot the RK shape functions in one and two dimensions, see Section 10.5.1.

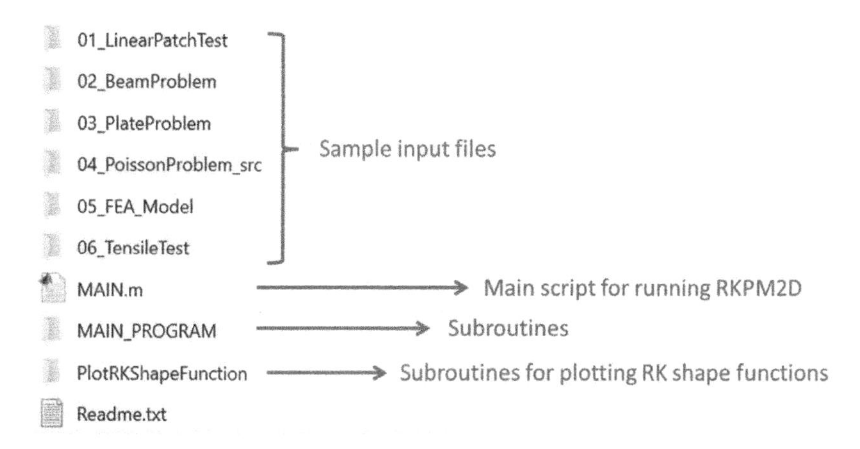

Figure 10.7 Folders containing input files and subroutines for meshfree analysis.

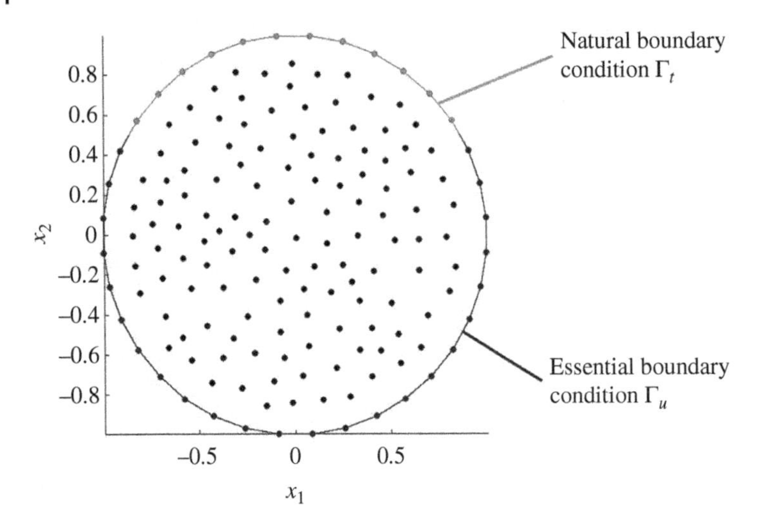

Figure 10.8 Domain discretization for model problem in 2D circular domain Ω. Source: Huang et al. [20], figure 9, p. 402 / Springer Nature.

10.4.1 Input File Generation

To illustrate how to set up an input file for RKPM2D, and the functionality of the associated subroutine `getInput`, consider a linear elasticity problem defined over the circular domain shown in Figure 10.8 with a manufactured solution of

$$\mathbf{u}^{\text{exact}} = \begin{bmatrix} 0.1 + 0.1x_1 + 0.2x_2 \\ 0.05 + 0.15x_1 + 0.1x_2 \end{bmatrix}. \tag{10.42}$$

Since the exact solution is linear, this problem is linear patch test.

The traction $\bar{\mathbf{t}}$, body force $\mathbf{b}$, displacement $\bar{\mathbf{u}}$ are prescribed based on the exact solution $\mathbf{u}^{\text{exact}}$ in (10.42), and the boundaries are either pure traction or pure displacement conditions:

$$\bar{\mathbf{t}} = \boldsymbol{\eta}^{\mathrm{T}}\mathbf{D}\mathcal{E}^{\text{exact}} \quad \forall \mathbf{x} \in \Gamma_{\mathbf{t}}, \tag{10.43}$$

$$\bar{\mathbf{u}} = \mathbf{u}^{\text{exact}} \quad \forall \mathbf{x} \in \Gamma_{\mathbf{u}}, \tag{10.44}$$

$$\mathbf{b} = -\begin{bmatrix} \sigma_{11,1}^{\text{exact}} + \sigma_{12,2}^{\text{exact}} \\ \sigma_{21,1}^{\text{exact}} + \sigma_{22,2}^{\text{exact}} \end{bmatrix} \quad \forall \mathbf{x} \in \Omega, \tag{10.45}$$

where $\boldsymbol{\eta}$ is the matrix of outward unit normals to the boundary, $\mathbf{D}$ is the matrix of the elastic tensor, $\mathcal{E}^{\text{exact}}$ is the exact strain vector in matrix form, and let $\boldsymbol{\Sigma}^{\text{exact}}$ be the exact stress vector in matrix form:

$$\boldsymbol{\eta} = \begin{bmatrix} n_1 & 0 \\ 0 & n_2 \\ n_2 & n_1 \end{bmatrix}, \tag{10.46}$$

$$\mathcal{E}^{\text{exact}} = \begin{bmatrix} u_{1,1}^{\text{exact}} \\ u_{2,2}^{\text{exact}} \\ u_{2,1}^{\text{exact}} + u_{1,2}^{\text{exact}} \end{bmatrix} = \begin{bmatrix} 0.1 \\ 0.1 \\ 0.35 \end{bmatrix}, \tag{10.47}$$

$$\boldsymbol{\Sigma}^{\text{exact}} = \begin{bmatrix} \sigma_{11}^{\text{exact}} \\ \sigma_{22}^{\text{exact}} \\ \sigma_{12}^{\text{exact}} \end{bmatrix} = \mathbf{D}\mathcal{E}^{\text{exact}}, \tag{10.48}$$

with

$$\begin{aligned} \mathbf{D} &= \frac{E}{(1+\nu)(1-2\nu)} \begin{bmatrix} 1-\nu & \nu & 0 \\ \nu & 1-\nu & 0 \\ 0 & 0 & 1-2\nu \end{bmatrix}, \quad \text{(plane strain)} \\ \mathbf{D} &= \frac{E}{(1-\nu^2)} \begin{bmatrix} 1 & \nu & 0 \\ \nu & 1 & 0 \\ 0 & 0 & (1-\nu)/2 \end{bmatrix}, \quad \text{(plane stress)} \end{aligned} \tag{10.49}$$

where E and ν are Young's modulus Poisson's ratio respectively.

The input file for this problem is created by the function `getInput`, with three data structures: `RK`, `Quadrature`, and `Model`, which define the RK shape functions, quadrature rules, and numerical parameters (such as penalty parameters, elastic moduli, etc.), respectively. A sample input file for the above linear patch test is given in Listing 10.8, with the values assigned to the data structures as described below.

Listing 10.8 Input file of the linear patch test.

```
function [RK,Quadrature,Model] = getInput()

%% INPUT FILE
% Sample Input File for Patch Test
%% (1) Material
% Linear Elasticity
% Lame Parameters for Young's modulus and Poisson ratio
Model.E = 2.1E11; Model.nu = 0.3;
Model.Condition = 'PlaneStress'; % PlaneStress, or PlaneStrain
Model.ElasticTensor = getElasticTensor(Model.E,Model.nu,Model.
Condition);
Model.DOFu = 2;                       % two dimensional problem

%% (2) Geometry
theta = [0:pi/18:2*pi-pi/18]'; r = ones(size(theta));
[x1_vertices,x2_vertices] = pol2cart(theta,r);
% ensure the boundary segments to be counter clockwise
[x1_vertices, x2_vertices] = poly2ccw(x1_vertices, x2_vertices);
Model.xVertices = [x1_vertices, x2_vertices];
Model.DomainArea = polyarea(x1_vertices,x2_vertices);
```

(Continued)

Listing 10.8 (Continued)

```
%% (3) Boundary condition
% If an edge is not specified, natural BC with zero traction is
imposed.
Model.CriteriaEBC = @(x1,x2) find(x2<=0.5); % user input
Model.CriteriaNBC = @(x1,x2) find(x2>0.5);  % user input
% beta parameter for Nitches Method
Model.Beta_Nor = 1E2;
% For verification purposes, provide the exact displacement solution
syms x1 x2  % use x1 and x2 as x- & y- coordinates exact solution
Model.ExactSolution.u_exact =[0.1 + 0.1*x1 + 0.2*x2;
                              0.05 + 0.15*x1 + 0.1*x2;];
[Model.ExactSolution.S,...
 Model.ExactSolution.g,...
 Model.ExactSolution.t,...
 Model.ExactSolution.b] = getBoundaryConditions(Model);
 Model.ExactSolution.Exist = 1;

%% (4) Discretization Method,
% (...A) MATLAB built-in FE mesh generator: Default
Model.Discretization.Method = 'A';
Model.Discretization.Hmax = 0.1; % max nodal distance

%% (5) RK shape function parameters
RK.KernelFunction = 'SPLIN3';        % SPLIN3
RK.KernelGeometry = 'CIR';           % CIR, REC
RK.NormalizedSupportSize = 2.01;     % suggested order n + 1;
RK.Order = 'Linear';                 % Constant, Linear, Quadratic

%% (6) Quadrature rules
Quadrature.Integration = 'SCNI';        % GAUSS, SCNI, DNI
Quadrature.Stabilization = 'N';         % M, N
Quadrature.Option_BCintegration = 'NODAL'; % NODAL OR GAUSS
Quadrature.nGaussPoints = 6; % #Gauss Points per cell
Quadrature.nGaussCells = 10; % #GaussCells on the short side of
domain

end
```

10.4.1.1 Model

This data structure contains the elastic constants, boundary conditions, Nitsche parameter, and domain geometry, which are defined using the following fields:

- `E`, `nu`: Young's modulus E, Poisson ratio ν.
- `DOFu`: the number of nodal degrees of freedom, `DOFu=2` for two-dimensional elasticity.

- `BetaNormalized`: normalized penalty parameter β_{nor}, see Section 10.1.2.
- `xVertices`: coordinates of domain vertices.
- `CriteriaEBC`: function handle to define the essential boundaries.
- `CriteriaNBC`: function handle to define the natural boundaries.

With reference to Listing 10.8, in this example, we set $E = 2.1 \times 10^{11}$ (`Model.E`) and $\nu = 0.3$ (`Model.nu`). The traction $\bar{\mathbf{t}}$ is imposed on Γ_t: $(x_1, x_2) \in \Gamma$, $x_2 > 0.5$ (`Model.CriteriaNBC`), the essential boundary condition $\bar{\mathbf{u}}$ is enforced on Γ_u: $(x_1, x_2) \in \Gamma$, $x_2 \le 0.5$ (`Model.CriteriaEBC`), and the body force is $\mathbf{b} = \mathbf{0}$ in this case according to the manufactured solution (10.42). The two-dimensional condition (`Model.Condition`) is set to plane stress, and the circular domain is generated using a few lines of code and finally assigned to the array `Model.xVertices`. The area of the domain (`Model.DomainArea`) is calculated automatically using `polyarea` based on these vertices. Nitsche's parameter (`Model.Beta_Nor`) is set to $\beta_{\mathrm{nor}} = 100$. We assign two to the number of degrees of freedom per node (`Model.DOFu`), which should be used by default unless the code has been modified to solve a problem other than two-dimensional linear elasticity.

To implement the values of traction $\bar{\mathbf{t}}$, body force $\mathbf{b}$, displacement $\bar{\mathbf{u}}$ based on the exact solution $\mathbf{u}^{\mathrm{exact}}$, and the switch matrix $\mathbf{S}$ from the boundary conditions, the symbolic feature in MATLAB is employed. As shown in Listing 10.9, for any given exact displacement field $\mathbf{u}^{\mathrm{exact}}(\mathbf{x})$ in symbolic form, the functions for $\bar{\mathbf{t}}$, $\mathbf{b}$, $\bar{\mathbf{u}}$, and $\mathbf{S}$ can be obtained. Listing 10.10 shows how these are calculated in `getBoundaryConditions` using Eqs. (10.43)–(10.48) with $\mathbf{u}^{\mathrm{exact}}$ as an input variable. These variables are saved as function handles under the structure `Model.ExactSolution` using the `matlabFunction` command (note: a function handle is a data type in MATLAB to store an association to a function).

Alternatively, if the exact solution $\mathbf{u}^{\mathrm{exact}}$ is not specified, then one can manually define these as individual functions, as shown in Listing 10.11. The structure `Model.ExactSolution` contains the following fields:

- `t`: return the traction $\bar{\mathbf{t}}$.
- `b`: return the body force vector $\mathbf{b}$.
- `g`: return the imposed displacement $\bar{\mathbf{u}}$.
- `S`: return the switch matrix $\mathbf{S}$.

Listing 10.9 Command lines of defining boundary conditions by providing an analytical expression of displacement $\mathbf{u}^{\mathrm{exact}}$ or defining imposed traction $\bar{\mathbf{t}}$, body force $\mathbf{b}$, displacement $\bar{\mathbf{u}}$, and switch matrix $\mathbf{S}$.

```
% give the exact solution of the displacement
syms x1 x2 % please use x1 and x2 as coordinates
Model.ExactSolution.u_exact =[0.1 + 0.1*x1 + 0.2*x2;
                              0.05 + 0.15*x1 + 0.1*x2;];

% give the expression of function handle of Switch S, essential
% boundary conditions g, traction t, and body force b
```

(Continued)

Listing 10.9 (Continued)

```
if isfield(Model,'ExactSolution') % if given analytical
displacement

    [Model.ExactSolution.S,Model.ExactSolution.g,...
     Model.ExactSolution.t,Model.ExactSolution.b] = ...
    getBoundaryConditions(Model);

else % if S, g, t, b are defined in functions

    Model.ExactSolution.S = @getSebc; % function getSebc
    Model.ExactSolution.g = @getGebc; % function getGebc
    Model.ExactSolution.t = @getTraction; % function getTraction
    Model.ExactSolution.b = @getBodyForce; % function getBodyForce

end
```

Listing 10.10 Command lines of the function that generates the exact traction $\bar{\mathbf{t}}$, body force $\mathbf{b}$, imposed displacement $\bar{\mathbf{u}}$, and switch matrix $\mathbf{S}$ through MATLAB symbolic operation.

```
function [function_S,function_g,function_traction,function_b] =
getBoundaryConditions(Model)

syms x1 x2 n1 n2

% function handle for essential boundary condition g
u = Model.ExactSolution.u_exact;
C = Model.ElasticTensor;
function_g = matlabFunction(u);

% function handle for stress
epsilon_x1 = diff(u(1),x1);
epsilon_x2 = diff(u(2),x2);
epsilon_x12 = (diff(u(1),x2)+diff(u(2),x1));
stress = C*[epsilon_x1;epsilon_x2;epsilon_x12];

% function handle for traction t
eta = [n1 0 n2; 0 n2 n1;];
traction = eta*stress;
function_traction = matlabFunction(traction,'Vars',[x1 x2 n1 n2]);
```

(Continued)

Listing 10.10 (Continued)

```
% function handle for body force b
b = [divergence([stress(1),stress(3)],[x1,x2]);
     divergence([stress(3),stress(2)],[x1,x2]);];
function_b = matlabFunction(b,'Vars',[x1 x2]);

% function handle for switch S
function_S = matlabFunction(sym(diag([1 1])),'Vars',[x1 x2]);

end
```

Listing 10.11 Functions that define imposed traction $\bar{\mathbf{t}}$, body force $\mathbf{b}$, displacement $\bar{\mathbf{u}}$, and switch matrix S.

```
function [ t ] = getTraction(x1,x2,n1,n2)

% Input: x1,x2: Cartesian coordinate
% Output: t: a 2 by 1 vector for the traction

E = 2.1E11; nu = 0.3;
Condition = 'PlainStress'; % PlaneStress, or PlaneStrain
C = getElasticTensor(E,nu,Condition);
Strain_exact = [0.1; 0.1; 0.35];
stress = (C*Strain_exact);
eta = [n1 0 n2; 0 n2 n1;];
t = eta*stress;

end

function [ SEBC ] = getSebc(x1,x2)

% Input:  x1,x2: coordinates in 1,2
% Output: SEBC: a 2 by 2 matrix for the switch matrix on EBC

SEBC = diag([1 1]);

End

function [ gEBC ] = getGebc(x1,x2)
```

(Continued)

Listing 10.11 (Continued)

```
% Input:   x1,x2: Cartesian coordinate
% Output:  gEBC: a 2 by 1 vector of prescribed displacement on EBC

gEBC =[0.1 + 0.1*x1 + 0.2*x2;
       0.05 + 0.15*x1 + 0.1*x2];

End

function [ b ] = getBodyForce(x1,x2)

% Input:  x1,x2: Cartesian coordinate
% Output: b: a 2 by 1 vector for the body force

b = [0; 0;];

end
```

10.4.1.2 RK

This data structure contains the following parameters for the reproducing kernel approximation to be employed in the simulation:

- `KernelFunction`: kernel functions with different levels of continuity, see Table 10.1 for the kernels available in RKPM2D.
- `KernelGeometry`: the nodal support shape where "`CIR`" and "`REC`" represent circular and rectangular supports, respectively. See Section 3.1.1 for more details.
- `NormalizedSupportSize`: normalized support size $\tilde{a}$ in the RK approximation, see Eq. (10.5).
- `Order`: the order of basis n: `Constant` $(n = 0)$, `Linear` $(n = 1)$, and `Quadratic` $(n = 2)$ are available in RKPM2D, although the subroutine can accommodate any n.

Note that the RK shape functions are computed at the beginning of the simulation based on the information in the structure `RK`. In this example, with reference to Listing 10.8, we have used cubic B-spline kernels (`RK.KernelFunction`) with circular supports (`RK.KernelGeometry`), a

Table 10.1 The abbreviation of the kernel functions used in the code.

Name	Continuity	Kernel function
HVSIDE	C^{-1}	Heaviside
SPLIN1	C^0	Linear B-Spline (tent)
SPLIN2	C^1	Quadratic B-Spline
SPLIN3	C^2	Cubic B-Spline
SPLIN4	C^3	Quartic B-Spline
SPLIN5	C^4	Quintic B-Spline

normalized support size of $\tilde{a}$=2.01 (RK.NormalizedSupportSize), and linear basis functions (RK.Order).

10.4.1.3 Quadrature

This data structure specifies how the numerical integration will be carried out and contains the following fields:

- Integration: basic quadrature rule, where the following options are provided: direct nodal integration "DNI", stabilized conforming nodal integration "SCNI", and Gauss integration "GAUSS".
- Stabilization: types of stabilization for nodal integration (DNI or SCNI only), where the symbol "N" represents naturally stabilized nodal integration and the symbol "M" represents modified nodal integration, as described in Section 10.2.3.
- Option_BCintegration: quadrature rules for boundary integrals, including "NODAL" and "GAUSS" options.
- nGaussPoints: the number of Gauss points $Ng \in \mathbb{N}$ in each background integration cell. E.g., $Ng = 6$ denotes 6×6 Gauss points in each cell for domain integration.
- nGaussCells: the parameter determines the number of background integration cells along x_1 or x_2 direction of the problem domain, depending on the shorter of the problem dimensions.

Note that the nGaussPoints and nGaussCells are used only for Gauss integration.

As seen in Listing 10.8, in this example we have chosen SCNI for domain integration (Quadrature.Integration) with natural stabilization (Quadrature.Stabilization), and nodal integration for boundary integrals (Quadrature.Option_BCintegration). The two remaining fields for Gauss integration are ignored by RKPM2D since "SCNI" was specified.

10.4.2 Executing RKPM2D

After the input file getInput.m is prepared, one can execute RKPM2D by running MAIN.m through the following steps:

1) Open MAIN.m and specify the path to the input folder, as shown in Figure 10.9.
2) Press "Run" to execute the RKPM2D code.
3) Once the simulation is finished, the post-processing figures will be generated, and one can perform further analysis of the results.

A successful execution of the meshfree analysis will show basic information for the simulation in the Command Window.

10.4.3 Post-Processing

When the simulation is finished, key parameters and data structures associated with the results are saved in MATLAB workspace, as shown in Figure 10.10.

The program's final step is to compute and visualize the computed displacement, strain and stress fields using the routine PostProcess. The following fields are computed and returned:

- uhI: matrix of size $NP \times 2$ that defines the displacements at RK nodes $\mathbf{u}^h(\mathbf{x}_I)$.
- Strain: double vector of size $NP \times 3$ that defines the strain at RK nodes $\mathcal{E} = [\varepsilon_{11} \quad \varepsilon_{22} \quad 2\varepsilon_{12}]^T$.
- Stress: double vector of size $NP \times 3$ that defines the stress at RK nodes $\mathbf{\Sigma} = [\sigma_{11} \quad \sigma_{22} \quad \sigma_{12}]^T$.

01_LinearPatchTest
02_BeamProblem
03_PlateProblem
04_PoissonProblem_src
05_FEA_Model
06_TensileTest
MAIN_PROGRAM
MAIN
Readme
Main script
% add path to input file folder
% path_problem = '01_LinearPatchTest';
% path_problem = '02_BeamProblem';
% path_problem = '03_PlateProblem';
% path_problem '04_PoissonProblem' is the diffusion problem in paper;
% path_problem = '05_FEA_Model';
path_problem = '06_TensileTest';
rmpath(path_problem); addpath(path_problem)
%% (1) Pre-Process
% (1.1) Generate input files (RK parameter,Quadrature, Material, BC)
[RK,Quadrature,Model] = getInput();
% (1.2) Generate nodal discretization (three options available)
[Discretization] = Pre_GenerateDiscretization(Model);
% (1.3) Evaluate support size & output model set-up parameters
[RK,Quadrature] = Pre_Initialization(RK,Quadrature,Discretization);
% (1.4) Generate quadrature weight & evaluatioin point
[Quadrature] = Pre_GenerateQuadraturePoints(Quadrature,Discretization)
% (1.5) Evaluate RK shape functions
[Quadrature] = Pre_GenerateShapeFunction(RK,Quadrature);
%% (2) Main Solver
% (2.1) Assemble global stiffness matrix & force vector
[K,F] = MatrixAssembly(Quadrature,Model);
% (2.2) Solve linear algebraic equation
RK.dI = mldivide(K,F); % MATLAB built-in matrix solver i.e., dI = K\F
%% (3) Post-Process
% (3.1) visualization
[uhI,Strain,Stress] = PostProcess(RK,Quadrature,Model,Discretization);
% clean path
rmpath(path_MAINCODE)
rmpath(path_problem)
Add path to input file folder
Preprocess
Solving Eqns
Postprocess

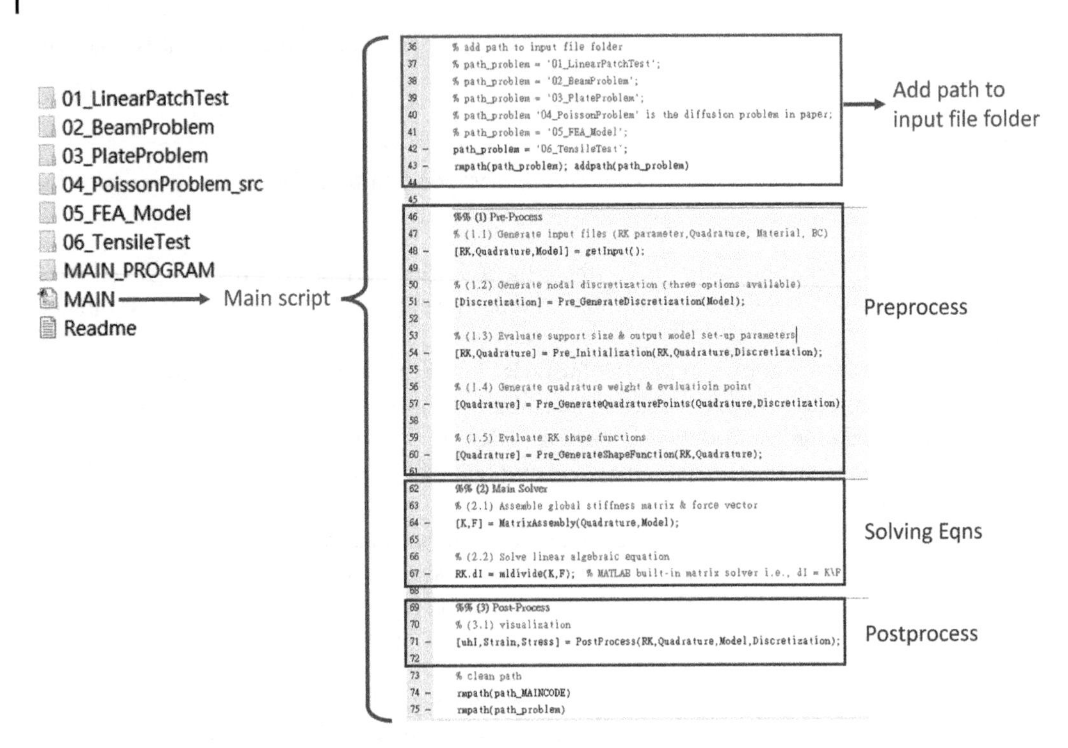

Figure 10.9 The main script (`MAIN.m`) for running meshfree analysis.

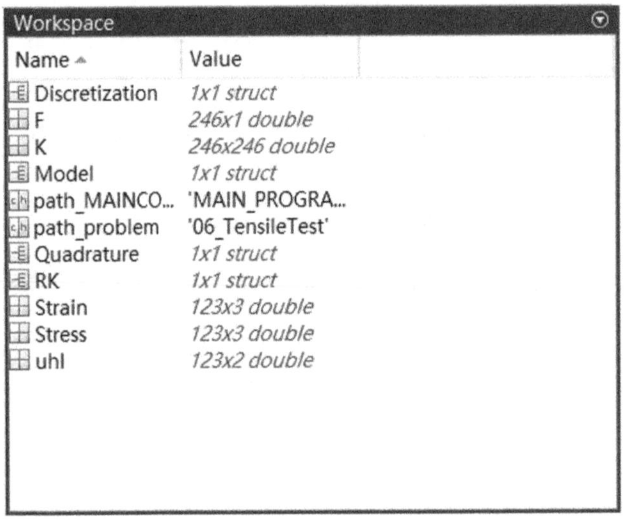

Workspace
Name
Value
Discretization
1x1 struct
F
246x1 double
K
246x246 double
Model
1x1 struct
path_MAINCO...
'MAIN_PROGRA...
path_problem
'06_TensileTest'
Quadrature
1x1 struct
RK
1x1 struct
Strain
123x3 double
Stress
123x3 double
uhl
123x2 double

Figure 10.10 Data structure and parameters saved in the MATLAB Workspace area.

For simplicity, the continuous fields of displacement, strain, and stress are plotted based on Delaunay triangulation of the whole domain by using the MATLAB built-in functions `delaunayTriangulation` and `trisurf` as shown in Listing 10.12. However, it should be noted that variables such as stress and strain are in reality, highly smooth in RKPM when a smooth kernel is adopted in the approximation.

Listing 10.12 Parts of the subroutine for Delaunay triangulation

```
% Using delaunay Triangulation to post processing the scattered data
% Index_BC is the index number of node on the boundary

tri = delaunayTriangulation(xI(:,1),xI(:,2) ,...
                  [Discretization.Index_BC',...
                   circshift(Discretization.Index_BC',[-1 0])]);

% plot u1 displacement
figure, trisurf(tri(tri.isInterior(),:), xI(:,1),xI(:,2),uhI(:,1));
```

10.5 Numerical Examples

10.5.1 Plotting the RK Shape Functions

In the first example, two scripts, `PlotRKShapeFunction1D` and `PlotRKShapeFunction2D`, shown in Figure 10.11 are employed to plot 1D and 2D RK shape functions, where the subroutines `getRKShapeFunction` and the corresponding 1D version `getRKShapeFunction1D` are utilized, respectively.

To plot the RK shape functions in 1D perform the following steps:

1) Open the folder PlotRKShapeFunction.
2) Open the MATLAB script `PlotRKShapeFunction1D.m`.
3) Specify the RK parameters, nodal distribution, and output figure type.

With reference to Listing 10.13, the definitions of RK parameters are identical to the discussions given in Section 10.4.1, and the variable `output_figure` specifies the types of output figures, and `randomness_number` imposes the magnitude of nodal coordinate perturbation if one wants to test non-uniform nodal discretizations.

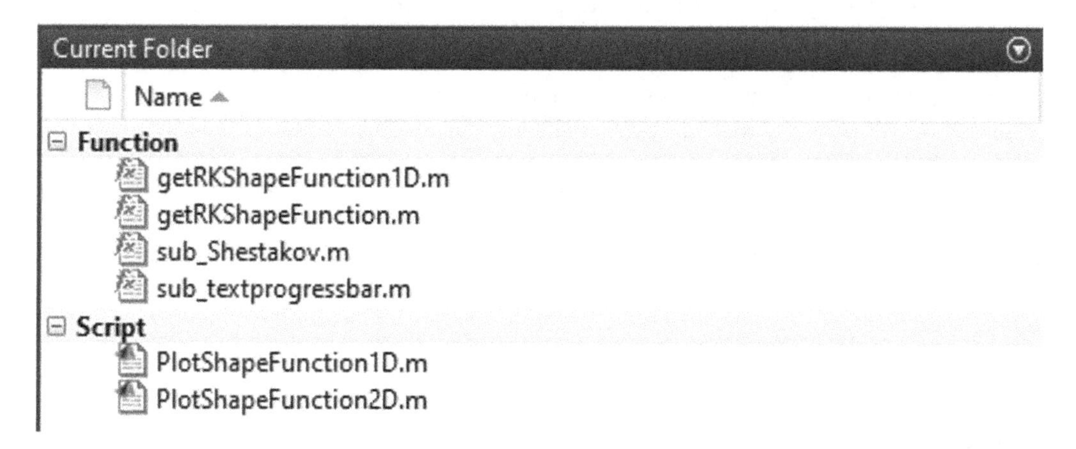

Figure 10.11 MATLAB scripts for plotting RK shape functions in 1D and 2D.

Listing 10.13 RK parameters and nodal distribution set up in PlotRKShapeFunction1D.m

```
%% RKPM Setting Area
RK.KernelFunction = 'SPLIN3';          % SPLIN1-5, HVSIDE
RK.KernelGeometry = 'CIR';             % CIR, REC
RK.Order = 'Linear';                   % Constant, Linear, Quadratic
RK.nP = 11;                            % # of nodes

% Support Size
NormalzieSupportSize1 = 1.5;

% Choose the output
output_figure = 'Shape_Function';
% Shape_Function, RK_condition, RK_condition_error,
Support_Comparison

%% Discretize RK nodes
xmin=0; xmax= 1;
nP = RK.nP;
L = xmax-xmin;

% randomness
randomness_number = 0.5;
```

Running the script `PlotRKShapeFunction1D.m`, one can observe how the kernel function controls the smoothness of the approximation, as shown in Figure 10.12, where the C^0 tent kernel function (`RK.KernelFunction = 'HVSIDE'`) is compared with the C^2 cubic B-spline kernel function (`RK.KernelFunction = 'SPLIN3'`). Figure 10.12a and b shows some results for a uniform discretization (`randomness_number = 0.0`), and Figure 10.12c and d shows some results for random nodes (`randomness_number = 0.5`).

The steps for plotting the RK shape functions in 2D are nearly identical to the 1D case:

1) Open the folder PlotRKShapeFunction.
2) Open the MATLAB script `PlotRKShapeFunction2D.m`.
3) Specify the RK parameters, nodal distribution, and output figure type.

With reference to Listing 10.14, the definitions of RK parameters are again identical to the discussion given in Section 10.4.1, and the variable `output_figure` determines the types of output figures. In this example, the Shestakov meshing method is used to generate the nodal distributions, as discussed in [20], where `nc` determines the refinement level and `randomness` determines the level of distortion.

10.5.2 Patch Test

In this example, the linear patch test is provided to verify the accuracy of RKPM2D using linear basis in the RK approximation with variationally consistent integration. The exact solution is defined as a linear polynomial function:

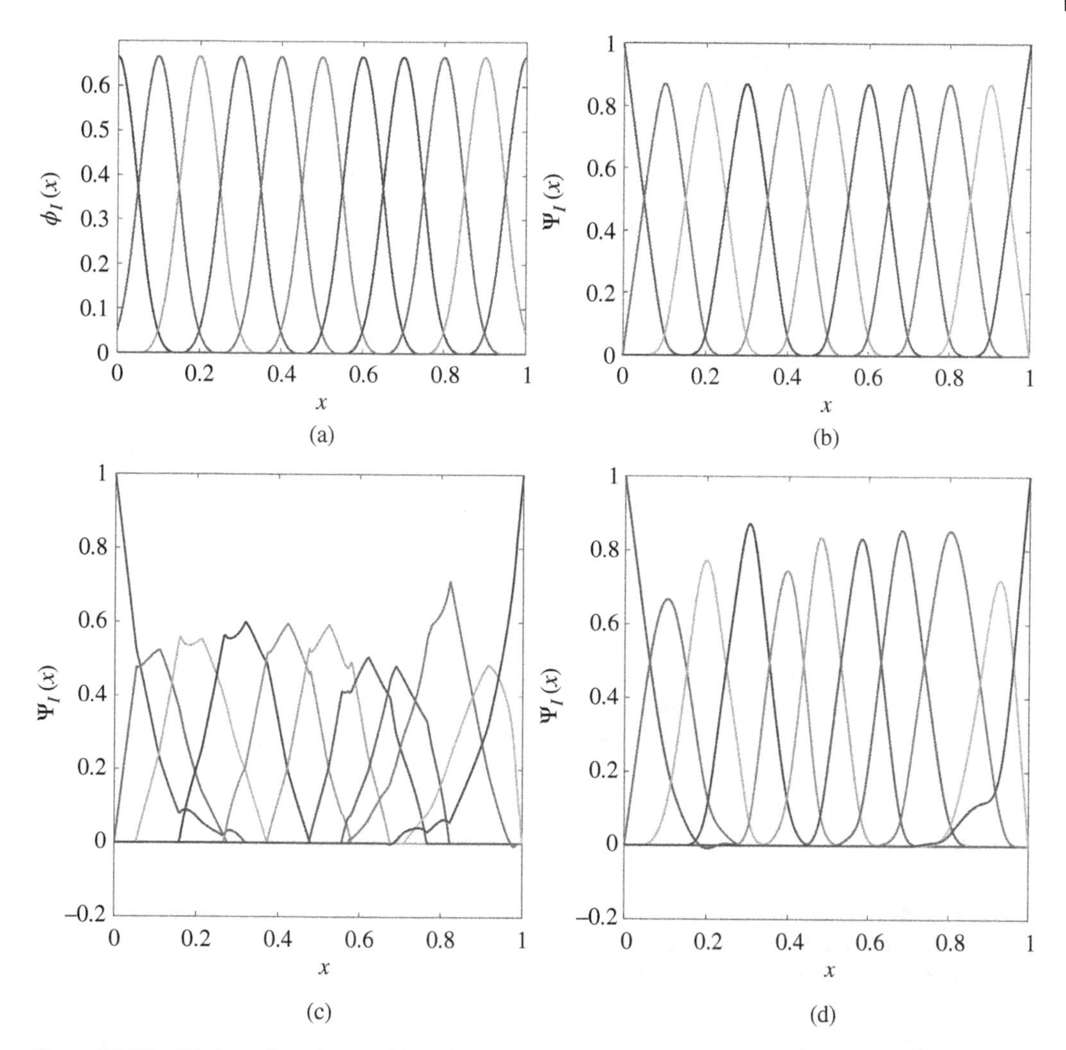

Figure 10.12 RK shape functions and kernels with linear basis and support size $\tilde{a} = 1.5$: (a) cubic B-Splines kernels for uniform nodes, (b) shape functions for uniform nodes with cubic B-Spline kernels, (c) shape functions for random nodes with tent kernels, and (d) shape functions for random nodes with cubic B-Spline kernels.

$$
\mathbf{u}^{\text{exact}} = \begin{bmatrix} 0.1 + 0.1x_1 + 0.2x_2 \\ 0.05 + 0.15x_1 + 0.1x_2 \end{bmatrix}. \tag{10.50}
$$

Accordingly, the traction is calculated as $\overline{\mathbf{t}} = \boldsymbol{\eta}^{\mathrm{T}}\mathbf{D}\boldsymbol{\varepsilon}^{\text{exact}}$ and is imposed here on $\Gamma_t : (x_1, x_2) \in \Gamma$, $x_2 > 0.5$, where $\boldsymbol{\eta}$ is the collection of outward unit normal vector of the boundary surface, $\mathbf{D}$ is the matrix of elastic moduli with Young's modulus $E = 2.1 \times 10^{11}$ and Poisson's ratio $\nu = 0.3$, $\boldsymbol{\varepsilon}^{\text{exact}} = [0.1,\ 0.1,\ 0.35]^{\mathrm{T}}$ is the exact strain; $\overline{\mathbf{u}} = \mathbf{u}^{\text{exact}}$ is enforced on $\Gamma_u : (x_1, x_2) \in \Gamma, x_2 \leq 0.5$, and the body force is $\mathbf{b} = \mathbf{0}$. In this problem, RKPM2D automatically calculates these quantities using (10.50) in conjunction with (10.43)–(10.45) as discussed in Section 10.4.1.

The input file for solving the linear patch test is provided in the folder "01_LinearPatchTest." A total of four other geometries are provided (they need to be uncommented for use), denoted as geometry 1, 2, 3, and 4 in the input file, and are depicted in Figure 10.13. For demonstration purposes, we use "geometry 1" here (the results for the others can be found in [20], although

Listing 10.14 Set-up of RK parameters and nodal distribution in PlotRKShapeFunction2D.m

```
%% Define RK Shape Function
RK.KernelFunction = 'SPLIN3';             % HVSIDE,SPLIN1-5
RK.KernelGeometry = 'CIR';                % CIR, REC
RK.NormalizedSupportSize = 1.5;           %
RK.Order = 'Linear';                      % Constant, Linear, Quadratic

%% define discrete RK nodes by Shestakov mesh
% we use the test distorted mesh generator to test
nc = 2; % the total number of nodes will be (2^nc+1)*(2^nc+1)
randomness = 0.1; % 0~0.5, where 0.5 is uniform case

% Choose the output
output_figure = 'Shape_Function'; % Shape_Function, RK_condition,
RK_condition_error
```

the user is encouraged to try them), and discretization method `'A'` (MATLAB Built-In Mesh Generator). NSCNI is used as the quadrature rule, and circular, cubic B-spline kernel with a normalized support size of 2.01 is used for the RK shape function construction. Detailed relevant information for setting up the model can be found in Listing 10.15. Since the exact solution is known, the boundary conditions and body force will be computed automatically using (10.43)–(10.45) according to (10.50).

As seen in Figure 10.14, NSCNI passes patch test as both the displacement and strain energy errors are on the order of machine precision. Changing `Quadrature.Stabilization` to `'M'` one observes that MSCNI does as well. Setting `Quadrature.Integration='DNI'` shows that DNI does not pass the patch test. These two cases are also shown in the figure for reference. The user can verify that both SCNI-based methods will indeed pass the patch test for all of the geometries given and any other geometries (such as the alternative ones in Figure 10.13 provided in the comments of the input file).

10.5.3 Cantilever Beam Problem

A cantilever beam problem shown in Figure 10.15 is considered next. The exact displacement solution to this problem is

$$
\begin{aligned}
u_1^{\text{exact}} &= \frac{Px_2}{6EI}\left[(6L-3x_2)x_1 + (2+\nu)\left(x_2^2 - \frac{W^2}{4}\right)\right],\\
u_2^{\text{exact}} &= \frac{-P}{6EI}\left[3\nu x_2^2(L-x_1) + (4+5\nu)\frac{W^2 x_1}{4} + (3L-x_1)x_1^2\right],
\end{aligned}
\tag{10.51}
$$

in which the Young's modulus $E = 1$ and Poisson's ratio $\nu = 0.3$ are selected, and the problem setup is shown in Figure 10.15. The exact displacement solution is prescribed on the left wall as the essential boundary condition, i.e., $\mathbf{u} = \bar{\mathbf{u}}$, and the corresponding traction is imposed on the right-side surface as the natural boundary condition using (10.43). The body force is zero which is commensurate with (10.45).

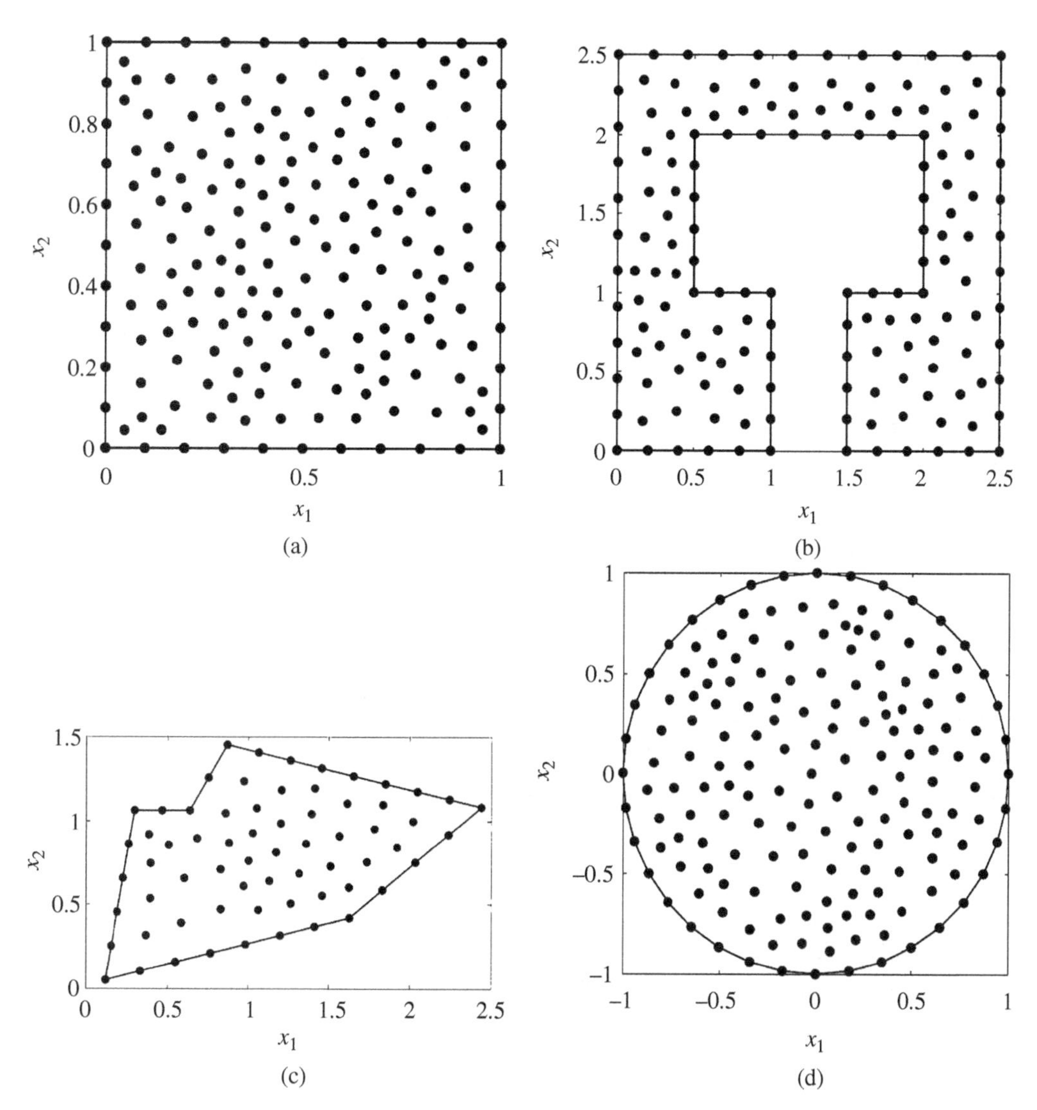

Figure 10.13 (a) Geometry 1, (b) geometry 2, (c) geometry 3, and (d) geometry 4 provided as patch test examples. Source: Huang et al. [20], figure 11, p. 419 / Springer Nature.

The input file is provided in the folder "02_BeamProblem." For demonstration purposes, we use discretization method `'B'` (rectangular domain) with randomness number 0.5. The RK shape function construction uses a Cubic B-Spline kernel with circular support (normalized support size 2.01). Detailed information for setting up the model can be found in Listing 10.16. The input file uses NSCNI to obtain a highly effective meshfree solution, although users are encouraged to use different quadrature rules to solve this problem to examine their performance, as is done here, the same way as the patch test input was modified in the previous example. As shown in Figure 10.16, both MSCNI and NSCNI perform well in the strain energy error, whereas DNI shows comparatively much larger error as well as spurious oscillations in the stress field.

Listing 10.15 Input file for solving a linear patch test with a square domain.

```
function [RK,Quadrature,Model] = getInput()
%% INPUT FILE
% Sample Input File for Patch Test

%% (1) Material
% Linear Elasticity
% Lame Parameters for Young's modulus and Poisson ratio
Model.E = 2.1E11; Model.nu = 0.3;
Model.Condition = 'PlaneStress'; % PlaneStress, or PlaneStrain
Model.ElasticTensor = getElasticTensor(Model.E,Model.nu,Model.
Condition);
Model.DOFu = 2;                    % two dimensional problem

%% (2) Geometry
x1_vertices = [0 1 1 0]';
x2_vertices = [0 0 1 1]';

% ensure the boundary segments to be counter clockwise
[x1_vertices, x2_vertices] = poly2ccw(x1_vertices, x2_vertices);
Model.xVertices = [x1_vertices, x2_vertices];
Model.DomainArea = polyarea(x1_vertices,x2_vertices);

%% (3) Boundary condition
% If an edge is not specified, natural BC with zero traction is
imposed.
Model.CriteriaEBC = @(x1,x2) find(x2<=0.5); % user input
Model.CriteriaNBC = @(x1,x2) find(x2>0.5); % user input

% beta parameter for Nitches Method
Model.Beta_Nor = 1E2;

% For verification purpose, provide the exact displacement solution
syms x1 x2        % use x1 and x2 as x- & y- coordinates

% exact solution
Model.ExactSolution.u_exact =[0.1 + 0.1*x1 + 0.2*x2;
                              0.05 + 0.15*x1 + 0.1*x2;];
[Model.ExactSolution.S,...
 Model.ExactSolution.g,...
 Model.ExactSolution.t,...
 Model.ExactSolution.b] = getBoundaryConditions(Model);
 Model.ExactSolution.Exist = 1;

%% (4) Discretization Method
% For general purpose, one can always use A
```

(Continued)

Listing 10.15 (Continued)

```
Model.Discretization.Method = 'A';
Model.Discretization.Hmax = 0;

%% (5) RK shape function parameters
RK.KernelFunction = 'SPLIN3';          % SPLIN3
RK.KernelGeometry = 'CIR';             % CIR, REC
RK.NormalizedSupportSize = 2.01;       %
RK.Order = 'Linear';                   % Constant, Linear, Quadratic

%% (6) Quadrature rule
Quadrature.Integration = 'SCNI';          % GAUSS, SCNI, DNI
Quadrature.Stabilization = 'N';           % M, N
Quadrature.Option_BCintegration = 'NODAL'; % NODAL OR GAUSS
Quadrature.nGaussPoints = 6;
Quadrature.nGaussCells = 10; % nGaussCells on the short side of the
domain

end
```

10.5.4 Plate With a Hole Problem

The problem of a plate with a circular hole is considered in the input folder "03_PlateProblem" which is a standard benchmark test.

The plate is subjected to far-field traction T_1 in the x_1 direction, and due to symmetry, only a quarter of the plate is modeled as illustrated in Figure 10.17, where length $L = 4$ and inner radius $R = 1$ are chosen. There is no body force, and the traction imposed on the finite domain boundary is defined using (10.43), calculated from the exact displacement field $\mathbf{u}^{\text{exact}}$:

$$\begin{aligned} u_1^{\text{exact}} &= \frac{T_1 r}{8\mu}\left[\frac{r(\kappa+1)\cos\theta}{R} + \frac{2R((1+\kappa)\cos\theta + \cos 3\theta)}{R} - \frac{2R^3\cos 3\theta}{r^3}\right], \\ u_2^{\text{exact}} &= \frac{T_1 r}{8\mu}\left[\frac{r(\kappa-1)\sin\theta}{R} + \frac{2R((1-\kappa)\cos\theta + \cos 3\theta)}{R} - \frac{2R^3\sin 3\theta}{r^3}\right], \end{aligned} \tag{10.52}$$

$$\begin{aligned} \sigma_{11}^{\text{exact}} &= T_1 - \frac{T_1 R^2}{r^2}\left[\frac{3}{2}(\cos 2\theta + \cos 4\theta)\right] + \frac{3T_1 R^4}{2r^4}\cos 4\theta, \\ \sigma_{22}^{\text{exact}} &= -\frac{T_1 R^2}{r^2}\left[\frac{1}{2}(\cos 2\theta - \cos 4\theta)\right] - \frac{3T_1 R^4}{2r^4}\cos 4\theta, \\ \sigma_{12}^{\text{exact}} &= -\frac{T_1 R^2}{r^2}\left[\frac{1}{2}(\sin 2\theta + \sin 4\theta)\right] + \frac{3T_1 R^4}{2r^4}\sin 4\theta, \end{aligned} \tag{10.53}$$

where (r, θ) denote polar coordinates, $T_1 = 10$, $\mu = E/2(1+\nu)$, $\kappa = (3-\nu)/(1+\nu)$, in which Young's modulus and Poisson ratio are taken as $E = 2.1 \times 10^{11}$ and $\nu = 0.3$, respectively.

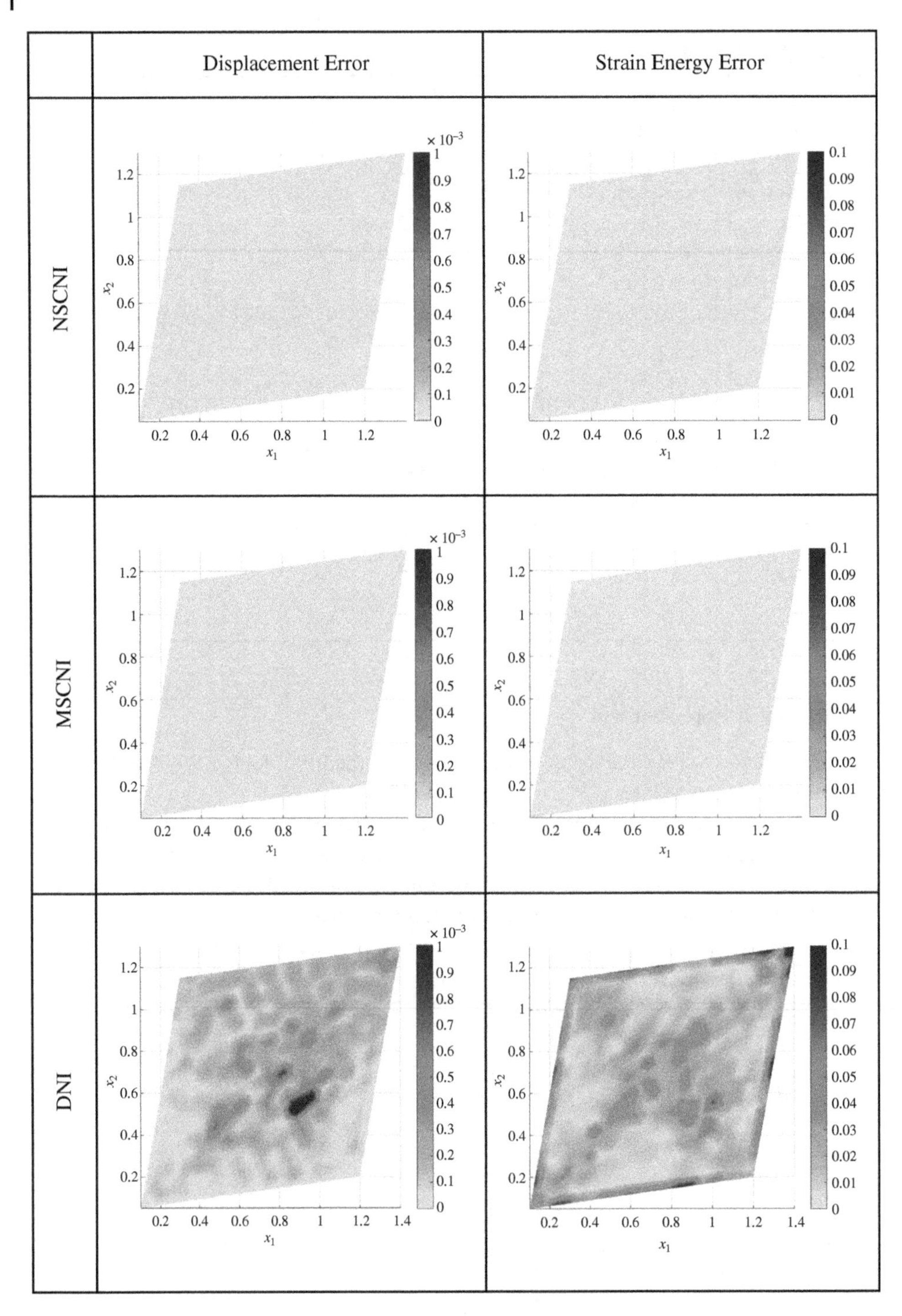

Figure 10.14 Results for the linear patch test using NSCNI, MSCNI, and DNI: error in the displacement and strain energy.

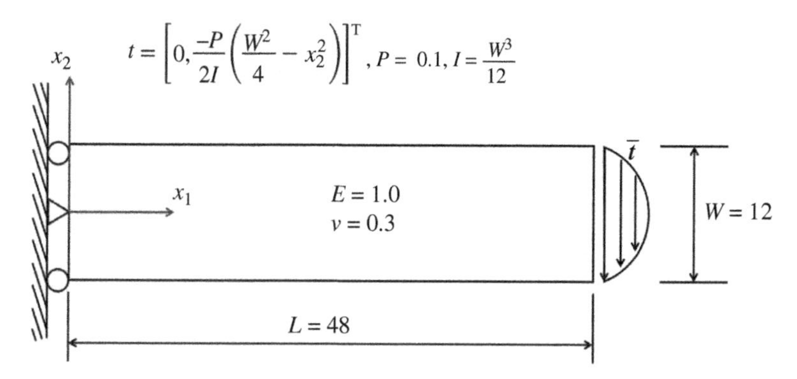

Figure 10.15 Problem setting of a cantilever beam under a shear force. Source: Huang et al. [20], figure 13, p. 422 / Springer Nature.

Listing 10.16 Sample input file for solving cantilever beam problem.

```
function [RK,Quadrature,Model] = getInput()
% Sample Input File for Beam Problem
%% (1) Material
% Linear Elasticity
% Lame Parameters for Young's modulus and poisson ratio
Model.E = 1.0E0; Model.nu = 0.3;  % user input
Model.Condition = 'PlaneStress';  % user input: 'PlaneStress' or
'PlaneStrain'
Model.ElasticTensor = getElasticTensor(Model.E,Model.nu,Model.
Condition);
Model.DOFu = 2;                   % two dimensional problem

%% (2) Geometry
% rectangular polygon
x1_vertices = [0 48 48 0]';  % user input
x2_vertices = [-6 -6 6 6]';  % user input
% ensure the boundary segments to be counter clockwise
[x1_vertices, x2_vertices] = poly2ccw(x1_vertices, x2_vertices);
Model.xVertices = [x1_vertices, x2_vertices];
Model.DomainArea = polyarea(x1_vertices,x2_vertices);

%% (3) Boundary condition
% If an edge is not specified, natural BC with zero traction is
imposed.
Model.CriteriaEBC = @(x1,x2) find(x1<=0+1E-7);    % user input
Model.CriteriaNBC = @(x1,x2) find(x1>=48-1E-7);   % user input
% beta parameter for Nitches Method
```

(Continued)

Listing 10.16 (Continued)

```
Model.Beta_Nor = 1E2;
% For verification purpose, provide the exact displacement solution
syms x1 x2        % use x1 and x2 as x- & y- coordinates
H = 12; L = 48; D = H; trac = 0.1;    % user input
I_inertia = (H^3)/12;    % user input
E = Model.E; nu = Model.nu;
% exact solution
u1 = (trac*x2/(6*E*I_inertia)).*((6*L-3*x1).*x1+(2+nu)*(x2.^2-
(D^2)/4));
u2 = -(trac/(6*E*I_inertia)).*(3*nu*x2.^2.*(L-x1)+...
                                  (4+5*nu).*((D^2*x1)/4)+(3*L-x1).*x1.^2);

Model.ExactSolution.u_exact =[u1;u2;];
     [Model.ExactSolution.S,...
      Model.ExactSolution.g,...
      Model.ExactSolution.t,...
      Model.ExactSolution.b] = getBoundaryConditions(Model);
      Model.ExactSolution.Exist = 1;

%% (4) Discretization Method
% For general purpose, one can always use A
Model.Discretization.Method = 'B';
% (...B) Uniform/Non-uniform discretization for rectangular domain:
% if A is chosen,
% define nx and ny: nx*ny is total number of nodes
% Randomness can be introduced to nodal distribution.
Model.Discretization.nx1 = 32;
Model.Discretization.nx2 = 8;
Model.Discretization.Randomness = 0.5;  % 0~1

%% (5) RK shape function parameter
RK.KernelFunction = 'SPLIN3';          % SPLIN3
RK.KernelGeometry = 'CIR';             % CIR, REC
RK.NormalizedSupportSize = 2.01;       % suggested order n + 1;
RK.Order = 'Linear';                   % Constant, Linear, Quadratic

%% (6) Quadrature rule
Quadrature.Integration = 'SCNI';         % GAUSS, SCNI, DNI
Quadrature.Stabilization = 'N';          % M, N, WO
Quadrature.Option_BCintegration = 'NODAL'; % NODAL OR GAUSS
Quadrature.nGaussPoints = 6; % nGaussPoints per cell
Quadrature.nGaussCells = 5; % nGaussCells on the short side of the
domain
```

(Continued)

Listing 10.16 (Continued)

```
%% (7) Plot the variables one prefers
Model.Plot.Discretization = 0;   % plot discretization, nodal
representative domain, or Gauss cell
Model.Plot.Displacement = 0;     % plot displacement
Model.Plot.Strain = 0;           % plot strain
Model.Plot.Stress = 1; % plot stress
Model.Plot.DeformedConfiguration = 0;  % plot deformed
configuration
Model.Plot.Error = 1; % plot absolute error

end
```

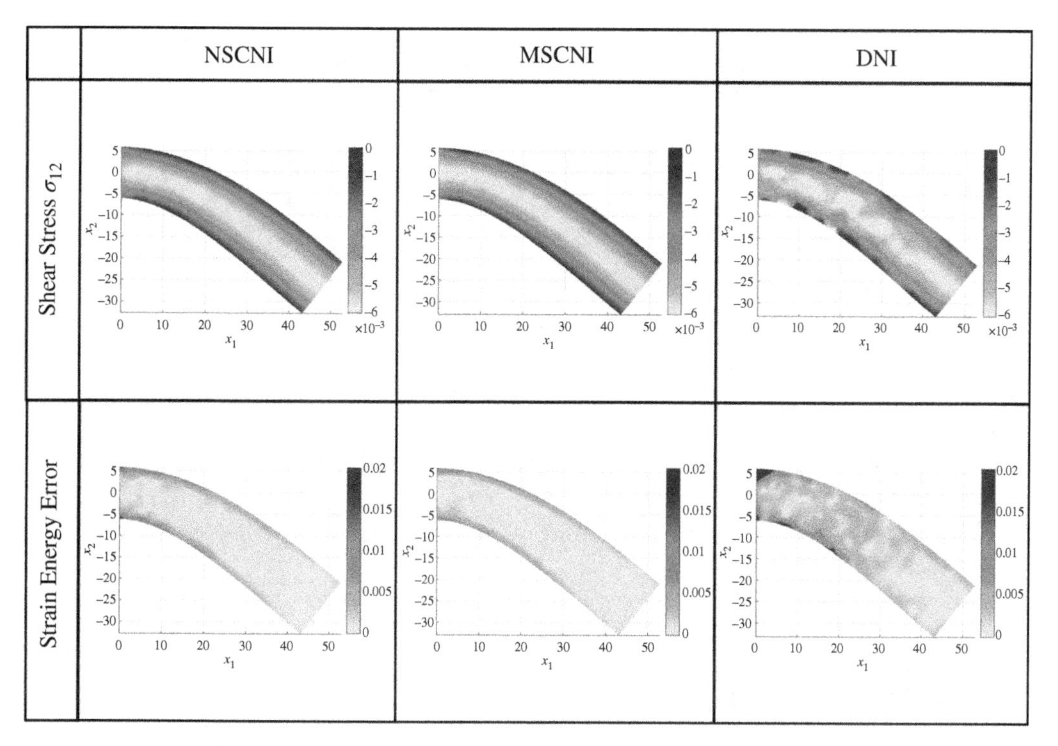

Figure 10.16 Stress fields and strain energy error of the cantilever beam problem under a nonuniform discretization.

With reference to Listing 10.17, NSCNI is chosen as the quadrature rule, with linear basis and a cubic circular B-spline kernel with a normalized support of 2.01. A Shestakov distorted discretization is chosen with the associated parameter set to 0.1. Figure 10.18 shows the results by NSCNI, along with MSCNI and DNI by modifying the input file as in the previous two examples: highly accurate and stable solutions are obtained for the variationally consistent stabilized methods. As demonstrated in [20], these methods are highly effective solution strategies even under highly severe non-uniform discretizations; the user is encouraged to verify.

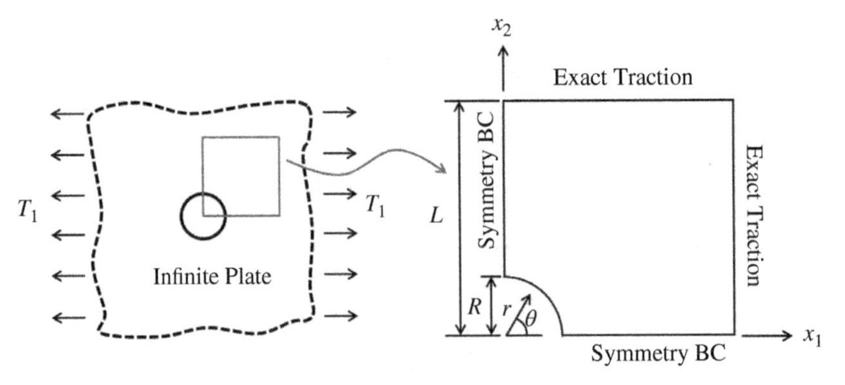

Figure 10.17 Problem settings for plate with a hole under far-field traction. Source: Huang et al. [20], figure 19, p. 427 / Springer Nature.

Listing 10.17 Sample input file for solving plate with hole problem.

```
function [RK,Quadrature,Model] = getInput()
%% GIVE INPUT PARAMETER
% Sample Input File for Beam Problem

%% (1) Material
% Linear Elasticity
% Lame Parameters for Young's modulus and poisson ratio
Model.E = 1.0E5; Model.nu = 0.3;  % user input
Model.Condition = 'PlaneStress';  % user input: 'PlaneStress' or
'PlaneStrain'
Model.ElasticTensor = getElasticTensor(Model.E,Model.nu,Model.
Condition);
Model.DOFu = 2;                   % two dimensional problem

%% (2) Geometry
theta = [0.5*pi:-pi/18:pi/18]; r = ones(size(theta));
[xI_e4,yI_e4] = pol2cart(theta,r);
x1_vertices = [1 4 4 0 xI_e4]';
x2_vertices = [0 0 4 4 yI_e4]';
% ensure the boundary segments to be counter clockwise
[x1_vertices, x2_vertices] = poly2ccw(x1_vertices, x2_vertices);
Model.xVertices = [x1_vertices, x2_vertices];
Model.DomainArea = polyarea(x1_vertices,x2_vertices);

%% (3) Boundary condition
% If an edge is not specified, natural BC with zero traction is
imposed.
Model.CriteriaEBC = @(x1,x2) find(x1<=4+1E-6 & x2<=4+1E-6);% user
input
Model.CriteriaNBC = @(x1,x2) find(x1>=4-1E-6 | x2>=4-1E-6); % user
```

(Continued)

Listing 10.17 (Continued)

```
input
% beta parameter for Nitches Method
Model.Beta_Nor = 1E2;
% For verification purpose, provide the exact displacement solution
syms x1 x2        % use x1 and x2 as x- & y- coordinates
theta = real(atan2(x2,x1)+eps); r = sqrt(x1.^2 + x2.^2);
R = 1;
Tx = 10;
mu = Model.E/(2*(1+Model.nu)); kappa = (3-Model.nu)/(1+Model.nu);
u1 =  ((Tx*R)./(8*mu)).*((r/R).*(kappa+1).*cos(theta) + (2*R./r).*
((1+kappa).*cos(theta) + cos(3*theta)) - (2*R^3./r.^3).*cos
(3*theta));
u2 =  ((Tx*R)./(8*mu)).*((r/R).*(kappa-3).*sin(theta) + (2*R./r).*
((1-kappa).*sin(theta) + sin(3*theta)) - (2*R^3./r.^3).*sin
(3*theta));

Model.ExactSolution.u_exact =[u1;
                              u2;];
     [Model.ExactSolution.S,...
      Model.ExactSolution.g,...
      Model.ExactSolution.t,...
      Model.ExactSolution.b] = getBoundaryConditions(Model);
      Model.ExactSolution.Exist = 1;

%% (4) Discretization Method
Model.Discretization.Method = 'C';
% (...C) Shestakov distorted discretization for the plate problem:
Model.Discretization.nc = 3;              % >=1
Model.Discretization.Distortion = 0.1;    % 0~0.5

%% (5) RK shape function parameter
RK.KernelFunction = 'SPLIN3';        % SPLIN3
RK.KernelGeometry = 'CIR';           % CIR, REC
RK.NormalizedSupportSize = 2.01;     % suggested order p + 0.5;
RK.Order = 'Linear';                 % Constant, Linear, Quadratic

%% (6) Quadrature rule
Quadrature.Integration = 'SCNI';        % GAUSS, SCNI, DNI
Quadrature.Stabilization = 'N';         % M, N
Quadrature.Option_BCintegration = 'NODAL'; % NODAL OR GAUSS
Quadrature.nGaussPoints = 6;
Quadrature.nGaussCells = 10; % nGaussCells on the short side of the
domain

%% (7) Plot the variables one prefers
```

(Continued)

Listing 10.17 (Continued)

```
Model.Plot.Discretization = 0;  % plot discretization, nodal
representative domain, or Gauss cell
Model.Plot.Displacement = 0;    % plot displacement
Model.Plot.Strain = 0;          % plot strain
Model.Plot.Stress = 1; % plot stress
Model.Plot.DeformedConfiguration = 0;  % plot deformed
configuration
Model.Plot.Error = 1; % plot absolute error
end
```

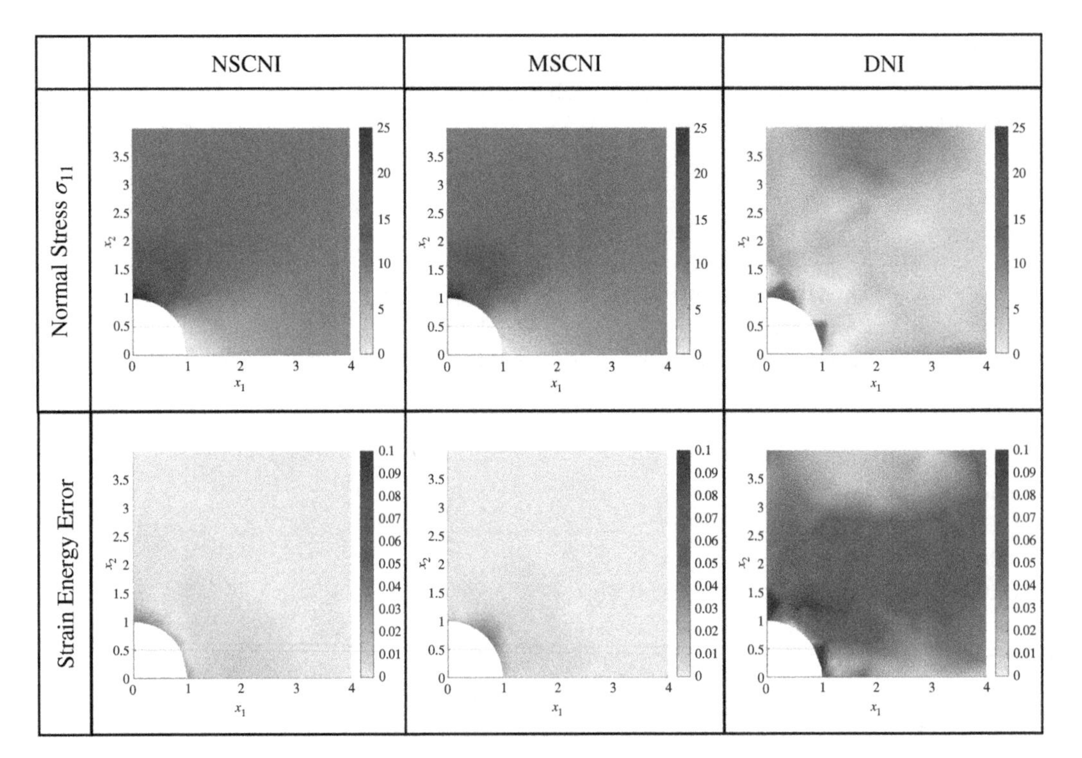

Figure 10.18 Stress fields and strain energy error of the plate with hole problem under a distorted discretization.

10.A Appendix

In this section, a diffusion problem is used to illustrate how to modify RKPM2D to solve different types of PDEs. The steady-state isotropic diffusion equation is considered as follows (see Chapter 2):

$$
\begin{aligned}
&(ku_{,i})_{,i} + s = 0 && \text{on} \quad \Omega, \\
&ku_{,i}n_i = -q_i n_i = \overline{q} && \text{on} \quad \Gamma_q, \\
&u = \overline{u} && \text{on} \quad \Gamma_u,
\end{aligned}
\tag{10.54}
$$

where u is a scalar field, k is the diffusivity, s is the source term, and $\overline{q}$ and $\overline{u}$ are the prescribed boundary flux and boundary values of u on Γ_q and Γ_u, respectively. Following Chapter 4, the strong form (10.54) can be recast into the following matrix equations using Nitsche's method:

$$\sum_{J\in S}(K_{IJ} + K_{IJ}^{\mathrm{N}})u_J = f_I + f_I^{\mathrm{N}} \quad \forall I \in S, \tag{10.55}$$

where

$$K_{IJ} = \int_{\Omega} \mathbf{B}_I^{\mathrm{T}}(\mathbf{x})\mathbf{D}\mathbf{B}_J(\mathbf{x})d\Omega, \tag{10.56}$$

$$f_I = \int_{\Omega} \Psi_I s d\Omega + \int_{\Gamma_q} \Psi_I \overline{q} d\Gamma, \tag{10.57}$$

$$K_{IJ}^{\mathrm{N}} = -\int_{\Gamma_u} (\Psi_{I,n} k \Psi_J + \Psi_I k \Psi_{J,n} - \beta \Psi_I \Psi_J) d\Gamma, \tag{10.58}$$

$$f_I^{\mathrm{N}} = \int_{\Gamma_u} (-k\Psi_{I,n}\overline{u} + \beta \Psi_I \overline{u}) d\Gamma, \tag{10.59}$$

with

$$\begin{aligned} &\mathbf{B}_I(\mathbf{x}) = \begin{bmatrix} \Psi_{I,1}(\mathbf{x}) \\ \Psi_{I,2}(\mathbf{x}) \end{bmatrix}, \quad \text{(direct derivatives, DNI)} \\ &\mathbf{B}_I(\mathbf{x}) \to \tilde{\mathbf{B}}_I(\mathbf{x}_L) = \begin{bmatrix} \tilde{b}_{1I}^L \\ \tilde{b}_{2I}^L \end{bmatrix}, \quad \text{(smoothed derivatives, SCNI)} \\ &\mathbf{D} = \begin{bmatrix} k & 0 \\ 0 & k \end{bmatrix}, \end{aligned} \tag{10.60}$$

where $\mathbf{D}$ is the diffusivity tensor, and k is the diffusion coefficient. Let us consider the diffusion problem (10.54) with a manufactured solution

$$u^{\mathrm{exact}} = 0.1 + 0.1x_1 + 0.2x_2, \tag{10.61}$$

in a circular domain $\Omega \subset \mathbb{R}^2$ like the one shown in Figure 10.8 from Section 10.4.1. The flux $\overline{q} = 0.1n_1 + 0.2n_2$ is imposed on Γ_t: $(x_1, x_2) \in \Gamma$, $x_2 > 0.5$ where n_1, and n_2 are the unit normal vector components. The solution $\overline{u}$ is enforced on Γ_u: $(x_1, x_2) \in \Gamma$, $x_2 \leq 0.5$, and the body source is $s = 0$.

The input file for this problem is generated in the function `getInput`. Compared to Listing 10.8, the following changes need to be made:

- Remove `Model.nu` and `Model.Condition`, as Poisson ratio and plane-stress/strain condition are not required for diffusion problem.
- Replace `Model.E` with `Model.d` (i.e., change the definition of Young's modulus E to be the diffusion coefficient d).

- Replace `Model.ElasticTensor` with `Model.DiffusiveTensor` (i.e., change the definition of elastic tensor **D** to be the diffusive tensor).
- Set `Model.DiffusiveTensor = diag([Model.d,Model.d])` to define the diffusive tensor $\mathbf{D} = \begin{bmatrix} k & 0 \\ 0 & k \end{bmatrix}$.
- Set `Model.DOFu = 1` to change the nodal degrees of freedom `DOFu` from 2 to 1.
- Set `u_exact = 0.1+0.1*x1+0.2*x2` to define the exact solution u^{exact}.

In addition, we also need to modify the subroutine `getBoundaryConditions` to generate the exact boundary flux $\overline{q} = kn_i u_{,i}^{\text{exact}}$, source term $s = -(ku_{,i}^{\text{exact}})_{,i}$, essential boundary conditions $\overline{u} = u^{\text{exact}}$, and switch matrix $S = 1$ based on a given expression of the exact solution u^{exact} in a symbolic form, as shown in Listing 10.18.

Listing 10.18 Command lines of function to generate exact heat flux q, heat sources s, imposed temperature $\overline{u}$, and switch matrix S for diffusion problem.

```
function [function_S,function_g,function_traction,function_b] =
getBoundaryConditions(Model)

syms x1 x2 n1 n2

% function handle for essential boundary condition g
u = Model.ExactSolution.u_exact;
D = Model.DiffusiveTensor;
function_g = matlabFunction(u);

% function handle for diff(u)
dudx1 = diff(u,x1);
dudx2 = diff(u,x2);
flux = D*[dudx1; dudx2;];

% function handle for surface flux (traction)
eta = [n1; n2;];
surf_flux = eta'*flux;
function_traction = matlabFunction(surf_flux,'Vars',[x1 x2 n1 n2]);

% function handle for source b
b = [diff(flux(1),x1)+ diff(flux(2),x2)];
function_b = matlabFunction(b,'Vars',[x1 x2]);

% function handle for switch S
function_S = matlabFunction(sym(1),'Vars',[x1 x2]);

end
```

Due to the change of dimensionality in the **B** and $\boldsymbol{\Psi}$ matrices, modifications are made to `MatrixAssmebly` (Listing 10.6) as follows

- Set `d = Model.d` to define the diffusivity coefficient.
- Set `D = Model.DiffusiveTensor` to define the diffusivity from input files.
- Replace `E` with `d` (i.e., replace the Young's modulus `E` with diffusion coefficient `d`).
- Replace `C` with `D` (i.e., replace elastic tensor `C` with diffusive tensor `D`).
- Set `B = sparse(2,nP*DOFu)`.
- Set `PSI = sparse(1,nP*DOFu)`.
- Modify the allocation of the `B` and `PSI` from shape function `SHP` and derivative `SHPDX1`, `SHPDX2` as:
  - `PSI = SHP(idx_nQuad,:);`
  - `B(1,:) = SHPDX1(idx_nQuad,:);`
  - `B(2,:) = SHPDX2(idx_nQuad,:);`
- Set `ETA = [n1; n2]` to define the surface normal vector $\boldsymbol{\eta}$.

RKPM2D is converted to a code for solving a diffusion problem with the above-mentioned modifications. Comparing the original code for the elasticity problem with the modified code for the diffusion problem, it is seen that very minimal code modification is required. This capability of easy code extension is a unique feature of RKPM2D.

References

1 MATLAB Release 2023a, The Mathworks Inc., Natick, MA.

2 Lu, Y.Y., Belytschko, T., and Gu, L. (1994). A new implementation of the element free Galerkin method. *Comput. Methods Appl. Mech. Eng.* 113 (3–4): 397–414.

3 Belytschko, T., Lu, Y.Y., and Gu, L. (1994). Element-free Galerkin methods. *Int. J. Numer. Methods Eng.* 37: 229–256.

4 Dolbow, J. and Belytschko, T. (1999). Numerical integration of the Galerkin weak form in meshfree methods. *Comput. Mech.* 23 (3): 219–230.

5 Chen, J.-S., Wu, C.-T., and Yoon, S. (2001). A stabilized conforming nodal integration for Galerkin mesh-free methods. *Int. J. Numer. Meth. Eng.* 50 (2): 435–466.

6 Chen, J.-S., Hillman, M., and Rüter, M. (2013). An arbitrary order variationally consistent integration for Galerkin meshfree methods. *Int. J. Numer. Methods Eng.* 95 (5): 387–418.

7 Chen, J.-S., Hu, W., Puso, M.A. et al. (2007). Strain smoothing for stabilization and regularization of galerkin meshfree methods. *Lect. Notes Comput. Sci. Eng.* 57: 57–75.

8 Puso, M.A., Chen, J.-S., Zywicz, E., and Elmer, W. (2008). Meshfree and finite element nodal integration methods. *Int. J. Numer. Methods Eng.* 74 (3): 416–446.

9 Hillman, M. and Chen, J.-S. (2016). An accelerated, convergent, and stable nodal integration in Galerkin meshfree methods for linear and nonlinear mechanics. *Int. J. Numer. Methods Eng.* 107: 603–630.

10 Sievers, J. (2020). VoronoiLimit(varargin). MATLAB Cent. File Exch.

11 Bentley, J.L. (1975). Multidimensional binary search trees used for associative searching. *Commun. ACM.* 18 (9): 509–517.

12 Perkins, E. and Williams, J.R. (2001). A fast contact detection algorithm insensitive to object sizes. *Eng. Comput.* 18.

13 Barbieri, E. and Meo, M. (2012). A fast object-oriented Matlab implementation of the reproducing kernel particle method. *Comput. Mech.* 49 (5): 581–602.

14 Cartwright, C., Oliveira, S., and Stewart, D.E. (2006). Parallel support set searches for meshfree methods. *SIAM J. Sci. Comput.* 28 (4): 1318–1334.

15 Parreira, G.F., Fonseca, A.R., Lisboa, A.C., Silva, E.J., Mesquita, R.C. (2006). Efficient algorithms and data structures for element-free Galerkin method. *IEEE Trans. Magn.* 42 (4): 659–662.

16 Olliff, J., Alford, B., and Simkins, D.C. (2018). Efficient searching in meshfree methods. *Comput. Mech.* 62 (6): 1461–1483.

17 Steger, C. (1996). On the calculation of arbitrary moments of polygons. Munchen Univ., Munchen, Ger. Tech. Rep. FGBV-96-05.

18 Saad, Y. (2003). *Iterative Methods for Sparse Linear Systems.* SIAM.

19 Shestakov, A.I., Kershaw, D.S., and Zimmerman, G.B. (1990). Test problems in radiative transfer calculations. *Nucl. Sci. Eng.* 105 (1): 88–104.

20 Huang, T.H., Wei, H., Chen, J.S., and Hillman, M. (2020). RKPM2D : Open-source implementation of nodally integrated reproducing kernel particle method for solving partial differential equations. *Comput. Part. Mech.* 7: 393–433.

Index

Note: **Boldface** indicates a definition or introduction to a concept; *italics* indicates figures; numbers followed by a, b, c, k, and t, indicates algorithms, boxes, command and function listings, key take-aways, and tables, respectively.

Meshfree and Particle Methods: Fundamentals and Applications, First Edition.
Ted Belytschko, J. S. Chen, and Michael Hillman.

C

d

e

W

Z